Recent Titles in This Series

167 **V. V. Lychagin, Editor,** The Interplay between Differential Geometry and Differential Equations
166 **O. A. Ladyzhenskaya, Editor,** Proceedings of the St. Petersburg Mathematical Society, Volume III
165 **Yu. Ilyashenko and S. Yakovenko, Editors,** Concerning the Hilbert 16th Problem
164 **N. N. Uraltseva, Editor,** Nonlinear Evolution Equations
163 **L. A. Bokut′, M. Hazewinkel, and Yu. G. Reshetnyak, Editors,** Third Siberian School "Algebra and Analysis"
162 **S. G. Gindikin, Editor,** Applied Problems of Radon Transform
161 **Katsumi Nomizu, Editor,** Selected Papers on Analysis, Probability, and Statistics
160 **K. Nomizu, Editor,** Selected Papers on Number Theory, Algebraic Geometry, and Differential Geometry
159 **O. A. Ladyzhenskaya, Editor,** Proceedings of the St. Petersburg Mathematical Society, Volume II
158 **A. K. Kelmans, Editor,** Selected Topics in Discrete Mathematics: Proceedings of the Moscow Discrete Mathematics Seminar, 1972–1990
157 **M. Sh. Birman, Editor,** Wave Propagation. Scattering Theory
156 **V. N. Gerasimov, N. G. Nesterenko, and A. I. Valitskas,** Three Papers on Algebras and Their Representations
155 **O. A. Ladyzhenskaya and A. M. Vershik, Editors,** Proceedings of the St. Petersburg Mathematical Society, Volume I
154 **V. A. Artamonov et al.,** Selected Papers in K-Theory
153 **S. G. Gindikin, Editor,** Singularity Theory and Some Problems of Functional Analysis
152 **H. Draškovičová et al.,** Ordered Sets and Lattices II
151 **I. A. Aleksandrov, L. A. Bokut′, and Yu. G. Reshetnyak, Editors,** Second Siberian Winter School "Algebra and Analysis"
150 **S. G. Gindikin, Editor,** Spectral Theory of Operators
149 **V. S. Afraĭmovich et al.,** Thirteen Papers in Algebra, Functional Analysis, Topology, and Probability, Translated from the Russian
148 **A. D. Aleksandrov, O. V. Belegradek, L. A. Bokut′, and Yu. L. Ershov, Editors,** First Siberian Winter School "Algebra and Analysis"
147 **I. G. Bashmakova et al.,** Nine Papers from the International Congress of Mathematicians, 1986
146 **L. A. Aĭzenberg et al.,** Fifteen Papers in Complex Analysis
145 **S. G. Dalalyan et al.,** Eight Papers Translated from the Russian
144 **S. D. Berman et al.,** Thirteen Papers Translated from the Russian
143 **V. A. Belonogov et al.,** Eight Papers Translated from the Russian
142 **M. B. Abalovich et al.,** Ten Papers Translated from the Russian
141 **H. Draškovičová et al.,** Ordered Sets and Lattices
140 **V. I. Bernik et al.,** Eleven Papers Translated from the Russian
139 **A. Ya. Aĭzenshtat et al.,** Nineteen Papers on Algebraic Semigroups
138 **I. V. Kovalishina and V. P. Potapov,** Seven Papers Translated from the Russian
137 **V. I. Arnol′d et al.,** Fourteen Papers Translated from the Russian
136 **L. A. Aksent′ev et al.,** Fourteen Papers Translated from the Russian
135 **S. N. Artemov et al.,** Six Papers in Logic
134 **A. Ya. Aĭzenshtat et al.,** Fourteen Papers Translated from the Russian
133 **R. R. Suncheleev et al.,** Thirteen Papers in Analysis
132 **I. G. Dmitriev et al.,** Thirteen Papers in Algebra
131 **V. A. Zmorovich et al.,** Ten Papers in Analysis
130 **M. M. Lavrent′ev, K. G. Reznitskaya, and V. G. Yakhno,** One-dimensional Inverse Problems of Mathematical Physics
129 **S. Ya. Khavinson,** Two Papers on Extremal Problems in Complex Analysis

(See the AMS catalog for earlier titles)

The Interplay between Differential Geometry and Differential Equations

American Mathematical Society

TRANSLATIONS

Series 2 • Volume 167

Advances in the Mathematical Sciences — 24

(*Formerly Advances in Soviet Mathematics*)

The Interplay between Differential Geometry and Differential Equations

V. V. Lychagin
Editor

American Mathematical Society
Providence, Rhode Island

1991 *Mathematics Subject Classification.* Primary 17B37, 35Axx, 58Axx, 58Gxx.

ABSTRACT. The purpose of the book is to emphasize the advantage of algebraic geometry approach to nonlinear differential equations, including applications of symplectic methods and the discussion of quantization problems. One of the common features for the majority of papers in the book is the systematic use of geometry of jet spaces. The book is useful to researchers and graduate students who are interested in nonlinear differential equations, differential geometry, quantum groups, and their applications.

Library of Congress Card Number 91-640741
ISBN 0-8218-0428-6
ISSN 0065-9290

Printed in the United States of America.

∞ The paper used in this book is acid-free and falls within the guidelines
established to ensure permanence and durability.
♻ Printed on recycled paper.
This volume was typeset by the authors using $\mathcal{A}\mathcal{M}\mathcal{S}$-TeX,
the American Mathematical Society's TeX macro system.

10 9 8 7 6 5 4 3 2 1 00 99 98 97 96 95

Contents

Foreword

This collection presents work concentrated mainly around the differential geometry approach to the theory of nonlinear differential equations.

Actually, differential geometry and differential equations are so closely related that it is practically impossible to draw a clear delimiting line between these two branches of mathematics.

Thus, whereas the connections between linear differential equations and differential geometry were few and far between, as were the connections between analytic geometry (in the elementary sense of the term) and geometry in the sense of Felix Klein, in contrast the theory of nonlinear differential equations is clearly a geometric theory, based on the special geometry of jet spaces.

Moreover, differential geometry, in its turn, may be presented as the part of the theory of nonlinear differential equations that studies differential equations *that have no solutions*. From this point of view, all the basic notions of differential geometry acquire their natural meaning. For example, various curvatures and torsion arise as obstructions to the solvability of the appropriate differential equations. In this sense the theory of differential equations in its classical understanding is the geometry of flat objects.

The papers comprising this collection may be conditionally subdivided into three groups.

The first group consists of papers dealing with general questions of the geometric theory of differential equations and the methods of differential geometry used in their study (I. Krasilshchik and P. Kersten, A. Kushner, A. Prastaro. A. Samokhin, N. Shananin, D. Tunitsky, L. Zilbergleit).

The second group deals with applications of the geometric theory of differential equations to the study of specific problems (A. Khor′kova and A. Verbovetsky, V. Chetverikov and A. Kudryavtsev, V. Shemarulin, L. Zilbergleit).

Finally, the third group is devoted to quantization problems (I. Donin and D. Gurevich, B. Enriques, S. Khoroshkin, A. Radul, A. Rosly, and V. Rubtsov).

The appearance of this last group of papers in a collection mainly concerned with differential equations understood in the classical sense, and not in the quantum one, nevertheless seems quite logical to me, since the true understanding of the problems of differential equations must unavoidably pass through the quantum domain.

V. Lychagin

Moscow, August 1994

Translated by A. B. SOSSINSKY

Amer. Math. Soc. Transl.
(2) Vol. **167**, 1995

Modeling Integro-Differential Equations and a Method for Computing their Symmetries and Conservation Laws

V. N. CHETVERIKOV AND A. G. KUDRYAVTSEV

ABSTRACT. A method for computing symmetries and conservation laws of integro-differential equations is proposed. The method consists in reducing an integro-differential equation to a system of boundary differential equations and in computing symmetries and conservation laws for this system. We also discuss methods used before by other authors as well as the foundations of the geometry of integro-differential and boundary differential equations. Results of computations of symmetries and conservation laws of the Smoluchowski coagulation equation and of an equation for nonuniform Brownian coagulation are given as examples.

Introduction

Classical infinitesimal symmetries of integro-differential equations were studied by many authors (for instance, see [1–8]). Different methods were used for this. In this paper we suggest one more method. It is distinguished by the fact that it works for any integro-differential equation (or system) and yields not only classical (that is, contact or Lie-point), but higher (that is, generalized or Lie–Bäcklund) symmetries as well as conservation laws. This is achieved by increasing the number of variables and thus complicating the computation. But the high dimensionality of the problems did not prevent us from computing the classical symmetries and some conservation laws of coagulation equations of two types (see equations (2.5) and (5.1), symmetries (4.2)–(4.4) and (5.3)–(5.4), conservation law (6.10)).

A symmetry is a geometric concept. Therefore, in order to define a symmetry of something it is enough to represent it as some geometric object (model). Then a transformation of the model is its symmetry. In the case of integro-differential equations, their analogy with differential equations can also be used. That is why we shall consider models of differential equations here and of integro-differential equations in the next sections.

Several levels of geometric models of differential equations can be noted (cf. [9–12]). At the first level, one considers the set of all solutions of the considered

1991 *Mathematics Subject Classification*. Primary 45K05, 58G35; Secondary 58A20, 58G20.
Key words and phrases. Integro-differential equations, boundary differential equations, functional differential equations, jet spaces, symmetries, conservation laws.

equation. Usually, it is impossible to find all solutions of a nonlinear differential equation or even to describe smooth or topological structures of the space of solutions. But the analogy "set of solutions—smooth manifold" can be used for an intuitive (inexact) introduction of new concepts. Thus by a symmetry of a differential equation one means a transformation of the space of its solutions, by an infinitesimal symmetry an infinitesimal transformation or a vector field on this space, by a conservation law a differential form. But at the first level it is impossible to define these concepts exactly.

A submanifold of a k-jet space is a geometric model of the next level for a certain differential equation of the kth order. Classical symmetries of the equation are the transformations of this submanifold mapping solutions of the equation into solutions.

At the third level one considers the infinite prolongation of a differential equation as an infinite-dimensional submanifold of the space of infinite jets. The transformations of this submanifold preserving the set of solutions are higher symmetries. At this level conservation laws are also defined, as cohomology classes of a certain complex. There exist models of differential equations of subsequent levels, each new level containing more informative models than the preceding ones (see [11–13]).

The aim of this paper is to construct in a similar way models of the third level for integro-differential equation. The first section reviews some works on symmetries of integro-differential equations and gives geometric models for them. Examples of the construction of geometric models in our sense is described in the second section. Our approach is based on A. M. Vinogradov's idea to use the notion of a covering (see [13]). That is, introducing nonlocal variables, we eliminate integrals and transform the original equation to a system of boundary (or functional) differential equations. In the third section, we define such generalized jet spaces so that boundary differential equations can be interpreted as submanifolds of these spaces. A geometric theory of boundary differential equations similar to the geometry of nonlinear differential equations presented in [11, 12] is constructed here. Sections 4 and 5 contain applications of this theory. The last section includes a definition and a computation scheme for conservation laws of integro-differential equations and the result of this computation for the coagulation kinetic equation.

§1. Methods previously used for computing symmetries of integro-differential equations

In recent years, two methods for the computation have gained ground. Here we consider two works where these methods had been applied.

In the paper [1], V. B. Taranov investigated the system of equation that describes longitudinal one-dimensional high-frequency motions of collisionless plasma when the ion constituent forms a stationary homogeneous background. In dimensionless variables, the distribution function $f(t, x, v)$ for electrons and the self-consistent electric field $E(t, x)$ satisfy the equations

$$(1.1)\qquad f_t + vf_x - Ef_v = 0, \qquad E_v = 0,$$
$$E_t = \int_{-\infty}^{\infty} vf\,dv, \qquad E_x = 1 - \int_{-\infty}^{\infty} f\,dv.$$

If we multiply the first equation from (1.1) by v^k (k is an arbitrary natural number) and integrate it with respect to v, we obtain the system

$$\begin{gathered} \partial_t M_k + \partial_x M_{k+1} + kEM_{k-1} = 0, \qquad k = 0, 1, \dots, \\ \partial_t E = M_1, \qquad \partial_x E = 1 - M_0, \end{gathered} \tag{1.2}$$

where $M_k(t, x) = \int_{-\infty}^{\infty} v^k f(t, x, v)\, dv$, $k = 0, 1, \dots$, are the moments of the distribution function, $\partial_t = \partial/\partial t$, $\partial_x = \partial/\partial x$.

To compute symmetries of the infinite system (1.2) for any $k = 1, 2, \dots$, V. B. Taranov computed symmetries of the finite parts of the system (1.2) for $k \leqslant n$ with an arbitrary n and passed to the limit with $n \to \infty$. The transformation group so obtained is generated by the infinitesimal transformations

$$X_1 = \partial_t, \quad X_2 = \partial_x, \quad X_3 = x\partial_x + E\partial_E + \sum_{k=1}^{\infty} kM_k \partial_{M_k},$$

$$X_4 = \cos t\,(\partial_x + \partial_E) - \sum_{k=1}^{\infty} kM_{k-1} \sin t\, \partial_{M_k},$$

$$X_5 = \sin t\,(\partial_x + \partial_E) + \sum_{k=1}^{\infty} kM_{k-1} \cos t\, \partial_{M_k}.$$

These vector fields on the space of variables t, x, E, M_k, $k = 0, 1, \dots$, correspond the following fields on the space (t, x, v, E, f)

$$\begin{gathered} X_1 = \partial_t, \quad X_2 = \partial_x, \quad X_3 = x\partial_x + v\partial_v + E\partial_E - f\partial_f, \\ X_4 = \cos t\,(\partial_x + \partial_E) - \sin t\,\partial_v, \quad X_5 = \sin t\,(\partial_x + \partial_E) + \cos t\,\partial_v, \end{gathered}$$

which generate the desired group of symmetries.

This result was used by V. B. Taranov [1] for reducing system (1.1) and for finding invariant solutions. This method was also employed for solving analogous problems for other equations in [2–4].

To get the geometric model which can be used to interpret this method in the case of system (1.1), it is enough to consider the 1-jet space with infinite number of dependent variables E, M_0, M_1, ... and with two independent variables x, t. In this space equations (1.2) cut out a submanifold which will be the model.

The jet space mentioned above can be defined in the following way. Let $\pi\colon \mathbb{R}^5 \to \mathbb{R}^3$ be the trivial bundle with coordinates t, x, v on the base and E, f on the fibers. Consider the sections for which the component E is independent of the variable v. Introduce an equivalence relation on the set of those sections. Two sections $\{f^1(t, x, v), E^1(t, x)\}$ and $\{f^2(t, x, v), E^2(t, x)\}$ are (l, n)-equivalent at the point $P(t; x)$ if the Taylor polynomials of degree l at the point P for the functions $E^1(t, x)$ and $E^2(t, x)$, $M_k^1(t, x)$ and $M_k^2(t, x)$, where

$$M_k^i(t, x) = \int_{-\infty}^{\infty} v^k f^i(t, x, v)\, dv, \qquad i = 1, 2, \; k = 0, 1, \dots, n,$$

coincide in pairs. Denote the set of all the equivalence classes for fixed numbers l and n by J^l_n. We have the chain of projections

$$J^l_0 \leftarrow J^l_1 \leftarrow \cdots \leftarrow J^l_n \leftarrow J^l_{n+1} \leftarrow \cdots .$$

The inverse limit of this chain is the l-jet space for (1.2).

Equations (1.2) define submanifolds in J^1_n for every $n \geqslant 1$; we denote this submanifold by $\mathcal{Y}_n$. The inverse limit of the chain

$$\mathcal{Y}_1 \leftarrow \mathcal{Y}_2 \leftarrow \cdots \leftarrow \mathcal{Y}_n \leftarrow \mathcal{Y}_{n+1} \cdots$$

is the desired geometric model $\mathcal{Y}$, which is an infinite-dimensional manifold. By a symmetry of this model (or of the system (1.1)) we mean its transformation preserving the set of its solutions. (Concerning the definition of transformations of infinite-dimensional manifolds, see [11, 12]). This definition is obviously a formal version of the concept used by V. B. Taranov (see above).

The second method was used by S. V. Meleshko in the paper [5] to find the symmetry group of the system of equations that describes one-dimensional motion of a viscoelastic (hereditary) medium. The simplest version of this system, which we shall consider here, is

$$v_t = \sigma_x , \quad e_t = v_x , \quad \varphi(e) = \sigma + \int_0^t K(t, \tau)\sigma(\tau)\, d\tau , \tag{1.3}$$

where time t and distance x are independent variables, stress σ, velocity v, and strain e are dependent variables $\varphi' \neq 0$.

We try to find infinitesimal classical symmetries in the form

$$X = \overset{\sigma}{\zeta}\partial_\sigma + \overset{v}{\zeta}\partial_v + \overset{e}{\zeta}\partial_e + \xi\partial_t + \eta\partial_x ,$$

where $\overset{\sigma}{\zeta}$, $\overset{v}{\zeta}$, $\overset{e}{\zeta}$, ξ, η are functions of t, x, σ, v, and e; $\partial_\sigma = \partial/\partial\sigma$ and so on. As in the differential case [9], in order to obtain the determining equations in $\overset{\sigma}{\zeta}$, $\overset{v}{\zeta}$, $\overset{e}{\zeta}$, ξ, η, the author used the fact that the flow of the vector field corresponding to an infinitesimal symmetry consists of local symmetries of the equation. The determining equations for (1.3) are

$$\begin{gathered} [D_t(\nu) - D_x(\psi)]_{(1.3)} = 0 , \quad [D_t(\psi) - D_x(\theta)]_{(1.3)} = 0 , \\ \left[\varphi'\theta - \psi - \int_0^t K(t, \tau)\psi(\tau)\, d\tau\right]\Bigg|_{(1.3)} = 0 , \end{gathered} \tag{1.4}$$

$$\nu = \overset{v}{\zeta} - \xi v_t - \eta v_x , \quad \psi = \overset{\sigma}{\zeta} - \xi\sigma_t - \eta\sigma_x , \quad \theta = \overset{e}{\zeta} - \xi e_t - \eta e_x ,$$
$$D_t = \partial_t + \sigma_t\partial_\sigma + v_t\partial_v + e_t\partial_e , \quad D_x = \partial_x + \sigma_x\partial_\sigma + v_x\partial_v + e_x\partial_e ;$$

here the expression $\psi(\tau)$ coincides with ψ but the variable t in ψ is replaced by the variable τ in $\psi(\tau)$, an equality of the form $A|_{(1.3)} = 0$ means that the expression A vanishes if the functions $\sigma(t, x)$, $v(t, x)$, $e(t, x)$ involved in A form a solution of (1.3).

The complete set of solutions of (1.4) is found under the assumption that there exist a solution of the Cauchy problem along x for with any initial condition at the point $x = x_0$: $\sigma(x_0, t) = \sigma_0(t)$, $v(x_0, t) = v_0(t)$. Consider system (1.4) for $x = x_0$ and substitute into it the following values of derivatives: $v_t = \sigma_x = v_0'$, $\sigma_t = \sigma_0'$, $v_x = e_t = g_1/\varphi'$, $e_x = g_2/\varphi'$, where

$$g_1 = \sigma_0' + K(t, t)\sigma_0 + \int_0^t K_t'(t, \tau)\sigma_0(\tau)\, d\tau,$$

$$g_2 = v_0' + \int_0^t K(t, \tau) v_0'(\tau)\, d\tau.$$

The obtained expressions are polynomials in v_0, v_0', and g_2 with coefficients containing the variables t, σ_0, σ_0', e, g_1. Therefore, if the variables v_0, v_0', and g_2 are functionally independent, then one can split the obtained equations, i.e., equate the coefficients of various powers of v_0, v_0', g_2 to zero and to obtain a new system of equations.

The functional independence of the variables v_0, v_0', and g_2 is proved in the following way. Consider an initial function of the form

$$v_0(t) = a_1 + a_2(t - t_0) + a_3(t_0 - t)^{n+1}/(n+1), \qquad n \geqslant 1.$$

For $t = t_0 > 0$ we have

$$v_0(t_0) = a_1, \quad v_0'(t_0) = a_2,$$

$$g_2(t_0) = a_2\left(1 + \int_0^{t_0} K(t_0, \tau)\, d\tau\right) - a_3 \int_0^{t_0} K(t_0, \tau)(t_0 - \tau)^n d\tau.$$

There exists a natural number n such that

$$\int_0^{t_0} K(t_0, \tau)(t_0 - \tau)^n d\tau \neq 0.$$

Indeed, $K(t_0, \tau) \not\equiv 0$ and the set of the functions $\{(t_0 - t)^{n-1}, n = 1, 2, \ldots\}$ is complete in the space $L_2[0, t_0]$. Therefore, choosing a_1, a_2, a_3, we can obtain any values $v_0(t_0)$, $v_0'(t_0)$, $g_2(t_0)$.

The new system can be again split with respect to the new variables and so on. The final result of [5] describes the algebras of classical infinitesimal symmetries of system (1.3) for various functions $\varphi(e)$. For instance, when $\varphi = \alpha + \beta \exp(\gamma e)$, where α, $\beta \neq 0$, $\gamma \neq 0$ are constant, the algebra is generated by

$$X_1 = \partial_x, \quad X_2 = \partial_v, \quad X_3 = \gamma x \partial_x + \gamma v \partial_v + 2\gamma(\sigma - \alpha\mu(t))\partial_\sigma + 2\partial_e,$$

where the function $\mu(t)$ is a solution of the equation

$$1 = \mu(t) + \int_0^t K(t, \tau)\mu(\tau)\, d\tau.$$

The papers [6–8] give other applications of this method.

The geometric model of the method is the space of solutions of the given system ((1.3) here) with a particular description of the set of coordinate functions on it. Namely, the variables which are functionally independent on this space and allow to split the equations are found (here v_0, v_0', g_2, and etc.).

§2. Examples of the construction of models of integro-differential equations in boundary form

The fundamental theorem of calculus is a basic technique which we intend to use for constructing the geometric models.

EXAMPLE 1. Consider the one-dimensional nonlinear integral Gammerstein equation of the second kind:

$$u(x) = \int_a^b K(x, s, u(s))\, ds, \tag{2.1}$$

where $K(x, s, u)$ is a given function, $x \in [a, b]$, $s \in [a, b]$. Introduce a dependent variable v by

$$v'_s(x, s) = K(x, s, u(s)), \qquad v(x, a) = 0. \tag{2.2}$$

Equation (2.1) is equivalent to the system (2.2) together with the equation

$$u(x) = v(x, b). \tag{2.3}$$

The system (2.2)–(2.3) involves two independent variables x, s, two dependent variables u, v and the restrictions of the dependent variable v to the boundary sets $\{s = a\}$ and $\{s = b\}$. The dependent variable u appears as $u(x)$ and as $u(s)$. To transform the system further, we introduce the following maps of the manifold $M = [a, b] \times [a, b]$:

$$g_a : M \to M : (x, s) \mapsto (x, a),$$
$$g_b : M \to M : (x, s) \mapsto (x, b),$$
$$g_c : M \to M : (x, s) \mapsto (s, x).$$

In addition, let us think of $u(x)$ as of a function of x and s satisfying the equation $u'_s = 0$. Then, using the notation $v_1 = v'_x$, $v_2 = v'_s$, $u_c = g_c^*(u)$, etc., system (2.2)–(2.3) can be written as

$$u_2 = 0, \quad v_2 = K(x, s, u_c), \quad v_a = 0, \quad u = v_b, \tag{2.4}$$

where K is the same as in (2.1).

The symbols u, v, u_1, v_1, u_2, v_2, u_a, v_a, u_b, v_b, u_c, v_c, etc. can be regarded as symbols of the coordinates of some generalized jet space (see below). The subscripts 1 and 2 are symbols of the derivatives with respect to x and s respectively. The letters a, b, or c indicate the action of the homomorphisms g_a^*, g_b^*, or g_c^*.

As in the case of differential equations, one must learn how to prolong any system of the form (2.4). In order to find new (prolonged) equations we must use two kinds of transformations: (a) differentiations or (b) actions of homomorphisms. Thus for the system (2.4) the new equations will involve new variables of the form $u_{cb} = g_b^*[g_c^*(u)] = (g_c \circ g_b)^*(u)$, $u_{2c} = g_c^*(u'_s)$, $u_{c2} = [g_c^*(u)]'_s$, etc. Note that $u_{2c} \neq u_{c2}$, since from the first equation of (2.4) we have $u_{2c} = 0$, but $u_{c2} = [u(s, x)]'_s = u_1(s, x) = u_{1c}$. Hence it is necessary to consider the whole semigroup generated by the maps of the form g_a, g_b, g_c and the identity map g_e of M (the identity element of the semigroup) and to take into account some natural principle to rearrange the subscripts in u and v.

For equation (2.1), the semigroup consists of 14 elements

$$g_e, \; g_a, \; g_b, \; g_c, \; g_a \circ g_c, \; g_c \circ g_a, \; g_b \circ g_c, \; g_c \circ g_b, \; g_a \circ g_c \circ g_a,$$
$$g_a \circ g_c \circ g_b, \; g_b \circ g_c \circ g_a, \; g_b \circ g_c \circ g_b, \; g_c \circ g_a \circ g_c, \; g_c \circ g_b \circ g_c$$

with the relations

$$g_a \circ g_b = g_a, \quad g_b \circ g_a = g_b, \quad g_a^2 = g_a, \quad g_b^2 = g_b, \quad g_c^2 = g_e,$$
$$g_a \circ g_c \circ g_b \circ g_c = g_a \circ g_c \circ g_b, \quad g_a \circ g_c \circ g_a \circ g_c = g_a \circ g_c \circ g_a,$$
$$g_b \circ g_c \circ g_b \circ g_c = g_b \circ g_c \circ g_b, \quad g_b \circ g_c \circ g_a \circ g_c = g_b \circ g_c \circ g_a,$$
$$g_c \circ g_a \circ g_c \circ g_a = g_a \circ g_c \circ g_a, \quad g_c \circ g_b \circ g_c \circ g_b = g_b \circ g_c \circ g_b,$$
$$g_c \circ g_a \circ g_c \circ g_b = g_b \circ g_c \circ g_a, \quad g_c \circ g_b \circ g_c \circ g_a = g_a \circ g_c \circ g_b.$$

Let G_1 denote this semigroup.

REMARK. There is an exact antirepresentation of the semigroup G_1 into the semigroup T of maps of the set $S = \{a, x, s, b\}$. This means that there exists an injective map $\alpha\colon G_1 \to T$ such that $\alpha(g \circ h) = \alpha(h) \circ \alpha(g)$, where $g, h \in G_1$, the symbol "$\circ$" denotes the operation in the corresponding semigroup. Namely, to the element $g \in G_1$ mapping a point with coordinates (x, s) to the point with coordinates (z, t) there corresponds the map

$$\alpha(g)\colon S \to S\colon \qquad a \mapsto a, \; b \mapsto b, \; x \mapsto z, \; s \mapsto t.$$

Here (z, t) is any pair of letters x, s, a, or b except (x, x) and (s, s). This antirepresentation can be used to make the computation of symmetries of any system with semigroup G_1 automatic.

EXAMPLE 2. Consider the coagulation kinetic equation

$$\frac{\partial u(x, t)}{\partial t} = \frac{1}{2}\int_0^x K(x-z, z)u(x-z, t)u(z, t)\,dz - u(x, t)\int_0^\infty K(x, z)u(z, t)\,dz, \tag{2.5}$$

where $K(x, z)$ is a known function such that $K(z, x) = K(x, z)$ for any $x \geqslant 0$ and $z \geqslant 0$. As in Example 1, we can introduce the dependent variables $\widetilde{v}(x, z, t)$, $w(x, z, t)$ by means of the equations

$$\begin{aligned} &\widetilde{v}'_z(x, z, t) = K(x-z, z)u(x-z, t)u(z, t), \\ &\widetilde{v}(x, 0, t) = 0, \\ &w'_z(x, z, t) = K(x, z)u(z, t), \\ &w(x, 0, t) = 0. \end{aligned} \tag{2.6}$$

But then we must consider an infinite semigroup, since this semigroup must include the map $f\colon (x, z, t) \mapsto (x-z, z, t)$ and all its powers. That is why we introduce the other dependent variable $v(x, z, t)$ such that $v(x-z, z, t) = \widetilde{v}(x, z, t) - \widetilde{v}(x, x/2, t)$, where $x \geqslant z \geqslant 0$. Then $\widetilde{v}'_z = -v_1(x-z, z, t) + v_2(x-z, z, t) = K(x-z, z, t)u(x-z, t)u(z, t)$, or

$$v_2(x, z, t) - v_1(x, z, t) = K(x, z)u(x, t)u(z, t), \tag{2.7}$$

where $x \geqslant 0$, $z \geqslant 0$. The right-hand side of the last equality is symmetric with respect to the change $x \to z$, $z \to x$. Therefore,

$$\begin{aligned}(\partial_z - \partial_x)&(v(x, z, t) + v(z, x, t)) \\ &= v_2(x, z, t) - v_1(x, z, t) + v_1(z, x, t) - v_2(z, x, t) \equiv 0.\end{aligned}$$

This means that function $F(x, z, t) = v(x, z, t) + v(z, x, t)$ is constant on the lines $x + z = \text{const}$, $t = \text{const}$. Moreover,

$$v(x, x, t) = \widetilde{v}(2x, x, t) - \widetilde{v}(2x, x, t) \equiv 0,$$

i.e.,

$$F(x, x, t) = 0.$$

As a consequence, we have $F \equiv 0$, $v(z, x, t) = -v(x, z, t)$. In addition, $\widetilde{v}(x, 0, t) = v(x, 0, t) + \widetilde{v}(x, x/2, t) = 0$, i.e.,

$$\begin{aligned}\widetilde{v}(x, x/2, t) &= -v(x, 0, t), \\ \partial u(x, t)/\partial t &= \widetilde{v}(x, x, t)/2 - u(x, t)w(x, \infty, t) \\ &= (v(0, x, t) + \widetilde{v}(x, x/2, t))/2 - u(x, t)w(x, \infty, t) \\ &= (v(0, x, t) - v(x, 0, t))/2 - u(x, t)w(x, \infty, t) \\ &= -v(x, 0, t) - u(x, t)w(x, \infty, t).\end{aligned}$$

The convergence of the second integral from (2.5) gives

$$\lim_{z\to\infty} [zK(x, z)u(z, t)] = 0.$$

Using the same arguments and notation as in Example 1 we get the system

$$\begin{aligned} &v_2 - v_1 = Kuu_c, \quad v_c = -v, \quad w_2 = Ku_c, \quad w_a = 0, \\ &u_2 = 0, \quad u_3 = -v_a - w_b u, \quad (xKu)_{cb} = 0, \end{aligned} \tag{2.8}$$

where $g_a\colon (x, z, t) \mapsto (x, 0, t)$, $g_b\colon (x, z, t) \mapsto (x, \infty, t)$, $g_c\colon (x, z, t) \mapsto (z, x, t)$ are maps of the set $M = [0, \infty] \times [0, \infty] \times \mathbb{R}$. One can make the symbol ∞ meaningful by defining

$$\begin{aligned} f(x, \infty, t) &= \lim_{z\to\infty} f(x, z, t), \\ f(\infty, z, t) &= \lim_{x\to\infty} f(x, z, t), \\ f(\infty, \infty, t) &= \lim_{(x, z)\to\infty} f(x, z, t) \end{aligned}$$

and considering only the functions for which these limits exist. This is enough for the definition of g_b, because below we use only the action of g_b on functions, but not g_b itself.

Note that the semigroup generated by g_a, g_b, g_c, $g_e = \mathrm{id}_M$ coincides with the semigroup G_1 from Example 1.

REMARK. In the next section we shall show how the geometric model of boundary type for an integro-differential equation (IDE) is constructed. Here we only gave examples of the reduction of IDE's to boundary differential equations (BDE's). The second example demonstrates that there are many BDE's for the same IDE, although solutions of each BDE are in one-to-one correspondence with solutions of the IDE. The form of the BDE, the semigroup of boundary maps and even the group of classical symmetries of BDE depend on the choice of nonlocal variables (see Remark from §5). But the group of nonlocal symmetries is independent of this choice since the change of BDE is a nonlocal transformation. Therefore, in this approach it is correct to seek a symmetry group consisting of nonclassical but nonlocal symmetries.

§3. The generalized jet spaces, boundary differential equations and their symmetries

In this section we generalize certain concepts from the geometry of differential equations to the case of boundary differential equations. Our exposition follows the corresponding parts of the book [11], and we only give an outline of the geometry of boundary differential equations.

Let $\pi\colon E \to M$ be a smooth bundle, $\Gamma(\pi)$ the set of all sections of the bundle π, G a semigroup (or a monoid) of certain maps of the base manifold M, such that the identity map of M is an element of G: $\mathrm{id}_M = g_e \in G$. We say that two sections $h_1\colon M \to E$ and $h_2\colon M \to E$ are (k, G)-*equivalent* (in particular, for $k = \infty$) at a point $x \in M$, if for any map $g \in G$ the submanifolds $h_1 \circ g(M)$ and $h_2 \circ g(M)$ are tangent to each other with order $\geqslant k$ at the point $h_1 \circ g(x) = h_2 \circ g(x) \in E$. The set of all sections that are (k, G)-equivalent to a section h at a point x is called the (k, G)-*jet* of h at the point x and denoted by $[h]_x^{(k,G)}$. The set of all the (k, G)-jets of sections of a bundle π is a smooth manifold with singularities. We denote this manifold by $J^k(\pi; G)$ and call it the *generalized jet space*.

REMARK. If the semigroup is trivial, i.e., $G = \{\mathrm{id}_M\}$, then $J^k(\pi; G)$ is the ordinary jet space $J^k\pi$.

The manifolds M and $J^k(\pi; G)$, $k = 0, 1, \dots, \infty$, are connected by natural maps $\pi_{k,s}\colon J^k(\pi; G) \mapsto J^s(\pi; G)$, $k > s$, and $\pi_k\colon J^k(\pi; G) \to M$, where $\pi_{k,s}([h]_x^{(k,G)}) = [h]_x^{(s,G)}$, $\pi_k([h]_x^{(k,G)}) = x$.

Each section $h \in \Gamma(\pi)$ defines a section $j_k(h)\colon M \to J^k(\pi; G)$ of the bundle $\pi_k\colon J^k(\pi; G) \to M$, namely, $j_k(h)(x) = [h]_x^{(k,G)}$. Every nonsingular point $x_{k+1} = [h]_x^{(k+1,G)} \in J^{k+1}(\pi; G)$ defines a subspace $L(x_{k+1}) \subset T_{x_k}(J^k(\pi; G))$ which is the tangent plane at the point $x_k = \pi_{k+1,k}(x_{k+1})$ to the graph of the (k, G)-jet of h: $L(x_{k+1}) = T_{x_k}(j_k(h)(M))$. Let $\mathcal{C}(x_k) \subset T_{x_k}(J^k(\pi; G))$ be the linear span of the subspaces of the form $L(x_{k+1})$, $x_{k+1} = \pi_{k+1,k}^{-1}(x_k)$. The distribution $\mathcal{C}\colon x_k \mapsto \mathcal{C}(x_k)$ thus obtained will be called the *Cartan distribution* on

$J^k(\pi; G)$. If $k = \infty$, then $\mathcal{C}(x_\infty) = L(x_\infty)$ and the maximal integral manifolds of the distribution $\mathcal{C}$ on $J^\infty(\pi; G)$ in some neighborhood of a nonsingular point have the form $j_\infty(h)(M)$, $h \in \Gamma(\pi)$.

By a *system of boundary partial differential equations* (or simply an *equation*) of order $\leqslant k$ for sections of π with a semigroup G, we mean a submanifold $\mathcal{Y} \subset J^k(\pi; G)$. A section $h \in \Gamma(\pi)$ is a *solution* of the equation $\mathcal{Y} \subset J^k(\pi; G)$, if $j_k(h)(M) \subset \mathcal{Y}$. The set of all jets $[h]_x^{(k+s, G)}$ such that for any map $g \in G$ the submanifold $(j_k(h) \circ g)(M)$ is tangent to $\mathcal{Y}$ at the point $[h]_x^{(k, G)}$ with order $\geqslant s$, is denoted by $\mathcal{Y}^{(s)}$ and is called the sth *prolongation* of the equation $\mathcal{Y}$. We have $\pi_{k+l, k+s}(\mathcal{Y}^{(l)}) \subset \mathcal{Y}^{(s)}$ for $l \geqslant s \geqslant 0$. Define the *infinite prolongation* of the equation $\mathcal{Y}$ as the inverse limit of the chain of maps

$$\mathcal{Y}^{(0)} \xleftarrow{\pi_{k+1, k}} \mathcal{Y}^{(1)} \longleftarrow \cdots \longleftarrow \mathcal{Y}^{(s-1)} \xleftarrow{\pi_{k+s, k+s-1}} \mathcal{Y}^{(s)} \longleftarrow \cdots,$$

which is denoted by $\mathcal{Y}^\infty$.

REMARKS. 1. If the semigroup G is nontrivial, i.e., $G \neq \{\mathrm{id}_M\}$, then $J^0(\pi; G) \neq E$ and it is possible that $\mathcal{Y}^{(0)} \neq \mathcal{Y}$, in contrast to the differential case. That is why we introduce the projection $\pi_G \colon J^\infty(\pi, G) \to E \colon [h]_x^{(\infty, G)} \mapsto h(x)$ in the boundary case.

2. If $\mathcal{Y} = J^0(\pi; G)$, then $\mathcal{Y}^{(s)} = J^s(\pi; G)$ and $\mathcal{Y}^\infty = J^\infty(\pi; G)$.

The direct limit of the chain of injections

$$C^\infty(M) \xrightarrow{\pi_0^*} \mathcal{F}_0(\pi; G) \xrightarrow{\pi_{1,0}^*} \cdots \longrightarrow \mathcal{F}_k(\pi; G) \xrightarrow{\pi_{k+1,k}^*} \mathcal{F}_{k+1}(\pi; G) \longrightarrow \cdots,$$

where $\mathcal{F}_k(\pi; G) = C^\infty(J^k(\pi; G))$, is an $\mathbb{R}$-algebra, denoted by $\mathcal{F}(\pi; G)$. By *smooth functions* on $J^\infty(\pi; G)$ (and on $\mathcal{Y}^\infty$) we mean elements of this algebra $\mathcal{F}(\pi; G)$. Any vector field X on the manifold M and any map g from G can be uniquely lifted to a vector field $\widehat{X}$ on $J^\infty(\pi; G)$ and a map g of this space respectively by means of the following formulas

$$j_\infty(h)^* \circ \widehat{X} = X \circ j_\infty(h)^*, \qquad j_\infty(h)^* \circ g^* = g^* \circ j_\infty(h)^*, \tag{3.1}$$

where h is an arbitrary section of the bundle π, and the left hand sides of the equalities contain the derivation $\widehat{X}$ and the homomorphism g^* of $\mathcal{F}(\pi; G)$.

To define the linearization of boundary differential operator, we consider the bundles $\pi \colon E_\pi \to M$, $\xi \colon E_\xi \to M$ and denote the algebra of sections of the bundle $\pi_k^*(\xi)$, where $\pi_k \colon J^k(\pi; G) \to M$, by $\mathcal{F}_k(\pi, \xi; G)$. From the equality $\pi_s \circ \pi_{k,s} = \pi_k$ for any $k \geqslant s$ we obtain the injection $\mathcal{F}_s(\pi, \xi; G) \to \mathcal{F}_k(\pi, \xi; G)$. Denote $\mathcal{F}(\pi, \xi; G) = \lim\operatorname{dir}_{s \to \infty} \mathcal{F}_s(\pi, \xi; G)$.

Every element $\varphi \in \mathcal{F}_k(\pi, \xi; G)$ can be identified with a nonlinear boundary differential operator Δ_φ of degree $\leqslant k$ mapping sections of the bundle π into sections of the bundle ξ according to the rule

$$\Delta_\varphi(h) = j_k(h)^*(\varphi), \tag{3.2}$$

where $h \in \Gamma(\pi)$ is an arbitrary section. Conversely, for any boundary differential operator Δ there exist a semigroup G and a single section $\varphi_\Delta \in \mathcal{F}_k(\pi, \xi; G)$ satisfying the condition (3.2) for any $h \in \Gamma(\pi)$.

Let $\psi \in \mathcal{F}(\pi, \xi; G)$, $\Delta = \Delta_\psi$ the corresponding operator, $\{\varphi_t \mid t \in (-\varepsilon, \varepsilon), \varepsilon > 0\}$ a smooth family of sections of the bundle $\pi_s^*(\pi)$ such that $\nabla_0 = \mathrm{id}$, and $\nabla = (\partial\nabla_t/dt)|_{t=0}$. Denote $\Delta(\nabla) = (d/dt)(\Delta_\psi \circ \nabla_{\varphi_t})|_{t=0}$ and $l_\psi(\varphi_\nabla) = \varphi_{\Delta(\nabla)}$ (or $l_\Delta(\varphi_\nabla)$). We obtain the mapping $l_\psi \colon \mathcal{F}(\pi, \pi; G) \to \mathcal{F}(\pi, \xi; G)$ which is called the *universal linearization operator* for the operator $\Delta = \Delta_\psi$.

REMARK. If $\Delta = X$ is a vector field, then $l_\Delta = \widehat{X}$, since

$$(d/dt)(X \circ \nabla_{\varphi_t})|_{t=0} = X \circ (d/dt)\nabla_{\varphi_t}|_{t=0} = X \circ \nabla, \qquad \varphi_{X\circ\nabla} = \widehat{X}(\varphi_\nabla)$$

(see (3.1) and (3.2) with $k = \infty$). If $\Delta = g^*$, $g \in G$, then similar relations imply $l_\Delta = g^*$.

The formula

$$l_{f\psi} = f l_\psi + \psi l_f, \tag{3.3}$$

where $f \in \mathcal{F}(\pi; G)$, $\psi \in \mathcal{F}(\pi, \xi; G)$, can be proved as in the differential case (see [11, 3.2.4]). Therefore, for any ξ, the relation $\mathfrak{D}_\varphi(\psi) = l_\psi(\varphi)$ defines a derivation $\mathfrak{D}_\varphi$ of the $\mathcal{F}(\pi; G)$-module $\mathcal{F}(\pi, \xi; G)$, which we call the *evolutionary derivation.* In particular, if ξ is the trivial bundle $\mathbb{R}\times M \to M$, then $\mathcal{F}(\pi, \xi; G) = \mathcal{F}(\pi; G)$ and $\mathfrak{D}_\varphi$ is a derivation of the $\mathbb{R}$-algebra $\mathcal{F}(\pi; G)$.

Denote $\Lambda^i(\mathcal{Y}^\infty) = \lim\operatorname{dir}\Lambda^i(\mathcal{Y}^s)$ and

$$\mathcal{C}\Lambda^i(\mathcal{Y}^\infty) = \{\omega \in \Lambda^i(\mathcal{Y}^\infty) \mid \omega|_{L(x_\infty)} = 0,\ x_\infty \in \mathcal{Y}^\infty\}.$$

A derivation X of the algebra $\mathcal{F}(\pi; G)$ is called a *$\mathcal{C}$-field* if $X(\mathcal{C}\Lambda^i(\mathcal{Y}^\infty)) \subset \mathcal{C}\Lambda^i(\mathcal{Y}^\infty)$. Any $\mathcal{C}$-field has the form

$$\mathfrak{D}_\varphi + \sum_i a_i \widehat{X}_i, \tag{3.4}$$

where $\varphi \in \mathcal{F}(\pi, \pi; G)$, $a_i \in \mathcal{F}(\pi; G)$, X_i is a vector field on M for each i. To prove this statement, it suffices to repeat the proof of Theorem 3.3.5 from [11] for the boundary case. Any field of the form $\sum_i a_i \widehat{X}_i$ is tangent to any solution on the infinitely prolonged equation. Hence, these fields are trivial components of $\mathcal{C}$-fields and the trajectories of a $\mathcal{C}$-field of the form (3.4) in the space of solutions are defined by the evolutionary component $\mathfrak{D}_\varphi$ of this $\mathcal{C}$-field. To find the trajectory starting on the solution $h(x)$, it is necessary to solve the evolution equation $\partial h_t(x)/\partial t = \nabla_\varphi(h_t(x))$ with initial condition $h_0(x) = h(x)$. Therefore, following [11, 12], we call this evolutionary derivation $\mathfrak{D}_\varphi$ (or the corresponding section $\varphi \in \mathcal{F}(\pi, \pi; G)$) a *higher symmetry* of a boundary differential equation $\mathcal{Y}$ if $\mathfrak{D}_\varphi$ is tangent to $\mathcal{Y}^\infty$.

A classical infinitesimal symmetry of $\mathcal{Y}$ is a $\mathcal{C}$-field that is tangent to $\mathcal{Y}^\infty$ and preserves the filtration $\cdots \subset \mathcal{F}_k(\pi; G) \subset \mathcal{F}_{k+1}(\pi; G) \subset \cdots$ of the algebra $\mathcal{F}(\pi; G)$. As in the differential case [11, 12], one can show that a classical infinitesimal symmetry is defined by a vector field on E when the dimension of the fibers of the bundle $\pi \colon E \to M$ is more than one, and on $J^1(\pi; G)$, when this dimension is one.

If a system of equations $\mathcal{Y}$ has the form $\Delta(h) = 0$, where Δ is a boundary differential operator, then the equation for a symmetry φ can be written as

$$l_{\Delta}^{\mathcal{Y}}(\varphi) = 0, \tag{3.5}$$

where $l_{\Delta}^{\mathcal{Y}}$ is the restriction of the operator l_{Δ} to the manifold $\mathcal{Y}^{\infty}$ (cf. [11, 12]). To compute the operator l_{Δ}, it suffices to use the remark above and formulas (3.3), $l_{\Delta_1+\Delta_2} = l_{\Delta_1} + l_{\Delta_2}$, $l_{X\circ\Delta} = \widehat{X} \circ l_{\Delta}$, $l_{g^*\circ\Delta} = g^* \circ l_{\Delta}$.

§4. Group analysis of the coagulation kinetic equation

The theory presented above will be applied here and in the following sections. We shall find the symmetry groups of two integro-differential equations and shall use the symmetries for reducing the equations.

The Smoluchowski equation (2.5) describes the time evolution of the size distribution of particles coagulating by two-body collisions. This equation was first applied to small suspended particles which collide and coagulate by virtue of their Brownian motion and has subsequently been applied to interacting polymers and to other physical systems.

To compute the symmetries of this equation, we replace it by system (2.8) and use formula (3.5), where $\mathcal{Y}$ is the submanifold defined in the jet space $J^1(\pi; G_1)$ by equations (2.8), π and G_1 are the bundle and the semigroup from Example 2 of §2. In this case equation (3.5) takes the form

$$\begin{gathered} D_2V - D_1V = K(Uu_c + uU_c), \quad V_c = -V, \quad D_2W = KU_c, \\ W_a = 0, \quad D_2U = 0, \quad D_3U = -V_a - W_bu - w_bU, \quad (xKU)_{cb} = 0, \end{gathered} \tag{4.1}$$

where U, V, W, $U_c = g_c^*(U)$, $V_a = g_a^*(V)$, etc. are components of the symmetry $\mathfrak{I}_{\varphi} = U\partial_u + V\partial_v + W\partial_w + U_c\partial_u + \dots$, D_i is the restriction of a total derivative to $\mathcal{Y}^{\infty}$ (i.e., $D_1 = \widehat{(\partial/\partial x)}|_{\mathcal{Y}^{\infty}}$, etc., see (3.1)), and the maps g_a, g_b, g_c are described in Example 2.

The symmetries of the first order are defined by the components of the form

$$\begin{aligned} U = U(&x, z, t, u, u_c, u_{ca}, u_{cb}, u_1, u_{1c}, u_{1ca}, u_{1cb}, v, v_a, v_b, \\ &v_{ac}, v_{bc}, v_{bca}, v_1, v_{1a}, v_{1b}, v_{1ac}, v_{1bc}, v_{1bca}, v_3, v_{3a}, v_{3b}, \\ &v_{3ac}, v_{3bc}, v_{3bca}, w, w_c, w_b, w_{ca}, w_{cb}, w_{bc}, w_{cac}, w_{bca}, w_{bcb}, \\ &w_{cbc}, w_1, w_{1c}, w_{1b}, w_{1ca}, w_{1cb}, w_{1bc}, w_{1cac}, w_{1bca}, w_{1bcb}, \\ &w_{1cbc}, w_3, w_{3c}, w_{3b}, w_{3ca}, w_{3cb}, w_{3bc}, w_{3cac}, w_{3bca}, w_{3bcb}, w_{3cbc}), \end{aligned}$$

and similarly for V, W.

The computations are performed in two simplified cases.

1. If we consider only the variables that appear in equations (2.8), i.e., assume that U, V, W are functions of x, z, t, u, u_c, u_1, u_{1c}, v, v_a, v_1, v_{1a}, v_3, v_{3a}, w, w_b, w_1, w_{1b}, w_3, w_{3b}, then we obtain

$$\begin{gathered} U = \eta - \xi^1 u_1 - \xi^3 u_3, \\ V = \eta^3 - \xi_c^1 K u u_c - (\xi^1 + \xi_c^1) v_1 - \xi^3 v_3 = \eta^3 - \xi^1 v_1 - \xi_c^1 v_2 - \xi^3 v_3, \\ W = \eta^4 - \xi_c^1 K u_c - \xi^1 w_1 - \xi^3 w_3 = \eta^4 - \xi^1 w_1 - \xi_c^1 w_2 - \xi^3 w_3, \end{gathered} \tag{4.2}$$

where the functions ξ^1, ξ^3, η, η^3, η^4 are defined by the function $K(x, z)$. For an arbitrary K we have

$$\begin{aligned} \xi^1 = 0, \quad \xi^3 = -C_1 t + C_2, \\ \eta = C_1 u, \quad \eta^3 = 2C_1 v, \quad \eta^4 = C_1 w. \end{aligned} \tag{4.3}$$

For a homogeneous function K, i.e., in this case $K(\lambda x, \lambda z) = \lambda^\sigma K(x, z)$, we have

$$\begin{aligned} \xi^1 = Cx, \quad \xi^3 = -[C_1 + C(1+\sigma/2)]t + C_2, \\ \eta = (C_1 - C\sigma/2)u, \quad \eta^3 = (2C_1 + C)v, \quad \eta^4 = [C_1 + C(1+\sigma/2)]w \end{aligned} \tag{4.4}$$

with arbitrary constants C, C_1, C_2.

For $K(x, z) = F(x)F(z)$, we have the additional solution

$$U = V = 0, \quad W = f(F'(x)w - F(x)w_1, x, t, u, v_a, w_b, u_1, v_{1a}, w_{1b}),$$

where the function f satisfies the condition

$$f(0, x, t, u, v_a, w_b, u_1, v_{1a}, w_{1b}) = 0.$$

2. If we look for a the solution in the form (4.2), where ξ^1, ξ^3, η, η^3, η^4 are functions of x, z, t, u, u_c, u_{ca}, u_{cb}, v, v_a, v_b, v_{ac}, v_{bc}, v_{bca}, w, w_c, w_b, w_{ca}, w_{cb}, w_{bc}, w_{cac}, w_{bca}, w_{bcb}, w_{cbc}, then we obtain the same result.

The case of a homogeneous kernel K is of particular interest to specialists [14]. For such a kernel, the symmetry algebra is spanned by the operators

$$\begin{aligned} X_1 &= \partial_t, \\ X_2 &= t\partial_t - (u\partial_u + u_c\partial_{u_c} + \dots) \\ &\qquad - 2(v\partial_v + v_a\partial_{v_a} + \dots) - (w\partial_w + w_c\partial_{w_c} + \dots), \\ X_3 &= x\partial_x + z\partial_z - (1+\sigma/2)t\partial_t - \sigma(u\partial_u + u_c\partial_{u_c} + \dots)/2 \\ &\qquad + (v\partial_v + v_a\partial_{v_a} + \dots) + (1+\sigma/2)(w\partial_w + w_c\partial_{w_c} + \dots). \end{aligned}$$

We shall now use this algebra to reduce the coagulation equation (2.5) to equations involving the independent variable alone.

To reduce an equation along a classical infinitesimal symmetry means to extract solutions of the equation that are mapped into themselves by translations along the infinitesimal symmetry (i.e., invariant solutions). If two symmetries generate conjugated subalgebras of the symmetry algebra, then the reductions along them are equivalent (see [9]). Therefore, it is necessary first to classify conjugacy classes of one-dimensional subalgebras under the adjoint action. The classification for the algebra considered here gives the one-dimensional subalgebras that are generated by the fields X_1, X_2, $-aX_2 + X_3$ (where $a + 1 + \sigma/2 \neq 0$, $a \in \mathbb{R}$) and $\pm X_1 - (1+\sigma/2)X_2 + X_3$.

For $X_1 = \partial_t$, invariant solutions have the form $u = \Phi(x)$, and the reduced equation is

$$\frac{1}{2}\int_0^x K(x-z, z)\Phi(x-z)\Phi(z)\,dz - \Phi(x)\int_0^\infty K(x, z)\Phi(z)\,dz = 0. \tag{4.5}$$

This equation was considered in [15].

For $X_2 = t\partial_t - u\partial_u$, the reduction is obtained by the substitution $u = t^{-1}\Phi(x)$. The reduced equation is

$$-\Phi(x) = \frac{1}{2}\int_0^x K(x-z, z)\Phi(x-z)\Phi(z)\,dz - \Phi(x)\int_0^\infty K(x, z)\Phi(z)\,dz. \tag{4.6}$$

For $-aX_2 + X_3 = x\partial_x - (a+1+\sigma/2)\,t\partial_t + (a-\sigma/2)\,u\partial_u$, the invariant solutions have the form $u = t^{-(a-\sigma/2)/\omega}\Phi(x^{1/\omega}t)$, where $\omega = a+1+\sigma/2$, and the reduced equation is

$$\begin{aligned}\omega^{-1}[\xi\Phi'(\xi) - (a-\sigma/2)\Phi(\xi)] &= \frac{1}{2}\int_0^\xi K(\xi-s, s)\Phi(\xi-s)\Phi(s)\,ds \\ &\quad - \Phi(w)\int_0^\infty K(\xi, s)\Phi(s)\,ds,\end{aligned} \tag{4.7}$$

where $\xi = xt^{1/\omega}$. In the case $\sigma/2 - a = 2$, this equation and the invariant solutions $u = t^{-2/(1-\sigma)}\Phi(xt^{-1/(1-\sigma)})$ have been obtained in [14, 16] using a scaling hypothesis.

For $X_1 - (1+\sigma/2)X_2 + X_3 = \partial_t + x\partial_x - (1+\sigma)u\partial_u$, the reduction is obtained by the substitution $u = e^{-(1+\sigma)t}\Phi(xe^{-t})$. The reduced equation is

$$\begin{aligned}-[\xi\Phi'(\xi) + (1+\sigma)\Phi(\xi)] &= \frac{1}{2}\int_0^\xi K(\xi-s, s)\Phi(\xi-s)\Phi(s)\,ds \\ &\quad - \Phi(\xi)\int_0^\infty K(\xi, s)\Phi(s)\,ds,\end{aligned} \tag{4.8}$$

where $\xi = xe^{\mp t}$.

REMARK. The reduction of the system (2.8) gives systems of boundary differential equations, which are equivalent to the reduced integro-differential equations (4.5)–(4.8).

§5. Classical symmetries of the equation for Brownian coagulation of aerosol particles in a stochastic medium

We shall now consider the equation

$$\begin{aligned}&(\partial/\partial t - \mathcal{D}(v_1)\nabla_{\mathbf{r}}^2)u(v_1, \mathbf{r}, t) \\ &= \frac{1}{2}\int_0^{v_1}[K(v_1-v_2, v_2)u(v_1-v_2, \mathbf{r}, t)u(v_2, \mathbf{r}, t) \\ &\qquad + L(v_1-v_2, v_2)\nabla_{\mathbf{r}}u(v_1-v_2, \mathbf{r}, t)\nabla_{\mathbf{r}}u(v_2, \mathbf{r}, t)]\,dv_2 \\ &\quad - \int_0^\infty [K(v_1, v_2)u(v_1, \mathbf{r}, t)u(v_2, \mathbf{r}, t) \\ &\qquad + L(v_1, v_2)\nabla_{\mathbf{r}}u(v_1, \mathbf{r}, t)\nabla_{\mathbf{r}}u(v_2, \mathbf{r}, t)]\,dv_2,\end{aligned} \tag{5.1}$$

where u is a dependent variable, v_1, $\mathbf{r} = (x, y, z)$, t are independent variables, K, L, $\mathcal{D}$ are known functions, K, L are symmetric, and $\nabla_{\mathbf{r}} = (\partial_x, \partial_y, \partial_z)$. This equation describes mean fields of nonuniform size distribution functions of particles under Brownian coagulation in a stochastic medium with small velocity fluctuations [17]. In the uniform case ($\nabla_{\mathbf{r}} u = 0$), the equation (5.1) takes the form the Smoluchowski coagulation equation (2.5).

As for equation (2.5), introduce an independent variable v_1 and dependent variables p and q as

$$\begin{aligned}(d/dv_2)\,p(v_1 - v_2, \mathbf{r}, t) &= K(v_1 - v_2, v_2)u(v_1 - v_2, \mathbf{r}, t)u(v_2, \mathbf{r}, t)\\ &\quad + L(v_1 - v_2, v_2)\nabla_{\mathbf{r}} u(v_1 - v_2, \mathbf{r}, t)\nabla_{\mathbf{r}} u(v_2, \mathbf{r}, t),\\ (d/dv)\,q(v_1 - v_2, \mathbf{r}, t) &= K(v_1, v_2)u(v_1, \mathbf{r}, t)u(v_2, \mathbf{r}, t)\\ &\quad + L(v_1, v_2)\nabla_{\mathbf{r}} u(v_1, \mathbf{r}, t)\nabla_{\mathbf{r}} u(v_2, \mathbf{r}, t).\end{aligned}$$

Using the notation of the preceding sections, we write the boundary differential form for the equation (5.1) as

$$\begin{aligned}&p_2 - p_1 = Kuu_c + L\nabla_{\mathbf{r}} u \nabla_{\mathbf{r}} u_c, \qquad p_c = -p,\\ &q_2 = Kuu_c + L\nabla_{\mathbf{r}} u \nabla_{\mathbf{r}} u_c, \qquad q_a = 0,\\ &u_2 = 0, \quad u_3 - \mathcal{D}(v_1)\Delta_{\mathbf{r}} u = -p_a - q_b,\\ &(v_2 K u_c)_b = 0, \quad (v_2 L \nabla_{\mathbf{r}} u \nabla_{\mathbf{r}} u_c)_b = 0,\end{aligned} \tag{5.2}$$

where $g_a\colon (v_1, v_2, \mathbf{r}, t) \mapsto (v_1, 0, \mathbf{r}, t)$, $g_b\colon (v_1, v_2, \mathbf{r}, t) \mapsto (v_1, \infty, \mathbf{r}, t)$, $g_c\colon (v_1, v_2, \mathbf{r}, t) \mapsto (v_2, v_1, \mathbf{r}, t)$ are generators of the same semigroup as in §2.

The symmetry algebra is generated by the infinitesimal symmetries

$$\begin{aligned}&X_1 = \partial_t, \quad X_2 = \partial_x, \quad X_3 = \partial_y, \quad X_4 = \partial_z,\\ &X_5 = y\partial_x - x\partial_y, \quad X_6 = x\partial_z - z\partial_x, \quad X_7 = z\partial_y - y\partial_z.\end{aligned} \tag{5.3}$$

If the functions K, L, $\mathcal{D}$ are homogeneous, i.e.,

$$\begin{aligned}K(\lambda v_1, \lambda v_2) &= \lambda^{\sigma_K} K(v_1, v_2),\\ L(\lambda v_1, \lambda v_2) &= \lambda^{\sigma_L} L(v_1, v_2),\\ \mathcal{D}(\lambda v) &= \lambda^{\sigma_{\mathcal{D}}} \mathcal{D}(v),\end{aligned}$$

then the algebra contains also the operator

$$\begin{aligned}X_8 &= 2v\partial_v + 2(\sigma_L - \sigma_K - \sigma_{\mathcal{D}})t\partial_t\\ &\quad + (\sigma_L - \sigma_K)(x\partial_x + y\partial_y + z\partial_z) + 2(\sigma_{\mathcal{D}} - \sigma_K - 1)u\partial_u.\end{aligned} \tag{5.4}$$

For physical applications it is interesting to consider the case $\sigma_L = \sigma_K + 1/3$, $\sigma_{\mathcal{D}} = 0$ [17]. Then we have the following form of invariant solutions $u =$

$t^{-(4+3\sigma_K)}\Phi(w_1, \rho)$, where $w_1 = v/t^3$, $\rho = (x/\sqrt{t}, y/\sqrt{t}, z/\sqrt{t})$. The corresponding reduced equation is

$$\begin{aligned} &-(4+3\sigma_K)\Phi - 3w_1\frac{\partial\Phi}{\partial w_1} - \frac{1}{2}\rho\nabla_\rho\Phi - \mathcal{D}(w_1)\Delta_\rho\Phi \\ &\quad = \frac{1}{2}\int_0^{w_1}[K(w_1-w_2, w_2)\Phi(w_1-w_2, \rho)\Phi(w_2, \rho) \\ &\qquad\qquad + L(w_1-w_2, w_2)\nabla_\rho\Phi(w_1-w_2, \rho)\nabla_\rho\Phi(w_2, \rho)]\,dw_2 \\ &\qquad - \int_0^\infty [K(w_1, w_2)\Phi(w_1, \rho)\Phi(w_2, \rho) \\ &\qquad\qquad + L(w_1, w_2)\nabla_{\mathbf{r}}\Phi(w_1, \rho)\nabla_{\mathbf{r}}\Phi(w_2, \rho)]\,dw_2 . \end{aligned}$$

REMARK. Consider another model for equation (5.1). Namely, replace the variable q in (5.2) by variables q^1 and q^2 such that $q = q^1 + q^2$, the third and the fourth equations of (5.2) by the four equations

$$q^1_2 = Kuu_c + L(u_4u_{4c} + u_5u_{5c}), \quad q^2_2 = Lu_4u_{4c}, \quad q^1_a = 0, \quad q^2_a = 0,$$

where the indices 4, 5, 6 denote derivatives with respect to x, y, z. Unlike system (5.2), the new system (with q^1 and q^2) does not have the symmetry $X_7 = z\partial_y - y\partial_z$. But if we introduce one more (nonlocal) dependent variable q^3 as $q^3_2 = L(u_5u_{6c} + u_6u_{5c})$, the new system has this symmetry. Hence, groups of classical symmetries for integro-differential equations depend on the choice of models: for the model (5.2) of equation (5.1), the symmetry X_7 is classical, but for the model with q^1 and q^2 it is nonlocal.

§6. Conservation laws of boundary differential equations

First consider how conserved quantities are obtained from conservation laws in the differential case (see [10, 12]). Let the base manifold M for the differential equation have the form $M = \mathbb{R} \times M_0$, where the coordinate on $\mathbb{R}$ corresponds to the time variable and the $(n-1)$-dimensional manifold M_0 has the boundary ∂M_0. Let an $(n-1)$-dimensional differential form $\omega(u(x))$ on M depend on solutions $u(x)$ of the equation and be closed for any $u(x)$. Consider the submanifold $N = [t_0, t_1] \times M_0 \subset M$ and the integral of $d\omega$ $(=0)$ over this submanifold. Using the Stokes theorem and the identity $\partial N = \{t_0\} \times M_0 \cup [t_0, t_1] \times \partial M_0 \cup \{t_1\} \times M_0$, we obtain

$$0 = \int_N d\omega = \int_{\partial N}\omega = \int_{M_0}\omega|_{t=t_0} + \int_{N_1}\omega - \int_{M_0}\omega|_{t=t_1}, \tag{6.1}$$

where $N_1 = [t_0, t_1] \times \partial M_0$. The next to last integral vanishes if any solution $u(x)$ of the differential equation vanishes on the submanifold $\mathbb{R} \times \partial M_0 \supset N_1$ and the differential form $\omega(u)$ equals zero when $u = 0$. Note that these restrictions are exactly the restrictions used to obtain conserved quantities (see [10]). In this case from (6.1) it follows that the value of the integral $\int_{M_0}\omega(u)$ is independent of the time t.

When the manifold M_0 has no boundary, it suffices to consider a collection of submanifolds $M_0^\tau \subset M_0$ depending on a parameter $\tau \geqslant \tau_0$ such that $M_0^\tau \to M_0$ as $\tau \to \infty$, and to pass to the limit as $\tau \to \infty$ in (6.1) with M_0^τ instead of M_0.

Thus if the differential form $\omega(u)$ satisfies the conditions

$$d\omega(u) = 0, \qquad \omega(u)|_{\mathbb{R}\times\partial M_0} = 0, \tag{6.2}$$

for any solution u, then

$$\int_N d\omega(u) = 0, \qquad \int_{N_1} \omega(u) = 0, \tag{6.3}$$

and $\int_{M_0} \omega(u)$ is a conserved quantity for the differential equation. If $\omega(u) = d\omega_1(u) = 0$ and $\omega_1(u)|_{\mathbb{R}\times\partial M_0} = 0$, then $\int_{M_0} \omega = \int_{\partial M_0} \omega_1 = 0$ and we have a trivial conserved quantity. That is why the class of forms $\{\omega + d\omega_1 \mid \omega_1|_{\mathbb{R}\times\partial M_0} = 0\}$, where the form ω satisfies (6.3), is called a conservation law.

In the boundary differential case, in order to obtain the first equation of (6.3) for any solution $u(x)$, one can use relations of the form

$$\int_N [\widetilde{\omega} - I_f f^*(\widetilde{\omega})] = 0, \tag{6.4}$$

where f is a diffeomorphism of the orientable manifold M belonging to the semigroup of the boundary differential equation and mapping the submanifold $N = [t_0, t_1]\times M_0 \subset M$ into itself, the number I_f equals -1 if the diffeomorphism f reverses the orientation of M and $+1$ otherwise. Below we call the number I_f the *orientation* of the diffeomorphism f.

Similar equalities can be also used to prove the second equation in (6.3). In addition, if g and h are maps taking M to ∂M and $g^2 = g$, $h^2 = h$, $g \circ h = g$, $h \circ g = h$, then $g\colon h(M) \to g(M)$ is a diffeomorphism and $(g|_{h(M)})^{-1} = h|_{g(M)}$. Hence, when M is a compact manifold,

$$\int_{h(M)} h^*(\omega) = \int_M h^*(h^*(\omega)) = \int_M h^*(\omega) = \int_M g^*(h^*(\omega)) = \int_{g(M)} h^*(\omega)$$

for any form ω.

The last relation and those of the form (6.4) allow us to compare integrals over different parts of the boundary of M and over M without computing their values. Now we define conservation laws in the boundary differential case, using these equalities and the homological language (cf. [12]). Let $\mathcal{Y}$ be a boundary differential equation with semigroup G, $M = \mathbb{R}\times M_0$ the manifold of independent variables of this equation, and M_0 a compact manifold or a limit of compact manifolds. Consider diffeomorphisms of M and projections (i.e., $g^2 = g$) from M to $\mathbb{R}\times\partial M_0 \subset M$ that belong to the semigroup G and map lines of the form $\mathbb{R}\times\{x\}$ to the same lines and preserve the fibers $\{t\}\times M_0$, $t \in \mathbb{R}$. These maps are defined by diffeomorphisms of M_0 or by projections from M_0 into ∂M_0 respectively. Complete, if necessary, this set of diffeomorphisms to a group G_0. Extend the semigroup G so that $G_0 \subset G$. In addition, note that if g is a projection from M_0 to ∂M_0 and f is a diffeomorphism of M_0, then $f^{-1}\circ g\circ f$ is also a projection from M_0 to ∂M_0 since any diffeomorphism of M_0 is simultaneously a diffeomorphism of its boundary ∂M_0.

Let the group G_0 have $l+1$ elements: $f_0=g$ (the identity element), f_1, $\dots$, f_l, and let $g_1,\dots,g_k$, $h_1,\dots,h_k$ be projections from M into ∂M satisfying the above-mentioned conditions and having $(n-1)$-dimensional images, where n is the dimension of M. Moreover, let $g_i\circ h_i=g_i$, $h_i\circ g_i=h_i$, $i=1,\dots,k$, and the images of the projections $f_j^{-1}\circ g_i\circ f_j$, $f_j^{-1}\circ h_i\circ f_j$, $j=0,1,\dots,l$, $i=1,\dots,k$, cover the whole set $\mathbb{R}\times\partial M_0$ and intersect only in sets of lesser dimension. If necessary, complete the semigroup G so that G has the elements f_j, g_i, h_i. Consider the bicomplex

$$\begin{array}{ccccccccccc}
\longrightarrow & \Lambda^i & \xrightarrow{d} & \Lambda^{i+1} & \longrightarrow & \cdots & \longrightarrow & \Lambda^{n-1} & \xrightarrow{d} & \Lambda^n & \longrightarrow 0\\
& \uparrow{\scriptstyle\partial_1} & & \uparrow{\scriptstyle\partial_1} & & & & \uparrow{\scriptstyle\partial_1} & & \uparrow{\scriptstyle\partial_1} & \\
\longrightarrow & \Lambda^i & \xrightarrow{d} & \Lambda^{i+1} & \longrightarrow & \cdots & \longrightarrow & \Lambda^{n-1} & \xrightarrow{d} & \Lambda^n & \longrightarrow 0\\
& \uparrow{\scriptstyle\partial_0} & & \uparrow{\scriptstyle\partial_0} & & & & \uparrow{\scriptstyle\partial_0} & & \uparrow{\scriptstyle\partial_0} & \\
\longrightarrow & C^i & \xrightarrow{d} & C^{i+1} & \longrightarrow & \cdots & \longrightarrow & C^{n-1} & \xrightarrow{d} & C^n & \longrightarrow 0,
\end{array} \tag{6.5}$$

where the first and the second lines are the de Rham complex of M, and the last line is the sum of l copies of this complex, i.e.,

$$C^i=\bigoplus_{j=1}^{l}\Lambda^i \text{ etc.},\qquad \partial_0\left(\bigoplus_{j=1}^{l}\omega_j\right)=\sum_{j=1}^{l}(\omega_j-I_{f_j}f_j^*(\omega_j)),$$

I_{f_j} is the orientation of f_j,

$$\partial_1(\omega)=\sum_{i=1}^{k}\left[h_i^*\left(\sum_{s=0}^{l}I_{f_s}f_s^*(\omega)\right)-g_i^*\left(\sum_{s=0}^{l}I_{f_s}f_s^*(\omega)\right)\right].$$

We have $\partial_1\circ\partial_0=0$. Indeed, $I_{f_s}I_{f_j}=I_{f_j\circ f_s}$ and

$$\sum_{s=0}^{l}I_{f_s}f_s^*\left[\sum_{j=1}^{l}\omega_j-I_{f_j}f_j^*(\omega)\right]$$
$$=\sum_{j=1}^{l}\left[\sum_{s=0}^{l}I_{f_s}f_s^*(\omega_j)-\sum_{s=0}^{l}I_{f_j\circ f_s}(f_j\circ f_s)^*(\omega_j)\right]=0,$$

since the maps $f_0,f_1,\dots,f_l$ constitute the group G_0 and $f_j\circ G_0=G_0$.

Let the differential forms from the bicomplex (6.5) depend on solutions of the equation $\mathcal{Y}$ and d be the differential for these forms. This means that if $x_1,\dots,x_n$ are coordinates on M, $D_1,\dots,D_n$ are the corresponding total derivatives (i.e., $D_i=\widehat{\partial/\partial x_i}$, see (3.1)), and ω is a differential form depending on the solutions of $\mathcal{Y}$, then $d\omega = \sum_{i=1}^{n} dx_i \wedge D_i(\omega)|_{\mathcal{Y}^\infty}$, where the result of the differentiation is restricted to $\mathcal{Y}^\infty$. Consider the modules $A^i = \ker(\partial_1|_{\Lambda^i})/\operatorname{im}(\partial_0|_{C^i})$. The cohomology classes of the complex

$$\longrightarrow A^i \xrightarrow{d} A^{i+1} \longrightarrow \cdots \longrightarrow A^{n-1} \xrightarrow{d} A^n \longrightarrow 0 \tag{6.6}$$

at the term A^{n-1} will be called *conservation laws* of the boundary differential equation $\mathcal{Y}$.

To compute the conservation laws of differential equations, the notion of the conjugate operator is used. In order to define this notion for the boundary case, fix an element $\omega \in A^n$ generating A^n. It exists since Λ^n, and hence A^n, are one-dimensional modules. We say that the linear operator $\Delta^+ : C^\infty(M) \to C^\infty(M)$ is *conjugate* to a given linear operator $\Delta : C^\infty(M) \to C^\infty(M)$ for the form ω, if for any functions φ, ψ on M the class of the form $(\psi\Delta^+(\varphi) - \varphi\Delta(\psi))\omega$ in A^n belongs to the image of the differential d. For instance, if Δ is the multiplication operator, then $\Delta^+ = \Delta$. Also, $D^+ = -D$, $(f^*)^+ = I_f\alpha(f^{-1})^*$, where D is the total derivative, $f \in G$ is a diffeomorphism of M, α is a function on M such that $(f^{-1})^*(\omega) = \alpha\omega$. Further, $(\Delta_1 \circ \Delta_2)^+ = \Delta_2^+ \circ \Delta_1^+$ provided that both Δ_1 and Δ_2 having conjugate operators. But there is no conjugate operator to the operator g^*, if g is a projection from M to ∂M.

Let the system of equations $\mathcal{Y}$ have the form $F_i(u) = 0$, $i = 1, \dots, m$, and let the differential form $\Omega(u) \in \Lambda^{n-1}$ specify a conservation law for $\mathcal{Y}$. According to the definition, $d\Omega(u) = 0$ for any solution $u(x)$ of the system $\mathcal{Y}$. Since the form ω generates A^n, there exist functions ψ^i, $i = 1, \dots, m$, such that

$$d\Omega(u) = \sum_{i=1}^{m} \psi^i F_i(u)\omega \tag{6.7}$$

for any section $u(x)$ (not only for solutions). Acting on (6.7) by an arbitrary evolutionary derivation ∂_φ, we obtain

$$\partial_\varphi(d\Omega(u)) = d\partial_\varphi(\Omega(u)) = \sum_{i=1}^{m} [F_i(u)\partial_\varphi(\psi^i\omega) + \psi^i\partial_\varphi(F_i(u))\omega].$$

From this it follows that the differential form $\sum_{i=1}^{m} \psi^i \partial_\varphi(F_i(u))\omega$ or

$$\sum_{i=1}^{m} \psi^i l_{F_i(u)}(\varphi)\omega \tag{6.8}$$

is exact for any solution $u(x)$ of the equation $\mathcal{Y}$. Using this fact, the concept of conjugate operator, and the arbitrariness of the section φ, we can obtain a system of equations of the form (3.5) for the functions ψ^i, $i = 1, \dots, m$. Solving it, we find the functions ψ^i. Then equation (6.7) gives a conservation law $\Omega(u)$. The corresponding conserved quantity is obtained as $\int_{M_0} \Omega(u)$.

This computation scheme is applicable in the boundary differential case as well as in the differential case. But the differential case differs from the boundary case in that the conjugate operator always exists. Therefore, in (6.8) we only can consider the determining equations $F_i(u) = 0$ of the system $\mathcal{Y}$ but not their prolongations of the form $\Delta_i(F_i(u)) = 0$, where Δ_i is some linear differential operator. In fact, using properties of the linearization $l_{F_i(u)}$ (see the end of §3) and the concept of conjugate operator, we can pass from the expression $\sum_{i=1}^{m} \psi^i l_{\Delta_i \circ F_i(u)}(\varphi)$ (see (6.8)) to the expression $\sum_{i=1}^{m} \Delta_i^+(\psi^i) l_{F_i(u)}(\varphi)\omega$ and seek functions $\widetilde{\psi}^i = \Delta_i^+(\psi^i)$. Moreover, we can replace the differential form (6.8) by the form

$$\sum_{i=1}^{m} l^+_{F_i(u)}(\psi^i)\varphi\omega . \tag{6.9}$$

The form (6.9) must be exact for any section φ. Hence in the differential case we obtain the equation $\sum_{i=1}^{m} l^+_{F_i(u)}(\psi^i) = 0$ [12].

In the boundary case, there is no conjugate operator to the operator g^* if $g \in G$ is not a diffeomorphism. Therefore, we must consider not only the determining equation $F_i(u) = 0$ of the system $\mathcal{Y}$, but also its prolongations $(g^* \circ F_i)(u) = 0$, $(g^* \circ D \circ F_i)(u) = 0$, $(g^* \circ f^* \circ F_i)(u) = 0$, etc., where D is the total derivative, $f \in G$ is a diffeomorphism. For the same reason, we cannot change (6.8) to the form (6.9). But using the arbitrariness of φ, we can choose a collection of differential forms whose linear combination can be an exact form only if all coefficients of this combination equal zero. This gives a system of equations in ψ^i, $i = 1, \dots, m$, like (6.9). Recall that a similar argument was used in [5] for finding symmetries (see above).

The authors used this method to compute conservation laws of the kinetic coagulation equation (2.5) when the unknown form $\Omega(u)$ (see (6.7)) is independent of higher order derivatives of $u(x)$. The integro-differential equation (2.5) is transformed to the system of boundary differential equations (2.8) with semigroup G_1 (see §2). The operators ∂_0 and ∂_1 from the bicomplex (6.5), where $n = 3$, $C^i = \Lambda^i$, are defined for this semigroup as

$$\partial_0 = g_e^* + g_c^*, \quad \partial_1 = g_b^* - g_a^* - g_b^* \circ g_c^* + g_a^* \circ g_c^* .$$

We now illustrate by a simple example how equations in the coefficients ψ^i are deduced in this case. Consider the evolutionary derivatives of the form $\ni_\varphi = W\partial_w + \dots$, where the coefficients at ∂_u and ∂_v equal zero. After this simplification the right-hand side of the expression (6.7) with $\omega = dx \wedge dz \wedge dt$ takes the form $[\psi^1 D_2(W) + \psi^2(g_a^* \circ D_2)(W) + \psi^3 g_a^*(W)]\omega$. Using $D_2^+ = -D_2$ we obtain the differential form $\omega_1 = [-D_2(\psi^1)\, W + \psi^2(g_a^* \circ D_2)(W) + \psi^3 g_a^*(W)]\omega$. This form must be exact for any function W. If $W = z^2 W_0$, then the last form equals $\omega_1 = -D_2(\psi^1)\, z^2 W_0 dx \wedge dz \wedge dt$.

Assume that we have already found functions ψ^1, ψ^2, ψ^3 such that the form ω_1 is exact for any W_0. Let $D_2(\psi^1)$ be a function on the k-jet space. Put $W_0 = D_1(p)$, where $p = u_{1\dots 1}$ ($k+1$ units), i.e., this is the derivative of order $(k+1)$ of the dependent variable u with respect to x. Let $\omega_1 = d\Omega$ for some form Ω. Comparing the coefficients of the highest derivatives for the forms ω_1 and $d\Omega$, we obtain $\Omega = -D_2(\psi^1)\, z^2 p\, dz \wedge dt$. In addition, the function $\alpha = D_2(\psi^1)\, z^2$ is constant.

On the other hand, if $W_0 = uu_c$, then $\omega_1 = -\alpha uu_c\, dx \wedge dz \wedge dt$. The equality $\omega_1 = d\Omega$ is possible only if $\alpha = 0$, since the form $uu_c\, dx \wedge dz \wedge dt$ cannot be exact for any solution u of system (2.8). Hence, $D_2(W^1) = 0$. Setting $W = zW_0$ and acting similarly, we obtain also $\psi^2 = 0$ and $\psi^3 = 0$. These are the desired equation in ψ^i.

Using arguments of this sort, the authors found a single conservation law. It is specified by the form

$$\Omega = xu\theta' dz \wedge dx + (xwu - x\theta(v_a + w_b u) - xv - \tfrac{1}{2}zv)\, dt \wedge dx - \tfrac{1}{2}xv\, dt \wedge dz,$$

where θ is a function of z such that $\theta(\infty) - \theta(0) = 1$, and θ' is its derivative. This form satisfies the conditions $\partial_1(\Omega) = 0$ and $d\Omega \in \operatorname{Im}(\partial_0|_{\Lambda^3})$, when $(zv)_b = 0$. Integrating the form Ω over the domain $0 \leqslant x < +\infty$, $0 \leqslant z < +\infty$ and using the condition on θ, we obtain the conservation law for mass

$$\int_0^\infty xu(t, x)\, dx \tag{6.10}$$

for the coagulation kinetic equation (2.5). As is known, this conservation law holds when the integral

$$\int_0^\infty \int_0^\infty K(x, z) u(t, x) u(t, z)\, dx\, dz$$

converges. The condition $(zv)_b = \lim_{z \to \infty}(zv(x, z, t)) = 0$ follows from the convergence of this integral. This fact conforms with the above-mentioned result.

Concluding remarks

The method used here to construct geometric models of integro-differential equations involves two steps: 1) the elimination of integrals by introducing potentials and using the fundamental theorem of calculus; 2) the representation of the obtained system of boundary differential equations as a submanifold of a generalized jet space. Models of differential equations (see the Introduction) is closer to the obtained model than to the other models considered here (see §1). This analogy allows us to extend the concepts of higher symmetries, conservation laws, and, apparently, many other concepts (see [9–13]) to the case of integro-differential equations.

Our approach is not free from drawbacks. One of them lies in the necessity of a suitable selection of the model. We keep in mind the following. Step 2) is made in a unique way (see §3), but the execution of step 1) depends on the choice of the potential variables. After a change of potential, the finite semigroup of boundary differential equations can become infinite (see Example 2 from §2) and a classical symmetry can become nonlocal (see the Remark from §5). Moreover, in order to find conservation laws, it is sometimes necessary to extend or to modify the corresponding semigroup (see §6).

This drawback is compensated, we think, by the range of applicability of this method. In addition, one easily synthesizes algorithms for all computations (of symmetries, conservation laws, etc.). Therefore, computational difficulties can be overcome by using computers.

Acknowledgement. This work was supported in part by the program "Russian University", project 1.4.11. We are grateful to A. M. Vinogradov for his suggestion to study these problems and for his advise to use coverings.

References

1. V. B. Taranov, *On symmetry of one-dimensional high-frequency motions of collisionless plasma*, Zh. Tekh. Fiz. **46** (1976), no. 6, 1271–1277. (Russian)
2. A. I. Bunimovich and A. V. Krasnoslobodtsev, *Group invariant solutions of kinetic equations*, Izv. Akad. Nauk SSSR Mekh. Zhidk. Gaza **1982**, no. 4, 135–140; English transl. in Fluid Dynamics **17** (1982).
3. ______, *On some invariant transformations of kinetic equations*, Vestnik Moskov. Univ. Ser. I Mat. Mekh. **1983**, no. 4, 69–72; English transl. in Moscow Univ. Math. Bull. **38** (1983).
4. A. V. Bobylev, *Exact solutions of the nonlinear Boltzmann equations and the theory of relaxation of the Maxwell gas*, Teoret. Mat. Fiz. **60** (1984), no. 2, 280–310; English transl. in Theoret. and Math. Phys. **60** (1984).
5. S. V. Meleshko, *Group properties of equations of motions of a viscoelastic medium*, Model. Mekh. **2 (19)** (1988), no. 4, 114–126. (Russian)
6. Yu. N. Grigor′ev and S. V. Meleshko, *Group theoretical analysis of kinetic Boltzmann equation and its models*, Arch. Mech. **42** (1990), no. 6, 693–701.
7. S. I. Senashov, *Group classification of an equation of a viscoelastic rod*, Model. Mekh. **4 (21)** (1990), no. 1, 69–72. (Russian)
8. V. F. Kovalev, S. V. Krivenko, and V. V. Pustovalov, *Symmetry group for kinetic equations for collisionless plasma*, Pis′ma Zh. Èksper. Teoret. Fiz. **55** (1992), no. 4, 256–259; English transl. in JETP Lett. (1992).
9. L. V. Ovsyannikov, *Group analysis of differential equations*, "Nauka", Moscow, 1978; English transl., Academic Press, New York, 1982.
10. N. H. Ibragimov, *Group transformations in the mathematical physics*, "Nauka", Moscow, 1983; English transl., Reidel, Boston, 1985.
11. I. S. Krasil′shchik, V. V. Lychagin, and A. M. Vinogradov, *Geometry of jet spaces and nonlinear partial differential equations*, Gordon and Breach, New York, 1986.
12. A. M. Vinogradov, *Local symmetries and conservation laws*, Acta Appl. Math. **2** (1984), no. 1, 21–78.
13. I. S. Krasil′shchik and A. M. Vinogradov, *Nonlocal symmetries and the theory of coverings*, Acta Appl. Math. **2** (1984), no. 1, 79–96.
14. R. L. Drake, *A general mathematical survey of the coagulation*, Topics in Current Aerosol Research, Part 2, International Reviews in Aerosol Physics and Chemistry, Vol. 3 (G. M. Hidy and J. R. Brock, eds.), Pergamon, New York, 1972.
15. L. I. Vinokurov and A. V. Kac, *Power solutions of the kinetic equation for the stationary coagulation of atmospheric aerosols*, Izv. Akad. Nauk SSSR Ser. Fiz. Atmosfer. i Okeana **16** (1980), no. 6; English transl. in Izv. Acad. Sci. USSR Atmospher. Ocean. Phys. **16** (1980).
16. A. A. Lushnikov, *Evolution of coagulating systems*, Journal of Colloid and Interface Science **45** (1973), no. 3, 549–556.
17. A. G. Kudryavtsev and S. D. Traitak, *Equation for Brownian coagulation of aerosol particles in a stochastic medium*, Zh. Èksper. Teoret. Fiz. **99** (1991), no. 1, 115–126; English transl., Soviet Phys. JETP (1991), no. 1, 63–69.

Translated by THE AUTHORS

Department of Applied Mathematics, Moscow State Technical University, ul. 2-ya Baumanskaya 5, 107005 Moscow, Russia

Department of Theoretical Problems, Russian Academy of Sciences, ul. Vesnina 12, 121002 Moscow, Russia

Amer. Math. Soc. Transl.
(2) Vol. **167**, 1995

Braiding of the Lie Algebra $sl(2)$

J. DONIN AND D. GUREVICH

ABSTRACT. We construct a (flat) braided deformation of the enveloping algebra $U(sl(2))$. The deformed algebra differs from the quantum group $U_q(sl(2))$ and lives instead in the category of $U_q(sl(2))$-modules. We consider the space generating this deformed enveloping algebra as a braided version of the Lie algebra $sl(2)$. We also construct quantum counterparts of $SL(2)$ orbits in $sl(2)^*$ and discuss the problem of defining "braided vector fields" on "quantum orbits".

§0. Introduction

It is not a big exaggeration to say that the most popular object connected with the quantum Yang-Baxter equation (QYBE) is the so-called quantum group $U_q(\mathfrak{g})$. It is well known that this object has a Hopf algebra structure, which is a deformation of the usual one of $U(\mathfrak{g})$. Nevertheless there exists another type of deformation arising from the YBE, namely *braiding*. Let us give some examples of braided or, in a more general context, *twisted* objects[1].

In [G1] one of the authors introduced the notion of generalized Lie algebra (called S-Lie algebra in some papers) assuming S to be an involutive ($S^2 = \mathrm{id}$) solution of the QYBE. The enveloping algebra of an S-Lie algebra is a twisted object, i.e., it has a twisted Hopf structure. This means that the compatibility of the multiplication μ and the comultiplication Δ can be expressed by means of the standard relation

$$\Delta\mu(a \otimes b) = \mu(\Delta a \otimes \Delta b),$$

but the multiplication in the r.h.s. of this relation is defined via the operator S (as in the definition of a super-Hopf algebras).

As far as we know, twisted Hopf algebras were first introduced by S. MacLane.

The dual object of the enveloping algebra mentioned above can be regarded as a twisted analog of a formal (co)group. Global group-like twisted objects of GL and SL type were considered in the paper [G3].

Sh. Majid [M1] has introduced the notion of braided group and discovered a process of transmutation converting the quantum group $U_q(\mathfrak{g})$ into a braided

1991 *Mathematics Subject Classification*. Primary 17B37.

[1]According to a tradition we use the term *braided* to denote the objects connected with a noninvolutive (quasitriangular) solution of the QYBE. The term S-*Lie algebras* is used only for involutive S and the term *twisted* is used in both cases.

group. The braided groups constructed by Sh. Majid have a braided Hopf structure.

Note that there exist solutions of the QYBE that cannot be obtained by means of a deformation from the usual permutation $S = \sigma$ $(\sigma(x \otimes y) = y \otimes x)$, for example a super-permutation. A new class of nondeformational solutions of the QYBE was constructed in [G2, G3] (some of them were independently discovered by M. Dubois-Violette and G. Launer [DL]).

All the above objects represent examples of twisted (in particular, braided) structures.

However in the present paper we use the term *braided* in a slightly different sense assuming a *braided* object to lie in the category $U_q(\mathfrak{g})$-Mod of $U_q(\mathfrak{g})$-modules. If this object is an algebra A, this means that the multiplication $\mu\colon A^{\otimes 2} \to A$ satisfies the condition

$$X\mu = \mu\Delta(X), \qquad X \in U_q(\mathfrak{g}), \tag{1}$$

where Δ is the coproduct in $U_q(\mathfrak{g})$, i.e., $U_q(\mathfrak{g})$ plays the role of the symmetry group of the algebra A.

Note that, as compared to $U_q(\mathfrak{g})$, which is defined by means of certain relations including analytic ones, the braided counterpart of the algebra $U(sl(2))$ considered in the paper is the so-called *quadratic-linear* algebra.

Since all such objects of this category are of deformational nature, it is reasonable to ask about the *flatness* of that *braiding* deformation. Roughly speaking, *flatness* means that the "quantity" of elements of the deformed object is stable under deformation. In particular, we call a space $V \in U_q(\mathfrak{g})$-Mod equipped with a bracket $[\ ,\ ]\colon V^{\otimes 2} \to V$ a *braided Lie algebra* if the bracket $[\ ,\ ]$ is a morphism in the category $U_q(\mathfrak{g})$-Mod and if its enveloping algebra is a flat deformation of the initial object.

Let us emphasize that the problem of defining the braided counterpart of Lie algebras (compared to S-Lie algebras) is the a subject of numerous papers. The main difficulty is to give a proper "braided" version of the Jacobi identity. Some papers reproduce the form given in [G1] for involutive S. However, a simplest example considered in the present paper shows that this definition is not reasonable.

We give another version of the Jacobi axiom ensuring the flatness of the deformation of the enveloping algebra (this version is a specialization of the Jacobi identity from [PP] to the braided case)[2].

Note that when we deform the enveloping algebra $U(\mathfrak{g})$, we simultaneously deform its graded adjoint algebra, i.e., the commutative algebra of polynomials on $\mathfrak{g}^*$. Therefore it is natural to consider the quasiclassical limit of this deformation. This is a so-called R-matrix Poisson bracket, i.e., the Poisson bracket generated by an R-matrix.

In the case under consideration, the corresponding R-matrix is of the form $R = \frac{1}{2}X \wedge Y$. It is a particular case of the following (modified) R-matrix defined

[2]When the paper was almost completed, we received the preprint [M2], where the author gives a definition of a braided Lie algebra based on the notion of braided groups but does not investigate the flatness of the braiding deformation. Note that the braided group is not a flat deformation of the initial group structure.

for any simple Lie algebra $\mathfrak{g}$

$$R = \frac{1}{2} \sum_{\alpha \in \Omega_+} X_\alpha \wedge X_{-\alpha} \in \wedge^2 \mathfrak{g}, \tag{2}$$

where $\{H_\alpha, X_\alpha, X_{-\alpha}\}$ is the Cartan–Weyl basis in the Chevalley normalization and Ω_+ is the set of positive roots of $\mathfrak{g}$.

There exists a natural way to assign to any R-matrix a bracket $f \otimes g \to \{f, g\}_R$ defined on any homogeneous space M and, in particular, on any orbit in $\mathfrak{g}^*$. However the bracket defined by means of the R-matrix (2) satisfies the Jacobi identity and therefore it is Poisson only under certain conditions on M (we say such a homogeneous space is of R-matrix type).

All R-matrix type orbits in $\mathfrak{g}^*$ for any simple Lie algebra $\mathfrak{g}$ over the field $k = \mathbb{C}$ have been classified in [GP]. We feel that the problem of quantization of all R-matrix brackets on R-matrix type orbits is of great interest. The result of such a quantization can be regarded as quantum or braided orbits.

From this point of view the algebra Lie $\mathfrak{g} = sl(2)$ plays an exceptional role, since the R-matrix bracket is Poisson on the whole $sl(2)^*$ (i.e., all orbits in $sl(2)^*$ are of R-matrix type). Therefore only for this algebra can a flat braided deformation of the functional space on whole space $\mathfrak{g}^*$ exist. We construct this deformation in §3. Simultaneously we construct a braided deformation of the enveloping algebra $U(sl(2))$.

In §4 we consider a braided counterpart of the quotient algebras of the algebra $U(sl(2))$ and prove the flatness of the corresponding deformation. We consider these quotient algebras of the deformed enveloping algebra as the quantum analog of the ordinary orbits (more precisely, of function spaces on them).

Let us emphasize that a similar deformation of the enveloping algebra arising from classical R-matrices was studied in [GRZ]. There a certain enlarged scheme of quantum mechanics was suggested. In particular "twisted vector fields" were used to define the simplest twisted models. In the case under consideration, it is not so clear what the proper definition of "braided vector fields" should be. We discuss this problem in the last section.

The present paper can be considered as a first step aiming to include the deformational algebras arising from quantization of some modified R-matrices in the scheme introduced in the paper [GRZ].

§1. Family of Poisson brackets

First of all we recall the definition of a flat deformation.

Let A_h be an associative algebra over the field $k[[h]]$, where k is the ground field, $k = \mathbb{R}$ or $k = \mathbb{C}$, and h is a formal parameter. This algebra is called a *flat deformation* of the algebra $A = A_0$ if A_h and $A[[h]]$ are isomorphic to each other as $k[[h]]$-modules and the multiplication μ_h in A_h is equal to

$$\mu_h = \mu_0 + h\mu_1 + h^2\mu_2 + \dots$$

for certain maps $\mu_i \colon A^{\otimes 2} \to A$, and where $\mu_0 = \mu$ is the multiplication in the initial algebra A (extended to $A[[h]]$). In a similar way a deformation of a coassociative structure can be defined.

It is well known that if A_h is a flat deformation of the algebra A, then μ_1 is a Hochschild cocycle on A. Assuming A to be commutative, we get more information about μ_1. Namely, the antisymmetric part of μ_1 usually is denoted

$$\{a, b\} = \mu_1(a, b) - \mu_1(b, a)$$

and is called the *Poisson bracket* satisfies the Leibniz and Jacobi identities.

A Poisson bracket defined in the algebra of (smooth) functions on a (smooth) manifold M can be described by means of certain skew bivector field on M. If such a Poisson bracket $\{\ ,\ \}$ is given, then a procedure of constructing an algebra A_h with the properties described above is called a *deformational quantization.*

If a Poisson bracket is nowhere degenerate (this means that the manifold has a symplectic structure), then a deformational quantization always exists (cf., for example, [Fed] and [dWL]). As far as we know, there does not exist any example of a nonquantizable (degenerate) bracket.

Let $\mathfrak{g}$ be a Lie algebra with structure constants c_{ij}^k in a fixed base $\{X_i\}$, i.e., $[X_i, X_j] = c_{ij}^k X_k$. Consider the linear Poisson–Lie[3] bracket $\{\ ,\ \}_{PL}$ defined in the space $\mathrm{Fun}(\mathfrak{g}^*)$ of polynomials on $M = \mathfrak{g}^*$ as follows:

$$\{f, g\}(\xi) = \langle [df, dg], \xi \rangle, \qquad \xi \in \mathfrak{g}^*.$$

Then the enveloping algebra $U(\mathfrak{g})$ with the parameter h introduced in the structure constants ($c_{ij}^k \to hc_{ij}^k$) is a quantization of the Poisson–Lie bracket (in what follows we use the notation $U^h(\mathfrak{g})$ for $U(\mathfrak{g})$ equipped with the parameter h as above). This follows from the PBW theorem.

The restriction of the Poisson–Lie bracket to any of symplectic leaves defines the so-called Kirillov–Kostant–Souriau bracket (we denote it by $\{\ ,\ \}_{KKS}$).

Let $R = r^{ij} X_i \otimes X_j \in \wedge^{\otimes 2}\mathfrak{g}$ be a classical or a modified R-matrix. Let us consider a new bracket

$$\{f, g\}_R = r^{ij}\{x_i, f\}\{x_j, g\}, \tag{3}$$

where $x = \langle X, \xi \rangle$, $\xi \in \mathfrak{g}^*$ (we call this the *R-matrix bracket*).

It is obvious that the bracket (3) can be restricted to any symplectic leaf of the Poisson–Lie bracket (the orbits of the corresponding group G). It is also clear that the bracket (3) is Poisson if R is a classical R-matrix. If R is a modified R-matrix, the bracket (3) is Poisson only after being restricted to some orbits in $\mathfrak{g}^*$ (we call them the R-matrix type orbits, cf. Introduction).

Hereafter we assume $\mathfrak{g}$ to be a simple Lie algebra and identify $\mathfrak{g}$ and $\mathfrak{g}^*$ via the Killing form.

We reproduce (partially) the result from [GP], where all R-matrix type orbits were classified (over the field $k = \mathbb{C}$).

[3] Nowadays this term is more often used to designate a quadratic Poisson bracket defined on the corresponding Lie group. We prefer to call the latter the Sklyanin (or Sklyanin–Drinfeld) bracket.

PROPOSITION 1.1. *Let* $\mathcal{O}$ *be the nonzero orbit in* $\mathfrak{g}^*$ *of an element* $x \in \mathfrak{g}^*$.

1. *Suppose the orbit* $\mathcal{O}$ *is of* R*-matrix type. Then* x *is either semisimple or nilpotent.*

2. *If* $\mathcal{O}$ *is semisimple, then* $\mathcal{O}$ *is of* R*-matrix type if and only if* $\mathcal{O}$ *is a symmetric space* (*this is true for* $k = \mathbb{R}$ *as well*).

We do not reproduce any description of R-matrix type nilpotent orbits since we do not need it. Let us only remark that the orbit of the highest weight vector is an orbit of such type (cf. also [DG1, DGM, DG2]).

It is clear that all orbits in $sl(2)^*$ apart from the highest weight orbit (the cone) are symmetric spaces. Therefore the R-matrix bracket is Poisson on all orbits in $sl(2)^*$, i.e., on the whole space $sl(2)^*$.

As to other simple Lie algebras $\mathfrak{g} \neq sl(2)$, for any of them there exist orbits in $\mathfrak{g}^*$ that are not of R-matrix type, for example orbits of semisimple elements that are not symmetric spaces.

Thus the Lie algebra $\mathfrak{g} = sl(2)$ is the only simple Lie algebra such that all orbits in $\mathfrak{g}^*$ are of R-matrix type and therefore the bracket (3) is Poisson on the whole space $\mathfrak{g}^*$.

Remark that it follows from the classification of R-matrix type orbits given in [GP] that all orbits of such type are multiplicity free. This means that in the decomposition $\mathrm{Fun}(\mathcal{O}) = \bigoplus V_i$ into direct sum of irreducible $\mathfrak{g}$-modules all $\mathfrak{g}$-modules V_i are pairwise nonisomorphic.

It is easy to see that the R-matrix bracket (assuming R to be a classical R-matrix) and the Poisson–Lie bracket are compatible. The brackets $\{\ ,\ \}_{KKS}$ and $\{\ ,\ \}_R$ have the same property for any modified R-matrix and any R-matrix type orbit. This means that all brackets of the family

$$\{\ ,\ \}_{a,b} = a\{\ ,\ \}_{KKS} + b\{\ ,\ \}_R \tag{4}$$

are Poisson.

Note that compatibility of the bracket $\{\ ,\ \}_{KKS}$ and the reduced Sklyanin bracket was investigated in [KRR] for orbits equipped with a Hermitian structure.

§2. Quantization and quadratic-linear algebras

The following is a problem of great interest: to quantize the whole family (4) simultaneously. The scheme of simultaneous quantization assuming R to be a nonmodified classical R-matrix was suggested in [GRZ]. If R is a modified R-matrix (2), then quantization is carried out only for certain brackets from the family (4) (cf. [DG1, DGM, DG2]).

Note that the brackets (4) are in general degenerate and therefore they cannot be quantized by the methods of the papers [Fed] and [dWL].

Let us reproduce the quantization scheme from [GRZ] in a short form, assuming R to be a classical R-matrix. Consider the element

$$F_\nu \in U(\mathfrak{g})^{\otimes 2}[[\nu]]$$

constructed by Drinfeld [D1] and satisfying the following relations

$$F_\nu = 1 \bmod \nu\,, \quad F_\nu - \sigma F_\nu = R \bmod \nu^2\,, \quad (\varepsilon \otimes \mathrm{id}) F_\nu = (\mathrm{id} \otimes \varepsilon) F_\nu = 1\,,$$
$$\Delta^{12} F_\nu F_\nu^{12} = \Delta^{23} F_\nu F_\nu^{23}\,,$$

where σ is the ordinary permutation and ε is the counit.

Deform now the initial multiplication

$$\mu_h\colon U^h(\mathfrak{g})^{\otimes 2} \to U^h(\mathfrak{g})$$

as follows

$$\mu_{h,\nu} = \mu_h(\rho \otimes \rho) F_\nu ,$$

where ρ is the extension of the representation $\rho_X(Y) = [X, Y]$ to $U^h(\mathfrak{g})$ and $(\rho \otimes \rho) F_\nu$ is the corresponding map from $U^h(\mathfrak{g})^{\otimes 2}[[\nu]]$ to itself.

The algebra $U^h(\mathfrak{g})$ equipped with the multiplication $\mu_{h,\nu}$ will be denoted by $U^{h,q}$ (we put $q = e^\nu$). It is a two-parameter quantization of the family (4). More precisely, this algebra equipped with the multiplication $\mu_{h,\nu}$, where $h = ah_1$, $\nu = bh_1$, is a quantization of the bracket $\{\ ,\ \}_{a,b}$.

Unfortunately, we cannot use similar methods to quantize the family (4) when R is the R-matrix (2), since there do not exist elements F_ν with the above properties. Nevertheless we will quantize the family (4) in the framework of the theory of quadratic and quadratic-linear algebras.

We shall reproduce some aspects of this theory, mainly using the paper [PP].

Let V be a fixed linear space and I be a subspace in $V^{\otimes 2}$. Then the algebra $\wedge_+(V, I) = T(V)/\{I\}$, where $\{I\}$ is the ideal in $T(V)$ generated by elements belonging to I, is called the *quadratic algebra* corresponding to I. Let $\wedge_+^l(V, I)$ denote its homogeneous component of degree l; introduce the linear spaces

$$\wedge_-^0 = k, \quad \wedge_-^1 = V,$$
$$\wedge_-^l(V, I) = \wedge_-^{l-1}(V, I) \otimes V \cap V^{\otimes(l-2)} \otimes I, \qquad l = 2, 3, \dots .$$

The quadratic algebra is called the *Koszul algebra* if the complex

$$\dots \xrightarrow{d} \wedge_+^k(V, I) \otimes \wedge_-^l(V, I) \xrightarrow{d} \wedge_+^{k+1}(V, I) \otimes \wedge_-^{l-1}(V, I) \xrightarrow{d} \dots$$

is exact (except from $k = l = 0$), where d is the natural differential:

$$d(a_1 \otimes \dots \otimes a_k) \otimes (b_1 \otimes \dots \otimes b_l) = (a_1 \otimes \dots \otimes a_k \otimes b_1) \otimes (b_2 \otimes \dots \otimes b_l).$$

The following statement is well known.

PROPOSITION 2.1. *If $\wedge_+(V, I)$ is a Koszul algebra, then the usual relation*

$$\mathcal{P}_-(t)\mathcal{P}_-(-t) = 1$$

for the Poincaré series $\mathcal{P}_\pm(t) = \sum \dim \wedge_\pm^k(V, I)t^k$ holds.

Given a map $[\ ,\ ]\colon I \to V$, define *the quadratic-linear* algebra (an analog of the enveloping algebra) in the natural way: $U(\mathfrak{g}) = T(V)/\{J\}$, where $\{J\} \subset T(V)$ is the ideal generated by elements $I - [\ ,\ ]I$. Since there exists a natural filtration in this algebra, we may consider the associated graded algebra $\operatorname{Gr} U(\mathfrak{g})$.

The following proposition proved in [PP] is a very useful generalization of the PBW theorem.

PROPOSITION 2.2. *Let us assume that the algebra* $\wedge_+(V, I)$ *is a Koszul algebra and that the following conditions*

$$([\ ,\ ]^{12} - [\ ,\ ]^{23}) \wedge_-^3 (V, I) \subset I, \tag{5}$$

$$[\ ,\ ]([\ ,\ ]^{12} - [\ ,\ ]^{23}) \wedge_-^3 (V, I) = 0 \tag{6}$$

are fulfilled. Then Gr $U(\mathfrak{g})$ *and* $\wedge_+(V, I)$ *are isomorphic as graded algebras.*

In the next section we shall use this proposition to construct a braiding of $sl(2)$.

To conclude this section, let us remark that for any triple $(V, I, [\ ,\ ])$ satisfying relations (5) and (6), it is not difficult to define a cohomology generalizing the usual cohomology of a Lie algebra. The corresponding complex can be constructed as follows

$$\dots \xrightarrow{d} \wedge_-^l(V, I) \xrightarrow{d} \wedge_-^{l-1}(V, I) \xrightarrow{d} \dots,$$

where a differential d is defined by

$$d(X_{i_1} \wedge \dots \wedge X_{i_l}) = \sum (-1)^j X_{i_1} \wedge \dots \wedge X_{i_{j-1}} \wedge [X_{i_j}, X_{i_{j+1}}] \wedge \dots \wedge X_{i_l}.$$

We leave it to the reader to verify that this complex is well defined (i.e.,

$$\operatorname{Im} d(\wedge_-^l(V, I)) \subseteq \wedge_-^{(l-1)}(V, I)$$

and the relation $d^2 = 0$ is satisfied).

It is clear that for ordinary Lie algebras the above differential coincides with the usual Chevalley–Eilenberg one (up to a factor).

§3. Braided Lie algebras: the $sl(2)$ case

First let us give the definition of a braided Lie algebra.

DEFINITION 3.1. Let $U_q(\mathfrak{g})$ be a quantum group and let V be a finite-dimensional object of the category $U_q(sl(2))$-Mod of all $U_q(sl(2))$-modules. Assume that $V^{\otimes 2}$ is a direct sum of two subspaces $V^{\otimes 2} = I \oplus \bar{I}$. We say that the space V is equipped with the structure of a *braided Lie algebra*, if there exists an operator $[\ ,\ ]\colon V^{\otimes 2} \to V$ satisfying the axioms

1. $[\ ,\ ]\bar{I} = 0$.
2. The relations of Proposition 2.2 are satisfied.
3. The spaces I, $\bar{I}$ are objects of the category $U_q(sl(2))$-Mod and the operator $[\ ,\ ]$ is a morphism in this category.

Let us note that this definition can be extended to a constant-linear bracket $[\ ,\ ]\colon V^{\otimes 2} \to V \oplus k$. It is possible to make this bracket purely linear by introducing a new central generator s and assuming $[\ ,\ ]V^{\otimes 2} \to V \oplus ks$ (we let $Xs = 0$ for any $X \in U_q(\mathfrak{g})$). Note also that Proposition 2.2 allows a generalization to the case of linear-constant brackets.

REMARK. In [G4] a definition of a braided Lie algebras with respect to more general monoidal category (not necessary of deformational type) was introduced.

In the general case, the property of the bracket $[\ ,\ ]$ (the subspaces I, $\overline{I}$) to be a morphism (objects) of the category can be formulated in term of an algebra of Reshetikhin–Takhtadzhyan–Faddeev type (cf. [G4] for details). In the case under consideration, we express these properties by means of the quantum group $U_q(\mathfrak{g})$. In addition, the axiom requiring "Koszulity" of the algebra $\wedge_+(V, I)$ was added in [G4]. We omit this axiom here, but we prove that all quadratic algebras under consideration are Koszul.

Let us restrict ourselves now to the case $\mathfrak{g} = sl(2)$. In this case the Hopf algebra structure in $U_q(sl(2))$ can be described as follows. This algebra is generated by the elements $\{H, X, Y\}$ satisfying the relations

$$[H, X] = 2X, \quad [H, Y] = -2Y, \quad [X, Y] = \frac{q^H - q^{-H}}{q - q^{-1}}.$$

The coproduct is defined by the following formulas

$$\Delta(X) = X \otimes 1 + q^{-H} \otimes X, \quad \Delta(Y) = 1 \otimes Y + Y \otimes q^H, \quad \Delta(H) = H \otimes 1 + 1 \otimes H$$

(we do not need the antipode).

Let V be a $U_q(sl(2))$-module of spin 1 with the base $\{u, v, w\}$ and of the following module structure:

$$Hu = 2u, \quad Hv = 0, \quad Hw = -2w, \quad Xu = 0, \quad Xv = -(q + q^{-1})u,$$
$$Xw = v, \quad Yu = -v, \quad Yv = (q + q^{-1})w, \quad Yw = 0.$$

The space $V^{\otimes 2}$ is the direct sum of three irreducible $U_q(sl(2))$-modules V_0, V_1, and V_2 of spin 0, 1, and 2 respectively. Let us describe them explicitly:

$$V_0 = \operatorname{span}((q^3 + q)uw + vv + (q + q^{-1})wu),$$
$$V_1 = \operatorname{span}(q^2uv - vu, (q^3 + q)(uw - wu) + (1 - q^2)vv, -q^2vw + wv),$$
$$V_2 = \operatorname{span}(uu, uv + q^2vu, uw - qvv + q^4wu, vw + q^2wv, ww)$$

(we omit the sign $\otimes$ if this does not lead to a misunderstanding). Note that the element $C_q = (q^3 + q)uw + vv + (q + q^{-1})wu$ is the q-analog of the Casimir element, i.e., it is $U_q(sl(2))$-invariant.

We define the subspaces $I = I_q$ and $\overline{I} = \overline{I}_q$ and the bracket $[\ ,\ ]$ as follows $I = V_1$, $\overline{I} = V_0 \oplus V_2$,

$$[\ ,\ ]\overline{I} = 0, \quad [\ ,\ ](q^2uv - vu) = -4hu,$$
$$[\ ,\ ]((q^3 + q)(uw - wu) + (1 - q^2)vv) = 4hv, \quad [\ ,\ ](-q^2vw + wv) = 4hw.$$

We leave it to the reader to verify that the bracket $[\ ,\ ]$ is a morphism of the category $U_q(sl(2))$-Mod. Indeed, it is possible to do this assuming the highest weight vectors of V and V_1 to be equal to each other and applying the decreasing operator Y to this equality.

Relations similar to the last three were constructed in [E], but the author of that paper does not define any bracket and does not investigate the flatness of the deformation of the enveloping algebra.

From the above relations let us find the quantities $[u, u]$, $[u, v]$, and so on:

$$[u, u] = 0, \quad [u, v] = -q^2 Mu, \quad [u, w] = (q+q^{-1})^{-1} Mv,$$
$$[v, u] = Mu, \quad [v, v] = (1-q^2)Mv, \quad [v, w] = -q^2 Mw,$$
$$[w, u] = -(q+q^{-1})^{-1} Mv, \quad [w, v] = Mw, \quad [w, w] = 0,$$

where $M = 4h(1+q^4)^{-1}$.

Let $U^{h,q}(sl(2))$ denote the enveloping algebra $T(V)/\{J\}$, where J is the linear space in $V^{\otimes 2} \oplus V$ spanning the elements

$$q^2 uv - vu + 2hu, \quad (q^3+q)(uw-wu) + (1-q^2)vv - 2hv, \quad -q^2 vw + wv - 2hw,$$

and $\{J\}$ is the corresponding ideal. Note that $U^{0,q}(sl(2)) = \wedge_+(V, I_q)$.

It is easy to see that the multiplication $\mu^{h,q}$ in the algebra $U^{h,q}(sl(2))$ satisfies condition (1).

In the sequel we assume $q-1$ to be small enough.

PROPOSITION 3.1. *The algebra* $\wedge_+(V, I_q)$ *is a Koszul algebra.*

PROOF. We fix the following ordering $v < u < w$ and introduce the lexicographic ordering in the family of monomials in u, v, w. Then the elements u^2, vu, uw, v^2, vw, w^2 cannot be expressed via smaller elements modulo the elements from I_q. It is clear that the family of elements

$$D_3 = \{v^a u^b w^c, \ a+b+c = 3\}$$

generates the homogeneous component $\wedge_+^3(V, I_q)$. Let us show now that this set is the base in the space $\wedge_+^3(V, I_q)$, i.e., the elements are independent.

Note first that the space $\wedge_-^3(V, I_q)$ is one-dimensional. Its generator is

$$\begin{aligned} Z &= -(q+q^{-1})w(q^2uv - vu) + v((q^3+q)(uw-wu) + (1-q^2)vv) \\ &\quad + (q^3+q)u(-q^2vw + wv) \\ &= -(q^3+q)(q^2uv - vu)w + ((q^3+q)(uw-wu) \\ &\quad + (1-q^2)vv)v + (q+q^{-1})(-q^2vw + wv)u. \end{aligned}$$

Hence the space I_q generates a 17-dimensional subspace $I_q \otimes V + V \otimes I_q \subset V^{\otimes 3}$ and $\wedge_+^3(V, I_q)$ has to be of dimension 10. On the other hand, the set D_3 consists of exactly 10 elements. Therefore the elements from D_3 are independent.

By the "diamond lemma" [B] we can state that the set

$$\{v^a u^b w^c \text{ for all integer } a, b, c \geqslant 0\}$$

is PBW base in $\wedge_+(V, I_q)$. Then by the Priddy theorem [Pri], the algebra $\wedge_+(V, I_q)$ is Koszul. This completes the proof.

Since $\dim \wedge_-^2(V, I_q) = 3$ and $\dim \wedge_-^3(V, I_q) = 1$, we see by Proposition 2.1 that $\mathcal{P}_+(t)$ does not change in the process of deformation, i.e., $\wedge_+(V, I_q)$ is a flat deformation of $\wedge_+(V, I_1)$.

It is easy to show that $([\ ,\ ]^{12} - [\ ,\ ]^{23})Z = 0$ and therefore the following data $(V, I, \overline{I}, [\ ,\ ])$ defines a braided Lie algebra[4]. By Proposition 2.2 $\operatorname{Gr} U^{h,q}(sl(2)) = \wedge_+(V, I_q)$ and therefore the algebra $\operatorname{Gr} U^{h,q}(sl(2))$ is a flat two-parameter deformation of the commutative algebra $\wedge_+(V, I_1)$.

Putting $q = e^{\nu/2}$ and computing the quasiclassical limit we find that the algebra $U^{0,q}(sl(2)) = \wedge_+(V, I_q)$ is a quantization of the R-matrix bracket

$$\{u, v\}_R = -uv, \quad \{u, w\}_R = v^2/2, \quad \{v, w\}_R = -vw,$$

where $R = \frac{1}{2}(u \otimes w - w \otimes u)$. Since the algebra $U^{h,1}(sl(2))$ is a quantization of the linear Poisson–Lie bracket, it is clear that the algebra $U^{h,q}(sl(2))$ is a quantization of the family (4).

Thus we have quantized the family (4) simultaneously. In the next section we investigate some quotient algebras of the algebras $U^{h,q}(sl(2))$.

§4. Quotient algebras of $U^{h,q}(sl(2))$

Now consider the associative algebra $A_c^{h,q} = U^{h,q}(sl(2))/\{C_q - c\}$ and let $\mu_c^{h,q}$ denote the multiplication in it. As usual $\{C_q - c\}$ denotes the ideal generated by the element $C_q - c$.

Let us emphasize that this algebra can be regarded as the braided counterpart of the algebra $A_c^{h,1}$, which was investigated by many authors from different points of view (cf. for example [Dix, Fei]). This last algebra is associative and it is a Lie algebra with respect to the natural bracket. As an $sl(2)$-module, $A_c^{h,1}$ is the direct sum of irreducible $sl(2)$-modules V_i, $0 \leqslant i < \infty$.

The braided counterpart $A_c^{h,q}$ of this algebra has a similar property. More precisely, for generic q the algebra $A_c^{h,q}$ can be decomposed into a direct sum of $U_q(sl(2))$-modules. This follows from the next Proposition

PROPOSITION 4.1. *The three-parameter deformation $A_c^{h,q}$ of the algebra $A_0^{0,1}$ is flat (recall that we always suppose $q - 1$ to be small enough).*

PROOF. Let us introduce a new decomposition of the space $V^{\otimes 2}$ into the direct sum of two subspaces, $V^{\otimes 2} = I \oplus \overline{I}$, where $I = I_q = V_0 \oplus V_1$, $\overline{I} = V_2$ (the spaces V_0, V_1, V_2 were introduced in §3). Introduce also a new bracket $[\ ,\ ]$ as above with the only difference that $[\ ,\ ](t) = c$, where t is the split q-Casimir, i.e., C_q regarded as an element of $V^{\otimes 2}$.

In the case under consideration, we have $\dim \wedge_-^3(V, I_q) = 4$. It is obvious that $Z \in \wedge_-^3(V, I_q)$, where Z was defined in the proof of Proposition 3.1. One can

[4] In the case under consideration, another form of the Jacobi identity is satisfied as well:

$$[\ ,\ ][\ ,\ ]^{12} \wedge_-^3 (V, I_q) = [\ ,\ ][\ ,\ ]^{23} \wedge_-^3 (V, I_q) = 0.$$

However it is not possible to rewrite the Jacobi identity in the more familiar form discussed in the last section because of the more complicated structure of the projector $V^{\otimes 3} \to \wedge_-^3 (V, I_q)$.

find the three other generators of this space by using the relations

$$u\otimes t-t\otimes u\in\{V_1\},\quad v\otimes t-t\otimes v\in\{V_1\},\quad w\otimes t-t\otimes w\in\{V_1\}.$$

Thus the following elements (together with Z) generate the space $\wedge^3_-(V, I_q)$:

$$\begin{aligned}
Z_1 &= ut-u((q^3+q)(uw-wu)+(1-q^2)vv)-q^{-2}v(q^2uv-vu)\\
&= tu+q^{-2}((q^3+q)(uw-wu)+(1-q^2)vv)u+(q^2uv-vu)v,\\
Z_2 &= vt+(q^3+q)u(-q^2vw+wv)+w(q^{-1}+q^{-3})(q^2uv-vu)\\
&= tv-(q^3+q)(q^2uv-vu)w-(q^{-1}+q^{-3})(-q^2vw+wv)u,\\
Z_3 &= wt+q^{-2}w((q^3+q)(uw-wu)+(1-q^2)vv)-v(-q^2vw+wv)\\
&= tw-((q^3+q)(uw-wu)+(1-q^2)vv)w+q^{-2}(-q^2vw+wv)v.
\end{aligned}$$

Now applying the operator $[\ ,\ ]([\ ,\ ]^{12}-[\ ,\ ]^{23})$ to the elements above, we get 0. Thus axioms of the definition of a braided Lie algebra are satisfied.

Now we state that the algebra $\wedge_+(V, I_q)$ is Koszul. In order to prove this, we use the same method as in Proposition 3.1 with the only difference that the role of the set D_3 is played by the set

$$\{v^3, v^2u, v^2w, vu^2, vw^2, u^3, w^3\},$$

i.e., the last set forms a base in the space $\wedge^3_+(V, I_q)$. Now applying Proposition 2.2, we complete the proof.

Note that the exactness of the Koszul complex related to the Hecke type solution of the QYBE (i.e., the solution with two eigenvalues) was proved in [G3]. The method of proof does not use any base (of PBW type) and is valid for nondeformational solutions of the QYBE as well. It would be very interesting to get a similar proof for the case under consideration.

As was noted above, for generic q the algebra $A_c^{h,q}$ can be decomposed into the direct sum of $U_q(sl(2))$-modules V_i, $i=0,1,2,\dots$. We fix a base $\{e^i_j, 1\leqslant j\leqslant \dim V_i\}$ in the space $V_i\subset V^{\otimes i}$ of spin i and regard the union of the bases as a base of the algebra $U_q(sl(2))$. More precisely, we put $e^i_j=Y^ju^{\otimes i}$ up to a factor. For example we have

$$\begin{gathered}
e^0_1=1,\quad e^1_1=u,\quad e^1_2=v,\quad e^1_3=w,\quad e^2_1=uu,\quad e^2_2=uv+q^2vu,\\
e^2_3=uw-qvv+q^4wu,\quad e^2_4=vw+q^2wv,\quad e^2_5=ww.
\end{gathered}$$

We use this base in the next section in order to define "braided vector fields". It would be very interesting to compute the multiplication table of the algebra $A_c^{h,q}$ in this base.

§5. On "braided vector fields"

We consider elements of the algebra $A_c^{0,q}$ as "functions" on "quantum orbits", since for $q=1$ the algebra $A_c^{0,1}$ is a function algebra either on the cone ($c=0$)

or on a hyperboloid ($c \neq 0$). A very interesting question is: what is the proper way to introduce vector fields on those quantum homogeneous spaces? We shall discuss this problem briefly (it will be considered elsewhere in more detail).

It is natural to use the above bracket to define "coadjoint braided vector fields" (we introduce "coadjoint vector fields" by means of the adjoint action, since we define their action on the "functions" on quantum orbits).

Consider three operators U, V, W acting on the generators u, v, w according to the above bracket. Namely, we put

$$Uu = 0, \quad Uv = -q^2 Mu, \quad Uw = (q+q^{-1})^{-1} Mv$$

and so on (recall that $M = 4h(1+q^4)^{-1}$). As to the zero component, we naturally assume $U1 = V1 = W1 = 0$.

We leave it to the reader to verify that the operators U, V, W satisfy the following relations

$$(q^3+q)uW + vV + (q+q^{-1})wU = 0, \quad q^2UV - VU = M(q^4-q^2+1)U,$$
$$(q^3+q)(UW - WU) + (1-q^2)V^2 = M(q^4-q^2+1)V,$$
$$-q^2VW + WV = -M(q^4-q^2+1)W.$$

Note that the last three relations differ from the ones between the generators u, v, w in the algebra $U^{q,h}(sl(2))$ by the factor (q^4-q^2+1).

Thus we have defined "braided vector fields" on "linear functions". How is it possible to extend them to higher powers? In the nondeformed case ($q = 1$) this is realized via the Leibniz rule. There exists also a so-called S-analog of the Leibniz rule for an involutive ($S^2 = 1$) solution S of the QYBE. It is connected with the notion of S-Lie algebra introduced in [G1].

Let us recall that an S-Lie algebra can be defined as a triple $(V, S\colon V^{\otimes 2} \to V^{\otimes 2}, [\ ,\ ]\colon V^{\otimes 2} \to V)$, where S is an involutive solution of the QYBE satisfying the following axioms:

1. $[\ ,\ ]S = -[\ ,\ ]$.
2. $[\ ,\ ][\ ,\ ]^{12}(\mathrm{id} + S^{12}S^{23} + S^{23}S^{12}) = 0$.
3. $S[\ ,\ ]^{12} = [\ ,\ ]^{23}S^{12}S^{23}$.

Let us introduce the subspaces I and $\bar{I} \in V^{\otimes 2}$ as follows

$$I = \mathrm{Im}(\mathrm{id} - S) = \mathrm{Ker}(\mathrm{id} + S), \quad \bar{I} = \mathrm{Im}(\mathrm{id} + S) = \mathrm{Ker}(\mathrm{id} - S).$$

They play the role of analogous subspaces from Definition 3.1. So the first axioms from the definition of a S-Lie algebra and from that of a braided Lie algebra correspond to each other. The last axiom from the definition of S-Lie algebra means that the bracket $[\ ,\ ]$ is a morphism of the tensor category generated by the space V (in this category the extended operator S plays the role of commutativity operator)[5].

However, the two axioms are different. And it is easy to check that the Jacobi identity for S-Lie algebra can be rewritten in the following form

$$[X, [Y, Z]] = [[X, Y], Z] - [\ ,\ ][\ ,\ ]^{12}S^{23}(X \otimes Y \otimes Z).$$

[5]It is easy to see that the algebra $\wedge(V, I)$ is Koszul (this is proved in [G3] in a more general context).

One can deduce from this fact that the adjoint operator $\mathrm{ad}_X(Y) = [X, Y]$ is a (left) representation of the S-Lie algebra.

However, this no longer true for a braided Lie algebra, i.e., the "adjoint representation" is not a representation at all. This explains why the relations between the operators U, V, W differ from the ones for the generators u, v, w.

As to the Leibniz rule, if S is involutive, it is natural to consider S-commutative algebras, i.e., algebras A equipped with the multiplication $\mu\colon A^{\otimes 2} \to A$ (which is a morphism in the corresponding category) and an involutive operator $S\colon A^{\otimes 2} \to A^{\otimes 2}$ satisfying the QYBE and such that $\mu S(a \otimes b) = \mu(a \otimes b)$. In this case the Leibniz rule can be expressed as follows

$$X\mu(a \otimes b) = \mu(X(a) \otimes b) + \mu((X \otimes \mathrm{id}) S(a \otimes b)).$$

The "functional spaces" $A_0^{0,q}$ on the "quantum orbits" are no longer S-commutative. However there exists an involutive operator $\widetilde{S}\colon A^{\otimes} \to A^{\otimes}$ such that this algebra is $\widetilde{S}$-commutative (cf. [DG1]). Therefore it is reasonable to substitute $\widetilde{S}$ in the last equality for S and extend "braided vector fields" to the higher components using this modified Leibniz rule. However, we shall do this in another way, namely by using the base $\{e_j^i\}$ constructed above.

Recall that the set $\{e_j^i\}$ is a base in the algebra $A_c^{h,q}$ and, in particular, in $A_c^{0,q}$. Considering the elements $\{e_j^i\}$ of the algebra $A_c^{0,q}$ as "symmetric", we can extend the differential operator X to any element of the base as follows

$$X(e_j^i) = i\mu(X \otimes \mathrm{id}_{i-1})(e_j^i),$$

where id_i is the identity operator in $\mathfrak{g}^{\otimes i}$ and μ is the multiplication in the algebra $A_c^{0,q}$ (here we regard e_j^i as elements of the space $\mathfrak{g}^{\otimes i}$). This definition in the nondeformed case ($q = 1$) agrees with the usual definition of vector fields.

Note that the "braided vector fields" extended to higher components do not satisfy the relations between generators in the algebra $A_c^{h,q}$ anymore (even up to a factor).

It is not difficult to verify that the operators U, V, W satisfy the relation

$$(q^3 + q)uW + vV + (q + q^{-1})wU = 0.$$

Thus it is natural to define "tangent braided vector fields" as the family of all linear combinations

$$aU + bV + cW, \qquad a, b, c \in A_c^{0,q}$$

modulo the elements $a((q^3 + q)uW + vV + (q + q{-}1)wU)$.

It is clear that all tangent braided vector fields satisfy the Leibniz rule in the form (7).

In conclusion we would like to note that braided vector fields introduced in this way are well defined, but their properties differ from those of ordinary vector fields. The main difference is that in the deformed case the "adjoint representation" is not a representation at all. It seems very plausible that S-Lie algebras are the most general objects that have the properties as in the classical case.

Acknowledgements. The authors are grateful to the Max-Planck-Institut für Mathematik and the Bar-Ilan University for their hospitality during the preparation of a preliminary version of the paper. We also wish to thank S. Khoroshkin, V. Rubtsov, and N. Zobin for stimulating discussions.

References

[B] G. Bergman, *The diamond lemma for ring theory*, Adv. in Math. **29** (1979), 178–218.

[dWL] M. de Wilde and P. B. A. Lecomte, *Existence of star-products revisited*, Note Mat. **X** (1990), no. 1, 205–210.

[Dix] J. Dixmier, *Quotients simples de l'algèbre enveloppante de* $sl(2)$, J. Algebra **24** (1973), 551–564.

[DG1] J. Donin and D. Gurevich, *Quasi-Hopf algebras and R-matrix structure in line bundles over flag manifolds*, Selecta Math. Soviet. **12** (1993), 37–48.

[DG2] ———, *Some Poisson structures associated to Drinfeld–Jimbo R-matrices and their quantization*, Israel J. Math. (to appear).

[DGM] J. Donin, D. Gurevich, and Sh. Majid, *R-matrix brackets and their quantization*, Ann. Inst. H. Poincaré **58** (1993), 235–246.

[D1] V. Drinfeld, *On constant quasiclassical solutions of the quantum Yang–Baxter equation*, Soviet Math. Dokl. **28** (1983), 667–671.

[D2] ———, *Quasi-Hopf algebra*, Leningrad Math. J. **1** (1990), 1419–1457.

[DL] M. Dubois-Violette and G. Launer, *The quantum group of a nondegenerated bilinear form*, Phys. Lett. B **245** (1990), 175–177.

[E] I. Egusquiza, *Quantum group invariance in quantum sphere valued statistical models*, Phys. Lett. B **276** (1992), 465–471.

[Fed] B. Fedosov, *Deformation quantization and asymptotic operator representation*, Functional Anal. Appl. **25** (1991), 184–194.

[Fei] B. Feigin, *The Lie algebras* $gl(\lambda)$ *and cohomology of Lie algebras of differential operators*, Russian Math. Surveys **43** (1988), 169–170.

[G1] D. Gurevich, *Generalized translation operators on Lie groups*, Soviet J. Contemporary Math. Anal. **18** (1983), 57–70.

[G2] ———, *Hecke symmetries and quantum determinants*, Soviet Math. Dokl. **38** (1989), no. 3, 555–559.

[G3] ———, *Algebraic aspects of the quantum Yang–Baxter equation*, Leningrad Math. J. **2** (1991), 801–828.

[G4] ———, *Hecke symmetries and braided Lie algebras*, Spinors, Twistors, Clifford Algebras and Quantum Deformation, Kluwer, Dordrecht, 1993, pp. 317–326.

[GP] D. Gurevich and D. Panyushev, *On Poisson pairs associated to modified R-matrices*, Duke Math. J. **73** (1994), no. 1.

[GRZ] D. Gurevich, V. Rubtsov, and N. Zobin, *Quantization of Poisson pairs: R-matrix approach*, J. Geom. Phys. **9** (1992), 25–44.

[M1] Sh. Majid, *Braided groups*, J. Pure Appl. Algebra **86** (1993), 187–221.

[M2] ———, *Quantum and braided Lie algebra*, Preprint DAMPT/93-4, Cambridge.

[KRR] S. Khoroshkin, A. Radul, and V. Rubtsov, *A family of Poisson structures on Hermitian symmetric spaces*, Comm. Math. Phys. **152** (1993), 299–316.

[PP] A. Polishchuk and L. Posicel′skiĭ, *On quadratic algebras*, Preprint (1991), Moscow.

[Pri] S. Priddy, *Koszul resolutions*, Trans. Amer. Math. Soc. **152** (1970), 39–60.

[RTF] N. Reshetikhin, L. Takhtadzhyan, and L. Faddeev, *Quantization of Lie groups and Lie algebras*, Leningrad Math. J. **1** (1990), 193–226.

Translated by THE AUTHORS

DEPARTMENT OF MATHEMATICS, BAR-ILAN UNIVERSITY, 52900 RAMAT-GAN, ISRAEL

CENTRE DE MATHÉMATIQUES, ECOLE POLYTECHNIQUE, F-91128 PALAISEAU CEDEX, FRANCE

Amer. Math. Soc. Transl.
(2) Vol. **167**, 1995

Poisson–Lie Aspects of Classical W-Algebras

B. ENRIQUEZ, S. KHOROSHKIN,
A. RADUL, A. ROSLY, AND V. RUBTSOV

ABSTRACT. We describe an infinite-dimensional Poisson–Lie group, related to classical W-algebras; its function algebra is a "universal W-algebra" since it gives the various $\mathcal{W}_n$ as quotients. We give two constructions of this group (it consists of extended Volterra operators), by logarithmic extension and by a polynomial continuation technique. The last approach allows us to construct new (parametrized by a complex number) Hamiltonian structures for the KP-hierarchy. We also give a short proof of the Kuperschmidt–Wilson theorem based on Poisson–Lie considerations.

Introduction

This paper is mainly concerned with two subjects that have been discussed extensively in recent times: W-algebras and (bi)-Hamiltonian integrable systems. Recently new interest in the subject appeared from the viewpoint of Poisson–Lie geometry ([27, 28, 42]).

V. Fateev and S. Lukyanov in the series of papers [15] considered different aspects of the relationship between W-algebras and Poisson–Lie groups. Their approach is strongly based on the classical paper of V. Drinfeld and V. Sokolov [11] and leads to a direct quantization of W-algebras [32] as well as to the very deep quantization scheme of B. Feigin and E. Frenkel [16].

Our position is more geometric and arises from the work of M. Semenov-Tyan-Shansky on dressing transformation geometry [45] and from related papers on quasi-classical geometry of quantum groups [36, 47]. The question of quantization in this approach is intimately connected with the problem of V. Drinfeld–T. Khovanova: to quantize in some way the "bialgebra" of differential operators on the circle [10]. One of us proposed a solution to this problem in terms of the vertex operator algebra (VOA) [12].

The geometry of Poisson–Lie groups was the object of considerable attention of mathematicians during these last years. Poisson–Lie groups appeared as the quasi-classical counterpart of quantum groups in the works of V. Drinfeld [9], but usually they were considered only in the framework of finite-dimensional simple or Kac–Moody Lie algebras. An additional interest to Poisson–Lie groups is due to

1991 *Mathematics Subject Classification*. Primary 17B65.

the identification of these groups with classical symmetries of (classical) chiral field theory, see [13, 14, 20].

We intend to relate the machinery of Poisson–Lie groups to another subject of mathematical physics, namely, bi-Hamiltonian structures. This notion was introduced in the context of integrable systems by F. Magri [37] and discussed very extensively by different authors ([21, 22, 31, etc.], it is impossible to mention all of them).

The pair of Poisson brackets $\{\ ,\ \}_1$ and $\{\ ,\ \}_2$ on a manifold M is called *compatible* or *coordinated* if any linear combination of these brackets $\alpha\{\ ,\ \}_1 + \beta\{\ ,\ \}_2$ is again a Poisson bracket and the manifold M is called *bi-Hamiltonian* in this case. The relevance of this notion for the theory of completely integrable systems is well known: in fact, almost all known examples of such systems are bi-Hamiltonian (that is, their phase spaces are endowed with such a structure, which leads to an infinite family of conserved quantities).

There is another reason to be interested in bi-Hamiltonian structures. Recently, a special type of such structures, for which one of the brackets is symplectic (nondegenerate) and the other is related to a constant solution of the classical Yang–Baxter equation (classical r-matrix), was studied from the point of view of deformation quantization [23] and Poisson–Lie geometry as well. The general quantization scheme of Poisson pairs [23] was thought be connected to the problem of rigorous quantization of the most celebrated Poisson pairs, the first and the second Gelfand–Dikiĭ (GD) brackets in the theory of Lax equations.

We do not know how to apply the quantization scheme of Poisson pairs to GD-pairs directly, because as the original work of M. Semenov-Tyan-Shansky [45, 46] implies, the r-matrices connected to the second GD brackets are in fact solutions of the so-called Modified Yang–Baxter Equation (MYBE). We think that the correct way to quantize these GD-brackets involves the BRST scheme (this was done by Feigin–Frenkel in the case of GD2). It is possible to give a natural candidate for the operation of adding a spectral parameter λ in this setting. But contrary to the classical case, it is unlikely that the family of Lie algebra structures obtained by the conjugation of quantum GD2 with this operation depends linearly on λ.

Poisson–Lie geometry of finite-dimensional bi-Hamiltonian manifolds with a pair of Poisson brackets, one of which is symplectic and the other relates to the Drinfeld–Jimbo solution of MYBE, was studied in [28]. The question of coordinating invariant symplectic structures on the orbit of the semisimple Poisson–Lie groups G with the reduced Drinfeld–Sklyanin bracket (see below) depends on the structure of the orbit. The answer is very simple and nice: the orbit is bi-Hamiltonian with respect to the pair of Poisson brackets of the type described above precisely when it is a Hermitian symmetric space. This restrictive property, together with the well-known relationship of the solution space of a scalar Lax equation with the infinite-dimensional Grassmannian [6, 39, 44], gave us some ideas and approaches for regarding the bi-Hamiltonian affine infinite-dimensional manifold $\overline{\mathcal{L}}_n$ introduced in [42] as a kind of homogeneous Poisson–Lie infinite-dimensional manifold.

It was a real surprise to learn that there exists a very natural procedure to endow all infinite-dimensional affine manifolds arising in this approach to integrable systems of Lax type (nth generalized KdV, Kadomtsev–Petviashvili hierarchy and their fractional generalizations) with the structure of Poisson–Lie submanifold for a certain Poisson–Lie group (extended Volterra group of integral operators).

This extension is realized by means of a certain cocycle, which was one of the most fascinating and beautiful discoveries of the last years in the theory of infinite-dimensional Lie groups and algebras. This "logarithmic cocycle" was introduced and studied in [25] and was generalized in [26]. It is related to the central charge of the Virasoro and Chern class of the line bundle over the moduli space of the complex curve [2], the Riemann–Roch theorem for DOP [17], and anomalous commutator relations for area-preserving Lie algebra diffeomorphisms [3].

The idea of extending the Lie algebra of formal pseudodifferential operators $\Psi\mathrm{DO}(S^1)$ on the circle yields an answer to a question posed by one of the authors (A. Radul) [42] and Morozov [38]: to describe a universal W-algebra, a Poisson bracket algebra (PBA) such that all classical W_n-algebras are obtained by reduction of this universal W. This question was discussed extensively in [18, 19].

We propose the following answer to this question. The Poisson–Lie structure on the extended Volterra group of integral operators on the circle (which is nothing but a certain kind of Drinfeld–Sklyanin bracket) gives a Poisson manifold such that all manifolds $\overline{\mathcal{L}}_n$ of Lax operators of nth degree (with the GD2 structure), as well as the KP (Kadomtsev–Petviashvili) phase space, are Poisson submanifolds of the obtained manifold. The dressing action of the Poisson–Lie algebra $\widetilde{\mathrm{DOP}}(S^1)$ (extended algebra of differential operators on the circle) on these manifolds coincides with the action defined in [42].

The Poisson–Lie structure of the extended Volterra group and fractional generalizations of Lax operators were studied independently by Khesin and Zakharevich in [27]. They used the logarithmic cocycle to define the Poisson–Lie structure on the group of extended Volterra operators. This idea is independently due to one of us (A. Rosly) and appears in [28]. In §3 we also present an alternative description of the construction of this Poisson–Lie structure, relying on the polynomiality in n of the structure constants of $\mathcal{W}_n$, established in [12], together with a short proof (using the notion of Poisson–Lie group) of the theorem of Kuperschmidt and Wilson on free field realization of $\mathcal{W}_n$ [31]. We remark that this gives a new proof of the result of [21].

A natural question that arises here concerns the quantization of the group structure on the group of Volterra operators. This question was discussed in [12] using the approach of Fateev and Lukyanov (involving quantum Miura coordinates). It would be interesting to find an interpretation of this question in the terms used by Feigin and Frenkel.

In §4, we discuss briefly a geometric reformulation of some of the considerations above. This reformulation may be useful for a possible interpretation in geometric terms of some ingredients of the theory of integrable systems. The most desirable for us would be to describe the τ-functions related to KP-hierarchies in a Poisson–Lie geometric way. F. Magri informed us that his work on relations between the τ-functions of KP and geometry of moment maps is in progress.

The elements of the manifold $\mathcal{L}$ can be expressed as operators containing complex powers of the derivation ∂. It is then natural to construct analogs of the KP flows and of their conserved densities on the phase space $\mathcal{L}$. This is done in §5, using techniques of polynomial continuation. (These results were also announced in [27].) We also generalize the bilinear identity which is at the origin of τ-functions in the usual KP theory.

Acknowledgements. This paper is a result of numerous discussions with many people from whom we benefited very much. Talking to M. A. Semenov-Tyan-Shansky and B. L. Feigin were extremely stimulating. We are also grateful to B. Khesin, F. Magri, T. Khovanova, and A. Reiman for very interesting discussions.

The first two parts of this paper are based on lectures given by one of us (V. Rubtsov) at the Centre de Mathématiques and at the Centre de Physique Théorique of Ecole Polytechnique, in June 1992. (This explains the expository character of these parts.) V. Rubtsov would like to thank Professor J. Lascoux for his interest in this work and Professors J.-P. Bourguignon and A. Guichardet for their hospitality during his visit at the Centre de Mathématiques of Ecole Polytechnique.

B. Enriquez would also like to thank D. Lebedev for his hospitality at ITEP in January 1993.

§1. Poisson–Lie groups. Generalities

1.1. We say that a Lie group G is a *Poisson–Lie group* if its algebra of smooth functions $C^\infty(G)$ is endowed with a Poisson bracket $\{\ ,\ \}$, the multiplication law $G \times G \to G$ is a Poisson morphism, and the inversion map $g \mapsto g^{-1}$ is a morphism of G to G with the opposite structure.

Multiplicativity of the Poisson structure on the Poisson–Lie group G may be easily generalized to the case of an action of a Poisson–Lie group G on a Poisson manifold M.

The left action σ of a Poisson–Lie group G with the bracket $\{\ ,\ \}_G$ on a Poisson manifold M with the Poisson bracket $\{\ ,\ \}_M$ is called a *Poisson–Lie action* if the map $\sigma\colon G \times M \to M$ is a Poisson morphism for the product Poisson structure on $G \times M$, or in the language of brackets:

$$\{\varphi\,,\,\psi\}_M(\sigma(g)x) = \{\varphi(\sigma(g))\,,\,\psi(\sigma(g))\}_M(x) + \{\varphi(\sigma_x)\,,\,\psi(\sigma_x)\}_G(g)$$

for $x \in M$, $g \in G$, $\varphi, \psi \in C^\infty(M)$ (we take $\sigma(g\,,\,x) = \sigma(g)(x) = \sigma_x(g)$).

1.2. At the level of Lie algebras, the notion of Poisson–Lie group is equivalent to the notion of *Lie bialgebra structure* on the Lie algebra $\mathfrak{g} = \mathrm{Lie}(G)$. This means that the dual space $\mathfrak{g}^*$ of $\mathfrak{g}$ is endowed with a Lie algebra structure $[\ ,\ ]_*\colon \Lambda^2\mathfrak{g}^* \to \mathfrak{g}^*$ such that the dual map $\delta\colon \mathfrak{g} \to \Lambda^2\mathfrak{g}$ (*cobracket*) is a 1-cocycle on $\mathfrak{g}$ with respect to the adjoint action of $\mathfrak{g}$ on $\Lambda^2\mathfrak{g}$:

$$\delta([x\,,\,y]) = [\delta(x)\,,\,y \otimes 1 + 1 \otimes y] + [x \otimes 1 + 1 \otimes x\,,\,\delta(y)]$$

for $x, y \in \mathfrak{g}$.

This notion is equivalent to the notion of *Manin triple*, which was introduced by Drinfeld [9]: the set of $\mathfrak{g}$, $\mathfrak{g}_1$, $\mathfrak{g}_2$ is a Manin triple if

a) $\mathfrak{g}$ is a Lie algebra;
b) $\langle\ ,\ \rangle\colon \mathfrak{g}^2 \to \mathbb{C}$ is a invariant nondegenerate pairing;
c) $\mathfrak{g}_1$, $\mathfrak{g}_2$ are two Lie algebras, isotropic with respect to $\langle\ ,\ \rangle$ and $\mathfrak{g} = \mathfrak{g}_1 \oplus \mathfrak{g}_2$ as a vector space.

If $\mathfrak{g}$ is a Lie bialgebra, then $(\mathfrak{g}\oplus\mathfrak{g}^*, \mathfrak{g}, \mathfrak{g}^*)$ is an example of a Manin triple if we endow the vector space $\mathfrak{g}\oplus\mathfrak{g}^*$ with the scalar product $\langle\ ,\ \rangle\colon (\mathfrak{g}\oplus\mathfrak{g}^*)^2\to k$

$$\langle x+\xi\,,\, y+\eta\rangle = \langle x\,,\,\eta\rangle + \langle y\,,\,\xi\rangle\,,$$

and the structure of Lie algebra such that the natural injections $\mathfrak{g}\to\mathfrak{g}\oplus\mathfrak{g}^*$, $\mathfrak{g}^*\to\mathfrak{g}\oplus\mathfrak{g}^*$ are Lie algebra morphisms and

$$[x\,,\,\xi] = \mathrm{ad}^*_x(\xi) - \mathrm{ad}^*_\xi(x)\,,$$

where $x, y\in\mathfrak{g}$, $\xi, \eta\in\mathfrak{g}^*$, ad^* is the coadjoint action (of two different Lie algebras!). It is obvious that $\mathfrak{g}$ and $\mathfrak{g}^*$ are isotropic via $\langle\ ,\ \rangle$ and that $\langle\ ,\ \rangle$ is an ad-invariant product. We denote this Manin triple by $\mathfrak{g}\bowtie\mathfrak{g}^*$. The set $(\mathfrak{g}^*\bowtie\mathfrak{g}, \mathfrak{g}^*, \mathfrak{g})$ is also a Manin triple, hence $\mathfrak{g}^*$ is a Lie bialgebra. Hence the notion of Manin triple is self-dual.

Taking exp of the entries of the Manin triple $(\mathfrak{g}^*\bowtie\mathfrak{g}, \mathfrak{g}^*, \mathfrak{g})$, we obtain a *dual* Poisson–Lie group G^* for the Poisson Lie group G. The dual group G^* acts on G by left and right actions: $G^*\times G\to G$ and $G\times G^*\to G$.

These actions are called *left* and *right dressing actions* or *dressing transformations*. A classical result of Semenov-Tyan-Shansky [45] claims that these actions do not preserve the Poisson structure on G but are Poisson morphisms in the Poisson–Lie sense (see 1.1).

1.3. Let us give an infinitesimal version of the action of a connected Poisson–Lie group G on a Poisson manifold M. The structure of Lie bialgebra on $\mathfrak{g}=\mathrm{Lie}(G)$ is defined by the Poisson–Lie structure on G, i.e., the multiplicative Poisson tensor field π (such that $\{f\,,\,g\}_G = \langle df\wedge dg\,,\,\pi\rangle$ for $f, g\in C^\infty(G)$, $\pi\in\Lambda^2(TG)$) defines a Lie algebra structure $[\ ,\ ]_*$ for $\mathfrak{g}^*$ [36]. Let $\rho\colon\mathfrak{g}\to\mathrm{Vect}(M)$ be a representation of the Lie algebra $\mathfrak{g}$ in vector fields of the manifold M. We say that the *Lie bialgebra* $\mathfrak{g}$ *acts on* M *in a Poisson–Lie way* if

$$\begin{aligned}\mathcal{L}_{\rho(x)}\{f\,,\,g\}_M(m) = \{\mathcal{L}_{\rho(x)}f\,,\,g\}_M(m) + \{f\,,\,\mathcal{L}_{\rho(x)}g\}_M(m)\\ + \langle x\,,\,[\rho^*df(m)\,,\,\rho^*dg(m)]_*\rangle\,.\end{aligned}$$

Here $x\in\mathfrak{g}$, $f, g\in C^\infty(M)$, $\{\ ,\ \}_M$ is the Poisson bracket on M, $[\ ,\ ]_*$ is the Lie bracket on $\mathfrak{g}^*$ and $\rho^*df(m)\in\mathfrak{g}^*$ such that

$$\langle x\,,\,\rho^*df(m)\rangle = \mathcal{L}_{\rho(x)}(f)(m)\,,\qquad m\in M\,.$$

1.4. Actually, every Poisson–Lie group structure on a semisimple connected Lie group G has the following form [9]:

$$\pi(g) = \ell_{g*}(r) - r_{g*}(r)\,,$$

where ℓ_{g*}, r_{g*} are differentials of left and right translations by $g\in G$ and $r\in\Lambda^2(\mathfrak{g})$ is a *modified classical* r-matrix, or a solution of the modified Yang–Baxter classical equation

$$[\![r\,,\,r]\!] := [r^{12}\,,\,r^{13}] + [r^{13}\,,\,r^{23}] + [r^{12}\,,\,r^{23}] \in \Lambda^3(\mathfrak{g})^{\mathfrak{g}}.$$

(A particular case of this condition is $[\![r, r]\!] = 0$, r is then called a *classical r-matrix*.) Here for $r = \sum_i a_i \otimes b_i$ we define elements $r^{12}, r^{23}, r^{13} \in (U_{\mathfrak{g}})^{\otimes 3}$ as follows:

$$r^{12} = \sum_i a_i \otimes b_i \otimes 1, \quad r^{23} = \sum_i 1 \otimes a_i \otimes b_i, \quad r^{13} = \sum_i a_i \otimes 1 \otimes b_i$$

and, for example,

$$[r^{13}, r^{23}] = \sum_{i,j} a_i \otimes a_j \otimes [b_i, b_j].$$

The bracket $[\![\ ,\ ]\!]$ is the Schouten–Nijenhuis bracket in the exterior algebra of $\mathfrak{g}$: $[\![\ ,\ ]\!]: \Lambda^i(\mathfrak{g}) \times \Lambda^j(\mathfrak{g}) \to \Lambda^{i+j-1}(\mathfrak{g})$.

The ad-invariance of $[\![r, r]\!]$ means that the bracket $[\ ,\ ]_*: \Lambda^2(\mathfrak{g}^*) \to \mathfrak{g}^*$ dual to $\delta: \mathfrak{g} \to \Lambda^2(\mathfrak{g})$, $\delta(x) = \mathrm{ad}_x(r)$ satisfies the Jacobi identity.

We call such a Poisson structure a *coboundary structure*.

The corresponding Poisson bracket on the group G may be expressed in coordinate form by means of this modified r-matrix and is usually called a *Drinfeld–Sklyanin* Poisson bracket. Let $r^{\mu\nu}$ be a r-matrix written in some basis x_i for $\mathfrak{g}$ and let L_μ, R_μ be the coordinates of left and right invariant fields on G. Then the Drinfeld–Sklyanin bracket has the form

$$\{f, g\} = r^{\mu\nu}(L_\mu f L_\nu g - R_\mu f R_\nu g).$$

1.5. We describe an important simple example of the Poisson–Lie structure due to M. Semenov-Tyan-Shansky. Let $(\mathfrak{g}, \mathfrak{g}^*)$ be a Lie bialgebra. Consider an r-matrix on $\mathfrak{g} \bowtie \mathfrak{g}^*$, $r \in \Lambda^2(\mathfrak{g} \bowtie \mathfrak{g}^*)$ corresponding to the skew-symmetric operator $R: \mathfrak{g} \bowtie \mathfrak{g}^* \to \mathfrak{g} \bowtie \mathfrak{g}^*$ such that $R = \mathrm{pr}_{\mathfrak{g}} - \mathrm{pr}_{\mathfrak{g}^*}$, where $\mathrm{pr}_{\mathfrak{g}}$ ($\mathrm{pr}_{\mathfrak{g}^*}$) is the projection on $\mathfrak{g}$ along $\mathfrak{g}^*$ (respectively, on $\mathfrak{g}^*$ along $\mathfrak{g}$). The Lie algebra $\mathfrak{g} \bowtie \mathfrak{g}^*$ is an example of a *Lie–Baxter algebra* in the sense of [46].

For a function $f \in C^\infty(G)$ ($\mathfrak{g} \bowtie \mathfrak{g}^* = \mathrm{Lie}(G)$), we define the element $\mathrm{grad}_\ell f \in \mathfrak{g} \bowtie \mathfrak{g}^*$ such that for each $\xi \in \mathfrak{g} \bowtie \mathfrak{g}^*$

$$\langle \mathrm{grad}_\ell f(x), \xi \rangle = (d/dt)|_{t=0} f(\exp(t\xi)x)$$

and the element $\mathrm{grad}_r f$:

$$\langle \mathrm{grad}_r f(x), \xi \rangle = (d/dt)|_{t=0} f(x \exp(t\xi)).$$

Then the Poisson bracket (Drinfeld–Sklyanin) corresponding to R (or r) is

$$\{f, g\}_{\text{D-S}} = \tfrac{1}{2}\langle R(\mathrm{grad}_r f), \mathrm{grad}_r g\rangle - \tfrac{1}{2}\langle R(\mathrm{grad}_\ell f), \mathrm{grad}_\ell g\rangle.$$

We can easily see that the first and the second parts of Drinfeld–Sklyanin bracket associated to a classical r-matrix are Poisson brackets too:

$$\{f, g\}_r = \tfrac{1}{2}\langle R(\mathrm{grad}_r f), \mathrm{grad}_r g\rangle, \qquad \{f, g\}_\ell = \tfrac{1}{2}\langle R(\mathrm{grad}_\ell f), \mathrm{grad}_\ell g\rangle.$$

We shall call them *right r-matrix bracket* and *left r-matrix bracket* respectively.

The brackets $\{\ ,\ \}_{\text{D-S}}$ are a particular case of a more general construction, which is usually referred to as the *Gelfand–Dikiĭ bracket* (following [46]). We shall consider these general brackets in infinite-dimensional cases, where they were first introduced by I. Gelfand and L. Dikiĭ [21].

§2. Poisson geometry of infinite-dimensional affine manifolds

2.1. We shall work within the framework of Gelfand's formal variational calculus. Let $\mathfrak{L}$ be one of the following infinite-dimensional affine varieties:

a) $\overline{\mathcal{L}}_n = \{L = \partial_x^n + u_1\partial_x^{n-1} + \cdots + u_n \mid u_i \in C^\infty(S^1)\}$ is the manifold of nth order differential operators on the circle S^1.

b) $\mathcal{L}_\alpha = \{L = \partial_x^\alpha + \sum_{i\geqslant 0} u_i\partial_x^{\alpha-i} \mid u_i \in C^\infty(S^1)\}$, for $\alpha \in \mathbb{C}$.

We shall call these varieties *Lax operators with complex powers of* ∂ and we explain below their appearance in this context. In the case $\alpha = 1$, $\mathcal{L}_\alpha$ is the set of Lax operators

$$L = \partial + u_0 + u_{-1}\partial^{-1} + \cdots + u_s\partial^{-s} + \dots$$

of the KP-hierarchy.

2.2. We intend to introduce certain Poisson structures on the manifolds $\mathfrak{L}$ in all cases above.

It is a well-known fact that the set of Lax operators $\overline{\mathcal{L}}_n$ in case a) is endowed with the Poisson structure constructed in [21]. Later, one of the authors (A. Radul) noted that the same procedure could be extended to manifolds of type b) with positive integer α [42]. Since then, this fact has been noticed by many authors ([5, 19]; the best source of references is the book [8]).

We begin with the description of the pairs of GD-brackets in the context of Poisson–Lie geometry for the manifolds $\overline{\mathcal{L}}_n$ of a).

The manifold $\overline{\mathcal{L}}_n$ will be endowed with an action of a certain Poisson–Lie formal group. Actually, we prefer to avoid all analytic and topological complications and will consider the infinitesimal version of this action, a Lie (bi)algebra action of the differential operators on the circle.

Let $\Psi\mathrm{DO}_N(S^1)$ be the set of symbols of order N:

$$\Psi\mathrm{DO}_N(S^1) = \left\{a(x, \partial) = \sum_{i=-\infty}^{N} a_i(x)\partial^i \mid a_i \in C^\infty(S^1)\right\}.$$

We equip the set $\Psi\mathrm{DO}(S^1) = \bigcup_{N\geqslant 0} \Psi\mathrm{DO}_N(S^1)$ with *symbolic multiplication*

$$a(x, \partial) \circ b(x, \partial) = \sum_{n\geqslant 0} \frac{1}{n!} a_\partial^{(n)} \cdot b_x^{(n)}.$$

This multiplication makes $\Psi\mathrm{DO}(S^1)$ an associative algebra. Let a_+ and a_- be the differential and integral parts of the symbol a. The differential symbols form a subalgebra in $\Psi\mathrm{DO}(S^1)$, which we denote by $\mathrm{DOP}(S^1)$. The symbolic multiplication coincides with the composition of differential operators.

Denote by $\mathrm{Tr}\colon \Psi\mathrm{DO}(S^1) \to \mathbb{C}$ the *Adler trace*

$$\mathrm{Tr} = \int_{S^1} \circ\, \mathrm{res}, \qquad \mathrm{res}(a) = a_{-1}.$$

This trace defines a scalar product

$$\langle\ ,\ \rangle\colon \Psi\mathrm{DO}(S^1)^2 \to \mathbb{C}, \qquad \langle a, b\rangle = \mathrm{Tr}(a \circ b)$$

and hence $\Psi\mathrm{DO}(S^1)$ is a *metrizable* Lie algebra with respect to the bracket $[\ ,\ ]\colon \Lambda^2(\Psi\mathrm{DO}(S^1)) \to \Psi\mathrm{DO}(S^1)$, $[a, b] = a \circ b - b \circ a$ and $\langle\ ,\ \rangle$ is ad-invariant with respect to this bracket.

Denote by $\mathrm{IOP}(S^1)$ the algebra of integral symbols on the circle:

$$\mathrm{IOP}(S^1) = \{a_{-1}\partial^{-1} + \cdots + a_{-s}\partial^{-s} + \cdots \mid a_i \in C^\infty(S^1)\}.$$

The pair $(\mathrm{DOP}(S^1), \mathrm{IOP}(S^1))$ is an example of an infinite-dimensional "almost" Manin triple or self-dual Lie bialgebra. As indicated in [10], this is not a good Manin triple because not every continuous linear functional on S^1 has the form $\langle\ , f\rangle$ for $f \in C^\infty(S^1)$ and hence $\mathrm{IOP}(S^1)$ is not identified with $\mathrm{DOP}(S^1)$ with respect to $\langle\ ,\ \rangle$. In fact, the computation of the cobracket of a given $a\partial^{-k} \in \mathrm{IOP}(S^1)$ gives an element of the form $\sum_{k,l,m,p} a^{(k)}(x)\delta^{(l)}(x-y)\partial_x^m\partial_y^p$ (the sum is finite; x and y are the variables of the first and second component of the tensor product $\mathrm{IOP}(S^1) \otimes \mathrm{IOP}(S^1)$). The occurrence of δ-functions in the Poisson cobracket (or in the Poisson bracket of the functions on the group) usually leads to VOA rules after quantization. In what follows, we shall forget about these problems and shall simply write "Lie bialgebra" $\mathrm{DOP}(S^1)$ or "Manin triple" $(\Psi\mathrm{DO}(S^1), \mathrm{DOP}(S^1), \mathrm{IOP}(S^1))$.

2.3. The *tangent space* of the manifold $\overline{\mathcal{L}}_n$ at the point L is a deformation space (within differential operators) and may identified with the set of differential operators of order $(n-1)$:

$$T_L\overline{\mathcal{L}}_n = \{\delta L = v_1(x)\partial^{n-1} + \cdots + v_n(x) \mid v_i \in C^\infty(S^1)\}.$$

Let $\mathrm{Fun}(\overline{\mathcal{L}}_n)$ be the space of functions on $\overline{\mathcal{L}}_n$ consisting of functionals

$$F(u) = \int_{S^1} f(\partial_x^{(j)}(u_i)) = \int_{S^1} f(u_i^{(j)}),$$

where $f(u_i^{(j)})$ is a polynomial with coefficients in $C^\infty(S^1)$.

To describe the cotangent space $T_L^*\overline{\mathcal{L}}_n$, we use the pairing

$$\langle\ ,\ \rangle\colon T_L\overline{\mathcal{L}}_n \times T_L^*\overline{\mathcal{L}}_n \to \mathbb{C}, \qquad \langle \delta L, X\rangle = \mathrm{Tr}(\delta L \circ X),$$

$$X \in T_L^*\overline{\mathcal{L}}_n = \left\{\sum_{i=1}^n f_i(x)\partial^{-i} \mid f_i \in C^\infty(S^1)\right\} = \Psi\mathrm{DO}_{-1}(S^1)/\Psi\mathrm{DO}_{-n}(S^1).$$

For example, for $F \in \mathrm{Fun}(\overline{\mathcal{L}}_n)$ we can define the differential dF at the point L which is a one-form on $\overline{\mathcal{L}}_n$ in L: $dF(L) = \sum_{i=0}^{n-1} \partial^{-i-1}\delta F/\delta u_i$ where $\delta/\delta u_i$ is the variational derivative with respect to u_i and the following relation holds

$$\delta F = \mathrm{Tr}(\delta L \circ dF) = \langle \delta L, \delta F\rangle = i_{\delta L} dF \quad \text{with } \delta L = \sum_{i=0}^{n-1} \delta u_i \partial^i.$$

2.4. To determine a Poisson (or Hamiltonian) structure on $\overline{\mathcal{L}}_n$, we define the Hamiltonian map $V\colon T^*\overline{\mathcal{L}}_n \to T\overline{\mathcal{L}}_n$ which, in turn, defines a Poisson bracket on $\mathrm{Fun}(\overline{\mathcal{L}}_n)$:

$$\{F, G\}_V = \langle V(dF), dG\rangle.$$

For a given $L \in \overline{\mathcal{L}}_n$ and $X \in T_L^*\overline{\mathcal{L}}_n$, $X = \sum_{i=0}^{n-1} \partial^{-i-1} \circ x_i$, take $V_L(X) = (LX)_+L - L(XL)_+$. This choice of V is usually called the *Adler map,* although the Poisson brackets $\{F, G\}_V(L) = \mathrm{Tr}((L\,dF)_+L\,dG - L(dFL)_+dG)$ are called the *second Gelfand–Dikiĭ brackets* [8, 21].

In the coordinates u_i of the manifold $\overline{\mathcal{L}}_n$ let us refer to these brackets as *fundamental* ones.

The expression of the Adler map $V_L(X)$ in a slightly different form $V_L(X) = L(XL)_- - (LX)_-L$ shows that $V_L(X)$ is a linear (on X) map to differential operators of order at most $n-1$. Hence we can write

$$V_L(X) = \sum_{i,j=0}^{n-1} H_{ij}(x_j)\partial^i ,$$

where H_{ij} are differential operators, with coefficients differential polynomials in u_k; now we can rewrite the Gelfand–Dikiĭ brackets $\{F, G\}_V$ as fundamental Poisson brackets between $u_i(x)$ and $u_j(y)$:

$$\{u_i(x), u_j(y)\}_V = -H_{ij}\delta(x-y).$$

The problem is to check that the bilinear operation defined above

$$\{\ ,\ \}_V \colon \mathrm{Fun}(\overline{\mathcal{L}}_n)^2 \to \mathrm{Fun}(\overline{\mathcal{L}}_n)$$

gives a Lie algebra structure on $\mathrm{Fun}(\overline{\mathcal{L}}_n)$, since the Jacobi identity for $\{\ ,\ \}_V$ is far from obvious.

The fundamental result of Gelfand–Dikiĭ was that this is indeed true and $(\mathrm{Fun}(\overline{\mathcal{L}}_n), \{\ ,\ \}_V)$ is a Lie algebra. Gelfand and Dikiĭ's computations are performed in [21]. Later Gelfand and Dorfman proposed a more conceptual scheme for the case of the first GD structure, using the Nijenhuis–Schouten bracket; then Semenov-Tyan-Shansky applied this scheme [46] to GD2. Drinfeld and Sokolov [11] also gave a proof using the Hamiltonian reduction.

This Lie algebra is said to be the *classical* $\mathcal{W}_n$*-algebra,* although this name is still discussed: in [43], this algebra was called the GD_n-algebra, while the name classical W_n-algebra was used for the related Lie algebra structure on the vector fields $\mathrm{Vect}(\overline{\mathcal{L}}_n)$ with the usual commutator $[[\ ,\ ]]$ such that

$$V_L(d\{F, G\}) = [[V_L(dF), V_L(dG)]].$$

Following [19], we call the algebras $(\mathrm{Fun}(\overline{\mathcal{L}}_n), \{\ ,\ \}_V)$ *classical realizations* of the Zamolodchikov–Fateev–Lukyanov W-algebras.

The last relation can be easily checked for example in the case of a $\mathcal{W}_2$-algebra if on the 1-forms we impose a restriction reflecting the Virasoro algebra constraints ($s\ell_2$-case) $u_1 = 0$ in L.

The matrix $\|H_{ij}\|$ in this case has the following form:

$$\|H_{ij}\| = \begin{pmatrix} -\partial^3 - u_1\partial^2 + \partial^2 u_1 - u_2\partial - \partial u_2 + u_1\partial u_1 & \partial^2 + u_1\partial \\ \partial^2 - \partial u_1 & 2\partial \end{pmatrix}.$$

After imposing the condition $u_1 = 0$, we obtain the realization of the Virasoro algebra

$$\begin{aligned}\{u_2(x), u_2(y)\} &= \tfrac{1}{2}(-\partial^3 - 2(u_2\partial + \partial u_2))\delta(x-y) \\ &= \tfrac{1}{2}\delta'''(x-y) + (u_2(x) + u_2(y))\delta'(x-y).\end{aligned}$$

The coefficient $1/2$ in the middle term is in good correspondence with the embedding of the Virasoro algebra as a subalgebra in $\mathcal{W}_n$:

$$\{u_2(x), u_2(y)\} = (u_2(x) + u_2(y))\delta'(x-y) + \tfrac{1}{12}(n^3 - n)\delta'''(x-y)$$

(relation in $\mathcal{W}_n$ reduced by $u_1 = 0$) [29]. Another interesting way (which is connected to the Poisson–Lie structure of the extended Volterra group) to prove that $(\mathrm{Fun}(\overline{\mathcal{L}}_n), \{\ ,\ \}_V)$ is indeed a Lie algebra was proposed in the work of B. Kuperschmidt and G. Wilson [31] and based on a "free field" realization of the manifold $\overline{\mathcal{L}}_n$. Let $L = (\partial - \varphi_n)(\partial - \varphi_{n-1})\cdots(\partial - \varphi_1)$ be "factorizable" ($\varphi_1, \dots, \varphi_n$ are called *Miura coordinates* for L). Define the fundamental brackets for φ_i by

$$\{\varphi_i(x), \varphi_j(x)\} = \delta_{ij}\delta'(x-y).$$

The theorem of Kuperschmidt and Wilson claims that the Poisson brackets between $u_i(x)$ induced by the Miura transformation coincide with the Gelfand–Dikiĭ brackets $\{\ ,\ \}_V$ above, that is, the embedding $\mathrm{Fun}(\overline{\mathcal{L}}_n) \to \mathrm{Fun}(\{\varphi_i\})$ is compatible with the brackets. Since the Jacobi identity is evident on the right-hand side, this proves that $\mathrm{Fun}(\overline{\mathcal{L}}_n)$ is a Lie algebra.

2.5. We shall consider another Hamiltonian map $V_L^{(1)}: T_L^*\overline{\mathcal{L}}_n \to T_L\overline{\mathcal{L}}_n$. Recall that $\overline{\mathcal{L}}_n$ is an affine manifold and hence we can consider the difference $V_{L+\lambda} - V_L$ of the Adler maps at two different points L and $L + \lambda$ ($\lambda \in \mathbb{C}$)

$$V_{L+\lambda}(X) - V_L(X) = \lambda[L, X]_+, \qquad X \in T^*\overline{\mathcal{L}}_n.$$

The map $V_L^{(1)}(X) = [L, X]_+$ is Hamiltonian and the corresponding Poisson bracket

$$\{F, G\}_{V^{(1)}}(L) = \langle [dF, L]_+ dG\rangle$$

is called the *first Gelfand–Dikiĭ bracket*.

We see that the first and second Gelfand–Dikiĭ brackets are compatible in the sense that was discussed in the introduction. There are many different realizations and interpretations of these two coordinated structures. Therefore it would be very desirable to clarify the compatibility for each realization of these structures.

The next section is devoted to one of the realizations of the first GD structure discovered by M. Adler [1], and by D. Lebedev and Yu. Manin [33].

2.6. Let G_- be the formal Volterra group defined by the exponentiation of the algebra $\mathrm{IOP}(S^1)$ considered in 2.2.

The properties of the map $\exp: \mathrm{IOP}(S^1) \to G_-$ have been discussed recently in the work of B. Khesin and I. Zakharevich [27]; they showed that this map is a surjection. The elements of G_- are expressions of the following type

$$1 + a_{-1}(x)\partial^{-1} + a_{-2}(x)\partial^{-2} + \dots, \qquad a_i \in C^\infty(S^1).$$

As we indicated earlier, the algebra $\mathrm{DOP}(S^1)$ can be considered in turn as dual to $\mathrm{IOP}(S^1)$ and hence we can define the coadjoint action of G_- on $\mathfrak{g}_-^*$:

$$\mathrm{Ad}_x^*(E) = (x^{-1}Ex)_+, \qquad x \in G_-, \ E \in \mathrm{DOP}(S^1).$$

Let $\mathcal{O}_E \subset \mathrm{DOP}(S^1)$ be the orbit of E under this action. Then we can consider the Kirillov structure on $\mathfrak{g}_-^* = \mathrm{DOP}(S^1)$. For $X, Y \in \mathrm{IOP}(S^1)$ we denote by ∂_X, ∂_Y elements of algebra $\mathrm{Vect}(\mathcal{O}_E)$ of vector fields on the orbit $\mathcal{O}_E$. Then the Kirillov symplectic form is defined by

$$\omega(\partial_X, \partial_Y)(E) = \langle E, [X, Y]\rangle.$$

It is clear that $\overline{\mathcal{L}}_n$ is a stable manifold under the action of $\mathrm{IOP}(S^1)$, hence it is possible to define on $\overline{\mathcal{L}}_n$ the Kirillov Poisson structure in the following way: take $L \in \overline{\mathcal{L}}_n$ and $X \in \mathrm{IOP}(S^1)$ such that

$$\exp(X) = x \in G_-, \quad X = \sum_{i=0}^{n-1} \partial^{-i-1} x_i = dF \quad \text{for some } F \in \mathrm{Fun}(\overline{\mathcal{L}}_n),$$

and take $Y \in \mathrm{IOP}(S^1)$ such that

$$\exp(Y) = y \in G_- \quad \text{and} \quad Y = \sum_{i=0}^{n-1} \partial^{-i-1} y_i = dG \quad \text{for } G \in \mathrm{Fun}(\overline{\mathcal{L}}_n).$$

We have

$$\begin{aligned}\omega(X_F, X_G)(L) &= \langle L, [dF, dG]\rangle = -\langle [L, dF]_+, dG\rangle \\ &= \langle [dF, L]_+, dG\rangle = \{F, G\}_{V^{(1)}}(L).\end{aligned}$$

Here $X_F, X_G \in \mathrm{Vect}(\overline{\mathcal{L}}_n)$ are Hamiltonian vector fields.

The result of [1] and of [33] is that the first GD structure is just an ordinary Lie–Poisson structure on the dual space $\mathrm{IOP}(S^1)^*$ of the Lie algebra $\mathrm{IOP}(S^1)$.

Notice that in our case we only have the inclusion $\mathrm{DOP}(S^1) \subset \mathrm{IOP}(S^1)^*$; the elements of $\mathrm{DOP}(S^1)$ form an invariant submanifold of $\mathrm{IOP}(S^1)^*$. Hence the first GD structure is invariant with respect to the action of the Volterra group G_-.

2.7. It seems reasonable to interpret the second GD structure in a similar way by using the orbit realization of the manifold $\overline{\mathcal{L}}_n$. But this approach meets very serious obstructions.

The dual Manin triple $(\Psi\mathrm{DO}(S^1), \mathrm{IOP}(S^1), \mathrm{DOP}(S^1))$ yields another Poisson structure on $\mathrm{IOP}(S^1)$, as in 1.5. Let us regard $r \in \Lambda^2(\Psi\mathrm{DO}(S^1))$ as an endomorphism of $\Psi\mathrm{DO}(S^1)$, skew symmetric with respect to $\langle\ ,\ \rangle$. Then it follows from [46] that there exists a Poisson structure on the algebra $\Psi\mathrm{DO}(S^1)$. Indeed, we identify $\Psi\mathrm{DO}(S^1)$ with $(\Psi\mathrm{DO}(S^1))^*$ using the scalar product $\langle\ ,\ \rangle$ and consider the operator $r = \mathrm{pr}_+ - \mathrm{pr}_-$ given by this skew-symmetric endomorphism; there is a Lie–Poisson bracket on $\Psi\mathrm{DO}(S^1)$ which defines a Lie–Baxter algebra structure on $\Psi\mathrm{DO}(S^1)$. The second Lie algebra structure is defined by r:

$$[x, y]_r = ([rx, y] + [x, ry])/2$$

(see [46]).

The Poisson–Lie bracket on $\Psi\mathrm{DO}(S^1)$ in this case coincides with the first GD-bracket and our manifold $\overline{\mathcal{L}}_n$ with Kirillov bracket is a Poisson submanifold in $\mathrm{IOP}(S^1)^*$, which in turn inherits the Lie Poisson structure of $\Psi\mathrm{DO}(S^1)$ mentioned above.

If we try to reduce the second GD structure (which, of course, also lives in $\Psi\mathrm{DO}(S^1)$ [47]), then we shall succeed in defining the Poisson–Lie group structure on the Volterra group of integral operators (this structure was called Benney–Hamilton structure in [34]).

Unfortunately, in fact we have no groups for the Lie algebras $\Psi\mathrm{DO}(S^1)$ and $\mathrm{DOP}(S^1)$, therefore it seems very hard to apply the Poisson–Lie group machinery of subsections 1.1–1.5 directly. But it is enough for us to have the action of the Volterra group on the Lie algebra $\Psi\mathrm{DO}(S^1)$. In terms of 1.4, we can say that the bialgebra structure on $\Psi\mathrm{DO}(S^1)$ relatively to r is *coboundary* and hence may be written as in 1.4.

Now we can apply results of [36] and [45], and define the Poisson–Lie structure on $\mathcal{D}$ ($\mathcal{D}$ is the formal group such that $\Psi\mathrm{DO}(S^1) = \mathrm{Lie}(\mathcal{D})$). We shall see in 3.3 that the Drinfeld–Sklyanin structure of the Volterra group G_- in given by

$$\{f, g\}(g_-) = \tfrac{1}{2}(\langle (R^*_{g_-}(df))_+, R^*_{g_-}(dg)\rangle - \langle (L^*_{g_-}(df))_+, L^*_{g_-}(dg)\rangle).$$

Here f and g belong to $\mathrm{Fun}(G_-)$, $g_- \in G_-$, $R^*_{g_-}$ and $L^*_{g_-}$ are differentials of right and left translations. (The r-matrix manifests itself by the sign subscripts $+$).

In 3.3, we shall see that this structure can be written as a structure of GD type corresponding the generalized Adler map

$$V\colon T^*_L G_- \to T_L G_-, \qquad V_L(X) = (LX)_+ L - L(XL)_+ .$$

The next stage is to try to induce this structure on the manifold $\overline{\mathcal{L}}_n$ in order to regard the latter as an infinite-dimensional example of a bi-Hamiltonian manifold in sense of F. Magri [37].

To do this, we could use the action (3.6) of G_- on $\overline{\mathcal{L}}_n$ if it were Poisson–Lie. However, as we noticed in [28], it is not. The reason is that the coadjoint action of 3.6 is not a Poisson–Lie group action of G_- but simply a Lie group action.

To define a more sophisticated (Poisson–Lie group) action on $\overline{\mathcal{L}}_n$, one of us proposed in [42] to consider an action of $\mathrm{DOP}(S^1)$ on $\overline{\mathcal{L}}_n$ as a Lie bialgebra action in the following way. Define a map $X \mapsto W_X$ from $\mathrm{DOP}(S^1)$ to $\mathrm{Vect}(\overline{\mathcal{L}}_n)$ by the following formula: at the point L,

$$W_X(L) = LX - (LXL^{-1})_+ = (LXL^{-1})_- L .$$

From the first equality it follows that W_X is a differential operator, and the second one implies that its order is $\leqslant n-1$.

THEOREM 1 [42]. *The map* $W\colon \mathrm{DOP}(S^1) \to \mathrm{Vect}(\overline{\mathcal{L}}_n)$, $X \mapsto W_X$ *is a Lie algebra homomorphism.*

Actually we have $W_X = V_L(XL^{-1})$ and a Lie algebra antihomomorphism $\mathrm{DOP}(S^1) \to \Omega^1(\overline{\mathcal{L}}_n)$ (obtained by composition), where the Lie algebra structure of $\Omega^1(\overline{\mathcal{L}}_n)$ is defined in the usual way for a Poisson manifold with Poisson tensor π:

$$[\![\omega_1, \omega_2]\!] = L_{\langle \pi, \omega_2\rangle}\omega_1 - L_{\langle \pi, \omega_1\rangle}\omega_2 - d\langle \pi, \omega_1 \wedge \omega_2\rangle .$$

In our case we have for two symbols X, $Y \in \Omega^1_L(\overline{\mathcal{L}}_n) \simeq T^*_L(\overline{\mathcal{L}}_n)$

$$[\![X, Y]\!] = L_{V(X)}Y + (X(LY)_- - (XL)_+Y)_- - L_{V(Y)}X - (Y(LX)_- + (YL)_+X)_- .$$

For example, for $X = dF$, $Y = dG$, we obtain $d\{F, G\}_V = [\![dF, dG]\!]$.

This sophisticated action has a simple geometric meaning, noticed also in [7, 30, 42]. We can understand the classical $\mathcal{W}_n$ as a symmetry of the nth order scalar differential equation $Lf = 0$. To deform the operator L and a solution f, we consider a pair of differential operators X and Y and choose them from the conditions:

$$\delta L = YL - LX \quad \text{and} \quad \delta L \in T_L\overline{\mathcal{L}}_n .$$

From this equation we obtain directly $Y = LXL^{-1} + \delta LL^{-1} = (LXL^{-1})_+$, $\delta f = Xf$ and so $\delta L = (LXL^{-1})_- L$, which is $-W_X(L)$.

We have another encouraging observation: the space of solutions of $Lf = 0$ has many features of the finite-dimensional orbits for Lie–Poisson groups which are bi-Hamiltonian with respect to the Kirillov bracket and the induced Drinfeld–Sklyanin brackets [28]. Namely, the space $\{Lf = 0\}$ can be easily mapped to the infinite-dimensional Grassmannian.

If we consider $\operatorname{Ker} L$ as a n-dimensional subspace of some Hilbert functional space $\mathcal{H}$, we obtain a point in $\operatorname{Gr}_n(\mathcal{H})$, the Grassmannian manifold of n-dimensional subspaces in $\mathcal{H}$. Moreover, it is easy to obtain an inverse map: for any point $G \in \operatorname{Gr}_n(\mathcal{H})$ with basis $G = \{f_1, \dots, f_n\}$, we consider the following differential operator of the nth order:

$$L(G)f = \frac{W_{n+1}(f, f_1, \dots, f_n)}{W_n(f_1, \dots, f_n)},$$

where $f \in \mathcal{H}$ and W_r is the Wronski determinant of order r. For example, this bijective correspondence gives $u_i(x)$ (the coefficients of L) as inhomogeneous Grassmannian coordinates. (We can remark that f_i are expressed simply in terms of Miura coordinates. Hence the Poisson brackets between them is of simple form, and can be quantized on the lattice using q-commutation relations [40].)

These geometric remarks lead to a description of the Poisson–Lie structure on infinite-dimensional Grassmannians and then to reduce them on different interesting objects (for example on $\overline{\mathcal{L}}_n$, cf. the work of I. Zakharevich [50]).

The situation with the first GD structure on $\overline{\mathcal{L}}_n$ is quite different. Namely, the following theorem holds

THEOREM 2. *The Lie algebra* $\operatorname{DOP}(S^1)$ *acts by* W *on the manifold* $\overline{\mathcal{L}}_n$ *with the second* GD*-structure in a Poisson–Lie way. The first* GD*-structure is not invariant under this action.*

The first statement is proved in [42], the second is more or less straightforward and is discussed in [28].

As a result, we have an infinite-dimensional bi-Hamiltonian manifold $\overline{\mathcal{L}}_n$ with two different types of actions: one of them (the coadjoint action of $\operatorname{IOP}(S^1)$) respects the first GD-bracket but not the second one, the second action (Theorem 1) is Poisson–Lie with respect to the second GD structure but not the first one.

To obtain an action which is "compatible" with both structures, we examine the dressing actions of M. Semenov-Tyan-Shansky mentioned above. The problem is to identify in some way the manifold $\overline{\mathcal{L}}_n$ with a submanifold in $\mathrm{IOP}(S^1)$, where, as we know, the action of $\mathrm{DOP}(S^1)$ may be considered as a "dressing".

The following observation became crucial in the description below. To make the analogies with finite-dimensional case more realistic, we shall describe the action of $\mathrm{DOP}(S^1)$ in terms of coadjoint action. This may be done if we extend $\mathrm{DOP}(S^1)$ with the help of a *logarithmic cocycle*.

§3. Cocycles on $\mathrm{DOP}(S^1)$, $\mathrm{IOP}(S^1)$, $\Psi\mathrm{DO}(S^1)$ and their extensions

3.1. It is well known that the Lie algebra of vector fields on a circle $\mathrm{Vect}(S^1)$ has a nontrivial extension, called the *Virasoro algebra*. This extension is given by the two-cocycle c ("Gelfand–Fuchs cocycle"):

$$c(f\partial, g\partial) = \frac{1}{6}\int_{S^1} f''\cdot g', \qquad \partial = \frac{d}{dx}, \; f(x), g(x) \in C^\infty(S^1).$$

This fact was generalized by Kac and Peterson [24], who discovered that the Lie algebra of differential operators $\mathrm{DOP}(S^1)$ has a unique central extension $\widehat{\mathrm{DOP}}(S^1)$ by means of the two-cocycle

$$c(f_m\partial^m, g_n\partial^n) = \frac{m!\,n!}{(m+n+1)!}\int_{S^1} f_m^{(n)}(x)\,g_n^{(m+1)}(x).$$

In [25] B. Khesin and O. Kravchenko gave a beautiful formula for this cocycle which allowed them to generalize these cocycles to the algebra $\Psi\mathrm{DO}(S^1)$. We recall the description of c in terms of the "logarithmic cocycle".

3.2. Consider the exterior derivation of the Lie algebra $\Psi\mathrm{DO}(S^1)$ given by

$$[\log\partial, a] = \sum_{n\geqslant 0}\frac{(-1)^{n-1}}{n}\,a_x^{(n)}\partial^{i-n},$$

where $a = \sum_{k=-\infty}^{N} a_k(x)\partial^k$ and $a_x^{(n)} = \sum_{k=-\infty}^{N} a_k^{(n)}(x)\partial^k \in \Psi\mathrm{DO}(S^1)$. The cocycle c can be expressed in terms of $\log\partial$ as

$$c(a, b) = \int_{S^1}\mathrm{res}([a, \log\partial]b) = \mathrm{Tr}([a, \log\partial]b).$$

Using this formula, all properties of $c(a, b)$ can be obtained in a straightforward way. Possible generalizations and properties of this logarithmic cocycle are discussed in [26]. It is worth noting in the W-algebra context that it is possible to construct a special base of $\mathrm{DOP}(S^1)$ such that this cocycle becomes "diagonal" [4]: this base is formed by the two-indexed generators $\{V_m^s\}$ ($m \in \mathbb{Z}$ is the Fourier mode number, $s \in \mathbb{Z}_+$ is the spin number) and the commutation relations between V_m^s, $V_n^{s'}$ give an isomorphism between $\mathrm{DOP}(S^1)$ and the algebra $\mathcal{W}_{1+\infty}$. "Diagonality" of c means that

$$c(V_m^s, V_n^{s'}) = -\frac{(m+s-1)!}{(m-s)!}\,\frac{B(s)^2}{2s-1}\,\delta_{s,s'}\delta_{m+n,0},$$

where $B(s) = 2^{s-3}(s-1)!/(2s-3)!!$.

We shall discuss below the isomorphism between $\mathrm{DOP}(S^1)$ and $\mathcal{W}_{1+\infty}$ from the viewpoint of Poisson geometry. Recall only that $\mathcal{W}_{1+\infty}$ is an infinite-dimensional Lie algebra that can be regarded as a possible "limit" of $\mathcal{W}_n$ $(n \to \infty)$ (with the appropriate precise definition of this notion).

3.3. Extend the algebra $\mathrm{IOP}(S^1)$ "co-centrally" by means of the logarithmic cocycle:

$$\widehat{\mathrm{IOP}}(S^1) = \left\{ \sum_{k=-\infty}^{-1} a_k(x)\partial^k + \lambda \log \partial \mid a_i \in C^\infty(S^1),\ \lambda \in \mathbb{C} \right\}.$$

Again, we want to form a "Manin triple" $(\widetilde{\Psi\mathrm{DO}}(S^1), \widehat{\mathrm{DOP}}(S^1), \widehat{\mathrm{IOP}}(S^1))$. To do this, we define an pairing between $\widehat{\mathrm{DOP}}(S^1)$ and $\widehat{\mathrm{IOP}}(S^1)$ by the formula:

$$\langle L + \alpha \log \partial , M + c\,1 \rangle = \langle L , M \rangle + \alpha c ,$$

$\alpha, c \in \mathbb{C}$, $L, M \in \Psi\mathrm{DO}(S^1)$. We then define a Lie algebra on $\widetilde{\Psi\mathrm{DO}}(S^1) = \widehat{\mathrm{DOP}}(S^1) \oplus \widehat{\mathrm{IOP}}(S^1)$ in the usual way. The following result holds.

PROPOSITION. *The triple* $(\widetilde{\Psi\mathrm{DO}}(S^1), \widehat{\mathrm{DOP}}(S^1), \widehat{\mathrm{IOP}}(S^1))$ *is a Manin triple and* $\widehat{\mathrm{DOP}}(S^1)$ *and* $\widehat{\mathrm{IOP}}(S^1)$ *are self-dual bialgebras.*

With the extended version $\widehat{\mathrm{IOP}}(S^1)$ of integral operators, the extension of the Volterra group exists as well. The elements of this extended Volterra group $\widehat{G}_-$ are expressions of the form:

$$\exp(\alpha \log \partial + u_{-1}\partial^{-1} + u_{-2}\partial^{-2} + \dots), \qquad \alpha \in \mathbb{C},\ u_i \in C^\infty(S^1).$$

The properties of this exponential were discussed in [27]. For example, the Baker–Campbell–Hausdorff formula shows that $\exp \colon \widehat{\mathrm{IOP}} \to \widehat{G}_-$ is a bijection. The group $\widehat{G}_-$ is identified with the group of expressions $\partial^\alpha + \sum_{k=1}^\infty \partial^{\alpha-k} u_k(x)$, $\alpha \in \mathbb{C}$, with relations $\partial^\alpha \partial^\beta = \partial^{\alpha+\beta}$,

$$\partial^\alpha f = \sum_{k \geqslant 0} \frac{\alpha(\alpha-1)\cdots(\alpha-k+1)}{k!} f^{(k)} \partial^{(\alpha-k)}.$$

We introduce a bivector on the manifolds $\mathcal{L}_\alpha$, $\alpha \in \mathbb{C}$. For $L \in \mathcal{L}_\alpha$, we identify $T_L\mathcal{L}_\alpha$ with $\{\sum_{i>0} \delta u_i \partial^{\alpha-i},\ u_i \in C^\infty(S^1)\}$ and (by means of the Adler trace) $T_L^*\mathcal{L}_\alpha$ with $\{\sum_{i\geqslant 0} x_i \partial^{-\alpha+i},\ x_i \in C^\infty(S^1)\}$. We then pose

$$V_X(L) = (LX)_+ L - L(XL)_+ ,$$

in clear analogy with the Gelfand–Dikiĭ formula. This defines a bivector on $\widehat{G}_-$, the function α (degree of the operator) being assumed central.

We show that this bivector defines a Poisson structure on $\widehat{G}_-$, which is opposite to the one defined by the Manin triple $(\widetilde{\Psi\mathrm{DO}}(S^1), \widehat{\mathrm{DOP}}(S^1), \widehat{\mathrm{IOP}}(S^1))$. Let us first make the following observation. Consider an arbitrary Manin triple $(\mathfrak{d}, \mathfrak{g}_1, \mathfrak{g}_2)$

with corresponding groups (D, G_1, G_2). Then ([47]) the natural mapping $G_2 \to D/G_1$ is Poisson. This means that if F and G are right G_1-invariant functions on D, then $\{F, G\}(g_2) = \{f, g\}(g_2)$ for any $g_2 \in G_2$ (f and g are the restrictions of F and G to G_2). The identity $F(g_2 g_1) = f(g_2)$ implies that $\langle R^*_{g_2} dF(g_2), X\rangle = \langle R^*_{g_2} df(g_2), X_2\rangle$, for any $X \in \mathfrak{d}$ (decomposed as $X_1 + X_2$, $X_i \in \mathfrak{g}_i$). This implies $R^*_{g_2} dF(g_2) = R^*_{g_2} df(g_2)$, since we know that $R^*_{g_2} df(g_2)$ belongs to $\mathfrak{g}_2$. Then, if V' denotes the bivector on $\widehat{G}_2$ defined by the Manin triple, we have

$$\begin{aligned}\langle V'_{df}(g_2), dg\rangle &= \{f, g\}(g_2) = \{F, G\}(g_2)\\ &= \tfrac{1}{2}\langle r, R^*_{g_2} dF(g_2) \wedge R^*_{g_2} dG(g_2) - L^*_{g_2} dF(g_2) \wedge L^*_{g_2} dG(g_2)\rangle\\ &= \tfrac{1}{2}\langle r, R^*_{g_2} df(g_2) \wedge R^*_{g_2} dg(g_2) - L^*_{g_2} df(g_2) \wedge L^*_{g_2} dg(g_2)\rangle,\end{aligned}$$

where r denotes the r-matrix of $\mathfrak{d}$.

We apply this result to the Manin triple $(\widetilde{\Psi\mathrm{DO}}(S^1), \widehat{\mathrm{DOP}}(S^1), \widehat{\mathrm{IOP}}(S^1))$. Put $g_2 = L$, $df(g_2) = X$, and $dg(g_2) = Y$. Then

$$\begin{aligned}\langle V'_X(L), Y\rangle &= \tfrac{1}{2}\langle r, XL \wedge YL - LX \wedge LY\rangle\\ &= \tfrac{1}{2}(\langle (XL)_+ - (XL)_-, YL\rangle - \langle (LX)_+ - (LX)_-, YL\rangle)\\ &= \tfrac{1}{2}(\langle L(XL)_+ - (LX)_+ L - L(XL)_- + (LX)_- L, Y\rangle)\\ &= \langle L(XL)_+ - (LX)_+ L, Y\rangle,\end{aligned}$$

where we set $g_- = L$ and recall that $\langle r, X \wedge Y\rangle = \langle X_+ - X_-, Y\rangle$ for $X, Y \in \Psi\mathrm{DO}(S^1)$. This proves our claim.

It is clear from the generalization of the GD formula that $\overline{\mathcal{L}}_n$ is a Poisson submanifold of $\widehat{G}_-$.

PROPOSITION. *The submanifold $\overline{\mathcal{L}}_n$, endowed with its second* GD-*structure, is a Poisson submanifold with respect to the generalized* GD-*structure of $\widehat{G}_-$.*

To prove this fact, we must check that every Hamiltonian vector field on $\overline{\mathcal{L}}_n$ is tangent to $\overline{\mathcal{L}}_n$. Indeed, let $L \in \overline{\mathcal{L}}_n$ and $X \in \mathrm{Ker}(T^*_L \widehat{G}_- \to T^*_L \overline{\mathcal{L}}_n)$. We want to verify that $V_X(L) = 0$. Recall that $T^*_L \widehat{G}_-$ is identified to $\{a_{-n}(x)\partial^{-n} + a_{-n+1}(x)\partial^{-n+1} + \dots\}$, $T^*_L \overline{\mathcal{L}}_n$ is identified to $\{a_{-n}(x)\partial^{-n} + \cdots + a_{-1}(x)\partial^{-1}\}$, so that $\mathrm{Ker}(T^*_L \widehat{G}_- \to T^*_L \overline{\mathcal{L}}_n)$ consists of differential operators. It is now clear that $V_X(L) = (LX)_+ L - L(XL)_+ = 0$, since both L and X are in $\mathrm{DOP}(S^1)$.

We can summarize our results as follows.

PROPOSITION. *For $\alpha \in \mathbb{C}$, $n \in \mathbb{N}$, $\mathcal{L}_\alpha$ and $\overline{\mathcal{L}}_n$ can be identified with Poisson submanifolds of $\widehat{G}_-$, which are therefore homogeneous subspaces for the "dressing action" of the extended bialgebra $\widehat{DOP}(S^1)$ on $\widehat{G}_-$.*

COROLLARY. *The Poisson bracket algebra of $\widehat{G}_-$ with generalized* GD *Poisson structure is a universal W-algebra in the sense of* [19, 38, 42].

Indeed, for every $n \in \mathbb{N}$, the $\mathcal{W}_n$-PBA is a Hamiltonian reduction of the $\mathcal{W}$-universal algebra, which is the PBA on $\widehat{G}_-$ with generalized GD structure.

Actually, we considered the extension of the Volterra group looking for an algebra for which every $\mathrm{Fun}(\overline{\mathcal{L}}_n)$ is a Poisson ideal. This point of view is not in contradiction with the discussion in [19]. There as a possible candidate for a universal W-algebra, $\mathrm{DOP}(S^1)$, is considered (as Lie algebra, not a PBA!). This algebra is in good relation with the Dirac reduction proposed in [19], but not with the Poisson reduction on the Poisson–Lie action, because the isomorphism $\mathrm{DOP}(S^1) \simeq W_{1+\infty}$ noted above means that this algebra $(\mathrm{DOP}(S^1))$ is the symmetry algebra for the KP-hierarchy, as proved in [48, 49]. Our geometric point of view avoids the "contradiction" discussed in [19]: $W_{1+\infty} \simeq \mathrm{DOP}(S^1)$ is a Hamiltonian structure algebra for the first GD structure on the KP phase space, regarded as a Poisson submanifold of $\Psi\mathrm{DO}(S^1)$, and the $\mathcal{W}_n$ are reductions of the second GD structure on $\widehat{G}_-$, the extended integral operators group. We shall discuss the Hamiltonian structure for the KP-hierarchy and generalizations of nth KdV-hierarchy in the next sections.

3.4. In this subsection we show how to recover the theorem of Kuperschmidt and Wilson directly. We define a bivector ϖ on $\mathcal{L} = \bigcup_{\alpha\in\mathbb{C}} \mathcal{L}_\alpha$ by the formula $V_L(X) = (LX)_+L - L(XL)_+$. Let us show that the multiplication map $\mathcal{L}\times\mathcal{L} \to \mathcal{L}$ sends $\varpi\otimes 1 + 1\otimes\varpi$ to ϖ, and that the inverse map $\mathcal{L}\to\mathcal{L}$ sends ϖ to $-\varpi$. We see that

$$\begin{aligned} V_{LM}(X) &= L[M(XLM)_+ - (MXL)_+M] + [L(MXL)_+ - (LMX)_+L]M \\ &= LV_M(XL) + V_L(XM)M, \end{aligned}$$

as required, since, for example, the left multiplication by L from M to LM acts from $T_M\mathcal{L}$ to $T_{LM}\mathcal{L}$ by the left multiplication by L and from $T^*_{LM}\mathcal{L}$ to $T^*_M\mathcal{L}$ by the right multiplication by L. Similarly, we have

$$V_{L^{-1}}(X) = -L^{-1}(V_L(L^{-1}XL^{-1}))L^{-1}.$$

Since the embedding $\overline{\mathcal{L}}_n \to \mathcal{L}$ has the property that for any $L\in\overline{\mathcal{L}}_n$, the element $\varpi(L)$ is the image of a bivector of $\overline{\mathcal{L}}_n$ (corresponding to the second GD structure), we see (putting $n=1$) that the mapping $\overline{\mathcal{L}}_1^{\,n} \to \overline{\mathcal{L}}_n$ given by the product sends $\varpi\otimes 1\otimes\ldots + 1\otimes\varpi\otimes\ldots$ to ϖ. The Poisson brackets on $\overline{\mathcal{L}}_1$ are easily computed, and this ends the proof of the Kuperschmidt–Wilson theorem.

As was noted in 2.4, this result implies that the second Gelfand–Dikiĭ structures (on $\overline{\mathcal{L}}_n$) are Poisson. To see directly that ϖ defines a Poisson structure on $\mathcal{L}$, we note the following. After restriction to $\mathcal{L}_\alpha$, the bracket $\{w_i(x), w_j(y)\}$ is expressed in the form

$$\sum_{p,q\leqslant\max(i,j),k} c^{pq}_{ij;k}(\alpha)\,w_p(x)\,w_q(y)\delta^{(k)}(x-y),$$

$c^{pq}_{ij;k}(\alpha)$ depending polynomially on α (as follows from a recurrence formula proved by using the Miura realization, cf. [12]). The Jacobi identity for this bracket is expressed by polynomial identities between $c^{pq}_{ij;k}(\alpha)$, which are satisfied for all α, being satisfied for all sufficiently large integers.

§4. Poisson–Lie geometry of differential operators

4.1. Let us define a partial symplectic structure on $\overline{\mathcal{L}}_n$, i.e., introduce a 2-form Ω on the image of the Hamiltonian map W by the formula:

$$\Omega(W_X, W_Y) = \langle X^{(+)}, Y^{(-)}\rangle, \qquad X, Y \in \mathrm{DOP}(S^1) \tag{4.1}$$

and $X^{(\pm)} := (LXL^{-1})_{\pm}$, W_X, W_Y are vector fields on $\mathcal{L}$. Let us justify this formula.

PROPOSITION. i) *The formula* (4.1) *defines a bilinear product* Ω *on the image of* W.

ii) Ω *is skew-symmetric.*

iii) Ω *is a closed 2-form.*

PROOF. i) We verify that if $(L(XL^{-1})_-)_+L = L((XL^{-1})L)_+$, then the pairing of $X^{(+)}$ with any $Y^{(-)}$ vanishes, i.e., $(LXL^{-1})_+ \in L(\mathrm{DOP}(S^1))L^{-1}$. In fact, we shall see that LXL^{-1} is differential. The assumption gives the formula $L^{-1}(L(XL^{-1})_-)_+L = X - (XL^{-1})_+L$, and by conjugation by L, $LXL^{-1} = (L(XL^{-1})_-)_+ + L(XL^{-1})_+$, as required. A similar verification for Y is straightforward.

ii) We have

$$\begin{aligned}\Omega(W_X, W_Y) + \Omega(W_Y, W_X) &= \mathrm{Tr}(X^{(+)}Y^{(-)}) + \mathrm{Tr}(X^{(-)}Y^{(+)})\\ &= \mathrm{Tr}((X^{(+)} + X^{(-)})(Y^{(+)} + Y^{(-)})) = \mathrm{Tr}(LXL^{-1}\cdot LYL^{-1})\\ &= \mathrm{Tr}(LXYL^{-1}) = \mathrm{Tr}(XY) = 0.\end{aligned}$$

iii) $d\Omega(W_X, W_Y, W_Z) = W_X\Omega(W_Y, W_Z) - \Omega([W_X, W_Y], W_Z) +$ (cyclic permutations of X, Y, Z) $= 0$ because of the identity:

$$\begin{aligned}&\mathrm{Tr}([X, Y]^{(+)}Z^{(+)} + [X, Y]^{(-)}Z^{(-)})\\ &\quad + \mathrm{Tr}(X^{(+)}[Y^{(+)}, Z^{(+)}] + X^{(-)}[Y^{(-)}, Z^{(-)}]) = 0.\end{aligned}$$

4.2. We see that the form $\Omega(V_X, V_Y) = \mathrm{Tr}(X^{(+)}Y^{(-)})$ may be rewritten as $\mathrm{Tr}(LXL^{-1}\cdot\delta_Y LL^{-1})$, where we have set $\delta_Y L = W_L(Y)$. The last expression can be understood as the value of the $\mathrm{IOP}(S^1)$-valued 1-form ω_W on the vector field V_Y: $L^{-1}\delta_Y L = i_{V_Y}\omega_W$, $\omega_W = L^{-1}dL \in \Lambda^1(T^*\overline{\mathcal{L}}_n)\otimes \mathrm{IOP}(S^1)$ and $\langle X, \omega_W\rangle = i_{V_X}\Omega$.

We can easily check the following

LEMMA. *The form* ω_W *satisfies the Maurer–Cartan equation*

$$d\omega_W + [\omega_W, \omega_W]_{\mathrm{IOP}(S^1)} = 0.$$

(The structure $[\ ,\ ]_{\mathrm{IOP}(S^1)}$ is "dual" to $[\ ,\ ]_{\mathrm{DOP}(S^1)}$ in the "Manin triple" of 3.3; see, for example, the corresponding finite-dimensional result in [35].)

4.3. We shall conclude with the definition of the *moment map* for our action (as a preparation for future considerations). To each $X \in \mathrm{DOP}(S^1)$ we associate a left-invariant 1-form $\Theta_X \in \Omega^1(G_-)$ ($\mathrm{DOP}(S^1) \subset \mathrm{IOP}(S^1)^*$). The 1-form $(XL^{-1})_- \in T^*_L\overline{\mathcal{L}}_n$ at the point L gives the "pre-momentum" map:

$$\lambda\colon \mathrm{DOP}(S^1) \to \Omega^1(\overline{\mathcal{L}}_n), \qquad \lambda(X)(L) = (XL^{-1})_-.$$

PROPOSITION. *Fix an element* L_0 *of* $\mathcal{L}_n$. *Then the map* $\mu : \overline{\mathcal{L}}_n \to G_-$, $\mu(L) = L_0^{-1}L$ *can be regarded as the moment map for the Poisson–Lie action* W (*following the finite-dimensional definition of* [35]) *since the following diagram*

$$\begin{array}{ccc} \mathrm{DOP}(S^1) & \xrightarrow{\;W\;} & \mathrm{Vect}(\overline{\mathcal{L}}_n) \\ \lambda\big\downarrow & & \big\uparrow V \\ \Omega^1(G_-) & \xrightarrow[\mu^*]{} & \Omega^1(\overline{\mathcal{L}}_n) \end{array}$$

commutes, i.e., the equalities $W(X) = V_{\mu^*\Theta_X} = V_{\lambda(X)(L)}$ *hold.*

§5. A generalization of the KdV and KP-hierarchies

5.1. In the context specified above, we study some equations associated with the *complex Lax operator*, i.e., an element of $\mathcal{L}_\alpha$:

$$L = \partial^\alpha + \sum_{i=1}^{\infty} u_i \partial^{\alpha - i}, \qquad \alpha \in \mathbb{C};$$

and define for it the *complex* KP *hierarchy*, with kth flow defined by the Lax operator $L_+^{k/\alpha}$:

$$\partial L/\partial t_k = [(L^{k/\alpha})_+, L] = [L, (L^{k/\alpha})_-]. \tag{5.1}$$

Here we put $L^{k/\alpha} = \exp((k/\alpha)\log L)$; $L^{k/\alpha}$ belongs to $\mathcal{L}_k$. In the special case $\alpha \in \mathbb{N}$, we can impose the KdV constraints $L_- = 0$.

5.2. We shall show that these flows are Hamiltonian with respect to the generalized GD structure on $\mathcal{L}_\alpha$. Recall the proof that on $\overline{\mathcal{L}}_n$, the differential of $\operatorname{Tr} L^{k/n}$ is $(k/n)L^{k/n-1}$. We have

$$\delta L = \sum_{s=0}^{n-1} L^{s/n}\delta(L^{1/n})L^{(n-s-1)/n}$$

and so

$$\begin{aligned} \delta(\operatorname{Tr} L^{k/n}) &= \sum_{u=1}^{k-1} L^{u/n}\delta(L^{1/n})L^{(k-1-u)/n} = k \operatorname{Tr} L^{(k-1)/n}\delta(L^{1/n}) \\ &= \frac{k}{n}\operatorname{Tr} L^{(k-n)/n}\sum_{s=0}^{n-1} L^{s/n}\delta(L^{1/n})L^{(n-1-s)/n} = \frac{k}{n}\operatorname{Tr} L^{(k-n)/n}\delta L, \end{aligned}$$

hence

$$X_{\operatorname{Tr} L^{k/n}} = (k/n)L^{k/n-1}.$$

Let us now prove that the functional $\operatorname{Tr} L^{k/\alpha}$ defined on $\mathcal{L}_\alpha$, has differential $(k/\alpha)L^{k/\alpha-1}$. Note that $\operatorname{Tr} L^{k/\alpha}$ is the integral over S^1 of a polynomial differential

P in w_i, $i \leqslant k+2$, depending rationally on α. On the other hand, $L^{k/\alpha-1}$ can be written as $\partial^{k-\alpha} + \sum_{i=1}^{\infty} z_i \partial^{k-\alpha-i}$, where z_i, being polynomial differentials in w_j, $j \leqslant i$, depending rationally on α. The desired identity then reads

$$\delta \int_{S^1} P(\alpha, w_1, \dots, w_{k+2}) = \frac{k}{\alpha} \operatorname{Tr}(\partial^{k-\alpha} + z_1 \partial^{k-\alpha-1} + \dots)(\delta w_1 \partial^{\alpha-1} + \dots)$$
$$= \frac{k}{\alpha} \sum_{i, j \leqslant k, r \geqslant 0} \pi_{i,j,r}(\alpha) \int_{S^1} z_i^{(r)} \delta w_j$$

($\pi_{i,j,r}(\alpha)$ are polynomials in α). Eliminating the integrals $\int_{S^1} \delta w_i$, we get identities between differential polynomials in the w_i ($i \leqslant k+2$), depending rationally on α. Since these identities hold for all α integer $\geqslant k+2$, they hold in general. We can conclude

$$X_{\operatorname{Tr} L^{k/\alpha}} = (k/\alpha) L^{k/\alpha-1}.$$

The Hamiltonian vector field on $\mathcal{L}_\alpha$ associated to $H_k = (\alpha/k) \operatorname{Tr} L^{k/\alpha}$ is

$$V_{X_{(\alpha/k)\operatorname{Tr} L^{k/\alpha}}}(L) = (LL^{k/\alpha-1})_+ L - L(L^{k/\alpha-1}L)_+ = [(L^{k/\alpha})_+, L];$$

it is the vector field of (5.1).

5.3. In this subsection we show that the Hamiltonians obtained in 5.2 commute. Indeed,

$$\{H_k, H_l\}(L) = \langle V_{H_k}, X_{H_l} \rangle = \langle [L^{k/\alpha}, L], (l/\alpha) L^{l/\alpha-1} \rangle = 0;$$

this implies that the flows defined by equations (5.1) commute.

REMARK. The commutation of the quantities $\operatorname{Tr} L^{k/\alpha}$ can be interpreted in the following way: as Fateev and Lukyanov remarked [15], $\bigcup_{n\geqslant 0} \mathcal{L}_n$ has an "r-matrix" $r(x, y) = \varepsilon(x-y)\sigma_{xy}$, σ_{xy} is the operator of permutation of x and y and ε is the Heaviside function (in this paragraph we assume that all operators live on the real line). Indeed, the second GD Poisson brackets can be encoded by the formula $\{L(x), L(y)\} = [r(x-y), L(x)L(y)]$. It can be interpreted in the following way: on the l.h.s., powers of ∂_x, ∂_y do not enter the bracket, and in the right-hand side σ_{xy} can be expanded as $\sum_{k\geqslant 0} (1/k!)(x-y)^k (\partial_y - \partial_x)^k$. This formula can be extended to the whole $\mathcal{L}$, by polynomiality arguments. On the other hand, we know [9] that on a Poisson–Lie group with coboundary structure, invariant functions commute with each other. And in the case of $\mathcal{L}$, the functions $\operatorname{Tr} L^{k/\alpha}$ are invariant.

5.4. Let us show that for any complex number β, the operator L^β satisfies the equations (5.1) associated to the manifold $\mathcal{L}_{\alpha\beta}$. This statement is clear for $\beta \in \mathbb{N}$. Let us write L^β in the form $\partial^\beta + \sum_{i=1}^{\infty} P_i(\beta, w_1, \dots, w_i) \partial^{\beta-i}$; for each i, P_i is a differential polynomial in w_j and a polynomial in β. The equations (5.1) for the operator L^β are $\partial_{t_k} L^\beta = [(L^{k/\alpha})_+, L^\beta]$. Expanding this identity, we get

$$\partial_{t_k} P_i(\beta, w_1, \dots, w_i)$$
$$= \sum_{\substack{1\leqslant j \leqslant k, \\ 1 \leqslant l \leqslant i+k, \\ s, t \geqslant 0}} \pi'_{j,l,s,t}(\beta) P_j(k/\alpha, w_1, \dots, w_j)^{(s)} P_l(\beta, w_1, \dots, w_l)^{(t)},$$

$\pi'_{j,l,s,t}(\beta)$ being polynomials in β. This identity holds for any integer $\beta \geqslant 0$ and is polynomial in β, so it holds for $\beta \in \mathbb{C}$.

It follows in particular that the operator $L^{1/\alpha}$ satisfies the equations of the usual KP-hierarchy. Hence, the interest of this complex KP-hierarchy is not in the equations themselves, but in the Poisson structures that they preserve. By the mapping $L \mapsto L^{1/\alpha}$, we obtain a family (indexed by $\alpha \in \mathbb{C}$) of (not compatible) Poisson structures on the KP phase space, such that the Hamiltonian of the KP-hierarchy commute for all these structures and define the same flows.

5.5. We shall describe a generalization of the bilinear identity which is at the origin of the τ-function in the KP theory (cf. [8]). We have

$$\begin{aligned}\{H_k\,,\, \operatorname{res} L^{l/\alpha}\}' &= (d/d\varepsilon)|_{\varepsilon=0} \operatorname{res}(L + \varepsilon[(L^{k/\alpha})_+\,,\, L])^{l/\alpha} \\ &= (d/d\varepsilon)|_{\varepsilon=0} \operatorname{res}(1 + \varepsilon(L^{k/\alpha})_+) L^{l/\alpha}(1 - \varepsilon(L^{k/\alpha})_+) \\ &= \operatorname{res}[(L^{k/\alpha})_+\,,\, L^{l/\alpha}]\end{aligned}$$

is symmetric in k and l. (We used the relation $(XLX^{-1})^\alpha = XL^\alpha X^{-1}$ for $X\,,\, L \in \mathcal{L}$, and $\alpha \in \mathbb{C}$, which can be established by polynomiality considerations.)

5.6. Here we construct generalizations of the first GD structures for the manifolds $\mathcal{L}_n$.

The translation in $\mathcal{L}_n$ by the complex parameter λ transforms the Poisson V structure into $V_X(L+\lambda) = V_X(L) + \lambda([X_+\,,\, L] + [L\,,\, X]_+)$. This proves that the bivector

$$V_X^{(1)}(L) = [X_+\,,\, L] + [L\,,\, X]_+$$

defines a Poisson structure on $\mathcal{L}_n$ (to check the Jacobi identity it is enough to consider the terms in λ^2 in the Jacobi identity of the translation by λ of V), which is compatible to the generalized Gelfand–Dikiĭ structure V.

The structure manifold $\overline{\mathcal{L}}_n$, together with it first GD structure, is a Poisson submanifold of $(\mathcal{L}_n\,,\, V^{(1)})$.

Remark. In the same way, one can treat the infinite family of bi-Hamiltonian structures indexed by integers for the KP-hierarchies invented in the first time in [41] and extensively discussed recently. All these structures are incompatible as it was shown in [41].

References

1. M. Adler, *On a trace functional for formal pseudodifferential operators and symplectic structure of the Korteweg–de Vries type equations*, Invent. Math. **50** (1979), no. 2, 219–248.
2. E. Arbarello, C. De Concini, V. G. Kac, and C. Procesi, *Moduli spaces of curves and representation theory*, Comm. Math. Phys. **117** (1988), 1–36.
3. I. Bakas, *Structure and representation of W_∞-algebra*, Comm. Math. Phys. **134** (1990), 487–508.
4. I. Bakas, E. Kiritsis, and B. Khesin, *The logarithm of the derivative operator and higher spin algebras of W_∞ type*, Comm. Math. Phys. **151** (1993), 233–243.
5. A. Das and J. Huang, *The Hamiltonian structures associated with a generalized Lax operator*, J. Math. Phys. **33** (1992), 2487–2497.
6. E. Date, M. Jimbo, M. Kashiwara, and T. Miwa, *Transformation groups for soliton equations*, Proc. RIMS, Symposia on nonlinear integrable systems—classical and quantum theories, Kyoto, Japan, May 1981 (T. Miwa, ed.), World Scientific, Singapore, 1982.

7. P. Di Francesco, C. Itzykson, and J.-B. Zuber, *Classical W-algebras*, Saclay preprint SPTH/90-149 (1990).
8. L. Dikiĭ, *Soliton equations and hamiltonian systems*, World Scientific, Singapore, 1992.
9. V. Drinfeld, *Quantum groups*, Proc. Intern. Congress Math. (Berkeley, 1986), vol. 1, Amer. Math. Soc., Providence, RI, 1988, pp. 789–820.
10. ———, *On some unsolved problems in quantum group theory*, Quantum Groups, Proceedings of the Leningrad Conference (P. P. Kulish, ed.), Lecture Notes in Math., vol. 1510, Springer-Verlag, Berlin and New York, 1990, pp. 1–8.
11. V. Drinfeld and V. Sokolov, *Lie algebras and equations of Korteweg–de Vries type*, J. Soviet Math. **30** (1985), 1975–2036.
12. B. Enriquez, *Complex parametrized W-algebras: the gl-case*, Lett. Math. Phys. **31** (1994), 15–33.
13. L. Faddeev, *On the exchange matrix for WZNW models*, Comm. Math. Phys. **132** (1990), 131–141.
14. F. Falceto and K. Gawedzki, *On quantum group symmetries of conformal field theories*, Preprint IHES, IHES/P91/59, Bures-sur-Yvette (1991).
15. V. Fateev and S. Lukyanov, *Additional symmetries in two-dimensional conformal field theory and exactly solvable models,* I. *Quantization of Hamiltonian structures*, Internat. J. Modern Phys. A **3** (1988), 507–548; II. *The representation of W-algebras*, Internat. J. Modern Phys. A **7** (1992), 7–34.
16. B. Feigin and E. Frenkel, *Affine Kac–Moody algebras at the critical level and Gelfand–Dikiĭ algebras*, Internat. J. Modern Phys. A **7** (1992), no. 1, 197–215.
17. B. Feigin and B. Tsygan, *Riemann–Roch theorem for differential operators*, Proceedings of the Winter School on Geometry and Physics (Crni, 1988), Rend. Circ. Mat. Palermo (2) Suppl. (1989), no. 21, 15–52.
18. J. M. Figueroa-O'Farrill, J. Mas, and E. Ramos, *Bihamiltonian structure of the KP hierarchy and W_{KP} algebra*, Phys. Lett. B **266** (1991), 298–302.
19. J. M. Figueroa-O'Farrill and E. Ramos, *Existence and uniqueness of the universal W-algebra*, J. Math. Phys. **33** (1992), 833–839.
20. K. Gawedzki, *Classical origin of quantum group symmetries in Wess–Zumino–Witten conformal field theory*, Comm. Math. Phys. **139** (1991), 201–213.
21. I. M. Gelfand and L. Dikiĭ, *A family of Hamiltonian structures connected with integrable nonlinear differential equations*, Preprint 136, Institute of Appl. Math. (1978), Moscow.
22. I. M. Gelfand and I. Ya. Dorfman, *Hamiltonian operators and associated algebraic structures*, Funktsional. Anal. i Prilozhen. **13** (1979), no. 3, 13–30; English transl., Functional Anal. Appl. **13** (1979), 248–262.
23. D. Gurevich, V. Rubtsov, and N. Zobin, *Quantization of Poisson pairs: R-matrix approach*, J. Geom. Phys. **9** (1992), 15–26.
24. V. G. Kac and D. Peterson, *Spin and wedge representations of infinite-dimensional Lie algebras and groups*, Proc. Nat. Acad. Sci. U.S.A. **78** (1981), 3308–3312.
25. B. Khesin and O. Kravchenko, *Central extension of the algebra of pseudodifferential symbols*, Funktsional. Anal. i Prilozhen. **25** (1991), no. 1, 83–85; English transl., Functional Anal. Appl. **25** (1991), no. 1, 78–79.
26. B. Khesin, *Generalized hierarchies of infinite-dimensional Lie algebras and logarithmic cocycle*, Preprint (1991), Berkeley.
27. B. Khesin and I. Zakharevich, *Lie–Poisson group of pseudodifferential symbols and fractional KP-KdV hierarchies*, C. R. Acad. Sci. Paris Ser. I **316** (1994), 621–626.
28. S. Khoroshkin, A. Radul, and V. Rubtsov, *A family of Poisson structures on Hermitian symmetric spaces*, Comm. Math. Phys. **152** (1993), 299–315.
29. T. Khovanova, *Lie algebras of Gelfand–Dickey and Virasoro algebra*, Funktsional. Anal. i Prilozhen. **20** (1986), no. 4, 89–90; English transl., Functional Anal. Appl. **20** (1986), no. 4, 322–324.
30. ———, *Lie superalgebra structure on eigenfunctions and jets of the resolvent Kernel near diagonal of nth order differential operator of Gelfand–Dickey and Virasoro algebra*, Funktsional. Anal. i Prilozhen. **20** (1986), no. 2, 88–89; English transl., Functional Anal. Appl. **20** (1986), no. 2, 151–153.
31. B. Kuperschmidt and G. Wilson, *Modifying Lax equations and the second Hamiltonian structure*, Invent. Math. **62** (1981), 403–436.
32. S. Lukyanov, *Quantization of Gelfand–Dikiĭ brackets*, Funktsional. Anal. i Prilozhen. **22** (1988), no. 4, 1–10; English transl. in Functional Anal. Appl. **22** (1988).

33. D. Lebedev and Yu. Manin, *The Gelfand–Dikiĭ Hamiltonian operator and the coadjoint representation of the Volterra group*, Funktsional. Anal. i Prilozhen. **13** (1979), no. 3, 40–46; English transl. in Functional Anal. Appl. **13** (1979).
34. ______, *Benney's long wave equations,* II. *The Lax representation and conservation laws*, Zap. Nauchn. Sem. Leningrad. Otdel. Mat. Inst. Steklov. (LOMI) **96** (1981), 168–178; English transl. in J. Soviet Math. **21** (1983), no. 5.
35. J. H. Lu, *Moment mappings and Poisson reduction*, Symplectic Geometry, Groupoids and Integrable Systems (P. Dazord and A. Weinstein, eds.), MSRI Publications, vol. 20, Springer-Verlag, New York, 1991, pp. 209–226.
36. J. H. Lu and A. Weinstein, *Poisson–Lie groups, dressing transformations and Bruhat decomposition*, J. Differential Geom. **31** (1990), 501–526.
37. F. Magri, *A simple model of the integrable Hamiltonian equation*, J. Math. Phys. **19** (1978), 1156–1162.
38. A. Morozov, *On the concept of universal W-algebra*, Yadernaya Fiz. **51** (1990), 1190–1198; English transl., Soviet J. Nuclear Phys. **51** (1990), 758–763.
39. M. Mulase, *Solvability of the super KP equations and a generalization of the Birkhoff decomposition*, Invent. Math. **92** (1988), 1–46.
40. Ya. Pugay, *Lattice W-algebras and Quantum Groups*, Talk presented at the 3rd International Conference on Mathematical Physics, String Theory and Quantum Gravity, Alushta, 1993, Landau Institute preprint.
41. A. Radul, *Two series of Hamiltonian structures for a hierarchy of Kadomtsev–Petviashvili equations*, Applied Methods in Nonlinear Analysis and Control (Mironov, Moroz, and Tchernyatin, eds.), Moscov. Gos. Univ., Moscow, 1987, pp. 149–157.
42. ______, *Lie algebras of differential operators their central extensions and W-algebras*, Funktsional. Anal. i Prilozhen. **25** (1991), 33–49; English transl., Functional Anal. Appl. **25** (1991), 86–91.
43. A. Radul and I. Vaysburd, *Differential operators and W-algebras*, Phys. Lett. B **275** (1991), 317–322.
44. G. Segal and G. Wilson, *Loop groups and equations of KdV type*, Inst. Hautes Études Sci. Publ. Math. **61** (1985), 5–65.
45. M. Semenov-Tyan-Shansky, *Dressing transformations and Poisson group actions*, Publ. Res. Inst. Math. Sci. Kyoto Univ. **21** (1985), 1237–1260.
46. ______, *What a classical r-matrix is*, Funktsional. Anal. i Prilozhen. **17** (1983); English transl., Functional Anal. Appl. **17** (1983), 17–33.
47. ______, *Poisson–Lie groups, quantum duality principle and quantum double*, Preprint, April 1993, hep-th 9304042.
48. K. Yamagishi, *A Hamiltonian structure of KP hierarchy, $W_{1+\infty}$ algebra and self-dual gravity*, Phys. Lett. B **259** (1991), 436–441.
49. F. Yu and Y.-S. Wu, *Hamiltonian structure, (anti) selfadjoint flows in KP hierarchy and the $W_{1+\infty}$ and W_∞ algebras*, Phys. Lett. B **236** (1991), 220–225.
50. I. Zakharevich, *Several notes on the geometry of Wess–Zumino–Witten model and Gelfand–Dikiĭ brackets*, Preprint (1991).

B. E.: Centre de Mathématiques, URA 169 du CNRS, Ecole Polytechnique, 91128 Palaiseau, France

S. Kh.: Institute Theor. Experim. Physics, Bol. Cheremushkinskaya, 25, 117259, Moscow, Russia

A. Rad.: Department of Mathematics, Massachsetts Institute of Technology, Cambridge, MA 21039 USA

A. Ros.: Institute Theor. Experim. Physics, Bol. Cheremushkinskaya, 25, 117259, Moscow, Russia

V. R.: Institute Theor. Experim. Physics, Bol. Cheremushkinskaya, 25, 117259, Moscow, Russia

Amer. Math. Soc. Transl.
(2) Vol. **167**, 1995

On Symmetry Subalgebras and Conservation Laws for the $k-\varepsilon$ Turbulence Model and the Navier–Stokes Equations

N. G. KHOR'KOVA AND A. M. VERBOVETSKY

ABSTRACT. The classical symmetries and conservation laws of the $k-\varepsilon$ turbulence model are calculated. All one-, two- and three-dimensional subalgebras of the infinite-dimensional symmetry algebra (which turns out to be isomorphic to the one for the Navier–Stokes equations) are classified under the adjoint action.

§1. Introduction

This paper deals with the symmetry algebra and conservation laws for the two systems of differential equations of incompressible viscous fluid mechanics: the Navier–Stokes equations and the $k-\varepsilon$ turbulence model. The latter system describes the motion of high Reynolds number turbulence flows and is derived from averages of the Navier–Stokes equations by introducing the k- and ε- equations in order to obtain a closed set of equations [1, 2]:

$$\begin{cases} \dfrac{\partial \overline{u}_j}{\partial x_j} = 0, \\ \dfrac{\partial \overline{u}_i}{\partial t} + \dfrac{\partial(\overline{u}_i\overline{u}_j)}{\partial x_j} = -\dfrac{1}{\rho}\dfrac{\partial \overline{p}}{\partial x_i} + \nu \dfrac{\partial^2 \overline{u}_i}{\partial x_j \partial x_j} - \dfrac{\partial}{\partial x_j}(\overline{u_i' u_j'}), \qquad i = 1, 2, 3, \\ \dfrac{\partial k}{\partial t} + \overline{u}_j \dfrac{\partial k}{\partial x_j} = \dfrac{\partial}{\partial x_j}\left[\left(\nu + \dfrac{c}{\sigma_k}\dfrac{k^2}{\varepsilon}\right)\dfrac{\partial k}{\partial x_j}\right] - \dfrac{\partial \overline{u}_i}{\partial x_j}\overline{u_i' u_j'} - \varepsilon, \\ \dfrac{\partial \varepsilon}{\partial t} + \overline{u}_j \dfrac{\partial \varepsilon}{\partial x_j} = \dfrac{\partial}{\partial x_j}\left[\left(\nu + \dfrac{c}{\sigma_\varepsilon}\dfrac{k^2}{\varepsilon}\right)\dfrac{\partial \varepsilon}{\partial x_j}\right] - c_1 \dfrac{\varepsilon}{k}\dfrac{\partial \overline{u}_i}{\partial x_j}\overline{u_i' u_j'} - c_2 \dfrac{\varepsilon^2}{k}, \end{cases}$$

where $\overline{u}_i$ is the mean velocity component in the x_i direction, $\overline{p}$ is the mean pressure, k is the turbulence kinetic energy, ε is the rate of dissipation of turbulence kinetic energy, $\nu = \text{const}$ is the viscosity, $\rho = \text{const}$ is the density,

1991 *Mathematics Subject Classification*. Primary 35A30, Secondary 35Q30, 35Q35, 58G35, 22E65.
Key words and phrases. Symmetries, conservation laws, adjoint action, the $k-\varepsilon$ turbulence model, the Navier–Stokes equations.

$$-\rho\overline{u_i'u_j'} = \rho\left[\frac{ck^2}{\varepsilon}\left(\frac{\partial \overline{u}_j}{\partial x_i} + \frac{\partial \overline{u}_i}{\partial x_j}\right) - \frac{2}{3}\delta_{ij}k\right]$$

is the Reynolds stress tensor, δ_{ij} being the Kronecker delta, and the five empirical constants that appear in the equations are assigned the values: $c = 0.09$, $c_1 = 1.44 - 1.59$, $c_2 = 1.9 - 2.0$, $\sigma_k = 1.0$, $\sigma_\varepsilon = 1.3 - 1.47$. Throughout, for repeated indices the summation convention is used, the indices running from 1 to 3.

Symmetries and conservation laws of the Navier–Stokes equations have been calculated in [3] (the classical symmetries were first calculated in [4]; see also [5]). In this paper we show that the $k - \varepsilon$ model has the same classical symmetries and conservation laws. Exact formulations are contained in Theorems 3.1 and 4.1.

Further our purpose is to give a complete ready-to-use list of symmetry subalgebras, which is necessary to construct invariant solutions of the equations under consideration. As we will see later, it is natural to classify the subalgebras under the adjoint representations. By now, certain results in this direction have been already obtained (see, e.g. [4, 6]). However, they concern only the eleven-dimensional symmetry algebra spanned by translations, Galilean boost, rotations, and scaling. Here we classify subalgebras of the complete symmetry algebra. At the same time it should be noted that the subalgebras in our list need not be unequivalent. A further classification does not seem to be useful, because the over-use of adjoint maps can complicate the list of subalgebras instead of simplifying it.

This paper is organized as follows. In §2 we bring together the necessary results from the theory of symmetries and conservation laws of partial differential equations on which this research is based, and fix the notations. The detailed exposition of the theory is contained in [7–9]. Then §§3–4 deal with the computation of symmetries and conservation laws of the $k - \varepsilon$ model. Finally, in §§5–7 we perform the classification of subalgebras of the symmetry algebra.

The authors are grateful to Professor V. V. Lychagin for stating the problem and for helpful discussions, and to Yu. R. Romanovsky for a number of valuable comments and suggestions. One of the authors (A. M. V.) would like to thank SMF, SMAI and CIMPA (France) for a financial help.

Notations. We use the following notations:

$\langle A, \dots, B\rangle$ denotes the vector space spanned by $A, \dots, B$;

$\dot{f}(t)$ is the derivative with respect to t of $f(t)$;

$\exp(tX)$ is the one-parameter transformation group associated with the vector field X,

Ad denotes the adjoint representation, and

$\mathbb{H}$ is the skew field of quaternions.

§2. Preliminaries

2.1. Suppose $\pi\colon E \to M$ is a linear fiber bundle, $\dim M = n$, $\dim E = n+m$, and $\Gamma(\pi)$ is the set of all local sections of the bundle π. Let $J^k(\pi)$ be the manifold of all k-jets of the bundle π. If $u \in \Gamma(\pi)$ and $U \subset M$ is the domain of u, we denote by $j_k(u)\colon U \to J^k(\pi)$ the map that takes each point $x \in U$ to the k-jet of the section u at the point x.

Consider the trivial bundle $\alpha\colon U \times \mathbb{R}^m \to U$, where U is a domain in $\mathbb{R}^n$. Let (x, u), $x = (x_1, \dots, x_n)$, $u = (u^1, \dots, u^m)$ be the corresponding coordinate system. Then on the manifold $J^k(\alpha)$ a *special coordinate system* (x_i, p_σ^j), $|\sigma| \leqslant k$, arises, where σ is a multi-index understood as a nonordering sequence of integers $(i_1, \dots, i_r)$, $1 \leqslant i_s \leqslant n$, $|\sigma| = r$. The functions p_σ^j are determined by the property

$$j_k(u)^*(p_\sigma^j) = \frac{\partial^{|\sigma|} u^j}{\partial x_{j_1} \cdots \partial x_{j_r}}.$$

Suppose $f\colon U \times \mathbb{R}^m \to E$ is a trivialization over the coordinate neighborhood $U \subset M$. Using the k-jet prolongation of f

$$f^{(k)}\colon J^k(\alpha) \to J^k(\pi),$$

we can carry over the special coordinates from $J^k(\alpha)$ to $\operatorname{Im} f^{(k)}$. The coordinates x and u are called *independent* and *dependent* variables respectively.

2.2. The inverse limit of the chain of natural maps

$$E = J^0(\pi) \longleftarrow J^1(\pi) \longleftarrow \cdots \longleftarrow J^k(\pi) \longleftarrow J^{k+1}(\pi) \longleftarrow \cdots$$

is denoted by $J^\infty(\pi)$. The local coordinates on $J^\infty(\pi)$ are x, u, p_σ^j, $|\sigma| < \infty$. The ring of smooth functions on $J^\infty(\pi)$ is, by definition, the direct limit of the chain of maps

$$C^\infty(J^0(\pi)) \to C^\infty(J^1(\pi)) \to \cdots \to C^\infty(J^k(\pi)) \to C^\infty(J^{k+1}(\pi)) \to \dots.$$

In other words, a function on $J^\infty(\pi)$ is a function on one of the manifolds $J^k(\pi)$, $k < \infty$. In this manner, any contravariant (respectively, covariant) object on $J^\infty(\pi)$ is understood as the inverse (respectively, direct) limit of the corresponding object on $J^k(\pi)$ as $k \to \infty$.

The manifold $J^\infty(\pi)$ can be endowed with a contact structure as follows. For any point $\theta \in J^\infty(\pi)$, consider the n-dimensional subspace $C_\theta = T_\theta(j_\infty(u)(M)) \subset T_\theta(J^\infty(\pi))$, where u is a section of π such that $\theta \in \operatorname{Im} j_\infty(u)$. The field $\theta \mapsto C_\theta$ of planes on $J^\infty(\pi)$ is called the *Cartan distribution* or infinite order contact structure on $J^\infty(\pi)$. In the special coordinates, the Cartan distribution is spanned by the so-called *total derivative operators*

$$D_i = \frac{\partial}{\partial x_i} + \sum_{\sigma, j} p_{\sigma i}^j \frac{\partial}{\partial p_\sigma^j}.$$

2.3. The k-order system of nonlinear partial differential equations (or simply "equation") imposed on the sections of the bundle π

$$\begin{cases} F_1(x, u, u_{(1)}, \dots, u_{(k)}) = 0, \\ \dots\dots\dots\dots\dots\dots\dots\dots \\ F_l(x, u, u_{(1)}, \dots, u_{(k)}) = 0 \end{cases}$$

(here $u_{(s)}$ is the totality of all derivatives $\partial^{|\sigma|} u^j / \partial x_\sigma$, $|\sigma| = s$) may be understood as a submanifold $\mathcal{Y} \subset J^k(\pi)$ given by the equations

$$\begin{cases} F_1(x, u, \dots, p_\sigma^j) = 0, \\ \dots\dots\dots\dots\dots\dots \\ F_l(x, u, \dots, p_\sigma^j) = 0. \end{cases} \tag{2.1}$$

It is natural to consider the *infinite prolongation* $\mathcal{Y}_\infty$ of the equation $\mathcal{Y}$ that is given in $J^\infty(\pi)$ by the following infinite system

$$F_s = 0, \quad D_i(F_s) = 0, \ \dots, \ D_\sigma(F_s) = 0, \ \dots, \tag{2.2}$$

where $D_\sigma = D_{i_1} \circ \cdots \circ D_{i_r}$ for $\sigma = (i_1, \dots, i_r)$. In fact, system (2.2) is the coordinate description of the infinite prolongation $\mathcal{Y}_\infty$, which can be defined purely geometrically. All geometric objects on $\mathcal{Y}_\infty$ can be constructed in the same way as on $J^\infty(\pi)$. Without loss of generality, it can be assumed that $\mathcal{Y}$ is a formally integrable equation. Then $\mathcal{Y}_\infty$ inherits the Cartan distribution: $\mathcal{C}_\theta \subset T_\theta(\mathcal{Y}_\infty)$, $\theta \in \mathcal{Y}_\infty$

2.4. Let us denote the restriction of a function $f \in C^\infty(J^\infty(\pi))$ by $\overline{f}$. It is obvious that a part of the coordinate functions $\overline{x}_i$, $\overline{p}^j_\sigma$ on $J^\infty(\pi)$ can be expressed through the others using equations (2.2). The maximal functionally independent part of the coordinates $\overline{x}_i$, $\overline{p}^j_\sigma$ on $\mathcal{Y}_\infty$ is said to be *internal* with respect to $\mathcal{Y}_\infty$. The remaining coordinates are called *external.* Clearly, internal and external coordinates can be chosen in many different ways.

2.5. A differential operator on $J^\infty(\pi)$ is called *$\mathcal{C}$-differential* if it can be restricted to every submanifold in $J^\infty(\pi)$ of the form $\mathcal{Y}_\infty$. In coordinates, $\mathcal{C}$-differential operators are matrices whose elements are operators of the form

$$\sum_\sigma a_\sigma D_\sigma, \qquad a_\sigma \in C^\infty(J^\infty(\pi)).$$

2.6. A vector field on $J^\infty(\pi)$ that preserves the Cartan distribution is called an (external infinitesimal) *symmetry* of equation $\mathcal{Y}$ if it is tangent to $\mathcal{Y}_\infty$. In coordinates, the fields on $J^\infty(\pi)$ preserving the Cartan distribution have the form

$$X = \exists_\varphi + \sum_{i=1}^n a_i D_i, \tag{2.3}$$

where $\varphi = (\varphi^1, \dots, \varphi^m)$, φ^j, a_i are arbitrary functions on $J^\infty(\pi)$ and

$$\exists_\varphi = \sum_{\sigma, j} D_\sigma(\varphi^j)\frac{\partial}{\partial p^j_\sigma}.$$

The operator $\exists_\varphi$ is said to be the *evolution differentiation* corresponding to the *generating function* φ. Vector fields of the form $\sum_{i=1}^n a_i D_i$ are tangent to every infinite prolonged equation $\mathcal{Y}_\infty$ and so are called *trivial.* It is natural to identify fields of the form (2.3) that have the common evolution part $\exists_\varphi$. The field $\exists_\varphi$ is tangent to $\mathcal{Y}_\infty$ if and only if

$$\overline{\exists_\varphi(F)} = 0,$$

where $F = (F_1, \dots, F_l)$ is the left-hand side of (2.1). This equality can be rewritten in the form

$$\bar{l}_F(\overline{\varphi}) = 0, \tag{2.4}$$

where

$$l_F = \begin{pmatrix} \sum_\sigma \frac{\partial F_1}{\partial p^1_\sigma} D_\sigma & \dots & \sum_\sigma \frac{\partial F_1}{\partial p^m_\sigma} D_\sigma \\ \vdots & \ddots & \vdots \\ \sum_\sigma \frac{\partial F_l}{\partial p^1_\sigma} D_\sigma & \dots & \sum_\sigma \frac{\partial F_l}{\partial p^m_\sigma} D_\sigma \end{pmatrix}$$

is a $\mathcal{C}$-differential operator called the *universal linearization operator*. The overbar denotes the restriction on $\mathcal{Y}_\infty$.

Let $\operatorname{Sym}\mathcal{Y}$ denote the vector space of all symmetries of an equation $\mathcal{Y}$. It may be identified with the solution space of equation (2.4). In the sequel we identify symmetries with their generating functions (to be precise, with restrictions of generating functions to $\mathcal{Y}_\infty$).

2.7. Commutators of evolution differentiations are also evolution differentiations. More exactly,

$$[\ni_{\varphi_1}, \ni_{\varphi_2}] = \ni_{[\varphi_1, \varphi_2]},$$

where $[\varphi_1, \varphi_2] = \ni_{\varphi_1}(\varphi_2) - \ni_{\varphi_2}(\varphi_1)$. If φ_1 and φ_2 are symmetries of $\mathcal{Y}$, then the commutator $[\varphi_1, \varphi_2]$ is also a symmetry of $\mathcal{Y}$, so that $\operatorname{Sym}\mathcal{Y}$ is a Lie algebra.

2.8. A *finite classical symmetry* of the system of partial differential equation $\mathcal{Y}$ (for a moment we are considering the case $m > 1$) is a transformation of $J^0(\pi)$ that maps solutions of $\mathcal{Y}$ into solutions. An (infinitesimal) *classical symmetry* of $\mathcal{Y}$ is a vector field X_0 on $J^0(\pi)$ such that the one-parameter transformation group $\exp(tX_0)$ associated with X_0 consists of finite symmetries of $\mathcal{Y}$. The field X_0 can be canonically lifted to a symmetry on $J^\infty(\pi)$. With

$$X_0 = \sum_{i=1}^n a_i(x, u)\frac{\partial}{\partial x_i} + \sum_{j=1}^m b^j(x, u)\frac{\partial}{\partial u^j},$$

the generating function of this symmetry has the form

$$\varphi^j = b^j(x, u) - \sum_{i=1}^n a_i(x, u)\, p^j_i$$

and the lifting of X_0 on $J^\infty(\pi)$ is given by the formula

$$X_\varphi = \ni_\varphi + \sum_{i=1}^n a_i D_i.$$

The set $\operatorname{Sym}_0\mathcal{Y}$ of all classical symmetries is a subalgebra in the Lie algebra $\operatorname{Sym}\mathcal{Y}$.

2.9. Let φ be a symmetry of an equation $\mathcal{Y}$. Consider the evolution equation

$$u_\tau = \varphi(x, u, u_{(1)}, \dots), \tag{2.5}$$

with τ a new independent variable. If the Cauchy problem for this equation is uniquely solvable, then a flow on the space of sections $\Gamma(\pi)\colon u(x) \mapsto u(x, \tau)$ is well defined, where $u(x, \tau)$ denotes the solution of the Cauchy problem (2.5) with initial data $u(x, 0) = u(x)$. In particular, this is the case for the classical symmetries. Further, if $u(x)$ is a solution of the equation $\mathcal{Y}$, then $u(x, \tau)$ is also a solution for every τ. For a classical symmetry X_0, this flow on the space of solutions coincides with $\exp(\tau X_0)$.

2.10. Let $\mathfrak{g}$ be a subalgebra of the Lie algebra $\mathrm{Sym}_0 \mathcal{Y}$. A solution of the equation $\mathcal{Y}$ is called $\mathfrak{g}$-*invariant* if it is a fixed point of the flows defined by the evolution equations (2.5) for every $\varphi \in \mathfrak{g}$. Clearly, $\mathfrak{g}$-invariant solutions of $\mathcal{Y}$ are solutions of the system

$$F = 0, \quad \varphi_1 = 0, \quad \dots, \quad \varphi_s = 0, \tag{2.6}$$

where $\{\varphi_1, \dots, \varphi_s\}$ is a basis of $\mathfrak{g}$. This system is overdetermined and the fact that φ_i are symmetries means that the system is compatible. Under some regularity conditions, the problem of solving system (2.6) is equivalent to that of solving a system with $n - s$ independent variables. So it is much easier to solve system (2.6) than the original system (2.1).

To find all the invariant solutions, we need a description of all s-dimensional subalgebras of the Lie algebra $\mathrm{Sym}_0 \mathcal{Y}$ for $s < n$. It is natural to identify those subalgebras which are equivalent under the adjoint representation, because if $\mathfrak{g}' = \mathrm{Ad}(F)(\mathfrak{g})$, where F is a finite classical symmetry, then every $\mathfrak{g}'$-invariant solution $u'(x)$ has the form $u' = F(u)$ for a $\mathfrak{g}$-invariant solution $u(x)$.

Assume that for the algebra $\mathrm{Sym}_0 \mathcal{Y}$ the following condition holds: for any symmetries φ_1 and φ_2 such that the composition $\exp(X_{\varphi_1}) \circ \exp(X_{\varphi_2})$ is defined, there exist a symmetry $\varphi_3 \in \mathrm{Sym}_0 \mathcal{Y}$ such that $\exp(X_{\varphi_3}) = \exp(X_{\varphi_1}) \circ \exp(X_{\varphi_2})$. Then the subalgebras $\mathfrak{g}$ and $\mathfrak{g}'$ are equivalent if and only if there exists a symmetry $\varphi \in \mathrm{Sym}_0 \mathcal{Y}$ such that $\mathfrak{g}' = \mathrm{Ad}(\exp(X_\varphi))(\mathfrak{g})$ (for details see [10]). We shall denote this equivalence of $\mathfrak{g}$ and $\mathfrak{g}'$ by $\mathfrak{g} \sim \mathfrak{g}'$.

For Lie algebras without any additional structure, sophisticated techniques for classifying subalgebras does not seem to be available in the literature. Roughly speaking, this problem is solved by subjecting a subalgebra to various adjoint transformations so as to simplify it as much as possible.

The following elementary proposition is useful for classifying subalgebras of dimensions more then one.

PROPOSITION ([10]). *Let* $\mathfrak{g} = \langle A_1, \dots, A_s \rangle$ *and* $\mathfrak{g}' = \langle A'_1, \dots A'_s \rangle$.

(1) *If there exists a symmetry* $\varphi \in \mathrm{Sym}_0 \mathcal{Y}$ *such that* $A'_i = \mathrm{Ad}(\exp(X_\varphi))(A_i)$, *then* $\mathfrak{g}' \sim \mathfrak{g}$.

(2) *If* $\mathfrak{g}' \sim \mathfrak{g}$, *then there exist a basis* $\langle B_1, \dots, B_s \rangle$ *of* $\mathfrak{g}'$ *and a symmetry* φ *such that* $B_i = \mathrm{Ad}(\exp(X_\varphi))(A_i)$.

2.11. Let $\mathcal{C}\Lambda^k(\mathcal{Y}) = \{\omega \in \Lambda^k(\mathcal{Y}_\infty) \mid Y_\theta \lrcorner \omega = 0, \forall Y_\theta \in C_\theta, \theta \in \mathcal{Y}_\infty\}$. Elements of the quotient space $\overline{\Lambda}^k(\mathcal{Y}) = \Lambda^k(\mathcal{Y}_\infty)/\mathcal{C}\Lambda^k(\mathcal{Y})$ are called *horizontal* k-*forms* on $\mathcal{Y}_\infty$. Since $d(\mathcal{C}\Lambda^k(\mathcal{Y})) \subset \mathcal{C}\Lambda^{k+1}(\mathcal{Y})$, the horizontal de Rham complex

$$0 \to \overline{\Lambda}^0 \to \overline{\Lambda}^1 \to \dots \to \overline{\Lambda}^{n-1} \to \overline{\Lambda}^n \to 0$$

is well defined. Its $(n-1)$-dimensional cohomology group is called the group of *conservation laws* of $\mathcal{Y}$. The $(n-1)$-dimensional horizontal cochains are said to be *conserved currents* and in coordinates are given by n-vectors $(\omega_1, \dots, \omega_n)$ such that $\sum_{i=1}^n \overline{D}_i(\overline{\omega}_i) = 0$. The latter equality means that there exists a $\mathcal{C}$-differential operator $A = (A_1, \dots, A_l)$ such that

$$\sum_{i=1}^{n} D_i(\omega_i) = \sum_{j=1}^{l} A_j(F_j).$$

The vector-function $\psi = (\psi_1, \dots, \psi_l)$, $\psi_j = A_j^*(1)$, is called the *generating function* of the conservation law associated with the current ω. Recall that if $\Delta = \sum_\sigma a_\sigma D_\sigma$ is a scalar operator, then

$$\Delta^* = \sum_\sigma (-1)^{|\sigma|} D_\sigma \circ a_\sigma.$$

For so-called normal equations, every conservation law is locally uniquely defined by its generating function. The key result in finding conservation laws is that the generating function ψ of a conservation law satisfies the equation

$$\overline{l}_F^*(\psi) = 0. \tag{2.7}$$

Not every solution ψ of equation (2.7) is a generating function of a conservation law. This is so if and only if the operator $\overline{l}_F + \overline{B}^*$ can be represented in the form $\overline{C} \circ \overline{l}_F$ with some self adjoint $\mathcal{C}$-differential operator C, where B is a $\mathcal{C}$-differential operator given by the equality $l_F^*(\psi) = B(F)$. (The operator Δ^* conjugated with the matrix $\mathcal{C}$-differential operator $\Delta = \|\Delta_{ij}\|$ is defined as $\Delta^* = \|\square_{ij}\|$, where $\square_{ij} = (\Delta_{ji})^*$.)

§3. Symmetries of the $k - \varepsilon$ model

In this and the next sections we shall use the following notation for coordinate functions on a jet space. Firstly, let t, x_1, x_2, x_3 be four independent variables and u^1, u^2, u^3, p, k, ε six dependent variables. Secondly, the special coordinates on the infinite jet space will be denoted by

$$u_\sigma^1,\ u_\sigma^2,\ u_\sigma^3,\ p_\sigma,\ k_\sigma,\ \varepsilon_\sigma,$$

where multi-indices σ have the form

$$\sigma = \underbrace{t\dots t}_{s\text{ times}}\ \underbrace{1\dots1}_{l_1\text{ times}}\ \underbrace{2\dots2}_{l_2\text{ times}}\ \underbrace{3\dots3}_{l_3\text{ times}}.$$

Put $|\sigma|_t = s$, $|\sigma|_i = l_i$, $i = 1, 2, 3$, $|\sigma| = s + l_1 + l_2 + l_3$.

Now the equation $\mathcal{Y}$ of the $k - \varepsilon$ model can be written as follows:

$$u_j^j = 0, \tag{3.1}$$

$$u_t^i + u^j u_j^i + p_i - \nu u_{jj}^i - D_j(\theta_{ij}) = 0, \qquad i = 1, 2, 3, \tag{3.2$_i$}$$

$$k_t + u^j k_j - D_j\left(\left(\nu + \frac{c}{\sigma_k}\frac{k^2}{\varepsilon}\right)k_j\right) - u_j^i\theta_{ij} + \varepsilon = 0, \tag{3.3}$$

$$\varepsilon_t + u^j\varepsilon_j - D_j\left(\left(\nu + \frac{c}{\sigma_\varepsilon}\frac{k^2}{\varepsilon}\right)\varepsilon_j\right) - c_1\frac{\varepsilon}{k}u_j^i\theta_{ij} + c_2\frac{\varepsilon^2}{k} = 0, \tag{3.4}$$

where

$$\theta_{ij} = \frac{ck^2}{\varepsilon}(u^i_j + u^j_i) - \frac{2}{3}\delta_{ij}k .$$

To start the computation of the algebra of classical symmetries $\mathrm{Sym}_0\,\mathcal{Y}$, we need internal coordinates on the infinitely prolonged equation $\mathcal{Y}_\infty$. It can be easily seen that the following system can be taken as a coordinate system on $\mathcal{Y}_\infty$

$$\begin{aligned} &u^1_\sigma, \quad |\sigma|_1 = 0, \qquad u^2_\sigma,\ u^3_\sigma, \quad |\sigma|_t = 0, \\ &p_\sigma, \quad |\sigma|_1 = 0, \qquad k_\sigma,\ \varepsilon_\sigma, \quad |\sigma|_t = 0. \end{aligned} \tag{3.5}$$

Indeed, using equations (3.1)–(3.4), one can get the expressions for u^1_1, u^2_t, u^3_t, p_1, k_t, ε_t:

$$\begin{aligned} u^1_1 &= \overline{u}^1_1(u^2_2, u^3_3) = -u^2_2 - u^3_3, \\ p_1 &= \overline{p}_1(u^i, u^1_t, u^i_j, u^i_{jl}, k, \varepsilon, k_i, \varepsilon_i) = -u^1_t - u^j u^1_j + \nu u^1_{jj} + D_j(\theta_{1j}), \\ u^j_t &= \overline{u}^j_t(u^i, u^i_l, u^i_{ls}, p_j, k, \varepsilon, k_i, \varepsilon_i) \\ &= -u^j u^i_j - p_j + \nu u^j_{ii} + D_i(\theta_{ji}), \qquad j = 2, 3, \\ k_t &= \overline{k}_t(u^i, u^i_j, k, \varepsilon, k_i, \varepsilon_i, k_{ii}, \varepsilon_{ii}) \\ &= -u^j k_j + D_j\left(\left(\nu + \frac{c}{\sigma_k}\frac{k^2}{\varepsilon}\right)k_j\right) + u^i_j\theta_{ij} - \varepsilon, \\ \varepsilon_t &= \overline{\varepsilon}_t(u^i, u^i_j, k, \varepsilon, k_i, \varepsilon_i, k_{ii}, \varepsilon_{ii}) \\ &= -u^j\varepsilon_j + D_j\left(\left(\nu + \frac{c}{\sigma_\varepsilon}\frac{k^2}{\varepsilon}\right)\varepsilon_j\right) + c_1\frac{\varepsilon}{k}u^i_j\theta_{ij} - c_2\frac{\varepsilon^2}{k} = 0. \end{aligned} \tag{3.6}$$

Differentiating (3.6), one can express an arbitrary derivation u^1_σ, u^2_σ, u^3_σ, p_σ, k_σ, ε_σ through (3.5).

Now let us write out equations (2.4) for the generating functions of symmetries

$$\overline{D}_i(\varphi^i) = 0, \tag{3.7}$$

(3.8$_i$)

$$\begin{aligned} &\overline{D}_t(\varphi^i) + u^j\overline{D}_j(\varphi^i) + u^i_j\varphi^j + D_i(\varphi^4) - \nu\overline{D}^2_j(\varphi^i) \\ &\quad - \overline{D}_j\left\{\frac{ck^2}{\varepsilon}(\overline{D}_j(\varphi^i) + \overline{D}_i(\varphi^j)) - \frac{2}{3}\delta_{ij}\varphi^5\right. \\ &\qquad \left. + \frac{2ck}{\varepsilon}(u^i_j + u^j_i)\varphi^5 - \frac{ck^2}{\varepsilon^2}(u^i_j + u^j_i)\varphi^6\right\} = 0, \qquad i = 1, 2, 3, \end{aligned}$$

(3.9)

$$\begin{aligned} &\overline{D}_t(\varphi^5) + u^i\overline{D}_i(\varphi^5) + \varphi^j k_j \\ &\quad - \overline{D}_j\left\{\left(\nu + \frac{c}{\sigma_k}\frac{k^2}{\varepsilon}\right)\overline{D}_j(\varphi^5) + 2\frac{c}{\sigma_k}\frac{k}{\varepsilon}k_j\varphi^5 - \frac{c}{\sigma_k}\frac{k^2}{\varepsilon^2}k_j\varphi^6\right\} - \theta_{ij}\overline{D}_j(\varphi^i) \\ &\quad - u^i_j\left\{\frac{ck^2}{\varepsilon}(\overline{D}_j(\varphi^i) + \overline{D}_i(\varphi^j)) - \frac{2}{3}\delta_{ij}\varphi^5\right. \\ &\qquad \left. + \frac{2ck}{\varepsilon}(u^i_j + u^j_i)\varphi^5 - \frac{ck^2}{\varepsilon^2}(u^i_j + u^j_i)\varphi^6\right\} + \varphi^6 = 0, \end{aligned}$$

(3.10)

$$\overline{D}_t(\varphi^6) + u^i\overline{D}_i(\varphi^6) + \varphi^j\varepsilon_j$$
$$- \overline{D}_j\left\{\left(\nu + \frac{c}{\sigma_\varepsilon}\frac{k^2}{\varepsilon}\right)\overline{D}_j(\varphi^6) + 2\frac{c}{\sigma_\varepsilon}\frac{k}{\varepsilon}\varepsilon_j\varphi^5 - \frac{c}{\sigma_\varepsilon}\frac{k^2}{\varepsilon^2}\varepsilon_j\varphi^6\right\}$$
$$- c_1\frac{\varepsilon}{k}\theta_{ij}\overline{D}_j(\varphi^i) + c_1\frac{\varepsilon}{k^2}u^i_j\theta_{ij}\varphi^5 - \frac{c_1}{k}u^i_j\theta_{ij}\varphi^6 + 2\frac{c_2\varepsilon}{k}\varphi^5 - c_2\frac{\varepsilon^2}{k^2}\varphi^6$$
$$- c_1\frac{\varepsilon}{k}u^i_j\left\{\frac{ck^2}{\varepsilon}(\overline{D}_j(\varphi^i) + \overline{D}_i(\varphi^j)) - \frac{2}{3}\delta_{ij}\varphi^5\right.$$
$$\left. + \frac{2ck}{\varepsilon}(u^i_j + u^j_i)\varphi^5 - \frac{ck^2}{\varepsilon^2}(u^i_j + u^j_i)\varphi^6\right\} = 0,$$

where $\varphi = (\varphi^1, \varphi^2, \varphi^3, \varphi^4, \varphi^5, \varphi^6)$ is a generating function.

To find symmetries of the $k - \varepsilon$ model, we must solve equations (3.7)–(3.10). Since we are interested in classical symmetries, we shall restrict ourselves to solutions of (3.7)–(3.10) of the following form

$$\begin{aligned}
\varphi^1 &= A\overline{u}^1_1 + Bu^1_2 + Cu^1_3 + Du^1_t + F,\\
\varphi^2 &= Au^2_1 + Bu^2_2 + Cu^2_3 + D\overline{u}^2_t + G,\\
\varphi^3 &= Au^3_1 + Bu^3_2 + Cu^3_3 + D\overline{u}^3_t + H,\\
\varphi^4 &= A\overline{p}_1 + Bp_2 + Cp_3 + Dp_t + L,\\
\varphi^5 &= Ak_1 + Bk_2 + Ck_3 + D\overline{k}_t + M,\\
\varphi^6 &= A\varepsilon_1 + B\varepsilon_2 + C\varepsilon_3 + D\overline{\varepsilon}_t + N,
\end{aligned} \tag{3.11}$$

where $A, B, C, D, F, G, H, L, M, N \in C^\infty(J^0(\pi))$.

Now we begin to solve system (3.7)–(3.10).

LEMMA 3.1. *If the functions* (3.11) *satisfy equation* (3.7), *then*

$$\begin{aligned}
F &= \left(\frac{\partial B}{\partial x_2} + \frac{\partial C}{\partial x_3} + \alpha(t)\right)u^1 - \frac{\partial A}{\partial x_2}u^2 - \frac{\partial A}{\partial x_3}u^3 + \widehat{F}(t, x_1, x_2, x_3),\\
G &= -\frac{\partial B}{\partial x_1}u^1 + \left(\frac{\partial A}{\partial x_1} + \frac{\partial C}{\partial x_3} + \alpha(t)\right)u^2 - \frac{\partial B}{\partial x_3}u^3 + \widehat{G}(t, x_1, x_2, x_3),\\
H &= -\frac{\partial C}{\partial x_1}u^1 - \frac{\partial C}{\partial x_2}u^2 + \left(\frac{\partial A}{\partial x_1} + \frac{\partial B}{\partial x_2} + \alpha(t)\right)u^3 + \widehat{H}(t, x_1, x_2, x_3),
\end{aligned} \tag{3.12}$$

where $\alpha = \alpha(t)$ *is an arbitrary function of* t *and*

$$\frac{\partial \widehat{F}}{\partial x_1} + \frac{\partial \widehat{G}}{\partial x_2} + \frac{\partial \widehat{H}}{\partial x_3} = 0 \tag{3.13}$$

while A, B, C *are arbitrary functions of* t, x_1, x_2, x_3 *and* $D = D(t)$ *is an arbitrary function of* t.

PROOF. Substituting (3.11) in (3.7) and using the fact that the functions

$$(u^1_t, \overline{u}^2_t, \overline{u}^3_t, p_t, k_t, \varepsilon_t) \quad \text{and} \quad (u^1_i, u^2_i, u^3_i, p_i, k_i, \varepsilon_i), \qquad i = 1, 2, 3,$$

are generating functions of translations along t and x_i respectively, we obtain the following equation for A, B, C, D, F, G, H:

$$
\begin{aligned}
&\overline{D}_1(A)\overline{u}_1^1 + \overline{D}_1(B)u_2^1 + \overline{D}_1(C)u_3^1 + \overline{D}_1(D)u_t^1 + \overline{D}_1(F) \\
&\quad + \overline{D}_2(A)u_1^2 + \overline{D}_2(B)u_2^2 + \overline{D}_2(C)u_3^2 + \overline{D}_2(D)\overline{u}_t^2 + D_2(G) \\
&\quad + \overline{D}_3(A)u_1^3 + \overline{D}_3(B)u_2^3 + \overline{D}_3(C)u_3^3 + \overline{D}_3(D)\overline{u}_t^3 + \overline{D}_3(H) = 0.
\end{aligned}
\tag{3.14}
$$

Now let us write out the expressions for $D_i(f)$, $i = 1, 2, 3$, with $f = f(t, x_1, x_2, x_3, u^1, u^2, u^3, p, k, \varepsilon)$

$$
\begin{aligned}
\overline{D}_1(f) &= \frac{\partial f}{\partial x_1} + \overline{u}_1^1 \frac{\partial f}{\partial u^1} + u_1^2 \frac{\partial f}{\partial u^2} + u_1^3 \frac{\partial f}{\partial u^3} + \overline{p}_1 \frac{\partial f}{\partial p} + \varepsilon_1 \frac{\partial f}{\partial \varepsilon} + k_1 \frac{\partial f}{\partial k}, \\
\overline{D}_2(f) &= \frac{\partial f}{\partial x_2} + u_2^i \frac{\partial f}{\partial u^i} + p_2 \frac{\partial f}{\partial p} + \varepsilon_2 \frac{\partial f}{\partial \varepsilon} + k_2 \frac{\partial f}{\partial k}, \\
\overline{D}_3(f) &= \frac{\partial f}{\partial x_3} + u_3^i \frac{\partial f}{\partial u^i} + p_3 \frac{\partial f}{\partial p} + \varepsilon_3 \frac{\partial f}{\partial \varepsilon} + k_3 \frac{\partial f}{\partial k}.
\end{aligned}
\tag{3.15}
$$

Here $\overline{u}_1^1$ and $\overline{p}_1$ are given by (3.6).

Equating to zero the coefficients of $(u_t^1)^2$, u_t^1, p_2, p_3 in (3.14), we get

$$
\begin{aligned}
&\frac{\partial A}{\partial p} = \frac{\partial B}{\partial p} = \frac{\partial C}{\partial p} = 0, \qquad D = D(t, x_1, x_2, x_3), \\
&F = \frac{\partial D}{\partial x_1} p + \widetilde{F}(t, x_i, u^j, k, \varepsilon), \\
&G = \frac{\partial D}{\partial x_2} p + \widetilde{G}(t, x_i, u^j, k, \varepsilon), \\
&H = \frac{\partial D}{\partial x_3} p + \widetilde{H}(t, x_i, u^j, k, \varepsilon).
\end{aligned}
\tag{3.16}
$$

Substituting (3.16) in (3.14) and equating to zero the coefficient of u_{jk}^i, we see that the function D is independent of x_1, x_2, x_3, hence $F = \widetilde{F}$, $G = \widetilde{G}$, $H = \widetilde{H}$. Equating to zero the coefficients of k_i, ε_i, $i = 1, 2, 3$, in (3.14), we see that the functions A, B, C, F, G, H are independent of k and ε. Equating to zero the coefficients of $u_j^i u_l^k$, we see that A, B, C depend only on t, x_1, x_2, x_3. To conclude the proof, it remains to equate to zero the coefficients of u_j^i and the absolute term to get (3.12) and (3.13) respectively.

LEMMA 3.2. *If the functions* (3.11) *satisfy equations* (3.7) *and* (3.9), *then*

$$(3.17)\quad \begin{aligned}
\varphi^1 &= A\overline{u}_1^1 + Bu_2^1 + Cu_3^1 + (\tfrac{2}{3}at + d)u_t^1 \\
&\quad + \tfrac{1}{3}au^1 - a_{12}u^2 - a_{13}u^3 - \dot{a}_{12}x_2 - \dot{a}_{13}x_3 - \dot{\widetilde{a}}\,, \\
\varphi^2 &= Au_1^2 + Bu_2^2 + Cu_3^2 + (\tfrac{2}{3}at + d)\overline{u}_t^2 \\
&\quad + a_{12}u^1 + \tfrac{1}{3}au^2 - a_{23}u^3 + \dot{a}_{12}x_1 - \dot{a}_{23}x_3 - \dot{\widetilde{b}}\,, \\
\varphi^3 &= Au_1^3 + Bu_2^3 + Cu_3^3 + (\tfrac{2}{3}at + d)\overline{u}_t^3 \\
&\quad + a_{13}u^1 + a_{23}u^2 + \tfrac{1}{3}au^3 + \dot{a}_{13}x_1 + \dot{a}_{23}x_2 - \dot{\widetilde{c}}\,, \\
\varphi^4 &= A\overline{p}_1 + Bp_2 + Cp_3 + (\tfrac{2}{3}at + d)\,p_t + L\,, \\
\varphi^5 &= Ak_1 + Bk_2 + Ck_3 + (\tfrac{2}{3}at + d)\overline{k}_t + \tfrac{2}{3}ak\,, \\
\varphi^6 &= A\varepsilon_1 + B\varepsilon_2 + C\varepsilon_3 + (\tfrac{2}{3}at + d)\overline{\varepsilon}_t + \tfrac{4}{3}a\varepsilon\,,
\end{aligned}$$

where

$$(3.18)\quad \begin{aligned}
A &= \tfrac{1}{3}ax_1 + a_{12}x_2 + a_{13}x_3 + \widetilde{a}\,, \\
B &= -a_{12}x_1 + \tfrac{1}{3}ax_2 + a_{23}x_3 + \widetilde{b}\,, \\
C &= -a_{13}x_1 - a_{23}x_2 + \tfrac{1}{3}ax_3 + \widetilde{c}\,.
\end{aligned}$$

Here a, d *are constants, while* a_{ij}, $i < j$, $\widetilde{a}$, $\widetilde{b}$, $\widetilde{c}$ *are functions of* t.

PROOF. Equating to zero the coefficients of p_t, u_t^1, p_2, p_3 in (3.9), we get

$$M = M(t, x_1, x_2, x_3, k, \varepsilon), \qquad N = N(t, x_1, x_2, x_3, u^1, u^2, u^3, k, \varepsilon).$$

Taking into account Lemma 3.1 and equating to zero the coefficients of k_{ij}, k_ik_j, ε_{ij}, $\varepsilon_i\varepsilon_j$, we obtain

$$(3.19)\quad \begin{aligned}
M &= Q(t, x_1, x_2, x_3)k + \widetilde{M}(t, x_1, x_2, x_3)\,, \\
N &= \frac{2\varepsilon}{k}M - \left(\nu\frac{\sigma_k}{c}\frac{\varepsilon^2}{k^2} + \varepsilon\right)\left(\dot{D} - 2\frac{\partial A}{\partial x_1}\right),
\end{aligned}$$

and

$$(3.20)\quad \begin{gathered}
\frac{\partial A}{\partial x_1} = \frac{\partial B}{\partial x_2} = \frac{\partial C}{\partial x_3}\,, \\
\frac{\partial A}{\partial x_2} + \frac{\partial B}{\partial x_1} = 0\,, \qquad \frac{\partial A}{\partial x_3} + \frac{\partial C}{\partial x_1} = 0\,, \qquad \frac{\partial B}{\partial x_3} + \frac{\partial C}{\partial x_2} = 0\,.
\end{gathered}$$

Now consider the coefficients of k_i, $i = 1, 2, 3$, in (3.9). Equating them to zero and combining these equations with (3.19), (3.20), we get

$$(3.21)\quad M = Q(t)k + \widetilde{M}(t)\,,$$

$$(3.22)\quad \begin{aligned}
A &= \tfrac{1}{3}(\dot{D} - \alpha)x_1 + a_{12}(t)x_2 + a_{13}(t)x_3 + \widetilde{a}(t)\,, \\
B &= -a_{12}(t)x_1 + \tfrac{1}{3}(\dot{D} - \alpha)x_2 + a_{23}(t)x_3 + \widetilde{b}(t)\,, \\
C &= -a_{13}(t)x_1 - a_{23}(t)x_2 + \tfrac{1}{3}(\dot{D} - \alpha)x_3 + \widetilde{c}(t)\,,
\end{aligned}$$

α, a_{ij}, $i<j$, $\widetilde{a}$, $\widetilde{b}$, $\widetilde{c}$ are functions of t,

$$\widehat{F} = -\frac{\partial A}{\partial t}, \quad \widehat{G} = -\frac{\partial B}{\partial t}, \quad \widehat{H} = -\frac{\partial C}{\partial t}. \tag{3.23}$$

Now consider equation (3.13). Combining it with (3.23) and (3.24), we can obtain expressions (3.18) for A, B, C and the corresponding expressions for F, G, H (see (3.23)).

Taking (3.19), (3.23) into account and equating to zero the coefficients of k^2/ε, k, ε, 1, ε, ε/k^2, ε^2/k^2, one can obtain

$$D = \tfrac{2}{3}at + d, \quad Q = \tfrac{2}{3}a, \quad M = \tfrac{2}{3}ak, \quad N = \tfrac{4}{3}a\varepsilon,$$

a, d are constants.

Lemma 3.2 is proved.

Now any function $\varphi = (\varphi^1, \varphi^2, \varphi^3, \varphi^4, \varphi^5, \varphi^6)$ that satisfies equations (3.7) and (3.9) can be written in the form

$$\begin{aligned} \varphi = {} & dX_0 + \tfrac{1}{3}aS + a_{12}(t)R_{12} + a_{13}(t)R_{13} + a_{23}(t)R_{23} \\ & + X_1(\widetilde{a}) + X_2(\widetilde{b}) + X_3(\widetilde{c}) + \widetilde{\varphi}, \end{aligned} \tag{3.24}$$

where

$$\begin{aligned} \widetilde{\varphi} = (&-\dot{a}_{12}x_2 - \dot{a}_{13}x_3, \dot{a}_{12}x_1 - \dot{a}_{23}x_3, -\dot{a}_{13}x_1 + \dot{a}_{23}x_2, \\ & L - \tfrac{2}{3}ap - (\ddot{\widetilde{a}}x_1 + \ddot{\widetilde{b}}x_2 + \ddot{\widetilde{c}}x_3), 0, 0) \end{aligned}$$

and

(3.25)

$$\begin{aligned} X_0 &= (u_t^1, \overline{u}_t^2, \overline{u}_t^3, p_t, \overline{k}_t, \overline{\varepsilon}_t), \\ X_i(f) &= (fu_i^1 - \delta_{i1}\dot{f}, fu_i^2 - \delta_{i2}\dot{f}, \\ &\quad fu_i^3 - \delta_{i3}\dot{f}, fp_i + \ddot{f}x_i, fk_i, f\varepsilon_i), \qquad i = 1, 2, 3, \ f = f(t), \\ R_{ij} &= (x_ju_i^1 - x_iu_j^1 + \delta_{1j}u^i - \delta_{1i}u^j, x_ju_i^2 - x_iu_j^2 + \delta_{2j}u^i - \delta_{2i}u^j, \\ &\quad x_ju_i^3 - x_iu_j^3 + \delta_{3j}u^i - \delta_{3i}u^j, x_jp_i - x_ip_j, x_jk_i - x_ik_j, x_j\varepsilon_i - x_i\varepsilon_j), \\ S &= (u_i^1x_i + 2tu_t^1 + u^1, x_iu_i^2 + 2t\overline{u}_t^2 + u^2, x_iu_i^3 + 2t\overline{u}_t^3 + u^3, \\ &\quad x_ip_i + 2tp_t + 2p, x_ik_i + 2t\overline{k}_t + 2k, x_i\varepsilon_i + 2t\overline{\varepsilon}_t + 4\varepsilon). \end{aligned}$$

It is not hard to show that if a function (3.24) satisfies equations (3.8_i), $i = 1, 2, 3$, then a_{ij} are constants, while

$$L = \tfrac{2}{3}ap + \ddot{\widetilde{a}}x_1 + \ddot{\widetilde{b}}x_2 + \ddot{\widetilde{c}}x_3 + \theta(t).$$

Thus, we have

$$\begin{aligned} \varphi = {} & dX_0 + \tfrac{1}{3}aS + a_{12}R_{12} + a_{13}R_{13} + a_{23}R_{23} \\ & + X_1(\widetilde{a}) + X_2(\widetilde{b}) + X_3(\widetilde{c}) + P(\theta), \end{aligned} \tag{3.26}$$

where a, d, a_{ij}, $1 \leqslant i < j \leqslant 3$ are constants, $\widetilde{a}$, $\widetilde{b}$, $\widetilde{c}$, θ are functions of t,

$$P(\theta) = (0, 0, 0, \theta(t), 0, 0), \tag{3.27}$$

while X_0, $X_i(f)$, $i = 1, 2, 3$, R_{ij}, S are given by (3.25).

It can be checked by direct computations that any function (3.26) also satisfies equation (3.10).

Thus we have proved the following theorem.

THEOREM 3.1. *The algebra of classical symmetries* $\mathrm{Sym}_0\,\mathcal{Y}$ *for the* $k - \varepsilon$ *model as a vector space over* $\mathbb{R}$ *is generated by the following functions:*

$$X_0 = (u_t^1, \overline{u}_t^2, \overline{u}_t^3, p_t, \overline{k}_t, \varepsilon_t),$$
$$X_i(f_i) = (f_i u_i^1 - \delta_{i1}\dot{f}_i, f_i u_i^2 - \delta_{i2}\dot{f}_i, f_i u_i^3 - \delta_{i3}\dot{f}_i, f_i p_i + \ddot{f}_i x_i, f_i k_i, f_i \varepsilon_i),$$
$$i = 1, 2, 3, \quad f_i = f_i(t),$$
$$S = (u_i^1 x_i + 2tu_t^1 + u^1, x_i u_i^2 + 2t\overline{u}_t^2 + u^2, x_i u_i^3 + 2t\overline{u}_t^3 + u^3,$$
$$x_i p_i + 2tp_t + 2p, x_i k_i + 2t\overline{k}_t + 2k, x_i\varepsilon_i + 2t\overline{\varepsilon}_t + 4\varepsilon),$$
$$P(\theta) = (0, 0, 0, \theta(t), 0, 0), \qquad \theta = \theta(t),$$
$$R_{ij} = (x_j u_i^1 - x_i u_j^1 + \delta_{1j}u^i - \delta_{1i}u^j, x_j u_i^2 - x_i u_j^2 + \delta_{2j}u^i - \delta_{2i}u^j,$$
$$x_j u_i^3 - x_i u_j^3 + \delta_{3j}u^i - \delta_{3i}u^j, x_j p_i - x_i p_j, x_j k_i - x_i k_j, x_j\varepsilon_i - x_i\varepsilon_j).$$

The Lie algebra structure of $\mathrm{Sym}_0\,\mathcal{Y}$ is given by Table 1.

The symmetry X_0 gives rise to the one-parameter group of time translation.

With $f_i(t)$, $i = 1, 2, 3$, arbitrary functions, the symmetry $X_i(f_i)$ corresponds to the transformation to a mobile coordinate system:

$$(t, x_i, u^j, p, k, \varepsilon)$$
$$\mapsto (t, x_i + \varepsilon f_i(t), u^j + \varepsilon f_i(t), p - \varepsilon f_i(t)x_i - \tfrac{1}{2}\varepsilon^2 f_i(t) f_i''(t), k, \varepsilon).$$

The origin traces an arbitrary path, the axes remaining parallel to themselves. The change of pressure compensates the inertial reaction produced by the acceleration of the frame. The functions $f_i(t)$ being constant or linear, this symmetry is assigned to space translations or Galilean boost respectively.

The symmetries R_{ij} generate the three-dimensional rotation group, the velocities rotating with the coordinates:

$$(t, x_i, u^j, p, k, \varepsilon) \mapsto (t, r_{ij}x_i, r_{ij}u^j, p, k, \varepsilon)$$

where $\|r_{ij}\|$ is an orthogonal 3×3 matrix.

The symmetry $P(\theta)$ realizes the one-parameter group of pressure changes:

$$(t, x_i, u^j, p, k, \varepsilon) \mapsto (t, x_i, u^j, p + \varepsilon\theta(t), k, \varepsilon).$$

The one-parameter group associated with the symmetry S consists of scale transformations:

$$(t, x_i, u^j, p, k, \varepsilon) \mapsto (\mu^2 t, \mu x_i, \mu^{-1}u^j, \mu^{-2}p, \mu^{-2}k, \mu^{-4}\varepsilon),$$

where $\mu = e^{\varepsilon}$ is a multiplicative parameter of the group.

TABLE 1

	X_0	S	R_{12}	R_{13}	R_{23}	$X_1(g)$	$X_2(g)$	$X_3(g)$	$P(\theta_1)$
X_0	0	$2X_0$	0	0	0	$X_1(\dot g)$	$X_2(\dot g)$	$X_3(\dot g)$	$P(\dot\theta_1)$
S	$-2X_0$	0	0	0	0	$X_1(2t\dot g-g)$	$X_2(2t\dot g-g)$	$X_3(2t\dot g-g)$	$P(2t\dot\theta_1+2\theta_1)$
R_{12}	0	0	0	$-R_{23}$	R_{13}	$-X_2(g)$	$X_1(g)$	0	0
R_{13}	0	0	R_{23}	0	$-R_{12}$	$-X_3(g)$	0	$X_1(g)$	0
R_{23}	0	0	$-R_{13}$	R_{12}	0	0	$-X_3(g)$	$X_2(g)$	0
$X_1(f)$	$-X_1(\dot f)$	$-X_1(2t\dot f-f)$	$X_2(f)$	$X_3(f)$	0	$P(\ddot f\dot g-\ddot g f)$	0	0	0
$X_2(f)$	$-X_2(\dot f)$	$-X_2(2t\dot f-f)$	$-X_1(f)$	0	$X_3(f)$	0	$P(\ddot f g-\ddot g f)$	0	0
$X_3(f)$	$-X_3(\dot f)$	$-X_3(2t\dot f-f)$	0	$-X_1(f)$	$-X_2(f)$	0	0	$P(\ddot f g-\ddot g f)$	0
$P(\theta_2)$	$-P(\dot\theta_2)$	$-P(2t\dot\theta_2+2\theta_2)$	0	0	0	0	0	0	0

§4. Conservation laws of the $k-\varepsilon$ model

The starting point of the calculation of conservation laws for the $k-\varepsilon$ model is the passage to new internal coordinates, which turn out to be extremely convenient. To introduce the coordinates, we need the following elementary consequence of (3.1) and (3.2):

$$p_{ij}+u^i_j u^j_i - D_{ij}(\Theta_{ij})=0. \tag{4.1}$$

Taking (4.1) into account, we see that

$$\begin{array}{ll} t\,,\ x_i & \text{for } i=1,2,3,\\ u^1_\sigma & \text{for } |\sigma|_t=|\sigma|_1=0,\\ u^2_\sigma\,,\ u^3_\sigma\,,\ k_\sigma\,,\ \varepsilon_\sigma & \text{for } |\sigma|_t=0,\\ p_\sigma\,,\ p_{1\sigma} & \text{for } |\sigma|_1=0 \end{array} \tag{4.2}$$

form an internal coordinate system on the infinite prolongation of the equation in question.

REMARK. This construction of the coordinate system closely follows [3].

Let $\mathcal{F}_r$, $r\geqslant 0$, denote the set of functions of the internal coordinates (4.2) for $|\sigma|\leqslant r$. By $\mathcal{F}_{-1}$ denote the set of function of t, x_i.

Let us write equation (2.7) for a generating function of a conservation law:

(4.3)
$$\overline{D}_i(\psi_i)=0,$$

(4.4_i)
$$\begin{aligned}&-\overline{D}_i(\psi_0)-[\overline{D}_t+u^j\overline{D}_j+\nu\overline{D}_{jj}+\overline{D}_j\circ ck^2\varepsilon^{-1}\overline{D}_j](\psi_i)\\ &\quad+u^j_i\psi_j-\overline{D}_j(ck^2\varepsilon^{-1}\overline{D}_i(\psi_j))+[\overline{D}_j\circ 2ck^2\varepsilon^{-1}(u^i_j+u^j_i)-\tfrac{2}{3}k\overline{D}_i+\tfrac{1}{3}k_i](\psi_4)\\ &\quad+[\overline{D}_j\circ 2cc_1k\,(u^i_j+u^j_i)-\tfrac{2}{3}c_1\varepsilon\overline{D}_i+(1-\tfrac{2}{3}c_1)\varepsilon_i](\psi_5)=0,\qquad i=1,2,3,\end{aligned}$$

(4.5)
$$\begin{aligned}&[2ck\varepsilon^{-1}(u^j_n+u^n_j)\overline{D}_n-\tfrac{2}{3}\overline{D}_j](\psi_j)\\ &\quad+\left[-\overline{D}_t-u^j\overline{D}_j-\overline{D}_j\circ\left(\nu+\frac{ck^2}{\sigma_k\varepsilon}\right)\overline{D}_j+\frac{2ckk_j}{\sigma_k\varepsilon}\overline{D}_j-\frac{2ck}{\varepsilon}u^n_j(u^n_j+u^j_n)\right](\psi_4)\\ &\quad+\left[\frac{2ck\varepsilon_j}{\sigma_k\varepsilon}\overline{D}_j-cc_1u^n_j(u^n_j+u^j_n)-c_2\frac{\varepsilon^2}{k^2}\right](\psi_5)=0,\end{aligned}$$

(4.6)
$$\begin{aligned}&-\frac{ck^2}{\varepsilon^2}(u^n_j+u^j_n)\overline{D}_n(\psi_j)+\left[-\frac{ck^2k_j}{\sigma_k\varepsilon^2}\overline{D}_j+\frac{ck^2}{\varepsilon^2}u^n_j(u^n_j+u^j_n)+1\right](\psi_4)\\ &\quad+\left[-\overline{D}_t-u^j\overline{D}_j-\overline{D}_j\circ\left(\nu+\frac{ck^2}{\sigma_\varepsilon\varepsilon}\right)\overline{D}_j-\frac{ck^2\varepsilon_j}{\sigma_\varepsilon\varepsilon^2}\overline{D}_j+\frac{2c_2\varepsilon}{k}\right](\psi_5)=0,\end{aligned}$$

where $\psi=(\psi_0,\psi_1,\psi_2,\psi_3,\psi_4,\psi_5)$ is a generating function.

Differentiating (4.4_i) with respect to x_i and summing over all i, we get

$$\overline{D}_{ii}(\psi_0) + a_j\overline{D}_{ii}(\psi_j) + \sum_{j=1}^{5}\Delta_j(\psi_j) = 0, \tag{4.7}$$

where $a_j \in \mathcal{F}_1$: Δ_j are linear $\mathcal{C}$-differential operators of order 1 for $j = 1, 2, 3$ and of order 2 for $j = 4, 5$, with the coefficients depending on u^i_σ, k_σ, ε_σ for $|\sigma| \leqslant 2$, $|\sigma|_t = 0$.

To find the conservation laws of the $k - \varepsilon$ model, we must solve equations (4.3)–(4.7). A key result here is the following.

LEMMA 4.1. *Let* $\psi = (\psi_0, \psi_1, \psi_2, \psi_3, \psi_4, \psi_5)$ *be a solution of* (4.3)–(4.6). *Then* $\psi_i \in \mathcal{F}_0$ *for* $0 \leqslant i \leqslant 3$ *and* $\psi_i \leqslant \mathcal{F}_{-1}$ *for* $i = 4, 5$.

PROOF. Assume that $\psi_i \in \mathcal{F}_r$. The proof is by induction over r. For $r = -1$ there is nothing to prove. For $r \geqslant 0$, first let us show that the components of ψ are independent of the variables p_σ for $|\sigma| + |\sigma|_t > r$. In the converse case, consider the set of multi-indices $\Omega = \{\sigma \mid \partial\psi_0/\partial p_\sigma \neq 0,\ |\sigma| + |\sigma|_t > r\}$ and its subsets $\Omega_1 = \{\sigma \in \Omega \mid |\sigma| + |\sigma|_t$ is maximal$\}$ and $\Omega_2 = \{\sigma \in \Omega \mid |\sigma|_2 + 2|\sigma|_t$ is maximal$\}$. Take $\sigma \in \Omega_2$. Let $\widehat{\sigma}$ be obtained from σ by substituting the couple of indices 22 for every index t. Then the left-hand side of (4.7) is linear with respect to the variable $u^1_{22\widehat{\sigma}}$. Equating to zero the coefficient of this variable, we obtain $\partial\psi_0/\partial p_\sigma = 0$. In the same way it can be proved that $\partial\psi_i/\partial p_\sigma = 0$ for all i, $|\sigma| + |\sigma|_t > r$.

Next, let us prove that $\psi_i \in \mathcal{F}_{r-1}$ for $1 \leqslant i \leqslant 5$. Conversely, assume that the set of multi-indices $\Omega = \{\sigma \mid \partial\psi_1/\partial u^1_\sigma \neq 0,\ |\sigma| \geqslant r\}$ is nonempty. (The other cases can be considered similarly.) Take a subset $\Omega_1 = \{\sigma \in \Omega \mid |\sigma|_2$ is maximal$\}$ and choose the multi-index σ such that $|\sigma|_3$ is maximal. Then the left-hand side of (4.4_1) is linear with respect to variable $u^1_{33\sigma}$. Equating to zero the coefficient of this variable, we get $\partial\psi_1/\partial u^1_\sigma = 0$.

Further, continuing the previous line of argument, we see that the components ψ_4, ψ_5 belong to $\mathcal{F}_{r-2}$.

After that, let us check that the component ψ_0 is independent of p_σ for $|\sigma| + |\sigma|_t = r$. Assume the converse. If $|\sigma|_t = 0$, then it follows easily from equation (4.4_2) that $\partial\psi_0/\partial p_\sigma = 0$. Otherwise, let us introduce the set of multi-indices $\Omega = \{\sigma \mid \partial\psi_0/\partial p_\sigma \neq 0,\ |\sigma| + |\sigma|_t = r,\ |\sigma|_t \neq 0\}$ and its subsets $\Omega_1 = \{\sigma \in \Omega \mid |\sigma|_3$ is maximal$\}$ and $\Omega_2 = \{\sigma \in \Omega \mid |\sigma|_2$ is maximal$\}$. Then for $\sigma \in \Omega_2$ the left-hand side of (4.7) is linear with respect to variable $p_{2\sigma}$. Equating to zero the coefficient of this variable, we obtain $\partial\psi_0/\partial p_\sigma = 0$.

Finally, arguing as above for ψ_i, $1 \leqslant i \leqslant 5$, we see that ψ_0 is independent of u^j_σ, k_σ, ε_σ for $|\sigma| = r$. This completes the proof.

Now let us solve equations (4.3)–(4.6) with ψ as in the Lemma 4.1. It follows from (4.3) that for $1 \leqslant i \leqslant 3$ $\psi_i = a(t)u^i + b_i$, where $b_i \in \mathcal{F}_{-1}$ and $\partial b_i/\partial x_i = 0$. Solving equations (4.5) and (4.6), we get

$$\psi_i \in \mathcal{F}_{-1} \quad \text{for } 1 \leqslant i \leqslant 3, \qquad \frac{\partial\psi_i}{\partial x_i} = 0,$$

$$\frac{\partial\psi_i}{\partial x_j} + \frac{\partial\psi_j}{\partial x_i} = 0 \quad \text{for } i < j,\ 1 \leqslant i, j \leqslant 3, \qquad \psi_4 = \psi_5 = 0.$$

Clearly, the left-hand sides of equations (4.4) are linear with respect to variables the k_j, $1 \leqslant j \leqslant 3$. Equating to zero the coefficients of these variables, we see that for every $1 \leqslant i \leqslant 3$ $\partial\psi_i/\partial x_i = 0$. Therefore,

$$\psi_1 = \alpha(t)x_2 + \beta(t)x_3 + a^1(t),$$
$$\psi_2 = -\alpha(t)x_1 + \gamma(t)x_3 + a^2(t),$$
$$\psi_3 = -\beta(t)x_1 - \gamma(t)x_2 + a^3(t).$$

Finally, using equations (4.4), we have

$$\alpha, \beta, \gamma = \text{const},$$
$$\psi_0 = a^i(t)u^i - \dot{a}^i(t)x_i + \alpha(x_2u^1 - x_1u^2) + \beta(x_3u^1 - x_1u^3) + \gamma(x_3u^2 - x_2u^3) + f(t).$$

Recapitulating, we have found the space of generating functions of conservation laws for the $k - \varepsilon$ model to be spanned by

$$\psi^1 = (x_2u^1 - x_1u^2, x_2, -x_1, 0, 0, 0),$$
$$\psi^2 = (x_3u^1 - x_1u^3, x_3, 0, -x_1, 0, 0),$$
$$\psi^3 = (x_3u^2 - x_2u^3, 0, x_3, -x_2, 0, 0),$$
$$\psi^4 = (a^1u^1 - \dot{a}^1x_1, a^1, 0, 0, 0, 0),$$
$$\psi^5 = (a^2u^2 - \dot{a}^2x_2, 0, a^2, 0, 0, 0),$$
$$\psi^6 = (a^3u^3 - \dot{a}^3x_3, 0, 0, a^3, 0, 0),$$
$$\psi^7 = (f, 0, 0, 0, 0, 0).$$

It is straightforward to calculate the conserved currents corresponding to the generating functions. In the physical notations of §1, the results obtained are the following

THEOREM 4.1. *The space of conservation laws for the $k - \varepsilon$ model is generated by the following conserved currents*:

$$\omega^i = [a^i\overline{u}_i, a^i\eta_{i1} - \dot{a}^ix_i\overline{u}_1, a^i\eta_{i2} - \dot{a}^ix_i\overline{u}_2, a^i\eta_{i3} - \dot{a}^ix_i\overline{u}_3], \qquad i = 1, 2, 3,$$
$$\omega^{ij} = \big[\rho(x_i\overline{u}_j - x_j\overline{u}_i), \rho(x_i\eta_{j1} - x_j\eta_{i1} + \nu(\overline{u}_j\delta_{i1} - \overline{u}_i\delta_{j1})), \rho(x_i\eta_{j2} - x_j\eta_{i2} + \nu(\overline{u}_i\delta_{i2} - \overline{u}_i\delta_{j2})), \rho(x_i\eta_{j3} - x_j\eta_{i3} + \nu(\overline{u}_i\delta_{i3} - \overline{u}_i\delta_{j3}))\big], \qquad 1 \leqslant i < j \leqslant 3,$$
$$\omega^7 = [0, f\overline{u}_1, f\overline{u}_2, f\overline{u}_3],$$

where

$$\eta_{ij} = \overline{u}_i\overline{u}_j + \overline{u_i'u_j'} + \frac{p}{\rho}\delta_{ij} - \nu\frac{\partial\overline{u}_i}{\partial x_j}.$$

The conservation laws corresponding to ω^i generalize the conservation laws of linear momentum ($a^i = \text{const}$) and the center-of-mass motion theorem ($\alpha^i(t) = \alpha^i t$, $\alpha^i = \text{const}$).

The currents ω^{ij} determine the conservation law of angular momentum, while ω^7 realizes the conservation law of mass (with $f = \rho$).

§5. Classification of one-dimensional subalgebras

In this and the two subsequent sections we classify subalgebras of the symmetry algebra $\mathrm{Sym}_0\,\mathcal{Y}$ calculated in §3.

To begin the classification process, we compute the adjoint representation using the Lie series

$$\mathrm{Ad}(\exp(X_{\varphi_1}))(\varphi_2)=\sum_{k=0}^{\infty}\frac{1}{k!}\underbrace{[\varphi_1,\dots[\varphi_1,\varphi_2]\dots]}_{k\text{ times}}.$$

As the result we obtain Table 2, with each entry giving $\mathrm{Ad}(\exp(\tau X_{\varphi_1}))(\varphi_2)$.

Let $A=\alpha_1X_0+\alpha_2S+\beta_{ij}R_{ij}+X_i(f_i)+P(\theta)$ be a symmetry. Our task is to simplify it by using suitable adjoint transformations.

Referring to Table 2, we can make the coefficients β_{13} and β_{23} vanish by acting on A by the rotations R_{ij}.

Further, assume that $\alpha_2\neq 0$. Scaling A if necessary, we can assume that $\alpha_2=1$. Using the rotation πR_{13}, we make the coefficient of R_{12} to be nonnegative. Acting on such an A by $-(\alpha_1/2)X_0$, we cancel the coefficient of X_0:

$$A\sim A'=S+\beta R_{12}+X_i(f_i')+P(\theta'),\qquad \beta\geqslant 0.$$

Then we can make the functions f_i' vanish by using the symmetry $X_i(F_i)$, where the functions F_i satisfy the system of equations

$$\begin{cases}2t\dot F_1-F_1+\beta F_2=f_1',\\ 2t\dot F_2-F_2-\beta F_1=f_2',\\ 2t\dot F_3-F_3=f_3'.\end{cases}$$

Thus $A'\sim A''=S+\beta R_{12}+P(\theta'')$. After that we reduce A'' to the symmetry $S+\beta R_{12}$ acting by $P(\Theta)$, with Θ a solution of $2t\dot\Theta+2\Theta=\theta''$.

If $\alpha_2=0$, while $\alpha_1\neq 0$, A is similarly seen to be equivalent to $A'=X_0+\beta R_{12}$, $\beta\geqslant 0$. If $\beta\neq 0$, we can further act on A' by a multiple of S to scale the coefficient of R_{12}: $A'\sim A''=X_0+R_{12}$.

If $\alpha_1=\alpha_2=0$, and $\beta_{12}\neq 0$, we scale to make $\beta_{12}=1$, and then act on A by $X_2(f_1)-X_1(f_2)$, so that A is equivalent to $R_{12}+X_3(f_3)+P(\theta')$.

Summarizing, we have obtained the list of one-dimensional subalgebras

(1) $\langle S+\beta R_{12}\rangle$, $\beta\geqslant 0$,
(2) $\langle X_0\rangle$,
(3) $\langle X_0+R_{12}\rangle$,
(4) $\langle R_{12}+X_3(f)+P(\theta)\rangle$,
(5) $\langle X_i(f_i)+P(\theta)\rangle$,

with the property that any other one-dimensional subalgebra is equivalent to a subalgebra in the list.

§6. Classification of two-dimensional subalgebras

Suppose $\mathfrak{g}=\langle A,B\rangle$ is a subalgebra, which we shall simplify through suitable adjoint maps.

TABLE 2

	X_0	S	R_{12}	R_{13}	R_{23}	$X_1(f_2)$	$X_2(g_2)$	$X_3(h_2)$	$P(\theta_2)$
X_0	X_0	$S + 2\tau X_0$	R_{12}	R_{13}	R_{23}	$X_1(f_2(t+\tau))$	$X_2(g_2(t+\tau))$	$X_3(h_2(t+\tau))$	$P(\theta_2(t +$
S	$e^{-2\tau} X_0$	S	R_{12}	R_{13}	R_{23}	$X_1(e^{-\tau} f_2(e^{2\tau} t))$	$X_2(e^{-\tau} g_2(e^{2\tau} t))$	$X_3(e^{-\tau} h_2(e^{2\tau} t))$	$P(e^{2\tau}\theta_2(e^2$
R_{12}	X_0	S	R_{12}	$R_{13}\cos\tau$ $-R_{23}\sin\tau$	$R_{23}\cos\tau$ $+R_{13}\sin\tau$	$X_1(f_2)\cos\tau$ $-X_2(f_2)\sin\tau$	$X_2(g_2)\cos\tau$ $+X_1(g_2)\sin\tau$	$X_3(h_2)$	$P(\theta_2)$
R_{13}	X_0	S	$R_{12}\cos\tau$ $+R_{23}\sin\tau$	R_{13}	$R_{23}\cos\tau$ $-R_{12}\sin\tau$	$X_1(f_2)\cos\tau$ $-X_3(f_2)\sin\tau$	$X_2(g_2)$	$X_3(h_2)\cos\tau$ $+X_1(h_2)\sin\tau$	$P(\theta_2)$
R_{23}	X_0	S	$R_{12}\cos\tau$ $-R_{13}\sin\tau$	$R_{13}\cos\tau$ $+R_{12}\sin\tau$	R_{23}	$X_1(f_2)$	$X_2(g_2)\cos\tau$ $-X_3(g_2)\sin\tau$	$X_3(h_2)\cos\tau$ $+X_2(h_2)\sin\tau$	$P(\theta_2)$
$X_1(f_1)$	$X_0 - \tau X_1(\dot f_1)$ $+\frac{\tau^2}{2}P(f_1\dddot f_1$ $-\dot f_1\ddot f_1)$	$S - \tau X_1(2t\dot f_1 - f_1)$ $+\tau^2 P(2f_1\ddot f_1$ $+t(f_1\dddot f_1 - \dot f_1\ddot f_1))$	R_{12} $+\tau X_2(f_1)$	R_{13} $+\tau X_3(f_1)$	R_{23}	$X_1(f_2)$ $+\tau P(\ddot f_1 f_2 - \ddot f_2 f_1)$	$X_2(g_2)$	$X_3(h_2)$	$P(\theta_2)$
$X_2(g_1)$	$X_0 - \tau X_2(\dot g_1)$ $+\frac{\tau^2}{2}P(g_1\dddot g_1$ $-\dot g_1\ddot g_1)$	$S - \tau X_2(2t\dot g_1 - g_1)$ $+\tau^2 P(2g_1\ddot g_2$ $+t(g_1\dddot g_1 - \dot g_1\ddot g_1))$	R_{12} $-\tau X_1(g_1)$	R_{13}	R_{23} $+\tau X_3(g_1)$	$X_1(f_2)$	$X_2(g_2)$ $+\tau P(\ddot g_1 g_2 - \ddot g_2 g_1)$	$X_3(h_2)$	$P(\theta_2)$
$X_3(h_1)$	$X_0 - \tau X_3(\dot h_1)$ $+\frac{\tau^2}{2}P(h_1\dddot h_1$ $-\dot h_1\ddot h_1)$	$S - \tau X_3(2t\dot h_1 - h_1)$ $+\tau^2 P(2h_1\ddot h_2$ $+t(h_1\dddot h_1 - \dot h_1\ddot h_1))$	R_{12}	R_{13} $-\tau X_1(h_1)$	R_{23} $-\tau X_2(h_1)$	$X_1(f_2)$	$X_2(g_2)$	$X_3(h_2)$ $+\tau P(\ddot h_1 h_2 - \ddot h_2 h_1)$	$P(\theta_2)$
$P(\theta_1)$	$X_0 - \tau P(\dot\theta_1)$	$S - \tau P(2t\dot\theta_1 + 2\theta_1)$	R_{12}	R_{13}	R_{23}	$X_1(f_2)$	$X_2(g_2)$	$X_3(h_2)$	$P(\theta_2)$

Let $\mathcal{L}$ be the algebra $\langle X_0, S, R_{ij}\rangle$, $\mathcal{N}$ the ideal $\langle X_i(f_i)+P(\theta)\rangle$, and ν: $\mathrm{Sym}_0\,\mathcal{Y}\to\mathcal{L}$ the projection along $\mathcal{N}$. There are three cases, depending on the dimension of the image of $\mathfrak{g}$ under ν.

Case 1. $\dim\nu(\mathfrak{g})=0$.

Obviously,

$$A=X_i(f_i)+P(\theta),\qquad B=X_i(g_i)+P(\eta),$$

and $[A, B]=0$. The latter relation means that $\ddot{f}_i g_i=\ddot{g}_i f_i$.

Case 2. $\dim\nu(\mathfrak{g})=1$.

Suppose that $\nu(\mathfrak{g})=\langle\nu(A)\rangle$. Put $B=X_i(g_i)+P(\eta)$. It follows easily from the results of §5 that A is equivalent to any of the following symmetries:

(1) $S+\beta R_{12}$, $\beta\geqslant 0$,

(2) X_0,

(3) X_0+R_{12},

(4) $R_{12}+X_3(f)+P(\theta)$.

Let us consider these four possibilities.

(1) $A=S+\beta R_{12}$.

Referring to Table 1, we have $[A, B]=\lambda B$, λ being a constant. Let us introduce the following notation. With $F=f_1+if_2$ a complex-valued function of t, by $X(F)$ we denote $X_1(f_1)+X_2(f_2)$. It is straightforward to check that

$$[R_{12}, X(F)]=X(-iF),\qquad \mathrm{Ad}(\exp(\tau X_{R_{12}}))(X(F))=X(e^{-i\tau}F).$$

Now we can write the condition $[A, B]=\lambda B$ in the following form

$$2t\dot{G}-G-i\beta G=\lambda G,\quad 2t\dot{g}_3-g_3=\lambda g_3,\quad 2t\dot{\eta}+2\eta=\lambda\eta,$$

where $G=g_1+ig_2$. Therefore,

$$G=\widetilde{a}t^{(1+\lambda+i\beta)/2},\quad g_3=ct^{(1+\lambda)/2},\quad \eta=dt^{(\lambda-2)/2},$$

where $\widetilde{a}\in\mathbb{C}$, $c, d\in\mathbb{R}$. Obviously, we can make the constant $\widetilde{a}$ real by acting on $\mathfrak{g}$ with a multiple of R_{12}:

$$B\sim X(at^{(1+\lambda+i\beta)/2})+X_3(ct^{(1+\lambda)/2})+P(dt^{(\lambda-2)/2}),$$

with a, c, d real constants. If $\lambda\neq 0$, while $a\neq 0$ or $c\neq 0$, we can guarantee that d vanish acting on $\mathfrak{g}$ by $X(\alpha t^{(1+i\beta)/2})+X_3(\gamma\sqrt{t})$, α, γ being constants.

(2) $A=X_0$.

Clearly, we have $[A, B]=\lambda B$, so $\dot{g}_i=\lambda g_i$, $\dot{\eta}=\lambda\eta$. Thus $g_i=a_ie^{\lambda t}$, $\eta=de^{\lambda t}$, where a_i, d are constants. Acting by a multiple of R_{12}, we cancel a_2. If $\lambda\neq 0$, while $a_1\neq 0$ or $a_3\neq 0$, we act on $\mathfrak{g}$ by $X_1(\alpha)+X_3(\gamma)$, α, γ are constants, to make d zero.

(3) $A=X_0+R_{12}$.

We evidently have $[A, B]=0$, therefore $\dot{G}-iG=\lambda G$, $\dot{g}_3=\lambda g_3$, $\dot{\eta}=\lambda\eta$, where $G=g_1+ig_2$. Hence $G=\widetilde{a}e^{(\lambda+i)t}$, $g_3=ce^{\lambda t}$, $\eta=de^{\lambda t}$, where $\widetilde{a}$ is a complex constant and c, d are real ones. As above, using a multiple of R_{12}, we can guarantee that $\widetilde{a}$ becomes real, so that

$$B \sim X(ae^{(\lambda+i)t}) + X_3(ce^{\lambda t}) + P(de^{\lambda t}), \qquad a = |\widetilde{a}|.$$

If $\lambda \neq 0$, while $a \neq 0$ or $c \neq 0$, we cancel d by using $X(\alpha e^{it}) + X_3(\gamma)$, α, γ being constants.

(4) $A = R_{12} + X_3(f) + P(\theta)$.

It is not hard to prove that $[A, B] = 0$, so that $g_1 = g_2 = 0$, $\ddot{f} g_3 = \ddot{g}_3 f$.

Case 3. $\dim \nu(\mathfrak{g}) = 2$.

First, let us classify the two-dimensional subalgebras of the algebra $\mathcal{L}$. Suppose $\langle X, Y\rangle \subset \mathcal{L}$ is a subalgebra,

$$X = \alpha_1 X_0 + \alpha_2 S + \alpha_{ij} R_{ij}, \qquad Y = \beta_1 X_0 + \beta_2 S + \beta_{ij} R_{ij}.$$

Since there exist no two-dimensional subalgebras in the Lie algebra $\langle R_{ij}\rangle$ (see, e.g., [11]), we can make α_{ij} equal to 0. Further, we clearly have

$$Y \sim \beta_1 X_0 + \beta_2 S + \beta R_{12}, \qquad \beta \geqslant 0.$$

Assume that $\alpha_2 \neq 0$, then $X \sim S$ and the subalgebra $\langle X, Y\rangle$ is spanned by S, X_0 or by S, R_{12}.

If $\alpha_2 = 0$, while $\beta_2 \neq 0$, then $Y \sim S + \beta R_{12}$, $\beta \geqslant 0$.

Thus a two-dimensional subalgebra in $\mathcal{L}$ is equivalent to any of the following subalgebras:

$$\langle S, R_{12}\rangle, \qquad \langle S + \beta R_{12}, X_0\rangle, \quad \beta \geqslant 0, \qquad \langle X_0, R_{12}\rangle.$$

Now, if $\nu(\mathfrak{g}) = \langle S, R_{12}\rangle$, then we can take $A = S$, $B = R_{12} + X_i(g_i) + P(\eta)$. The relation $[A, B] = 0$ gives

$$2t\dot{g}_i - g_i = 0 \implies g_i = a_i\sqrt{t}, \qquad 2t\dot{\eta} + 2\eta = 0 \implies \eta = d/t,$$

where a_i, d are constants. Acting on $\mathfrak{g}$ by a multiple of R_{12} we can guarantee that a_2 vanishes.

Further, if $\nu(\mathfrak{g}) = \langle S + \beta R_{12}, X_0\rangle$, then we can put

$$A = S + \beta R_{12}, \qquad B = X_0 + X(G) + X_3(g_3) + P(\eta).$$

From $[A, B] = -2B$ it follows that

$$\begin{aligned} 2t\dot{G} - G - i\beta G &= -2G \implies G = \widetilde{a} t^{(i\beta-1)/2}, \\ 2t\dot{g}_3 - g_3 &= -2g_3 \implies g_3 = c/\sqrt{t}, \\ 2t\dot{\eta} + 2\eta &= -2\eta \implies \eta = d/t^2, \end{aligned}$$

where $\widetilde{a} \in \mathbb{C}$, c, $d \in \mathbb{R}$. We act on $\mathfrak{g}$ by a multiple of R_{12} to make $\widetilde{a}$ real.

Finally, if $\nu(\mathfrak{g}) = \langle X_0, R_{12}\rangle$, we can choose $A = X_0$, $B = R_{12} + X_i(g_i) + P(\eta)$. The relation $[A, B] = 0$ yields $g_i = \text{const}$ and $\eta = \text{const}$. Acting on $\mathfrak{g}$ by a multiple of R_{12}, we cancel g_2.

Bringing together all the above, we obtain the following list of two-dimensional subalgebras:

(1) $\langle X_i(f_i) + P(\theta), X_i(g_i) + P(\eta)\rangle, \qquad \ddot{f}_i g_i = \ddot{g}_i f_i,$

(2) $\langle S + \beta R_{12}, X_1(at^{(1+\lambda)/2}\cos((\beta/2)\ln t))$
$\qquad + X_2(at^{(1+\lambda)/2}\sin((\beta/2)\ln t)) + X_3(ct^{(1+\lambda)/2})\rangle, \qquad \beta \geqslant 0, \ \lambda \neq 0,$

(3) $\langle S + \beta R_{12}, P(t^\lambda)\rangle, \qquad \beta \geqslant 0, \ \lambda \neq 2,$

(4) $\langle S + \beta R_{12}, X_1(a\sqrt{t}\cos((\beta/2)\ln t))$
$\qquad + X_2(a\sqrt{t}\sin((\beta/2)\ln t)) + X_3(c\sqrt{t}) + P(d/t)\rangle, \qquad \beta \geqslant 0,$

(5) $\langle X_0, X_1(ae^{\lambda t}) + X_3(ce^{\lambda t})\rangle, \qquad \lambda \neq 0,$

(6) $\langle X_0, P(e^{\lambda t})\rangle, \qquad \lambda \neq 0,$

(7) $\langle X_0, X_1(a) + X_3(c) + P(d)\rangle,$

(8) $\langle X_0 + R_{12}, X_1(ae^{\lambda t}\cos t) + X_2(ae^{\lambda t}\sin t) + X_3(ce^{\lambda t})\rangle,$

(9) $\langle X_0 + R_{12}, P(e^{\lambda t})\rangle,$

(10) $\langle X_0 + R_{12}, X_1(a\cos t) + X_2(a\sin t) + X_3(c) + P(d)\rangle,$

(11) $\langle R_{12} + X_3(f) + P(\theta), X_3(g) + P(\eta)\rangle, \qquad \ddot{f}g = \ddot{g}f,$

(12) $\langle S, R_{12} + X_1(a\sqrt{t}) + X_3(c\sqrt{t}) + P(d/t)\rangle,$

(13) $\langle S + \beta R_{12}, X_0 + X_1(at^{-1/2}\cos((\beta/2)\ln t))$
$\qquad + X_2(at^{-1/2}\sin((\beta/2)\ln t)) + X_3(ct^{-1/2}) + P(dt^{-2})\rangle, \qquad \beta \geqslant 0,$

(14) $\langle X_0, R_{12} + X_1(a) + X_3(c) + P(d)\rangle.$

§7. Classification of three-dimensional subalgebras

Now let $\mathfrak{g} = \langle A, B, C\rangle$ denote the three-dimensional subalgebra under consideration. There are four cases, depending on the dimension of the algebra $\nu(\mathfrak{g})$.

Case 1. $\dim \nu(\mathfrak{g}) = 0$.

Clearly,

$$A = X_i(f_i) + P(\theta), \quad B = X_i(g_i) + P(\eta), \quad C = X_i(h_i) + P(\xi).$$

It is not hard to prove that the dimension of the commutant $[\mathfrak{g}, \mathfrak{g}]$ equals either 1 or 0. In the former case we can put $h_i = 0$, $\xi = \ddot{f}_i g_i - \ddot{g}_i f_i$. In the second case we have

$$\ddot{f}_i g_i = \ddot{g}_i f_i, \quad \ddot{f}_i h_i = \ddot{h}_i f_i, \quad \ddot{g}_i h_i = \ddot{h}_i g_i.$$

Case 2. $\dim \nu(\mathfrak{g}) = 1$.

Suppose that $\nu(\mathfrak{g}) = \langle \nu(A)\rangle$. It is easy to see that $\operatorname{Ker}(\nu|_{\mathfrak{g}})$ is a commutative ideal in $\mathfrak{g}$ and the map $X \mapsto [A, X]$, $X \in \operatorname{Ker}(\nu|_{\mathfrak{g}})$, is a linear operator on it.

Choosing a basis $\{B, C\}$ in $\operatorname{Ker}(v|_{\mathfrak{g}})$, we can guarantee that the commutators of A, B, C have one of the following forms:

$$
\begin{aligned}
&\text{(a)} \quad [A, B] = \lambda B, && [A, C] = \mu C, && [B, C] = 0, \\
&\text{(b)} \quad [A, B] = \lambda B, && [A, C] = B + \lambda C, && [B, C] = 0, \\
&\text{(c)} \quad [A, B] = \lambda B - \mu C, && [A, C] = \mu B + \lambda C, && [B, C] = 0,
\end{aligned}
$$

where λ and μ are constants and in case (c) we have $\mu > 0$.

Letting $B = X(G)+X_3(g_3)+P(\eta)$, $C = X(H)+X_3(h_3)+P(\xi)$, $G = g_1+ig_2$, $H = h_1 + ih_2$, we consider these three subcases.

(a) Recall that A is equivalent to $S + \beta R_{12}$, $\beta \geqslant 0$, or to X_0, or to $X_0 + R_{12}$, or to $R_{12} + X_3(f) + P(\theta)$.

For $A = S + \beta R_{12}$, we get

$$2t\dot{G} - G - i\beta G = \lambda G, \quad 2t\dot{g}_3 - g_3 = \lambda g_3, \quad 2t\dot{\eta} + 2\eta = \lambda \eta.$$

Therefore,

$$G = \widetilde{a}_1 t^{(1+\lambda+i\beta)/2}, \quad g_3 = c_1 t^{(1+\lambda)/2}, \quad \eta = d_1 t^{(\lambda-2)/2}.$$

Here and below we denote constants by the letters a, b, c, d, α, β, γ, δ with or without indices; a tilde or a hat over the letter means that the constant is complex or quaternionic respectively. Similarly, we have

$$H = \widetilde{a}_2 t^{(1+\mu+i\beta)/2}, \quad h_3 = c_2 t^{(1+\mu)/2}, \quad \xi = d_2 t^{(\mu-2)/2}.$$

Acting on $\mathfrak{g}$ by a multiple of R_{12}, we make the constant $\widetilde{a}_1$ real: $\widetilde{a}_1 = a_1$. It is straightforward to verify that the condition $[B, C] = 0$ gives

$$(\lambda + \mu)((\lambda - \mu)(a_1 \operatorname{Re} \widetilde{a}_2 + c_1 c_2) + 2\widetilde{a}_1 \operatorname{Im} \widetilde{a}_2) = 0.$$

REMARK. Such computations are perhaps most easily done in the complex notation by using the obvious formula

$$[X(F_1), X(F_2)] = P(\operatorname{Re}(\ddot{F}_1 \overline{F}_2 - \overline{F}_1 \ddot{F}_2)).$$

For $A = X_0$ we obtain

$$
\begin{aligned}
&\dot{g}_i = \lambda g_i \implies g_i = a_i e^{\lambda t}, && \dot{\eta} = \lambda \eta \implies \eta = d_1 e^{\lambda t}, \\
&\dot{h}_i = \mu h_i \implies h_i = b_i e^{\mu t}, && \dot{\xi} = \mu \xi \implies \xi = d_2 e^{\mu t}.
\end{aligned}
$$

We can further act on $\mathfrak{g}$ by a multiple of R_{12} to make $a_2 = 0$. The equality $[B, C] = 0$ yields $(\lambda^2 - \mu^2)(a_1 b_1 + c_1 c_2) = 0$.

For $A = X_0 + R_{12}$, we have

$$\dot{G} - iG = \lambda G \implies G = \widetilde{a} e^{(\lambda+i)t},$$

$$\dot{g}_3 = \lambda g_3 \implies g_3 = c_1 e^{\lambda t}, \qquad \dot{\eta} = \lambda \eta \implies \eta = d_1 e^{\lambda t},$$

$$\dot{H} - iH = \mu H \implies H = \widetilde{b}^{(\mu+i)t},$$

$$\dot{h}_3 = \mu h_3 \implies h_3 = c_2 e^{\mu t}, \qquad \dot{\xi} = \mu \xi \implies \xi = d_2 e^{\mu t}.$$

Using R_{12}, we make $\widetilde{a}$ real: $\widetilde{a} = a$. From $[B, C] = 0$ it follows that

$$(\lambda + \mu)((\lambda - \mu)(a \operatorname{Re} \widetilde{b} + c_1 c_2) + 2a \operatorname{Im} \widetilde{b}) = 0.$$

For $A = R_{12} + X(f) + P(\theta)$, we easily see that

$$\lambda = \mu = 0, \quad G = H = 0,$$
$$\ddot{f} g_3 = \ddot{g}_3 f, \quad \ddot{f} h_3 = \ddot{h}_3 f, \quad \ddot{g}_3 h_3 = \ddot{h}_3 g_3.$$

(b) Let $A = S + \beta R_{12}$, $\beta \geqslant 0$. Then the condition $[A, B] = \lambda B$ gives

$$2t\dot{G} - G - i\beta G = \lambda G, \quad 2t\dot{g}_3 - g_3 = \lambda g_3, \quad 2t\dot{\eta} + 2\eta = \lambda \eta.$$

The condition $[A, C] = B + \lambda C$ yields

$$2t\dot{H} - H - i\beta H = G + \lambda H, \quad 2t\dot{h}_3 - h_3 = g_3 + \lambda h_3, \quad 2t\dot{\xi} + 2\xi = \eta + \lambda \xi.$$

Therefore,

$$G = \widetilde{a}_1 t^{(1+\lambda+i\beta)/2}, \quad g_3 = c_1 t^{(1+\lambda)/2}, \quad \eta = d_1 t^{(\lambda-2)/2},$$
$$H = \widetilde{a}_2 t^{(1+\lambda+i\beta)/2} + (\widetilde{a}_1/2)\, t^{(1+\lambda+i\beta)/2} \ln t,$$
$$h_3 = c_2 t^{(1+\lambda)/2} + (c_1/2)\, t^{(1+\lambda)/2} \ln t,$$
$$\xi = d_2 t^{(\lambda-2)/2} + (d_1/2)\, t^{(\lambda-2)/2} \ln t.$$

Using R_{12}, we make $\widetilde{a}_1$ real: $\widetilde{a}_1 = a_1$. The condition $[B, C] = 0$ has the form

$$\lambda(2\beta a_1 \operatorname{Im} \widetilde{a}_2 - a_1^2 - c_1^2) = 0.$$

Next set $A = X_0$. The reader will have no difficulty in showing that

$$G = a_1 e^{\lambda t}, \quad g_3 = c_1 e^{\lambda t}, \quad \eta = d_1 e^{\lambda t}, \quad H = \widetilde{a}_2 e^{\lambda t} + a_1 t e^{\lambda t},$$
$$h_3 = c_2 e^{\lambda t} + c_1 t e^{\lambda t}, \quad \xi = d_2 e^{\lambda t} + d_1 t e^{\lambda t}, \quad \lambda(a_1^2 + c_1^2) = 0.$$

In the same way, setting $A = X_0 + R_{12}$, we obtain

$$G = a_1 e^{(\lambda+i)t}, \quad g_3 = c_1 e^{\lambda t}, \quad \eta = d_1 e^{\lambda t}, \quad H = \widetilde{a}_2 e^{(\lambda+i)t} + a_1 t e^{(\lambda+i)t},$$
$$h_3 = c_2 e^{\lambda t} + c_1 t e^{\lambda t}, \quad \xi = d_2 e^{\lambda t} + d_1 t e^{\lambda t}, \quad \lambda(2a_1 \operatorname{Im} \widetilde{a}_2 - a_1^2 - c_1^2) = 0.$$

Finally, let $A = R_{12} + X_3(f) + P(\theta)$. It follows easily that

$$\lambda = 0, \quad g_1 = g_2 = g_3 = 0, \quad H = 0, \quad \eta = \ddot{f} h_3 - \ddot{h}_3 f.$$

(c) For the calculation in this case, it will be convenient to rewrite the relations $[A, B] = \lambda B - \mu C$ and $[A, C] = \mu B + \lambda C$ in the form

$$[A, D] = (\lambda + \mu j) D, \tag{7.1}$$

where $D = B + Cj$, $j \in \mathbb{H}$ is an imaginary unit. With $F = f_1 + f_2 i + f_3 j + f_4 k$ a quaternion-valued function of t, by $X(F)$ denote $X_1(f_1) + X_2(f_2) + X_1(f_3) j + X_2(f_4) j \in \mathrm{Sym}_0 \mathcal{Y} \oplus (\mathrm{Sym}_0 \mathcal{Y}) j$. It is obvious that

$$[R_{12}, X(F)] = X(-iF),$$
$$\mathrm{Ad}(\exp(\tau R_{12}))(X(F)) = X(e^{-i\tau} F), \quad jX(F) = X(Fj).$$

Now set $D = X(G) + X_3(g) + P(\eta)$, where $G = g_1 + g_2 i + h_1 j + h_2 k$,

$$g = g_3 + h_3 j, \quad X_3(g) = X_3(g_3) + X_3(h_3) j,$$
$$\eta = \zeta + \xi j, \quad P(\eta) = P(\zeta) + P(\xi) j.$$

As before we must consider four cases, depending on the form of symmetry A.

If $A = S + \beta R_{12}$, $\beta \geqslant 0$, then from (7.1) we immediately get

$$G = t^{(1+\lambda+i\beta)/2} \widehat{a}\, t^{\mu j/2}, \quad \widehat{a} = \widetilde{a}_1 + \widetilde{a}_2 j,$$
$$g = \widehat{c}\, t^{(1+\lambda+\mu j)/2}, \quad \eta = \widehat{d}\, t^{(\lambda-2+\mu j)/2}.$$

We can make the constant $\widetilde{a}_1$ real by using a multiple of R_{12}. It is straightforward to check that the condition $[B, C] = 0$ yields

$$\lambda(\mu(|\widehat{a}|^2 + |\widehat{c}|^2) + 2\beta a_1 \operatorname{Im} \widetilde{a}_2) = 0. \tag{7.2}$$

If $A = S + \beta R_{12}$, $\beta \geqslant 0$, $\beta = 0$ or 1, then we have

$$G = e^{(\lambda+i\beta)t} \widehat{a}\, e^{j\mu t}, \quad \widehat{a} = \widetilde{a}_1 + \widetilde{a}_2 j,$$
$$g = \widehat{c}\, e^{(\lambda+j\mu)t}, \quad \eta = \widehat{d}\, e^{(\lambda+j\mu)t}.$$

As above, we can make the constant $\widetilde{a}_1$ real and the condition $[B, C] = 0$ gives (7.2).

Finally, if $A = R_{12} + X_3(f) + P(\theta)$, then the reader will easily prove that

$$\lambda = 0, \quad \mu = 1, \quad G = g_1 + g_2 i - g_2 j + g_1 k, \quad g = 0, \quad \eta = 0.$$

The condition $[B, C] = 0$ implies $\ddot{g}_1 g_2 = \ddot{g}_2 g_1$.

Case 3. $\dim v(\mathfrak{g}) = 2$.

Let us choose a basis $\{A, B, C\}$ in $\mathfrak{g}$ so that $\mathrm{Ker}(v|_{\mathfrak{g}})$ is spanned by C. If the algebra $v(\mathfrak{g})$ is commutative, then we can guarantee that the commutators of A, B, C have one of the following forms:

(a) $[A, B] = 0$, $[A, C] = \lambda C$, $[B, C] = \mu C$,

(b) $[A, B] = C$, $[A, C] = 0$, $[B, C] = 0$.

Otherwise, these commutator relations can be reduced to

(c) $[A, B] = -2B$, $[A, C] = \lambda C$, $[B, C] = \mu C$,

(d) $[A, B] = -2B + C$, $[A, C] = -2C$, $[B, C] = 0$.

We consider these four subcases separately.

(a) We already know that if $\nu(\mathfrak{g})$ is commutative, then it is equivalent either to $\langle S, R_{12}\rangle$ or to $\langle X_0, R_{12}\rangle$. In the former case we can take

$$A = S, \quad B = R_{12} + X_i(f_i) + P(\theta), \quad C = X_i(g_i) + P(\eta).$$

Using the symmetry $X_2(f_1) - X_1(f_2)$, we cancel f_1 and f_2. The relation $[A, B] = 0$ yields

$$f_3 = c_1\sqrt{t}, \qquad \theta = d_1/t.$$

From the condition $[A, C] = \lambda C$ it follows that

$$g_i = a_i t^{(1+\lambda)/2}, \qquad \eta = d_2 t^{(\lambda-2)/2}.$$

Finally, the relation $[B, C] = \mu C$ gives $\mu = 0$, $a_1 = a_2 = 0$, $\lambda a_3 c_1 = 0$.

In the same way, for $\nu(\mathfrak{g}) = \langle X_0, R_{12}\rangle$ we get

$$\mu = 0, \quad A = X_0, \quad B = R_{12} + X_3(c_1) + P(d_1),$$
$$C = X_3(c_2 e^{\lambda t}) + P(d_2 e^{\lambda t}), \quad \lambda c_1 c_2 = 0.$$

(b) As in the case (a), we can put

$$A = S \text{ or } X_0, \quad B = R_{12} + X_3(f) + P(\theta), \quad C = X_i(g_i) + P(\eta).$$

For $A = S$ it can be easily checked that

$$g_1 = g_2 = 0, \quad g_3 = c_2\sqrt{t}, \quad \eta = d_2/t,$$
$$f = c_1\sqrt{t} + (c_2/2)\sqrt{t}\ln t, \quad \theta = d_1/t + (d_2/2t)\ln t.$$

For $A = X_0$ we see that

$$g_1 = g_2 = 0, \quad g_3 = c_2, \quad \eta = d_2, \quad f = c_1 + c_2 t, \quad \theta = d_1 + d_2 t.$$

(c) If $\nu(\mathfrak{g})$ is noncommutative, then $\nu(\mathfrak{g}) \sim \langle S + \beta R_{12}, X_0\rangle$, $\beta \geqslant 0$, so that we can set $A = S + \beta R_{12}$, $B = X_0$, $C = X_i(f_i) + P(\theta)$. From $[B, C] = \mu C$ it follows that

$$f_i = a_i e^{\mu t}, \quad \theta = d e^{\mu t}.$$

If $\beta \neq 0$, then using $[A, C] = \lambda C$ we obtain

$$\mu = 0, \quad a_1 = a_2 = 0, \quad a_3 = 0 \text{ for } \lambda \neq -1, \quad d = 0 \text{ for } \lambda \neq 2.$$

If $\beta = 0$, then we see that $C = X_i(a_i)$ for $\mu = 0$, $\lambda = -1$. Acting on $\mathfrak{g}$ by rotations we can make a_1 and a_2 vanish.

(d) Let $C = X(F) + X_3(f) + P(\theta)$. From the conditions $[A, C] = -2C$ and $[B, C] = 0$ it follows that

$$2t\dot{F} - F - i\beta F = -2F, \quad 2t\dot{f}_3 - f_3 = -2f_3, \quad \dot{F} = 0, \quad \dot{f}_3 = 0.$$

Hence $F = 0$ and $f_3 = 0$. Further, the same conditions yield

$$2t\dot{\theta} + 2\theta = -2\theta, \qquad \dot{\theta} = 0.$$

Hence $\theta = 0$ and this case is impossible.

Case 4. $\dim \nu(\mathfrak{g}) = 3$.

Let us consider the projection ν' of $\mathrm{Sym}_0 \mathcal{Y}$ onto $\langle R_{ij} \rangle$ along $\langle S, X_0 \rangle + \mathcal{N}$. There are two subcases, depending on the dimension of the image of $\mathfrak{g}$ under ν'.

If $\dim \nu'(\mathfrak{g}) = 3$, then we can put $A = R_{12} + A_1$, $B = R_{13} + B_1$, $C = R_{23} + C_1$, where A_1, B_1, C_1 belong to $\mathrm{Ker}\, \nu'$.

The obvious relations $[A, B] = -C$, $[A, C] = B$, $[B, C] = -A$ are easily seen to yield

$$\begin{aligned} A &= R_{12} + X_1(f) + X_2(g) + P(\theta_1), \\ B &= R_{13} + X_1(h) + X_3(g) + P(\theta_2), \\ C &= R_{23} + X_2(h) - X_3(f) + P(\theta_3). \end{aligned}$$

Acting on $\mathfrak{g}$ by the symmetry $-X_1(g) + X_2(f) + X_3(h)$, we reduce $\mathfrak{g}$ to $\langle R_{ij} \rangle$.

If $\dim \nu'(\mathfrak{g}) = 1$ then we can guarantee that

$$A = S, \quad B = R_{12} + X_i(f_i) + P(\theta), \quad C = X_0 + X_i(g_i) + P(\eta).$$

Acting on $\mathfrak{g}$ by $X_2(f_1) - X_1(f_2)$, we cancel f_1 and f_2. Further, it is clear that $[S, X_i(g_i) + P(\eta)] = -2X_i(g_i) - 2P(\eta)$, hence $g_i = a_i/\sqrt{t}$, $\eta = d/t^2$. So, if we act on $\mathfrak{g}$ by a symmetry of the form $X_i(\alpha_i \sqrt{t}) + P(\delta/t)$, we can reduce C to X_0, so that

$$\mathfrak{g} \sim \langle S, R_{12} + X_3(f_3) + P(\theta), X_0 \rangle.$$

It is obvious that $[S, X_3(f_3) + P(\theta)] = 0$ and $X_0, X_3(f_3) + P(\theta)] = 0$, whence $f_3 = 0$, $\theta = 0$.

Recapitulating, we have obtained the following list of the three-dimensional subalgebras:

(1) $\langle X_i(f_i) + P(\theta), X_i(g_i) + P(\eta), P(\ddot{f}_i g_i - \ddot{g}_i f_i) \rangle$.

(2) $\langle X_i(f_i) + P(\theta), X_i(g_i) + P(\eta), X_i(h_i) + P(\xi) \rangle$,

$$\ddot{f}_i g_i = \ddot{g}_i f_i, \quad \ddot{f}_i h_i = \ddot{h}_i f_i, \quad \ddot{g}_i h_i = \ddot{h}_i g_i.$$

(3) $\langle S + \beta R_{12}, X_1(a_1 t^{(1+\lambda)/2} \cos((\beta/2)\ln t)) + X_2(a_1 t^{(1+\lambda)/2} \sin((\beta/2)\ln t))$
$+ X_3(c_1 t^{(1+\lambda)/2}) + P(d_1 t^{(\lambda-2)/2}), X_1(a_2 t^{(1+\mu)/2} \cos((\beta/2)\ln t + \alpha))$
$+ X_2(a_2 t^{(1+\mu)/2} \sin((\beta/2)\ln t + \alpha)) + X_3(c_2 t^{(1+\mu)/2}) + P(d_2 t^{(\mu-2)/2}) \rangle$,

$$\beta \geqslant 0, \ (\lambda + \mu)((\lambda - \mu)(a_1 a_2 \cos\alpha + c_1 c_2) + 2a_1 a_2 \beta \sin\alpha) = 0.$$

(4) $\langle X_0, X_1(a_1 e^{\lambda t}) + X_3(c_1 e^{\lambda t}) + P(d_1 e^{\lambda t})$,
$X_1(a_2 e^{\mu t}) + X_2(b_2 e^{\mu t}) + X_3(c_2 e^{\mu t}) + P(d_2 e^{\mu t}) \rangle$,

$$(\lambda^2 - \mu^2)(a_1 a_2 + c_1 c_2) = 0.$$

(5) $\langle X_0 + R_{12}\,,\ X_1(a_1e^{\lambda t}\cos t) + X_2(a_1e^{\lambda t}\sin t) + X_3(c_1e^{\lambda t}) + P(d_1e^{\lambda t})\,,$
$X_1(a_2e^{\mu t}\cos(t+\alpha)) + X_2(a_2e^{\mu t}\sin(t+\alpha)) + X_3(c_2e^{\mu t}) + P(d_2e^{\mu t})\rangle\,,$
$(\lambda+\mu)((\lambda-\mu)(a_1a_2\cos\alpha + c_1c_2) + 2a_1a_2\sin\alpha) = 0\,.$

(6) $\langle R_{12} + X_3(f) + P(\theta)\,,\ X_3(g) + P(\eta)\,,\ X_3(h) + P(\xi)\rangle\,,$
$\ddot{f}g = \ddot{g}f\,,\ \ \ddot{f}h = \ddot{h}f\,,\ \ \ddot{g}h = \ddot{h}g\,.$

(7) $\langle S + \beta R_{12}\,,\ X_1(a_1t^{(1+\lambda)/2}\cos((\beta/2)\ln t))$
$+ X_2(a_1t^{(1+\lambda)/2}\sin((\beta/2)\ln t)) + X_3(c_1t^{(1+\lambda)/2}) + P(d_1t^{(\lambda-2)/2})\,,$
$X_1(t^{(1+\lambda)/2}(a_2\cos((\beta/2)\ln t+\alpha) + (a_1/2)\ln t\cos((\beta/2)\ln t+\alpha)))$
$+ X_2(t^{(1+\lambda)/2}(a_2\sin((\beta/2)\ln t+\alpha) + (a_1/2)\ln t\sin((\beta/2)\ln t+\alpha)))$
$+ X_3(c_2t^{(1+\lambda)/2} + (c_1/2)t^{(1+\lambda)/2}\ln t)$
$+ P(d_2t^{(\lambda-2)/2} + (d_1/2)t^{(\lambda-2)/2}\ln t)\rangle\,,$
$\beta \geqslant 0\,,\ \lambda(2a_1a_2\beta\sin\alpha - a_1^2 - c_1^2) = 0\,.$

(8) $\langle X_0\,,\ X_1(a_1) + X_3(c_1) + P(d_1)\,,$
$X_1(a_2 + a_1t) + X_2(b_2) + X_3(c_2 + c_1t) + P(d_2 + d_1t)\rangle\,.$

(9) $\langle X_0\,,\ P(e^{\lambda t})\,,\ X_1(a_2e^{\lambda t}) + X_2(b_2e^{\lambda t}) + X_3(c_2e^{\lambda t}) + P(d_2e^{\lambda t} + te^{\lambda t})\rangle\,,$
$\lambda \neq 0\,,$

(10) $\langle X_0 + R_{12}\,,\ X_1(a_1e^{\lambda t}\cos t) + X_2(a_2e^{\lambda t}\sin t) + X_3(c_1e^{\lambda t}) + P(d_1e^{\lambda t})\,,$
$X_1(e^{\lambda t}(a_2\cos(t+\alpha) + a_1t\cos t)) + X_2(e^{\lambda t}(a_2\sin(t+\alpha) + a_1t\sin t))$
$+ X_3(c_2e^{\lambda t} + c_1te^{\lambda t}) + P(d_2e^{\lambda t} + d_1te^{\lambda t})\rangle\,,$
$\lambda(2a_1a_2\sin\alpha - a_1^2 - c_1^2) = 0\,.$

(11) $\langle R_{12} + X_3(f) + P(\theta)\,,\ P(\ddot{f}h - \ddot{h}f)\,,\ X_3(h) + P(\xi)\rangle\,.$

(12) $\langle S + \beta R_{12}\,,\ X_1(a_1t^{(1+\lambda)/2}\cos((\beta/2)\ln t)\cos((\mu/2)\ln t)$
$- a_2t^{(1+\lambda)/2}\cos((\beta/2)\ln t+\alpha)\sin((\mu/2)\ln t))$
$+ X_2(a_1t^{(1+\lambda)/2}\sin((\beta/2)\ln t)\cos((\mu/2)\ln t)$
$- a_2t^{(1+\lambda)/2}\sin((\beta/2)\ln t+\alpha)\sin((\mu/2)\ln t))$
$+ X_3(ct^{(1+\lambda)/2}\cos((\mu/2)\ln t+\gamma))$
$+ P(dt^{(\lambda-2)/2}\cos((\mu/2)\ln t+\delta))\,,$
$X_1(a_1t^{(1+\lambda)/2}\cos((\beta/2)\ln t)\sin((\mu/2)\ln t)$
$+ a_2t^{(1+\lambda)/2}\cos((\beta/2)\ln t+\alpha)\cos((\mu/2)\ln t))$
$+ X_2(a_1t^{(1+\lambda)/2}\sin((\beta/2)\ln t)\sin((\mu/2)\ln t)$
$+ a_2t^{(1+\lambda)/2}\sin((\beta/2)\ln t+\alpha)\cos((\mu/2)\ln t))$
$+ X_3(ct^{(1+\lambda)/2}\sin((\mu/2)\ln t+\gamma)) + P(dt^{(\lambda-2)/2}\sin((\mu/2)\ln t+\delta))\rangle\,,$

$$\beta \geqslant 0\,,\;\; \mu > 0\,,\;\; \lambda(\mu(a_1^2 + a_2^2 + c^2) + 2\beta a_1 a_2 \sin\alpha) = 0\,.$$

(13) $\langle X_0 + \beta R_{12}\,,\, X_1(a_1 e^{\lambda t}\cos(\beta t)\cos(\mu t) - a_2 e^{\lambda t}\cos(\beta t + \alpha)\sin(\mu t))$
$+ X_2(a_1 e^{\lambda t}\sin(\beta t)\cos(\mu t) - a_2 e^{\lambda t}\sin(\beta t + \alpha)\sin(\mu t))$
$+ X_3(c e^{\lambda t}\cos(\mu t + \gamma)) + P(d e^{\lambda t}\cos(\mu t + \delta))\,,$
$X_1(a_1 e^{\lambda t}\cos(\beta t)\sin(\mu t) + a_2 e^{\lambda t}\cos(\beta t + \alpha)\cos(\mu t))$
$+ X_2(a_1 e^{\lambda t}\sin(\beta t)\sin(\mu t) + a_2 e^{\lambda t}\sin(\beta t + \alpha)\cos(\mu t))$
$+ X_3(c e^{\lambda t}\sin(\mu t + \gamma)) + P(d e^{\lambda t}\sin(\mu t + \delta))\rangle\,,$
$\beta = 0 \text{ or } 1\,,\;\; \mu > 0\,,\;\; \lambda(\mu(a_1^2 + a_2^2 + c^2) + 2\beta a_1 a_2 \sin\alpha) = 0\,.$

(14) $\langle R_{12} + X_3(f) + P(\theta)\,,\, X_1(g_1) + X_2(g_2)\,,\, X_2(g_1) - X_1(g_2)\rangle\,,$
$\ddot{g}_1 g = \ddot{g}_2 g_1\,.$

(15) $\langle S\,,\, R_{12} + P(d_1/t)\,,\, X_3(\sqrt{t}) + P(d_1/t)\rangle\,.$

(16) $\langle S\,,\, R_{12} + P(d_1/t)\,,\, X_3(t^\lambda)\rangle\,,\qquad \lambda \neq 1/2\,.$

(17) $\langle S\,,\, R_{12} + X_3(c_1\sqrt{t}) + P(d_1/t)\,,\, P(t^\lambda)\rangle\,.$

(18) $\langle X_0\,,\, R_{12} + P(d_1)\,,\, X_3(1) + P(d_2)\rangle\,.$

(19) $\langle X_0\,,\, R_{12} + P(d_1)\,,\, X_3(e^{\lambda t})\rangle\,,\qquad \lambda \neq 0\,.$

(20) $\langle X_0\,,\, R_{12} + X_3(c_1) + P(d_1)\,,\, P(e^{\lambda t})\rangle\,.$

(21) $\langle S\,,\, R_{12} + X_3(c_1\sqrt{t} + (c_2/2)\sqrt{t}\ln t)$
$+ P(d_1/t + (d_2/2t)\ln t)\,,\, X_3(c_2\sqrt{t}) + P(d_2/t)\rangle\,.$

(22) $\langle X_0\,,\, R_{12} + X_3(c_1 + c_2 t) + P(d_1 + d_2 t)\,,\, X_3(c_2) + P(d_2)\rangle\,.$

(23) $\langle S + \beta R_{12}\,,\, X_0\,,\, X_3(1)\rangle\,,\qquad \beta \geqslant 0\,.$

(24) $\langle S + \beta R_{12}\,,\, X_0\,,\, P(1)\rangle\,,\qquad \beta \geqslant 0\,.$

(25) $\langle R_{12}\,,\, R_{13}\,,\, R_{23}\rangle\,.$

(26) $\langle S\,,\, R_{12}\,,\, X_0\rangle\,.$

References

1. W. P. Jones and B. E. Launder, *The prediction of laminarization with a two-equation model of turbulence*, Internat. J. Heat Mass Transfer **15** (1972), no. 2, 301–314.
2. W. Kollmann (ed.), *Prediction methods for turbulent flows*, Hemisphere, Washington, 1980.
3. V. N. Gusyatnikova and V. A. Yumaguzhin, *Symmetries and conservation laws of Navier–Stokes equations*, Acta Appl. Math. **15** (1989), 65–81.
4. V. O. Bytev, *Invariant Solutions of the Navier–Stokes equations*, Ph.D. Thesis (1972), Novosibirsk State University, Novosibirsk. (Russian)
5. S. P. Lloyd, *The infinitesimal group of the Navier–Stokes equations*, Acta Mech. **38** (1981), 85–98.
6. L. F. Barannik and W. I. Fushchich, *Continuous subgroups of the generalized Schrödinger groups*, J. Math. Phys. **30** (1989), no. 2, 280–290.
7. I. S. Krasil′shchik, V. V. Lychagin, and A. M. Vinogradov, *Geometry of jet spaces and nonlinear partial differential equations*, Gordon and Breach, New York, 1986.
8. A. M. Vinogradov, *C-spectral sequence, Lagrangian formalism, and conservation laws*, J. Math. Anal. Appl. **100** (1984), 2–129.
9. P. J. Olver, *Applications of Lie groups to differential equations*, Academic Press, New York, 1986.

10. L. V. Kapitanskiĭ, *Group-theoretical analysis of the Navier–Stokes equations in the rotationally symmetric case and some new exact solutions*, J. Soviet Math. **21** (1980), 314–327.
11. L. V. Ovsyannikov, *Group analysis of differential equations*, "Nauka", Moscow, 1978; English transl., Academic Press, New York, 1982.

Translated by THE AUTHORS

MOSCOW STATE TECHNICAL UNIVERSITY AND THE SOPHUS LIE CENTER, MOSCOW, RUSSIA. CORRESPONDENCE TO: N. G. KHOR′KOVA, VOLOKOLAMSKOE SH. 16B-2-45, 125080 MOSCOW, RUSSIA

THE SOPHUS LIE CENTER, MOSCOW, RUSSIA. CORRESPONDENCE TO: A. M. VERBOVETSKY, PROFSOYUZNAYA 98-9-132, 117485 MOSCOW, RUSSIA

Amer. Math. Soc. Transl.
(2) Vol. **167**, 1995

Graded Frölicher–Nijenhuis Brackets and the Theory of Recursion Operators for Super Differential Equations

P. H. M. KERSTEN AND I. S. KRASIL'SHCHIK

ABSTRACT. The theory of the Frölicher–Nijenhuis bracket for n-graded commutative algebras is developed. On the basis of this theory, deformations and recursion operators are defined, both in local and nonlocal cases. New recursion operators are computed for super versions of KdV, NLS, and Boussinesq equations. These operators are shown to generate new series of symmetries for these equations.

Introduction

Let $\mathcal{E}$ be a differential equation and $\mathcal{E}^\infty$ be its infinite prolongation viewed at as a submanifold of the manifold $J^\infty(\pi)$ of infinite jets, $\pi\colon E \to M$ being a vector bundle [8]. In [7], a cohomology theory $H_{\mathcal{C}}^*(\mathcal{E})$ was constructed for any formally integrable $\mathcal{E}$. The basic notions for this theory are the Cartan connection $\mathcal{C}$ in $\mathcal{E}^\infty$ and the Frölicher–Nijenhuis bracket $[\![\ ,\ ]\!]$ in the tensors of the form $\Lambda^*(\mathcal{E}^{(\infty)}) \otimes D(\mathcal{E}^{(\infty)})$. In particular, elements of $H_{\mathcal{C}}^1(\mathcal{E})$ coincide with nontrivial infinitesimal deformations of the equation structure on $\mathcal{E}$, and in [9] it was shown that deformations of a certain special type are naturally identified with recursion operators for $\mathcal{E}$.

Here we generalize the theory of [6, 9] to the case of n-graded equations (super differential equations in particular). Just as in the classical case, equations defining recursion operators are, as a rule, solvable only in a nonlocal setting. This poses the problem of reconstructing real recursion operators from their so-called shadows (see [9, 10]). We show here that this reconstruction can be always fulfilled only in the universe of all graded equations (by infinitely applying the functor K that kills the group of horizontal cohomology $H_h^1(\mathcal{E})$ and the linearization functor T).

Finally, we apply the techniques of computing recursion operators for super versions of KdV [12] and for two super versions of NLS [15] equations. We

---

1991 *Mathematics Subject Classification*. (1991): 58F07, 58G07, 58H10, 58H15, 58G37, 58A50, 35Q53, 35Q55, 35Q58, 58Q35, 16W55.

Key words and phrases. Super differential equations, graded Frölicher–Nijenhuis brackets, cohomology, recursion operators, coverings, integrable systems, symmetries, super KdV equation, super NLS equation, super Boussinesq equation.

also derive a super version of the Boussinesq equation and in all cases obtain corresponding recursion operators.

§1. Graded Frölicher–Nijenhuis and Richardson–Nijenhuis brackets

Here we redefine the Frölicher–Nijenhuis bracket for the case of n-graded algebras. All definitions below are obvious generalizations of those from [6]. Proofs also follow the same lines and are omitted.

1.1. Graded polyderivations and forms. Let R be a commutative ring with unit $1 \in R$ and $\mathcal{A}$ be a commutative n-graded unitary algebra over R, i.e.

$$\mathcal{A} = \sum_{i \in \mathbb{Z}^n} \mathcal{A}_i, \qquad \mathcal{A}_i \mathcal{A}_j \subset \mathcal{A}_{i+j},$$

and

$$ab = (-1)^{a \cdot b} ba$$

for any homogeneous elements $a, b \in \mathcal{A}$. Here and below the notation $(-1)^{a \cdot b}$ means $(-1)^{i_1 j_1 + \cdots + i_n j_n}$, where $i = (i_1, \ldots, i_n)$, $j = (j_1, \ldots, j_n) \in \mathbb{Z}^n$ are the gradings of the elements a and b respectively. We also use the notation $a \cdot b$ for the scalar product of the gradings of the elements a and b. In what follows one can consider $\mathbb{Z}_2^n$-graded objects as well. We consider the category of n-graded (left)[1] $\mathcal{A}$-modules $\mathcal{M} = \mathcal{M}(\mathcal{A})$ and introduce the functors

$$D_i \colon \mathcal{M}(\mathcal{A}) \Rightarrow \mathcal{M}(\mathcal{A})$$

as follows (cf. [7, 9]): $D_0(P) = P$ for any $P \in \mathrm{Ob}(\mathcal{M})$, $P = \sum_{i \in \mathbb{Z}^n} P_i$, and

$$D_{1,j}(P) = \{\Delta \in \hom_R(A, P) \mid \Delta(\mathcal{A}_i) \subset P_{i+j},$$
$$\Delta(ab) = \Delta(a)b + (-1)^{\Delta \cdot a} a\Delta(b)\},$$

where $j = \mathrm{gr}(\Delta) = (j) = (j_1, \ldots, j_n) \in \mathbb{Z}^n$ is the grading of Δ; we set

$$D_1(P) = \sum_{j \in \mathbb{Z}^n} D_{1,j}(P).$$

Further, if $D_0, \ldots, D_s$ are defined, we set

$$D_{s+1,j}(P) = \{\Delta \in \hom_R(\mathcal{A}, D_s(P)) \mid \Delta(\mathcal{A}_i) \subset D_{s,i+j}(P),$$
$$\Delta(ab) = \Delta(a)b + (-1)^{\Delta \cdot a} a\Delta(b), \Delta(a, b) + (-1)^{a \cdot b}\Delta(b, a) = 0\},$$
$$D_{s+1}(P) = \sum_{j \in \mathbb{Z}^n} D_{s+1,j}(P).$$

Elements of $D_s(P)$ are called *graded P-valued s-derivations* of $\mathcal{A}$ and elements of $D_*(P) = \sum_{s \geqslant 0} D_s(P)$ are called *graded P-valued polyderivations* of $\mathcal{A}$.

[1] We also consider objects of $\mathcal{M}(\mathcal{A})$ as right $\mathcal{A}$-modules by setting $pa = (-1)^{a \cdot p} ap$ for any homogeneous $a \in \mathcal{A}$, $p \in P$.

PROPOSITION 1.1. *The functors* D_s, $s = 0, 1, 2, \dots$, *are representable in* $\mathcal{M}(\mathcal{A})$, *i.e., there exist n-graded modules* $\Lambda^0, \Lambda^1, \dots, \Lambda^s, \dots$, *such that*

$$D_s(P) = \hom_{\mathcal{A}}(\Lambda^s, P)$$

for all $P \in \mathrm{Ob}(\mathcal{M})$.

Our local target is the construction of *graded calculus* in the limits needed for what follows. By *calculus* we mean the set of basic operations related to the functors D_s and to the modules Λ^s together with the most important equalities connecting these operations. To simplify our considerations we shall suppose that

(i) $\mathcal{A}_{\mathbf{0}} = C^\infty(M)$ for some smooth manifold M, where $\mathbf{0} = (0, \dots, 0)$;
(ii) all homogeneous components P_i of the modules under consideration are projective $\mathcal{A}_{\mathbf{0}}$-modules of finite type or are filtered by such modules.

1.2. Wedge products.

PROPOSITION 1.2. (i) *There exists a derivation* $d\colon \mathcal{A} \to \Lambda^1$ *of the grading* $\mathbf{0}$ *such that for any* $\mathcal{A}$*-module* P *and any graded derivation* $\Delta\colon \mathcal{A} \to P$ *there exists a uniquely defined morphism* $f_\Delta\colon \Lambda^1 \to P$ *such that* $f_\Delta \circ d = \Delta$.

(ii) *The module* Λ^1 *is generated over* $\mathcal{A}$ *by the elements* $da = d(a)$, $a \in \mathcal{A}$, *with the relations*

$$d(ab) = (da)b + a\,db, \qquad a, b \in \mathcal{A}.$$

The j*th homogeneous component of* Λ^1 *is of the form*

$$\Lambda^1_j = \Big\{\sum a\,db \mid a, b \in A,\ \mathrm{gr}(a) + \mathrm{gr}(b) = j\Big\}.$$

(iii) *The modules* Λ^s *are generated over* $\mathcal{A}$ *by the elements of the form*

$$\omega_1 \wedge \dots \wedge \omega_s, \qquad \omega_1, \dots, \omega_s \in \Lambda^1,$$

with the relations

$$\omega \wedge \theta + (-1)^{\omega\cdot\theta}\theta \wedge \omega = 0, \quad \omega \wedge a\theta = \omega a \wedge \theta, \qquad \omega, \theta \in \Lambda^1,\ a \in \mathcal{A}.$$

The j*th homogeneous component of* Λ^s *is of the form*

$$\Lambda^s_j = \Big\{\sum \omega_1 \wedge \dots \wedge \omega_s \mid \omega_i \in \Lambda^1,\ \mathrm{gr}(\omega_1) + \dots + \mathrm{gr}(\omega_s) = j\Big\}.$$

(iv) *Let* $\omega \in \Lambda^s_j$, $j = (j_1, \dots, j_n)$. *Set* $\mathrm{gr}^1(\omega) = (j_1, \dots, j_n, s)$. *Then*

$$\Lambda^* = \sum_{s \geqslant 0} \Lambda^s = \sum_{s \geqslant 0} \sum_{j \in \mathbb{Z}^n} \Lambda^s_j$$

is an $(n+1)$*-graded commutative algebra with respect to the wedge product*

$$\omega \wedge \theta = \omega \wedge \dots \wedge \omega_s \wedge \theta_1 \wedge \dots \wedge \theta_r, \qquad \omega \in \Lambda^s,\ \theta \in \Lambda^r,\ \omega_\alpha, \theta_\beta \in \Lambda^1,$$

i.e.,

$$\omega \wedge \theta = (-1)^{\omega \cdot \theta + sr} \theta \wedge \omega ,$$

where $\omega \cdot \theta$ *in the power of* (-1) *denotes scalar product of gradings inherited by* ω *and* θ *from* $\mathcal{A}$.

REMARK. In fact, the definition of differential forms given above is not sufficient for geometrical considerations because, in this setting, for example, the form $d(\sin x) - \cos x \, dx$ is not trivial in $\Lambda^1 C^\infty(R)$. To tackle this problem one must restrict oneself to so-called geometrical modules, i.e., the modules P for which

$$\bigcap_{\mu \in \mathrm{Specm}} \mu P = 0 ,$$

where Specm is the space of maximal ideals of $C^\infty(M)$, see [8] for details.

A similar wedge product can be defined in $D_*(\mathcal{A})$. Namely for $a, b \in D_0(\mathcal{A}) = \mathcal{A}$ we set $a \wedge b = ab$ and then by induction define

$$(\Delta \wedge \nabla)(a) = \Delta \wedge \nabla(a) + (-1)^{\nabla \cdot a + r} \Delta(a) \wedge \nabla , \tag{1.1}$$

where $a \in \mathcal{A}$, $\Delta \in D_s(\mathcal{A})$, $\nabla \in D_r(\mathcal{A})$ and ∇ in the power of (-1) denotes the grading of ∇ inherited from $\mathcal{A}$.

PROPOSITION 1.3. (i) *Definition* (1.1) *determines a map*

$$\wedge \colon D_s(\mathcal{A}) \otimes_{\mathcal{A}} D_r(\mathcal{A}) \to D_{s+r}(\mathcal{A})$$

which is in agreement with the graded structure of polyderivations:

$$D_{s,i}(\mathcal{A}) \wedge D_{r,j}(\mathcal{A}) \subset D_{s+r,i+j}(\mathcal{A}) .$$

(ii) $D_*(\mathcal{A}) = \sum_{s \geqslant 0} \sum_{j \in \mathbb{Z}^n} D_{s,j}$ *is an* $(n+1)$*-graded commutative algebra with respect to the wedge product*:

$$\Delta \wedge \nabla = (-1)^{\Delta \cdot \nabla + rs} \nabla \wedge \Delta$$

for any $\Delta \in D_s(\mathcal{A})$, $\nabla \in D_r(\mathcal{A})$.[2]

(iii) $D_*(\mathcal{A})$ *is generated by* $D_0(\mathcal{A}) = \mathcal{A}$ *and* $D_1(\mathcal{A})$, *i.e., any* $\Delta \in D_s(\mathcal{A})$ *is a sum of the elements of the form*

$$a\Delta_1 \wedge \cdots \wedge \Delta_s , \qquad \Delta_i \in D_1(\mathcal{A}) , \ a \in \mathcal{A} .$$

REMARK. One can define a wedge product $\wedge \colon D_i(\mathcal{A}) \otimes_{\mathcal{A}} D_j(P) \to D_{i+j}(P)$ with respect to which $D_*(P)$ acquires the structure of an $(n+1)$-graded $D_*(\mathcal{A})$-module (see [7]), but it will not be needed below.

---

[2]This distinction between the first n gradings and the additional $(n+1)$st one will be preserved both for graded forms and graded polyderivations throughout the paper.

1.3. Contractions and graded Richardson–Nijenhuis bracket. We define the contraction of a polyderivation $\Delta \in D_s(\mathcal{A})$ into a form $\omega \in \Lambda^r$ in the following way

$$\begin{aligned} i_\Delta\omega &\equiv \Delta \lrcorner\, \omega = 0\,, && \text{if } s > r\,,\\ i_\Delta\omega &= \Delta(\omega)\,, && \text{if } s = r\,, \text{ due to the definition of } \Lambda^r\,,\\ i_a\omega &= a\omega\,, && \text{if } a \in \mathcal{A} = D_0(\mathcal{A})\,, \end{aligned}$$

and for $r > s$ we set by induction

$$i_\Delta(da \wedge \omega) = i_{\Delta(a)}(\omega) + (-1)^{\Delta\cdot a+s} da \wedge i_\Delta(\omega)\,. \tag{1.2}$$

PROPOSITION 1.4. (i) *For any $\Delta \in D_s(\mathcal{A})$ definition* (1.2) *determines an $(n+1)$-graded differential operator $i_\Delta\colon \Lambda^* \to \Lambda^*$ of order s.*

(ii) *In particular, if $\Delta \in D_1(\mathcal{A})$, then i_Δ is a graded derivation of Λ^*:*

$$i_\Delta(\omega \wedge \theta) = i_\Delta(\omega) \wedge \theta + (-1)^{\Delta\cdot\omega+r}\omega \wedge i_\Delta\theta\,, \qquad \omega \in \Lambda^r\,,\ \theta \in \Lambda^*.$$

Now we consider tensor products of the form $\Lambda^r \otimes_{\mathcal{A}} D_s(\mathcal{A})$ and generalize contraction and wedge product operations as follows

$$(\omega \otimes \Delta) \wedge (\theta \otimes \Delta) = (-1)^{\Delta\cdot\theta}(\omega \wedge \theta) \otimes (\Delta \wedge \nabla)\,,$$

$$i_{\omega\otimes\Delta}(\theta \otimes \nabla) = \omega \wedge i_\Delta(\theta) \otimes \nabla\,,$$

where $\omega\,, \theta \in \Lambda^*\,,\ \Delta\,, \nabla \in D_*(\mathcal{A})$. Define the Richardson–Nijenhuis bracket in $\Lambda^* \otimes D_s(\mathcal{A})$ by setting

$$[\![\Omega\,, \Theta]\!]^R_s = i_\Omega(\Theta) - (-1)^{(\omega+\Delta)\cdot(\Theta+\nabla)+(q-s)(r-s)} i_\Theta(\Omega)\,, \tag{1.3}$$

where $\Omega = \omega \otimes \Delta \in \Lambda^r \otimes D_s(\mathcal{A})\,,\ \Theta = \theta \otimes \nabla \in \Lambda^q \otimes D_s(\mathcal{A})$. In what follows we confine ourselves to the case $s = 1$ and introduce an $(n+1)$-graded structure into $\Lambda^* \otimes D_1(\mathcal{A})$ by setting

$$\mathrm{gr}(\omega \otimes X) = (\mathrm{gr}(\omega) + \mathrm{gr}(X)\,, r)\,, \tag{1.4}$$

where $\mathrm{gr}(\omega)$ and $\mathrm{gr}(X)$ are initial n-gradings of the elements $\omega \in \Lambda^r\,,\ X \in D_1(\mathcal{A})$. We also denote respectively by Ω and Ω_1 the first n and $(n+1)$st gradings of Ω in the powers of (-1).

PROPOSITION 1.5. (i) *For any two elements $\Omega\,, \Theta \in \Lambda^* \otimes D_1(\mathcal{A})$ one has*

$$[i_\Omega\,, i_\Theta] = i_{[\![\Omega\,,\Theta]\!]^R_1}.$$

Hence, the Richardson–Nijenhuis bracket $[\![\ ,\ ]\!]^R = [\![\ ,\ ]\!]^R_1$ determines in $\Lambda^* \otimes D_1(\mathcal{A})$ the structure of an $(n+1)$-graded Lie algebra with respect to the grading in which the $(n+1)$st component is shifted by 1 in (1.4), *i.e.,*

$$[\![\Omega\,, \Theta]\!]^R + (-1)^{\Omega\cdot\Theta+(\Omega_1+1)(\Theta_1+1)}[\![\Theta\,, \Omega]\!]^R = 0\,, \tag{ii}$$

$$\oint (-1)^{\Theta\cdot(\Omega+\Xi)+(\Theta_1+1)(\Omega_1+\Xi_1)}[\![[\![\Omega\,, \Theta]\!]^R\,, \Xi]\!]^R = 0\,, \tag{iii}$$

where $\oint$ denotes the sum of cyclic permutations.

(iv) *Moreover, if $\rho \in \Lambda^*$, then*

$$[\![\Omega\,, \rho \wedge \Theta]\!]^R = (\Omega \lrcorner\, \rho) \wedge \Theta + (-1)^{\Omega\cdot\rho+(\Omega_1+1)\rho_1} \wedge [\![\Omega\,, \Theta]\!]^R.$$

(v) *Finally, the composition of two contractions is expressed by*

$$i_\Omega \circ i_\Theta = i_{\Omega \lrcorner\, \Theta} + (-1)^{\Theta_1} i_{\Omega\wedge\Theta}\,.$$

1.4. De Rham complex and Lie derivatives. The de Rham differential $d\colon \Lambda^r \to \Lambda^{r+1}$ is defined as follows. For $r = 0$ it coincides with the derivation $d\colon \mathcal{A} \to \Lambda^1$ introduced in Proposition 1.2. For any $a\,db \in \Lambda^1$, $a, b \in \mathcal{A}$, we set

$$d\,(a\,db) = da \wedge db$$

and for a decomposable form $\omega = \theta \wedge \rho \in \Lambda^r$, $\theta \in \Lambda^{r'}$, $\rho \in \Lambda^{r''}$, $r > 1$, $r', r'' < r$, set

$$d\omega = d\theta \wedge \rho + (-1)^{\theta_1}\theta \wedge d\rho .$$

By definition, $d\colon \Lambda^* \to \Lambda^*$ is a derivative of the grading $(\mathbf{0}, 1)$ and, obviously, $d \circ d = 0$. Thus, one gets a complex

$$0 \longrightarrow \mathcal{A} \xrightarrow{d} \Lambda^1 \longrightarrow \cdots \longrightarrow \Lambda^r \xrightarrow{d} \Lambda^{r+1} \longrightarrow \ldots ,$$

which is called the *de Rham complex* of $\mathcal{A}$.

Let $X \in D_1(A)$ be a derivation. A *Lie derivation* $L_X\colon \Lambda^* \to \Lambda^*$ is defined as

(1.5) $$L_X = [i_X, d] = i_X \circ d + d \circ i_X .$$

Thus for any $\omega \in \Lambda^*$ one has

$$L_X\omega = X \lrcorner\, d\omega + d\,(X \lrcorner\, \omega).$$

The basic properties of L_X are described by

PROPOSITION 1.6. (i) *If* $\omega, \theta \in \Lambda^*$, *then*

$$L_X(\omega \wedge \theta) = L_X\omega \wedge \theta + (-1)^{X\cdot\omega}\omega \wedge L_X\theta ,$$

i.e., L_X *is a derivation of the grading* $(\mathrm{gr}\,(X), 0)$.

(ii) $$[L_X, d] = L_X \circ d - d \circ L_X = 0 .$$

(iii) *For any* $a \in \mathcal{A}$ *and* $\omega \in \Lambda^*$ *one has* $L_{aX}(\omega) = aL_X\omega + da \wedge i_X(\omega)$.

(iv) $$[L_X, i_Y] = [i_X, L_Y] = i_{[X, Y]} .$$

(v) $$[L_X, L_Y] = L_{[X, Y]} .$$

Now we extend the classical definition of Lie derivative to the elements of $\Lambda^* \otimes D_1(\mathcal{A})$ and for any $\Omega \in \Lambda^* \otimes D_1(\mathcal{A})$ define

$$L_\Omega = [i_\Omega, d] = i_\Omega \circ d + (-1)^{\Omega_1} d \circ i_\Omega .$$

If $\Omega = \omega \otimes X$, then

$$L_{\omega\otimes X} = \omega \wedge L_X + (-1)^{\omega_1} d\omega \wedge i_X .$$

PROPOSITION 1.7. (i) *For any* $\Omega \in \Lambda^* \otimes D_1(\mathcal{A})$,

$$L_\Omega(\rho \wedge \theta) = L_\Omega(\rho) \wedge \theta + (-1)^{\Omega \cdot \rho + \Omega_1 \rho_1} \rho \wedge L_\Omega \theta, \qquad \rho, \theta \in \Lambda^*,$$

i.e., L_Ω *is a derivation of* Λ^* *whose grading coincides with that of* Ω.

(ii) $$[L_\Omega, d] = L_\Omega \circ d - (-1)^{\Omega_1} d \circ L_\Omega = 0.$$

(iii) $$L_{\rho \wedge \Omega} = \rho \wedge L_\Omega + (-1)^{\rho_1 + \Omega_1} d\rho \wedge i_\Omega, \qquad \rho \in \Lambda^*.$$

To formulate properties of L_Ω similar to (iv)–(v) of Proposition 1.5, a new notion is needed.

1.5. Graded Frölicher–Nijenhuis bracket and associated complexes.

PROPOSITION 1.8. (i) *For any two elements* $\Omega, \Theta \in \Lambda^* \otimes D_1(\mathcal{A})$, *the commutator of the corresponding Lie derivatives* $[L_\Omega, L_\Theta]$ *is of the form* L_Ξ *for some* $\Xi \in \Lambda^* \otimes D_1(\mathcal{A})$.

(ii) *The correspondence* $L\colon \Lambda^* \otimes D_1(\mathcal{A}) \to D_1(\Lambda^*)$, $\Omega \mapsto L_\Omega$, *is injective and hence* Ξ *in* (i) *is defined uniquely. It is called the* (*graded*) *Frölicher–Nijenhuis bracket of the elements* Ω, Θ *and is denoted by* $\Xi = [\![\Omega, \Theta]\!]$. *Thus, by definition,*

$$[L_\Omega, L_\Theta] = L_{[\![\Omega, \Theta]\!]}.$$

(iii) *If* Ω *and* Θ *are of the form*

$$\Omega = \omega \otimes X, \quad \Theta = \theta \otimes Y, \qquad \omega, \theta \in \Lambda^*, \ X, Y \in D_1(\mathcal{A}),$$

then

$$\begin{aligned} [\![\Omega, \Theta]\!] &= (-1)^{X \cdot \theta} \omega \wedge \theta \otimes [X, Y] + \omega \wedge L_X \theta \otimes Y \\ &\quad + (-1)^{\Omega_1} d\omega \wedge (X \lrcorner \theta) \otimes Y - (-1)^{\Omega \cdot \Theta + \Omega_1 \Theta_1} \theta \wedge L_Y \omega \otimes X \\ &\quad - (-1)^{\Omega \cdot \Theta + (\Omega_1 + 1)\Theta_1} d\theta \wedge (Y \lrcorner \omega) \otimes X \\ &= (-1)^{X \cdot \theta} \omega \wedge \theta \otimes [X, Y] + L_\Omega(\theta) \otimes Y \\ &\quad - (-1)^{\Omega \cdot \Theta + \Omega_1 \Theta_1} L_\Theta(\omega) \otimes X. \end{aligned} \tag{1.6}$$

(iv) *If* $\Omega = X$, $\Theta = Y \in D_1(\mathcal{A}) = \Lambda^0 \otimes D_1(\mathcal{A})$, *then the graded Frölicher–Nijenhuis bracket of* Ω *and* Θ *coincides with the graded commutator of vector fields*:

$$[\![X, Y]\!] = [X, Y].$$

The main properties of the Frölicher–Nijenhuis bracket are described by

PROPOSITION 1.9. *For any* $\Omega, \Theta, \Xi \in \Lambda^* \otimes D_1(\mathcal{A})$ *and* $\rho \in \Lambda^*$,

(1.7) (i) $$[\![\Omega, \Theta]\!] + (-1)^{\Omega \cdot \Theta + \Omega_1 \Theta_1} [\![\Theta, \Omega]\!] = 0.$$

(1.8) (ii) $$\oint (-1)^{(\Omega + \Xi) \cdot \Theta + (\Omega_1 + \Xi_1)\Theta_1} [\![\Omega, [\![\Theta, \Xi]\!]]\!] = 0,$$

i.e., $[\![\;,\;]\!]$ *defines a graded Lie algebra structure on* $\Lambda^* \otimes D_1(\mathcal{A})$.

$$\text{(iii)}\quad [\![\Omega, \rho \wedge \Theta]\!] = L_\Omega(\rho) \wedge \Theta - (-1)^{\Omega\cdot(\Theta+\rho)+(\Omega_1+1)(\Theta_1+\rho_1)} d\rho \wedge i_\Theta \Omega + (-1)^{\Omega\cdot\rho+\Omega_1\rho_1} \rho \wedge [\![\Omega, \Theta]\!]. \tag{1.9}$$

$$\text{(iv)}\quad [L_\Omega, i_\Theta] + (-1)^{\Omega\cdot\Theta+\Omega_1(\Theta_1+1)} L_{\Theta \,\lrcorner\, \Omega} = i_{[\![\Omega,\Theta]\!]}. \tag{1.10}$$

$$\text{(v)}\quad i_\Xi[\![\Omega, \Theta]\!] = [\![i_\Xi\Omega, \Theta]\!] + (-1)^{\Omega\cdot\Xi+\Omega_1(\Xi_1+1)}[\![\Omega, i_\Xi\Theta]\!] + (-1)^{\Omega_1} i_{[\![\Xi,\Omega]\!]}\Theta - (-1)^{\Omega\cdot\Theta+(\Omega_1+1)\Theta_1} i_{[\![\Xi,\Theta]\!]}\Omega. \tag{1.11}$$

Now let U be an element of $\Lambda^1 \otimes D_1(\mathcal{A})$. Define

$$\partial_U = [\![U, \cdot\,]\!] \colon \Lambda^r \otimes D_1(\mathcal{A}) \to \Lambda^{r+1} \otimes D_1(\mathcal{A}). \tag{1.12}$$

Then from the definitions it follows that

$$\partial_U(U) = [\![U, U]\!] = (1 + (-1)^{U\cdot U}) L_U \circ L_U \tag{1.13}$$

and from (1.7) and (1.8) we get

$$(1 - (-1)^{U\cdot U})[\![U, U]\!] = 0,$$
$$(1 + (-1)^{U\cdot U})\partial_U(\partial_U\Omega) + (-1)^{U\cdot U}[\![\Omega, [\![U, U]\!]]\!] = 0$$

for any $\Omega \in \Lambda^* \otimes D_1(\mathcal{A})$.

We are interested in the case when (1.12) is a complex, i.e., $\partial_U \circ \partial_U = 0$, and give the following

DEFINITION. An element $U \in \Lambda^1 \otimes D_1(\mathcal{A})$ is said to be *integrable*, if
(i) $[\![U, U]\!] = 0$ and
(ii) $(-1)^{U\cdot U}$ equals to 1.

From the above it follows that for an integrable element U we have $\partial_U \circ \partial_U = 0$, and we can introduce the corresponding cohomology by

$$H_U^r(\mathcal{A}) = \frac{\ker(\partial_U \colon \Lambda^r \otimes D_1(\mathcal{A}) \to \Lambda^{r+1} \otimes D_1(\mathcal{A}))}{\operatorname{Im}(\partial_U \colon \Lambda^{r-1} \otimes D_1(\mathcal{A}) \to \Lambda^r \otimes D_1(\mathcal{A}))}.$$

The main properties of ∂_U are described by

PROPOSITION 1.10. *Let* $U \in \Lambda^1 \otimes D_1(\mathcal{A})$ *be an integrable element and* $\Omega, \Theta \in \Lambda^* \otimes D_1(\mathcal{A})$, $\rho \in \Lambda^*$. *Then*

(i) $\partial_U(\rho \wedge \Omega) = L_U(\rho) \wedge \Omega - (-1)^{U\cdot(\Omega+\rho)} d\rho \wedge i_\Omega U + (-1)^{U\cdot\rho+\rho_1} \rho \wedge \partial_U\Omega$,

(ii) $$[L_U, i_\Omega] = i_{\partial_U\Omega} + (-1)^{U\cdot\Omega+\Omega_1} L_{\Omega \,\lrcorner\, U},$$

(iii) $$[i_\Omega, \partial_U]\Theta + (-1)^{U\cdot\Theta} i_{[\![\Omega,\Theta]\!]} U = [\![i_\Omega U, \Theta]\!] + (-1)^{U\cdot\Omega+\Omega_1} i_{\partial_U\Omega}\Theta,$$

(iv) $$\partial_U[\![\Omega, \Theta]\!] = [\![\partial_U\Omega, \Theta]\!] + (-1)^{U\cdot\Omega+\Omega_1}[\![\Omega, \partial_U\Theta]\!].$$

From the last equality it follows that the Frölicher–Nijenhuis bracket is inherited by $H_U^*(\mathcal{A}) = \sum_{r\geqslant 0} H_U^r(\mathcal{A})$ and thus the latter forms an $(n+1)$-graded Lie algebra with respect to this bracket.

§2. Graded extensions, $\mathcal{C}$-cohomology, and recursion operators

In this section we adapt the cohomological theory of recursion operators [6, 9] to the case of polygraded (in particular, super) differential equations. Our first step is the appropriate definition of polygraded equations (cf. [14] and the literature cited there). In what follows we still suppose all the modules to be projective and of finite type over the main algebra $\mathcal{A}$ $(= C^\infty(M))$ or to be filtered by such modules in a natural way.

2.1. General construction. Let R be a commutative ring with a unit and $A_{-1} \subset A_0$ be two unitary associative commutative $\mathbb{Z}^n$-graded R-algebras. Let $\mathcal{D} = \mathcal{D}_0 \subset D(A_{-1}, A_0)$ be an A_0-submodule in

$$D(A_{-1}, A_0) = \{\partial \in \hom_R(A_{-1}, A_0) \mid \\ \partial(aa') = \partial a \cdot a' + (-1)^{a \cdot \partial} a \cdot \partial a', \ a, a' \in A_{-1}\}.$$

Define a $\mathbb{Z}^n$-graded A_0-algebra A_1 by the generators

$$[\partial, a], \quad a \in A_0, \ \partial \in D_0, \qquad \operatorname{gr}[\partial, a] = \operatorname{gr}(\partial) + \operatorname{gr}(a),$$

with the relations

$$\begin{aligned} [\partial, a] &= \partial a, \qquad a \in A_{-1}, \\ [\partial, a + a'] &= [\partial, a] + [\partial, a'], \\ [a'\partial' + a''\partial'', a] &= a'[\partial', a] + a''[\partial'', a], \\ [\partial, aa'] &= [\partial, a]a' + (-1)^{\partial \cdot a} a[\partial, a']. \end{aligned}$$

For any $\partial \in \mathcal{D}_0$ we can define a derivation $\partial^{(1)} \in D(A_0, A_1)$ by setting

$$\partial^{(1)}(a) = [\partial, a], \qquad a \in A_1.$$

Obviously, $\partial^{(1)} a = \partial a$ for $a \in A_0$. Denoting by $\mathcal{D}_1$ the A_1-submodule in $D(A_0, A_1)$ generated by the elements of the form $\partial^{(1)}$ one gets a triple

$$\{A_0, A_1, \mathcal{D}_1\}, \qquad A_0 \subset A_1, \ \mathcal{D}_1 \subset D(A_0, A_1),$$

which allows one to construct $\{A_1, A_2, \mathcal{D}_2\}$, etc. and to get two infinite sequences of embeddings

$$\begin{aligned} A_{-1} \to A_0 \to \cdots \to A_i \to A_{i+1} \to \cdots, \\ \mathcal{D}_0 \to \mathcal{D}_1 \to \cdots \to \mathcal{D}_i \to \mathcal{D}_{i+1} \to \cdots, \end{aligned}$$

where $A_{i+1} = (A_i)_1$, $\mathcal{D}_{i+1} = (\mathcal{D}_i)_1 \subset D(A_{i-1}, A_i)$, and $\mathcal{D}_i \to \mathcal{D}_{i+1}$ is a morphism of A_{i+1}-modules. Set

$$A_\infty = \lim_{i \to \infty} A_i, \qquad \mathcal{D}_\infty = \lim_{i \to \infty} \mathcal{D}_i.$$

Then $\mathcal{D}_\infty \subset D(A_\infty)$ and any element $\partial \in \mathcal{D}_0$ determines a derivation $\mathcal{D}(\partial) = \partial^{(\infty)} \in D(A_\infty)$. The correspondence $\mathcal{D}\colon \mathcal{D}_0 \to D(A_\infty)$ possesses the following properties

$$\mathcal{D}(X)(a) = X(a) \qquad \text{for } a \in A_{-1},$$
$$\mathcal{D}(aX) = a\mathcal{D}(X) \qquad \text{for } a \in A_0.$$

Moreover, by definition one has

$$[\mathcal{D}(X), \mathcal{D}(Y)](a) = \mathcal{D}(X)(Y(a)) - (-1)^{X\cdot Y}\mathcal{D}(Y)(X(a)),$$
$$a \in A_{-1}, \quad X, Y \in \mathcal{D}_0.$$

2.2. Connections. Let A, B be two n-graded algebras, $A \subset B$. Consider modules of derivations $D(A, B)$ and $D(B)$ and a B-linear map

$$\nabla\colon D(A, B) \to D(B).$$

The map ∇ is called a *connection* for the pair (A, B), or an (A, B)-*connection*, if

$$\nabla(X)|_A = X.$$

From the definition it follows that ∇ is of degree **0** and that for any $X, Y \in D(A, B)$ the element

$$\nabla(X) \circ Y - (-1)^{X\cdot Y}\Delta(Y) \circ X$$

again lies in $D(A, B)$. Thus one can define

$$R_\nabla(X, Y) = [\nabla(X), \nabla(Y)] - \nabla(\nabla(X) \circ Y - (-1)^{X\cdot Y}\nabla(Y) \circ X),$$

which is called the *curvature* of the connection ∇ and possesses the following properties

$$R_\nabla(X, Y) + (-1)^{X\cdot Y}R_\nabla(Y, X) = 0, \qquad X, Y \in D(A, B),$$
$$R_\nabla(aX, Y) = aR_\nabla(X, Y), \qquad a \in B,$$
$$R_\nabla(X, bY) = (-1)^{X\cdot b}bR_\nabla(X, Y), \qquad b \in B.$$

A connection ∇ is called *flat* if $R_\nabla(X, Y) = 0$ for all $X, Y \in D(A, B)$.

REMARK. Let M, N be two manifolds and $\xi\colon N \to M$ be a fiber bundle. Let $A = C^\infty(M)$, $B = C^\infty(N)$, A being embedded into B by means of ξ^*. Then the definitions above are equivalent to the ordinary definition of a connection and its curvature in the fiber bundle ξ.

2.3. Graded extensions of differential equations. Let now M be a smooth manifold and $\pi\colon E \to M$ be a smooth locally trivial fiber bundle over M. Let $\mathcal{E} \subset J^k(\pi)$ be a kth order differential equation represented as a submanifold in the manifold of k-jets for the bundle π. We consider $\mathcal{E}$ to be formally integrable and denote by $\mathcal{E}^{(\infty)} \subset J^\infty(\pi)$ its infinite prolongation.

Let $\mathcal{F}(\mathcal{E})$ be the algebra of smooth functions on $\mathcal{E}^{(\infty)}$ and $\mathcal{C}D(\mathcal{E}) \subset D(\mathcal{E}) = D(\mathcal{F}(\mathcal{E}))$ be the Lie algebra generated by total derivatives $\mathcal{C}X$, $X \in D(M)$, $\mathcal{C}\colon D(M) \to D(\mathcal{E})$ being the Cartan connection on $\mathcal{E}^{(\infty)}$ (for details see [8]).

Let $\mathcal{F}$ be an n-graded commutative algebra such that $\mathcal{F}_0 = \mathcal{F}(\mathcal{E})$. Denote by $\mathcal{C}D_0(\mathcal{E})$ the $\mathcal{F}$-submodule in $D(\mathcal{F}(\mathcal{E}), \mathcal{F})$ generated by $\mathcal{C}D(\mathcal{E})$ and consider the triple $(\mathcal{F}(\mathcal{E}), \mathcal{F}, \mathcal{C}D_0(\mathcal{E}))$ as a starting point for the construction from subsection 2.1. Then we shall get a pair $(\mathcal{F}_\infty, \mathcal{C}D_\infty(\mathcal{E}))$, where $\mathcal{C}D_\infty(\mathcal{E}) \stackrel{\text{def}}{=} (\mathcal{C}D_0(\mathcal{E}))_\infty$. We call the pair $(\mathcal{F}_\infty, \mathcal{C}D_\infty(\mathcal{E}))$ a *free differential $\mathcal{F}$-extension* of the equation $\mathcal{E}$.

The algebra $\mathcal{F}_\infty$ is filtered by its graded subalgebras $\mathcal{F}_i$, $i = -1, 0, 1, \dots$, and we consider its filtered graded $\mathcal{C}D_\infty(\mathcal{E})$-stable ideal I. Any vector field (derivation) $X \in \mathcal{C}D_\infty(\mathcal{E})$ determines a derivation $X_I \in D(\mathcal{F}_I)$, where $\mathcal{F}_I = \mathcal{F}/I$. Let $\mathcal{C}D_I(\mathcal{E})$ be the $\mathcal{F}_I$-submodule generated by such derivations. Obviously, it is closed with respect to the Lie bracket. We call the pair $(\mathcal{F}_I, \mathcal{C}D_I(\mathcal{E}))$ a *graded extension* of the equation $\mathcal{E}$, or a *graded $\mathcal{E}$-equation*, if (a) there exists an epimorphism $\varphi\colon \mathcal{F}_I \to \mathcal{F}(\mathcal{E})$ and (b) for any $X \in \mathcal{C}D(\mathcal{E})$ there exists $Y \in \mathcal{C}D_I(\mathcal{E})$ such that $\varphi X(a) = Y\varphi(a)$ for any $a \in \mathcal{F}_I$.

Let $\mathcal{F}_{-\infty} = C^\infty(M)$. In an appropriate algebraic setting (see above) the Cartan connection $\mathcal{C}\colon D(\mathcal{F}_{-\infty}) \to D(\mathcal{F}(\mathcal{E}))$ can be uniquely extended to a connection

$$\mathcal{C}_I\colon D(\mathcal{F}_{-\infty}, \mathcal{F}_I) \to \mathcal{C}D_I(\mathcal{E}) \subset D(\mathcal{F}_I).$$

In what follows we call graded $\mathcal{E}$-equations which admit such an extension *$\mathcal{C}$-natural.* From the flatness of the Cartan connection and from the definition of the algebra $\mathcal{C}D_\infty(\mathcal{E})$ (see 2.1) it follows that $\mathcal{C}_I$ is a flat connection as well, i.e., $R_{\mathcal{C}_I}(X, Y) = 0$, $X, Y \in D(\mathcal{F}_{-\infty}, \mathcal{F}_I)$, for any $\mathcal{C}$-natural graded $\mathcal{E}$-equation $(\mathcal{F}_I, \mathcal{C}D_I(\mathcal{E}))$.

2.4. The structural element and $\mathcal{C}$-cohomology. Consider a $\mathcal{C}$-natural graded $\mathcal{E}$-equation $(\mathcal{F}_I, \mathcal{C}D_I(\mathcal{E}))$ and define the homomorphism $U_I \in \hom_{\mathcal{F}_I}(D(\mathcal{F}_I), D(\mathcal{F}_I))$ by

$$U_I(X) = X - \mathcal{C}_I(X_{-\infty}), \qquad X \in D(\mathcal{F}_I),\ X_{-\infty} = X|_{\mathcal{F}_{-\infty}}. \tag{2.1}$$

The element U_I is said to be the *structural element* of $(\mathcal{F}_I, \mathcal{C}D_I(\mathcal{E}))$.

Due to the assumptions formulated above, U_I can be regarded as an element of $\Lambda^*(\mathcal{F}_I) \otimes D_1(\mathcal{F}_I)$.

THEOREM 2.1. *For any $\mathcal{C}$-natural graded $\mathcal{E}$-equation $(\mathcal{F}_I(\mathcal{E}), \mathcal{C}D_I(\mathcal{E}))$, $\mathcal{E}$ being formally integrable, its structural element is integrable*:

$$[\![U_I, U_I]\!] = 0.$$

PROOF. Let $X, Y \in D(\mathcal{F}_I)$ and consider $[\![U_I, U_I]\!]$ as an element of the module $\hom_{\mathcal{F}_I}(D_I(\mathcal{E}) \wedge D_I(\mathcal{E}), D_I(\mathcal{E}))$. Then applying (1.11) twice we get

$$\begin{aligned}[\![U_I, U_I]\!](X, Y) &= \varepsilon\{(-1)^{U\cdot Y}[U_I(X), U_I(Y)] \\ &\quad - (-1)^{U\cdot Y} U_I([U_I(X), Y]) - U_I([X, U_I(Y)]) + U_I^2([X, Y])\},\end{aligned} \tag{2.2}$$

where $\varepsilon = (-1)^{X\cdot Y}(1+(-1)^{U\cdot U})$. The expression (2.2) can be called the *graded Nijenhuis torsion* (cf. [5]).

From (2.1) if follows that the grading of U_I is $\mathbf{0}$, and thus (2.2) becomes

$$(2.3)\quad \begin{aligned}[][\![U_I\,,\,U_I]\!](X\,,\,Y) = (-1)^{X\cdot Y}2\{[U_I(X)\,,\,U_I(Y)] - U_I[U_I(X)\,,\,Y]\\ - U_I[X\,,\,U_I(Y)] + U_I^2[X\,,\,Y]\}.\end{aligned}$$

Now, using definition (2.1) of U_I, one gets from (2.3):

$$\begin{aligned}[][\![U_I\,,\,U_I]\!](X\,,\,Y) = (-1)^{X\cdot Y}2\{[\mathcal{C}_I(X_{-\infty})\,,\,\mathcal{C}_I(Y_{-\infty})] - \mathcal{C}_I([\mathcal{C}_I(X_{-\infty})\,,\,Y]_{-\infty})\\ - \mathcal{C}_I([X\,,\,\mathcal{C}_I(Y_{-\infty})]_{-\infty}) + \mathcal{C}_I(\mathcal{C}_I([X\,,\,Y]_{-\infty}))_{-\infty}\}.\end{aligned}$$

But for any vector fields $X\,,\,Y \in D(\mathcal{F}_I)$,

$$(\mathcal{C}_I(X_{-\infty}))_{-\infty} = X_{-\infty} \quad \text{and} \quad [X\,,\,Y]_{-\infty} = X\circ Y_{-\infty} - (-1)^{X\cdot Y}Y\circ X_{\infty}\,.$$

Hence,

$$\begin{aligned}[][\![U_I\,,\,U_I]\!](X\,,\,Y) &= (-1)^{X\cdot Y}\cdot 2\{[\mathcal{C}_I(X_{-\infty})\,,\,\mathcal{C}_I(Y_{-\infty})] - \mathcal{C}_I(\mathcal{C}_I(X_{-\infty})\circ Y_{-\infty}\\ &\qquad - (-1)^{X\cdot Y}\mathcal{C}_I(Y_{-\infty})\circ X_{-\infty})\}\\ &= (-1)^{X\cdot Y}2R_{\mathcal{C}_I}(X_{-\infty}\,,\,Y_{-\infty}) = 0\,.\end{aligned}$$

Therefore, with any $\mathcal{C}$-natural graded $\mathcal{E}$-equation, in an appropriate algebraic setting, one can associate a complex

$$(2.4)\quad \begin{aligned}0 \longrightarrow D(\mathcal{F}_I) \longrightarrow \Lambda^1(\mathcal{F}_I)\otimes D(\mathcal{F}_I) \longrightarrow \cdots\\ \longrightarrow \Lambda^r(\mathcal{F}_I)\otimes D(\mathcal{F}_I) \xrightarrow{\partial_I} \Lambda^{r+1}(\mathcal{F}_I)\otimes D(\mathcal{F}_I) \longrightarrow \dots\,,\end{aligned}$$

where $\partial_I(\Omega) = [\![U_I\,,\,\Omega]\!]$, $\Omega \in \Lambda^r(\mathcal{F}_I)\otimes D(\mathcal{F}_I)$, with corresponding cohomology modules.

As in [6], we confine ourselves to a subtheory of this cohomology theory.

2.5. Vertical subtheory.

DEFINITION. An element $\Omega \in \Lambda^*(\mathcal{F}_I)\otimes D(\mathcal{F}_I)$ is called *vertical*, if $L_\Omega(\varphi) = 0$ for any $\varphi \in \mathcal{F}_{-\infty} \subset \mathcal{F}_I = \Lambda^0(\mathcal{F}_I)$. Denote by $D^v(\mathcal{F}_I)$ the set of all vertical vector fields from $D(\mathcal{F}_I) = \Lambda^0(\mathcal{F}_I)\otimes D(\mathcal{F}_I)$.

PROPOSITION 2.2. *Let $(\mathcal{F}_I\,,\,\mathcal{C}D_I(\mathcal{E}))$ be a $\mathcal{C}$-natural graded $\mathcal{E}$-equation. Then*

(i) *The set of vertical elements in $\Lambda^r(\mathcal{F}_I)\otimes D(\mathcal{F}_I)$ coincides with $\Lambda^r(\mathcal{F}_I)\otimes D^v(\mathcal{F}_I)$.*

(ii) *$\Lambda^*(\mathcal{F}_I)\otimes D^v(\mathcal{F}_I)$ is closed with respect to the Nijenhuis bracket as well as with respect to the contraction operation*:

$$[\![\Lambda^r(\mathcal{F}_I)\otimes D^v(\mathcal{F}_I)\,,\,\Lambda^s(\mathcal{F}_I)\otimes D^v(\mathcal{F}_I)]\!] \subset \Lambda^{r+s}(F_I)\otimes D^v(\mathcal{F}_I)\,,$$
$$(\Lambda^r(\mathcal{F}_I)\otimes D^v(\mathcal{F}_I)) \lrcorner (\Lambda^s(\mathcal{F}_I)\otimes D^v(\mathcal{F}_I)) \subset \Lambda^{r+s-1}(\mathcal{F}_I)\otimes D^v(\mathcal{F}_I)\,.$$

(iii) *An element* $\Omega \in \Lambda^*(\mathcal{F}_I) \otimes D(\mathcal{F}_I)$ *lies in* $\Lambda^*(\mathcal{F}_I) \otimes D^v(\mathcal{F}_I)$ *iff*

$$i_\Omega(U_I) = \Omega .$$

(iv) *The structural element is vertical*: $U_I \in \Lambda^1(\mathcal{F}_I) \otimes D^v(\mathcal{F}_I)$.

From the last proposition it follows that the complex (2.4) can be restricted to

$$\begin{aligned} (2.5) \qquad & 0 \longrightarrow D^v(\mathcal{F}_I) \longrightarrow \Lambda^1(\mathcal{F}_I) \otimes D^v(\mathcal{F}_I) \longrightarrow \cdots \\ & \qquad \longrightarrow \Lambda^r(\mathcal{F}_I) \otimes D^v(\mathcal{F}_I) \xrightarrow{\partial_I} \Lambda^{r+1}(\mathcal{F}_I) \otimes D^v(\mathcal{F}_I) \longrightarrow \cdots . \end{aligned}$$

The cohomology

$$H_I^r(\mathcal{E}) = \frac{\ker(\partial_I : \Lambda^r(\mathcal{F}_I) \otimes D^v(F_I) \to \Lambda^{r+1}(\mathcal{F}_I) \otimes D^v(\mathcal{F}_I))}{\operatorname{Im}(\partial_I : \Lambda^{r-1}(\mathcal{F}_I) \otimes D^v(F_I) \to \Lambda^r(\mathcal{F}_I) \otimes D^v(\mathcal{F}_I))}$$

is called the *$\mathcal{C}$-cohomology* of a graded $\mathcal{E}$-equation. The basic properties of the differential ∂_I in (2.5) are corollaries of Propositions 1.9 and 2.2.

PROPOSITION 2.3. *Let* $(\mathcal{F}_I(\mathcal{E}), \mathcal{C}D_I(\mathcal{E}))$ *be a $\mathcal{C}$-natural graded $\mathcal{E}$-equation and denote by* L_I *the operator* L_{U_I}. *Then for any* $\Omega, \Theta \in \Lambda^*(\mathcal{F}_I) \otimes D^v(\mathcal{F}_I)$ *and* $\rho \in \Lambda^*(\mathcal{F}_I)$,

$$\begin{aligned} &\text{(i)} && \partial_I(\rho \wedge \Omega) = (L_I(\rho) - d\rho) \wedge \Omega + (-1)^{\rho_I} \rho \wedge \partial_I \Omega , \\ &\text{(ii)} && [L_I , i_\Omega] = i_{\partial_I \Omega} + (-1)^{\Omega_I} L_\Omega , \\ &\text{(iii)} && [i_\Omega , \partial_I]\Theta = (-1)^{\Omega_I} (\partial_I \Omega) \lrcorner \Theta , \\ &\text{(iv)} && \partial_I [\![\Omega , \Theta]\!] = [\![\partial_I \Omega , \Theta]\!] + (-1)^{\Omega_I} [\![\Omega , \partial_I \Theta]\!] . \end{aligned}$$

Let $d_h = d - L_I : \Lambda^*(\mathcal{F}_I) \to \Lambda^*(\mathcal{F}_I)$. From (1.13) and Proposition 1.6(ii) it follows that $d_h \circ d_h = 0$. We call d_h the *horizontal differential* of $(\mathcal{F}_I , \mathcal{C}D_I(\mathcal{E}))$ (cf. [8, 9]) and denote its cohomology by $H_h^*(\mathcal{E} ; I)$.

COROLLARY 2.4. *For any $\mathcal{C}$-natural graded $\mathcal{E}$-equation one has*

(i) $H_I^*(\mathcal{E}) = \sum_{r \geqslant 0} H_I^r(\mathcal{E})$ *is a graded* $H_h^*(\mathcal{E} ; I)$*-module.*

(ii) $H_I^*(\mathcal{E})$ *is a graded Lie algebra with respect to the Frölicher–Nijenhuis bracket inherited from* $\Lambda^*(\mathcal{F}_I) \otimes D^v(\mathcal{F}_I)$.

(iii) *From* $\Lambda^*(\mathcal{F}_I) \otimes D^v(\mathcal{F}_I)$ $H_I^*(\mathcal{E})$ *inherits the contraction operation*

$$H_I^r(\mathcal{E}) \lrcorner H_I^s(\mathcal{E}) \subset H_I^{r+s-1}(\mathcal{E}),$$

and $H_I^*(\mathcal{E})$, *with shifted* $(n+1)$*st grading, is a graded Lie algebra with respect to the inherited Richardson–Nijenhuis bracket.*

2.6. Symmetries and deformations. Skipping standard arguments (see [8]) we define infinitesimal symmetries of a graded $\mathcal{E}$-equation $(\mathcal{F}_I(\mathcal{E}), \mathcal{C}D_I(\mathcal{E}))$ as

$$D_{\mathcal{C}_I}(\mathcal{E}) = \{X \in D_I(\mathcal{E}) \mid [X\,,\, \mathcal{C}D_I(\mathcal{E})] \subset \mathcal{C}D_I(\mathcal{E})\}\,;$$

$D_{\mathcal{C}_I}(\mathcal{E})$ forms an n-graded Lie algebra, while $\mathcal{C}D_I(\mathcal{E})$ is its graded ideal consisting of trivial symmetries. Thus, the Lie algebra of nontrivial symmetries is

$$\operatorname{sym}_I \mathcal{E} = D_{\mathcal{C}_I}(\mathcal{E})/\mathcal{C}D_I(\mathcal{E})\,.$$

If the equation at hand is $\mathcal{C}$-natural, then, due to the connection $\mathcal{C}_I$, one has direct sum decompositions

$$D(\mathcal{F}_I) = D^v(\mathcal{F}_I) \oplus \mathcal{C}D_I(\mathcal{E})\,, \qquad D_{\mathcal{C}_I}(\mathcal{E}) = D^v_{\mathcal{C}_I}(\mathcal{E}) \oplus \mathcal{C}D_I(\mathcal{E})\,, \tag{2.6}$$

where

$$D^v_{\mathcal{C}_I}(\mathcal{E}) = \{X \in D^v_I(\mathcal{E}) \mid [X\,,\, \mathcal{C}D_I(\mathcal{E})] = 0\} = D^v(\mathcal{F}_I) \cap D_{\mathcal{C}_I}(\mathcal{E})\,,$$

and $\operatorname{sym}_I \mathcal{E}$ is identified with the first summand in (2.6).

Let $\varepsilon \in \mathcal{R}$ be a small parameter and $U_I(\varepsilon) \in \Lambda^1(\mathcal{F}_I) \otimes D^v(\mathcal{F}_I)$ be a smooth family such that

(i) $U_I(0) = U_I$,
(ii) $[\![U_I(\varepsilon)\,,\, U_I(\varepsilon)]\!] = 0$ for all ε.

Then $U_I(\varepsilon)$ is a (vertical) deformation of the graded $\mathcal{E}$-equation structure, and if

$$U_I(\varepsilon) = U_I + U_I^1 \cdot \varepsilon + O(\varepsilon^2)\,,$$

then U_I^1 is said to be the (*vertical*) *infinitesimal deformation* of U_I. Again, skipping motivations (see [2, 6, 9]) and literally repeating the corresponding proof from [6], we have

THEOREM 2.5. *For any $\mathcal{C}$-natural graded $\mathcal{E}$-equation $(\mathcal{F}_I\,,\, \mathcal{C}D_I(\mathcal{E}))$ we have*
(i) $H_I^0(\varepsilon) = \operatorname{sym}_I(\mathcal{E})$;
(ii) $H_I^1(\mathcal{E})$ *consists of the classes of nontrivial infinitesimal vertical deformations of the graded $\mathcal{E}$-equation structure U_I.*

The following result is an immediate consequence of the results of the previous subsection.

THEOREM 2.6. *Let $(\mathcal{F}_I\,,\, \mathcal{C}D_I(\mathcal{E}))$ be a graded $\mathcal{E}$-equation. Then*
(i) *$H_I^1(\mathcal{E})$ is an associative algebra with respect to contraction.*
(ii) *The map*

$$\mathcal{R}\colon H_I^1(\mathcal{E}) \to \operatorname{End}(H_I^0(\mathcal{E}))\,,$$

where

$$\mathcal{R}_\Omega(X) = X \lrcorner \Omega\,, \qquad X \in H_I^0(\mathcal{E})\,,\ \Omega \in H_I^1(\mathcal{E})$$

is a representation of this algebra. And consequently,

$$(\operatorname{sym}_I \mathcal{E}) \lrcorner H_I^1(\mathcal{E}) \subset \operatorname{sym}_I \mathcal{E}\,. \tag{iii}$$

2.7. Recursion operators. The first equality in (2.6) gives us the dual decomposition

(2.7) $$\Lambda^1(\mathcal{F}_I) = \mathcal{C}\Lambda^1(\mathcal{F}_I) \oplus \Lambda^1_h(\mathcal{F}_I),$$

where

$$\mathcal{C}\Lambda^1(\mathcal{F}_I) = \{\omega \in \Lambda^1(\mathcal{F}_I) \mid \mathcal{C}D_I(\mathcal{E}) \lrcorner\, \omega = 0\},$$
$$\Lambda^1_h(\mathcal{F}_I) = \{\omega \in \Lambda^1(\mathcal{F}_I) \mid D^v(\mathcal{F}_I) \lrcorner\, \omega = 0\}.$$

In fact, let $\omega = \sum_\alpha f_\alpha dg_\alpha$, $f_\alpha, g_\alpha \in \mathcal{F}_I$, be a 1-form. Then, since by definition $d = d_h + L_I$, we have

$$\omega = \sum_\alpha f_\alpha(d_h g_\alpha + L_I(g_\alpha)).$$

Let $X \in D^v(\mathcal{F}_I)$. Then from Proposition 2.3(ii) it follows that

$$X \lrcorner\, L_I(g) = -L_I(X \lrcorner\, g) + \partial_I(X) \lrcorner\, g + L_X(g) = X(g), \qquad g \in \mathcal{F}_I.$$

Hence,

$$X \lrcorner\, d_h g = X \lrcorner\, (d - L_I)g = X(g) - X(g) = 0.$$

On the other hand, $L_I(g) = U_I \lrcorner\, dg$, and if $Y \in \mathcal{C}D_I(\mathcal{E})$, then

$$Y \lrcorner\, L_I(g) = Y \lrcorner\, (U_I \lrcorner\, dg) = (Y \lrcorner\, U_I) \lrcorner\, dg$$

due to Proposition 1.5(v); but $Y \lrcorner\, U_I = 0$ for any $Y \in \mathcal{C}D_I(\mathcal{E})$. Thus, one has the decomposition

(2.8) $$\Lambda^r(\mathcal{F}_I) = \sum_{p+q=r} \mathcal{C}^p\Lambda(\mathcal{F}_I) \wedge \Lambda^q_h(\mathcal{F}_I),$$

where

$$\mathcal{C}^p\Lambda(\mathcal{F}_I) = \underbrace{\mathcal{C}\Lambda^1(\mathcal{F}_I) \wedge \cdots \wedge \mathcal{C}\Lambda^1(\mathcal{F}_I)}_{p \text{ times}}, \qquad \Lambda^q_h(\mathcal{F}_I) = \underbrace{\Lambda^1_h(\mathcal{F}_I) \wedge \cdots \wedge \Lambda^1_h(\mathcal{F}_I)}_{q \text{ times}},$$

and the wedge product $\wedge$ is taken in the graded sense (see 1.2).

REMARK. The summands in (2.8) can also be described in the following way:

$$\mathcal{C}^p\Lambda(\mathcal{F}_I) \wedge \Lambda^q_h(\mathcal{F}_I) = \big\{\omega \in \Lambda^{p+q}(\mathcal{F}_I) \mid X_1 \lrcorner\, \cdots \lrcorner\, X_{p+1} \lrcorner\, \omega = 0,$$
$$Y_1 \lrcorner\, \cdots \lrcorner\, Y_{q+1} \lrcorner\, \omega = 0$$
$$\text{for all } X_\alpha \in D^v(\mathcal{F}_I),\ Y_\beta \in \mathcal{C}D_I(\mathcal{E})\big\}.$$

PROPOSITION 2.7. *Let $(\mathcal{F}_I, \mathcal{C}D_I(\mathcal{E}))$ be a $\mathcal{C}$-natural equation. Then*

$$\partial_I(\mathcal{C}^p\Lambda(\mathcal{F}_I) \wedge \Lambda^q_h(\mathcal{F}_I) \otimes D^v(\mathcal{F}_I)) \subset \mathcal{C}^p\Lambda(\mathcal{F}_I) \wedge \Lambda^{q+1}_h(\mathcal{F}_I) \otimes D^v(\mathcal{F}_I)$$

for all $p, q \geqslant 0$.

The proof is based on two lemmas.

LEMMA 2.8. $d_h \mathcal{C}^1\Lambda(\mathcal{F}_I) \subset \mathcal{C}^1\Lambda(\mathcal{F}_I) \wedge \Lambda_h^1(\mathcal{F}_I)$.

PROOF. Due to the last remark, it is sufficient to show that

$$X^v \lrcorner Y^v \lrcorner d_h\omega = 0, \qquad X^v, Y^v \in D^v(\mathcal{F}_I), \tag{2.9}$$

$$X^h \lrcorner Y^h \lrcorner d_h\omega = 0, \qquad X^h, Y^h \in \mathcal{C}D_I(\mathcal{E}), \tag{2.10}$$

$\omega \in \mathcal{C}^1\Lambda(F_I)$. Obviously, we can restrict ourselves to the case $\omega = L_I(g)$, $g \in \mathcal{F}_I$:

$$\begin{aligned} Y^v \lrcorner d_h\omega &= Y^v \lrcorner d_h L_I(g) = -Y^v \lrcorner L_I d_h g \\ &= L_I(Y^v \lrcorner d_h g) + L_{Y^v}(d_h g) = d_h Y^v(g). \end{aligned}$$

Hence, $X^v \lrcorner Y^v \lrcorner d_h\omega = X^v \lrcorner d_h Y^v(g) = 0$, which proves (2.9). Now,

$$Y^h \lrcorner d_h\omega = -Y^h \lrcorner L_I d_h g = Y^h \lrcorner (d\,(U_I \lrcorner d_h g) - U_I \lrcorner d\,(d_h g)).$$

But U_I is a vertical element, i.e., $U_I \in \Lambda^1(\mathcal{F}_I) \otimes D^v(\mathcal{F}_I)$. Therefore, we have $U_I \lrcorner d_h g = 0$ and

$$\begin{aligned} Y^h \lrcorner d_h\omega &= -Y^h \lrcorner (U_I \lrcorner d\,(d_h g)) \\ &= -(Y^h \lrcorner U_I) \lrcorner d\,(d_h g) - (Y^h \wedge U_I) \lrcorner d\,(d_h g). \end{aligned}$$

The first summand in the right-hand side of the last equality vanishes, since, by definition, $Y^h \lrcorner U_I = 0$ for any $Y^h \in \mathcal{C}D_I(\mathcal{E})$. Hence,

$$\begin{aligned} X^h \lrcorner Y^h \lrcorner d_h\omega &= -X^h \lrcorner (Y^h \wedge U_I) \lrcorner d\,(d_h g) \\ &= -(X^h \lrcorner (Y^h \wedge U_I)) \lrcorner d\,(d_h g) - (X^h \wedge Y^h \wedge U_I) \lrcorner d\,(d_h g) \\ &= -(X^h \wedge Y^h \wedge U_I) \lrcorner d\,(d_h g). \end{aligned}$$

But $X^h \wedge Y^h \wedge U_I$ is a (form valued) 3-vector, while $d\,(d_h g)$ is a 2-form; hence

$$X^h \lrcorner Y^h \lrcorner d_h\omega = 0,$$

which finishes the proof of Lemma 2.8. □

LEMMA 2.9. $\partial_I D^v(\mathcal{F}_I) \subset \Lambda_h^1 \otimes D^v(\mathcal{F}_I)$.

PROOF. This lemma immediately follows from Proposition 2.3(iii).

PROOF OF PROPOSITION 2.7. This proposition follows from the previous lemmas and Proposition 2.3(i), which can be rewritten as

$$\partial_I(\rho \wedge \Omega) = -d_h(\rho) \wedge \Omega + (-1)^{\rho_1} \rho \wedge \partial_I(\Omega). \quad \square$$

Taking into account the last result, one has the following decomposition

$$H_I^r(\mathcal{E}) = \sum_{p+q=r} H_I^{p,q}(\mathcal{E}), \qquad H_I^{p,q}(\mathcal{E}) = \frac{\ker(\partial_I^{p,q})}{\operatorname{Im}(\partial_I^{p,q-1})},$$

where $\partial^{i,j}\colon \mathcal{C}^i\Lambda(\mathcal{F}_I) \wedge \Lambda_k^j(\mathcal{F}_I) \otimes D^v(\mathcal{F}_I) \to \mathcal{C}^i(\mathcal{F}_I) \wedge \Lambda_h^{j+1}(\mathcal{F}_I) \otimes D^v(\mathcal{F}_I)$.

In particular,

$$H_I^1(\mathcal{E}) = H_I^{0,1}(\mathcal{E}) \oplus H_I^{1,0}(\mathcal{E}). \tag{2.11}$$

Note now that from the point of view of $H_I^1(\mathcal{E})$-action onto $H_I^0(\mathcal{E}) = \operatorname{sym}_I \mathcal{E}$, the first summand in (2.11) is of no interest, since

$$D^v(\mathcal{F}_I) \lrcorner \Lambda_h^1(\mathcal{F}_I) = 0.$$

We call $H_I^{*,0}(\mathcal{E})$ the *Cartan part* of $H_I^*(\mathcal{E})$, while the elements of $H_I^{1,0}(\mathcal{E})$ are called the *recursion operators* for the equation $(\mathcal{F}_I, \mathcal{C}D_I(\mathcal{E}))$. We have the following

PROPOSITION. $H_I^{p,0}(\mathcal{E}) = \ker \partial_I^{p,0}$.

PROOF. In fact, Proposition 2.7 implies

$$\operatorname{Im}(\partial_I) \cap (\mathcal{C}^*\Lambda(\mathcal{F}_I) \otimes D^v(\mathcal{F}_I)) = 0,$$

which proves the result. □

Note that $H_I^{*,0}(\mathcal{E})$ inherits an associative graded algebra structure with respect to contraction, $H_I^{1,0}(\mathcal{E})$ being its subalgebra.

2.8. Commutativity theorem. In this subsection we prove the following

THEOREM 2.11. $[\![H_I^{1,0}(\mathcal{E}), H_I^{1,0}(\mathcal{E})]\!] \subset H_I^{2,0}(\mathcal{E})$.

The proof is based on the following

LEMMA. *For any* $\omega \in \mathcal{C}^1\Lambda(\mathcal{F}_I)$ *one has*

$$U_I \lrcorner\, \omega = \omega. \tag{2.12}$$

PROOF. It is sufficient to prove (2.12) for generators of $\mathcal{C}^1\Lambda(\mathcal{F}_I)$ of the form

$$\omega = L_I(g), \qquad g \in \mathcal{F}_I.$$

Now (1.10) implies $L_I \circ i_{U_I} - i_{U_I} \circ L_I + L_{U_I \lrcorner\, U_I} = i_{[\![U_I, U_I]\!]}$, or

$$L_I \circ i_{U_I} - i_{U_I} \circ L_I + L_I = 0. \tag{2.13}$$

Applying (2.13) to some $g \in \mathcal{F}_I$, one sees that

$$U_I \lrcorner\, L_I(g) = L_I(g). \quad \square$$

PROOF OF THEOREM 2.11. Let $\Omega, \Theta \in H_I^{1,0}(\mathcal{E})$, i.e., $\Omega, \Theta \in \mathcal{C}^1\Lambda(\mathcal{F}_I)$ and $\partial_I\Omega = \partial_I\Theta = 0$. Then from (1.11) it follows that

$$U_I \lrcorner\, [\![\Omega, \Theta]\!] = [\![U_I \lrcorner\, \Omega, \Theta]\!] + [\![\Omega, U_I \lrcorner\, \Theta]\!],$$

or, by Lemma 2.12, $U_I \lrcorner\, [\![\Omega, \Theta]\!] = 2[\![\Omega, \Theta]\!]$. Hence,

$$\begin{aligned}
[\![\Omega, \Theta]\!] &= \tfrac{1}{2} U_I \lrcorner\, [\![\Omega, \Theta]\!] = \tfrac{1}{4} U_I \lrcorner\, (U_I \lrcorner\, [\![\Omega, \Theta]\!]) \\
&= \tfrac{1}{4}((U_I \lrcorner\, U_I) \lrcorner\, [\![\Omega, \Theta]\!] - (U_I \wedge U_I) \lrcorner\, [\![\Omega, \Theta]\!]) \\
&= \tfrac{1}{4}(U_I \lrcorner\, [\![\Omega, \Theta]\!] - (U_I \wedge U_I) \lrcorner\, [\![\omega, \Theta]\!]) \\
&= \tfrac{1}{2}[\![\Omega, \Theta]\!] - \tfrac{1}{4}(U_I \wedge U_I) \lrcorner\, [\![\Omega, \Theta]\!],
\end{aligned}$$

or

$$[\![\Omega, \Theta]\!] = -\tfrac{1}{2}(U_I \wedge U_I) \lrcorner\, [\![\Omega, \Theta]\!].$$

But $U_I \in \mathcal{C}^1\Lambda(\mathcal{F}_I) \otimes D^v(\mathcal{F}_I)$, which finishes the proof. □

COROLLARY 2.13. *U_I is a unit of the associative algebra $H_I^{1,0}(\mathcal{E})$.*

PROOF. This follows from the definition of U_I and from Lemma 2.12.

COROLLARY 2.14. *If $H_I^{2,0}(\mathcal{E}) = 0$, then all recursion operators for the equation $(\mathcal{F}_I, \mathcal{C}D_I(\mathcal{E}))$ commute with respect to the Frölicher–Nijenhuis bracket.*

Let $\Omega \in H_I^{1,0}(\mathcal{E})$ be a recursion operator. Denote its action on $H_I^0(\mathcal{E}) = \mathrm{sym}_I(\mathcal{E})$ by $\Omega(X) = X \lrcorner \Omega$, $X \in H_I^0(\mathcal{E})$. Then, from (1.11) it follows that

$$\begin{aligned} Y \lrcorner X \lrcorner [\![\Omega, \Theta]\!] = (-1)^{X\cdot Y}\{&(-1)^{Y\cdot\Omega}[\Omega(X), \Theta(Y)] \\ &+ (-1)^{(Y+\Omega)\cdot\Theta}[\Theta(X), \Omega(Y)] \\ &- (-1)^{\Omega\cdot\Theta}\Omega((-1)^{Y\cdot\Theta}[\Theta(X), Y] + [X, \Theta(Y)]) \\ &- \Theta((-1)^{Y\cdot\Omega}[\Omega(X), Y] + [X, \Omega(Y)]) \\ &+ ((-1)^{\Omega\cdot\Theta}\Omega \circ \Theta + \Theta \circ \Omega)[X, Y]\}, \end{aligned} \tag{2.14}$$

$X, Y \in \mathrm{sym}_I(\mathcal{E})$, $\Omega, \Theta \in H_I^{1,0}(\mathcal{E})$.

COROLLARY 2.15. *If $H_I^{2,0}(\mathcal{E}) = 0$. For any symmetries $X, Y \in \mathrm{sym}_I(\mathcal{E})$ and recursion operators $\Omega, \Theta \in H_I^{1,0}(\mathcal{E})$ one has*

$$\begin{aligned} &(-1)^{Y\cdot\Omega}[\Omega(X), \Theta(Y)] + (-1)^{(Y+\Omega)\cdot\Theta}[\Theta(X), \Omega(Y)] \\ &\quad = (-1)^{\Omega\cdot\Theta}\Omega((-1)^{Y\cdot\Omega}[\Theta(X), Y] + [X, \Theta(Y)]) \\ &\qquad + \Theta((-1)^{Y\cdot\Omega}[\Omega(X), Y] + [X, \Omega(Y)]) + ((-1)^{\Omega\cdot\Theta}\Omega \circ \Theta + \Theta \circ \Omega)[X, Y]. \end{aligned} \tag{2.15}$$

In particular,

$$\begin{aligned} (1 + (-1)^{\Omega\cdot\Omega})\{&(-1)^{Y\cdot\Omega}[\Omega(X), \Omega(Y)] \\ &- (-1)^{Y\cdot\Omega}\Omega[\Omega(X), Y] - \Omega[X, \Omega(Y)] + \Omega^2[X, Y]\} = 0, \end{aligned}$$

and if $\Omega \cdot \Omega$ is even, then

$$\begin{aligned} &[\Omega(X), \Omega(Y)] \\ &\quad = \Omega([\Omega(X), Y] + (-1)^{Y\Omega}[X, \Omega(Y)] - (-1)^{Y\Omega}\Omega[X, Y]). \end{aligned} \tag{2.16}$$

§3. Nonlocal theory and the case of evolution equations

Here we extend the theory of coverings and that of nonlocal symmetries [10] to the case of graded equations (cf. [14]). We confine ourselves to evolution equations, although the results obtained, at least partially, are applicable to more general cases. For any graded equation the notion of its tangent covering (nonlocal linearization) is introduced; this reduces the computation of recursion operators to computations of special nonlocal symmetries. In this setting, we also solve the problem of extending "shadows" of recursion operators up to real operators. Finally, using algebraic techniques of previous sections, we get a rather complete description of the Lie algebra structure of $\mathrm{sym}(\mathcal{E})$ for equations $\mathcal{E}$ possessing a recursion operator.

3.1. GDE(M) **category.** Let M be a smooth manifold and $A = C^\infty(M)$. We define a GDE(M) category of *graded differential equations* over M as follows. The objects of GDE(M) are pairs $(\mathcal{F}, \nabla_\mathcal{F})$, where $\mathcal{F}$ is a commutative n-graded A-algebra (the case $n = \infty$ is included) endowed with a filtration

$$A = \mathcal{F}_{-\infty} \subset \cdots \subset \mathcal{F}_i \subset \mathcal{F}_{i+1} \subset \dots, \qquad \bigcup_i \mathcal{F}_i = \mathcal{F}, \tag{3.1}$$

while $\nabla_\mathcal{F}$ is a flat $(A, \mathcal{F})$-connection (see 2.2), i.e.,

(i) $\nabla_\mathcal{F} \in \hom_\mathcal{F}(D(A, \mathcal{F}), D(\mathcal{F}))$,
(ii) $\nabla_\mathcal{F}(X)(a) = X(a)$, $X \in D(A, \mathcal{F})$, $a \in A$,
(iii) $[\nabla_\mathcal{F}(X), \nabla_\mathcal{F}(Y)] = \nabla_\mathcal{F}(\nabla_\mathcal{F}(X) \circ Y - \nabla_\mathcal{F}(Y) \circ X)$, $X, Y \in D(A, \mathcal{F})$.

From the definition it follows that the grading of $\nabla_\mathcal{F}$ is $\mathbf{0}$, and we also suppose that for any $X \in \mathcal{D}(A, \mathcal{F})$ the derivation $\nabla_\mathcal{F}(X)$ agrees with the filtering (3.1), i.e.,

$$\nabla_\mathcal{F}(X)(\mathcal{F}_i) \subset \mathcal{F}_{i+s}$$

for some $s = s(X)$ and all i's large enough.

Let $(\mathcal{F}, \nabla_\mathcal{F})$, $(\mathcal{G}, \nabla_\mathcal{G})$ be two objects and $\varphi\colon \mathcal{F} \to \mathcal{G}$ be a graded filtered homomorphism. Then for any $X \in D(A, \mathcal{F})$ the composition $\varphi \circ X$ lies in $D(A, \mathcal{G})$. We say that it is a morphism of the objects $(\mathcal{F}, \nabla_\mathcal{F})$, $(\mathcal{G}, \nabla_\mathcal{G})$ if the diagram

$$\begin{array}{ccc} \mathcal{F} & \xrightarrow{\varphi} & \mathcal{G} \\ {\scriptstyle \nabla_\mathcal{F}(X)}\downarrow & & \downarrow{\scriptstyle \nabla_\mathcal{G}(\varphi \circ X)} \\ \mathcal{F} & \xrightarrow[\varphi]{} & \mathcal{G} \end{array}$$

is commutative for all $X \in D(A, \mathcal{F})$. If φ is a monomorphism, we say that it represents a covering of $(\mathcal{G}, \nabla_\mathcal{G})$ over $(\mathcal{F}, \nabla_\mathcal{F})$.

REMARKS. 1. Let $\mathcal{E}$ be an equation in some bundle over M. Then all graded $\mathcal{E}$-equations are obviously objects of GDE(M).

2. The theory of the previous section can be literally applied to the objects of GDE(M) as well.

3. Each "nongraded" object of GDE(M) carries a differential variety structure, and thus the nongraded part of GDE(M) can be regarded as a subcategory of Vinogradov's category of differential equations [18].

3.2. Local representation. In what follows we shall deal with the following kinds of objects of GDE(M):

(i) infinite prolongations of differential equations;
(ii) their superizations, i.e., 1-graded extensions;
(iii) coverings over (i) and (ii).

For specific applications, local versions of these objects will be considered. This means the following:

(i) in a neighborhood $\mathcal{O} \subset M$ local coordinates $x = (x_1, \dots, x_n)$ are chosen (independent variables);

(ii) the bundle $\pi\colon E \to M$ in which $\mathcal{E}$ is defined is assumed to be a vector bundle, and it trivializes over $\mathcal{O}$. If $(e_1, \dots, e_m)$ is a basis of local sections of π over $\mathcal{O}$, then $f = u^1 e_1 + \dots + u^m e_m$ for any $f \in \Gamma(\pi|_{\mathcal{O}})$, and $u^1, \dots, u^m$ play the role of dependent variables for the equation $\mathcal{E}$;

(iii) $\mathcal{E}$ is represented by a system of relations

$$\begin{cases} F_1(x, \dots, u^j_\sigma, \dots) = 0, \\ \dots\dots\dots\dots\dots\dots \\ F_1(x, \dots, u^j_\sigma, \dots) = 0, \end{cases}$$

where $u^j_\sigma = \partial^{|\sigma|} u^j / \partial x_\sigma$, $\sigma = (i_1, \dots, i_n)$, $|\sigma| = i_1 + \dots + i_n \leqslant k$, are coordinates in the manifold of k-jets $J^k(\pi)$, k being the order of $\mathcal{E}$;

(iv) a graded extension $\mathcal{F}$ of $\mathcal{F}(\mathcal{E})$ (see 2.3) is freely generated over $\mathcal{F}(\mathcal{E})$ by homogeneous elements $v^1, v^2, \dots$. This means that $\mathcal{F}_\infty$ is generated by v^j_τ, where $v^j_{\mathbf{0}} = v^j$ and

$$v^j_{(i_1, \dots, i_s+1, \dots, i_n)} = [D_s, v^j_{(i_1, \dots, i_n)}]$$

D_s being the total derivative on $\mathcal{E}^{(\infty)}$ corresponding to $\partial/\partial x_s$. In this setting any graded extension of $\mathcal{E}$ can be represented as

$$\text{(3.2)} \quad \begin{cases} F_1(x, \dots, u^j_\sigma, \dots) + \phi_1(x, \dots, u^j_\sigma, \dots, v^j_\tau, \dots) = 0, \\ \dots\dots\dots\dots\dots\dots\dots\dots\dots\dots\dots\dots \\ F_r(x, \dots, u^j_\sigma, \dots) + \phi_r(x, \dots, u^j_\sigma, \dots, v^j_\tau, \dots) = 0, \\ \dots\dots\dots\dots\dots\dots\dots\dots \\ \phi_{r+1}(x, \dots, u^j_\sigma, \dots v^j_\tau, \dots) = 0, \\ \dots\dots\dots\dots\dots\dots\dots\dots \\ \phi_{r+l}(x, \dots, u^j_\sigma, \dots v^j_\tau, \dots) = 0, \end{cases}$$

where $\phi_1, \dots, \phi_r$ are functions of degree $\mathbf{0}$ such that $\phi_1 = 0, \dots, \phi_{r+l} = 0$ for $v^j_\tau = 0$.

(v) for any covering $\varphi\colon \mathcal{F} \to \mathcal{G}$ of the graded extension $\mathcal{F}$ by an object $(\mathcal{G}, \nabla_{\mathcal{G}})$, we assume that $\mathcal{G}$ is freely generated over $\mathcal{F}$ by homogeneous elements $w^1, w^2, \dots$ and

$$\nabla_{\mathcal{G}}\left(\frac{\partial}{\partial x_i}\right) = D_i + \sum_s X^s_i \frac{\partial}{\partial w^s} \overset{\text{def}}{=} \widetilde{D}_i,$$

with

$$[\widetilde{D}_i, \widetilde{D}_j] = \left[D_i, \sum_s X^s_j \frac{\partial}{\partial w^s}\right] + \left[\sum_s X^s_i \frac{\partial}{\partial w^s}, D_j\right] + \left[\sum_s X^s_i \frac{\partial}{\partial w^s}, \sum_s X^s_j \frac{\partial}{\partial w^s}\right] = 0,$$

where $i, j = 1, \dots, n$, $X^s_i \in \mathcal{G}$, and $D_1, \dots, D_n$ are total derivatives extended onto $\mathcal{F}$: $D_i = \nabla_{\mathcal{F}}(\partial/\partial x_i)$. Elements $w^1, w^2, \dots$ are called *nonlocal variables* related to the covering φ, while the number of nonlocal variables is said to be the dimension of φ.

3.3. Evolution equations. Below we deal with super ($\mathbb{Z}_2$-graded) evolution equations $\mathcal{E}$ in two independent variables x and t:

$$
\begin{cases} u_t^1 = f^1(x, t, u^1, \dots, u^m, \dots, u_k^1, \dots, u_k^m), \\ \dots\dots\dots\dots\dots\dots\dots\dots\dots\dots \\ u_t^m = f^m(x, t, u^1, \dots, u^m, \dots, u_k^1, \dots, u_k^m), \end{cases} \tag{3.3}
$$

where $u^1, \dots, u^m$ are of either even or odd grading, and u_s^j denotes $\partial^s u^j/\partial x^s$. We take $x, t, u_0^1, \dots, u_0^m, \dots, u_i^1, \dots u_i^j, \dots$ for the internal coordinates on $\mathcal{E}^{(\infty)}$. The total derivatives D_x and D_t restricted to the infinite prolongation of (3.3) are of the form

$$
D_x = \frac{\partial}{\partial x} + \sum_{i=0}^{\infty}\sum_{j=1}^{m} u_{i+1}^j \frac{\partial}{\partial u_i^j}, \qquad D_t = \frac{\partial}{\partial t} + \sum_{i=0}^{\infty}\sum_{j=1}^{m} D_x^i(f^j) \frac{\partial}{\partial u_i^j}. \tag{3.4}
$$

In the chosen local coordinates the structural element $U = U_{\mathcal{E}}$ of the equation $\mathcal{E}$ is represented as

$$
U = \sum_{i=0}^{\infty}\sum_{j=1}^{m} (du_i^j - u_{i+1}^j dx - D_x^i(f^j)\, dt) \otimes \frac{\partial}{\partial u_i^j}. \tag{3.5}
$$

Then for a basis of the module $\mathcal{C}^1\Lambda(\mathcal{E})$ one can choose the forms

$$
\omega_i^j = L_U(u_i^j) = du_i^j - u_{i+1}^j\, dx - D_x^i(f^j)\, dt,
$$

while (3.5) is rewritten as

$$
U = \sum_{i=0}^{\infty}\sum_{j=1}^{m} \omega_i^j \otimes \frac{\partial}{\partial u_i^j}. \tag{3.6}
$$

Let

$$
\Theta = \sum_{i=0}^{\infty}\sum_{j=1}^{m} \theta_i^j \otimes \frac{\partial}{\partial u_i^j} \in \Lambda^p(\mathcal{E}) \otimes D^v(\mathcal{E}).
$$

Then from (1.6) one has

(3.7)

$$
\begin{aligned}
\partial_{\mathcal{E}}(\Theta) &= [\![U, \Theta]\!] \\
&= \sum_{i=0}^{\infty}\sum_{j=1}^{m} \Bigg(dx \wedge \{\theta_{i+1}^j - D_x(\theta_i^j)\} \\
&\qquad + dt \wedge \Bigg\{ \sum_{\beta=0}^{\infty}\sum_{\alpha=1}^{m} \theta_\beta^\alpha \frac{\partial D_x^i f^j}{\partial u_\beta^\alpha} - D_t(\theta_i^j) \Bigg\} \Bigg) \otimes \frac{\partial}{\partial u_i^j}.
\end{aligned}
$$

From (3.7) one easily gets the following

THEOREM. *Let $\mathcal{E}$ be an equation of the form* (3.3). *Then $H_{\mathcal{C}}^{p,0}(\mathcal{E})$ consists of the elements*

$$\ni_\theta = \sum_{i=0}^{\infty} \sum_{j=1}^{m} D_x^i(\theta^j) \otimes \frac{\partial}{\partial u_i^j},$$

where $\theta = (\theta^1, \dots, \theta^m)$, $\theta^j \in \mathcal{C}^p\Lambda(\mathcal{E})$, is a vector-valued form satisfying the equations

$$\sum_{i=0}^{k} \sum_{j=1}^{m} D_x^i(\theta^j) \frac{\partial f^l}{\partial u_i^j} = 0, \qquad l = 1, \dots, m, \tag{3.8}$$

or in short,

$$H_{\mathcal{C}}^{p,0}(\mathcal{E}) = \ker l_{\mathcal{E}}^{[p]},$$

$l_{\mathcal{E}}^{[p]}$ being the extension of the operator of universal linearization to the space $\mathcal{C}^p\Lambda(\mathcal{E}) \otimes_{\mathbb{R}} \mathbb{R}^m$:

$$(l_{\mathcal{E}}^{[p]}(\theta))^l = \sum_{i=0}^{k} \sum_{j=1}^{m} D_x^i(\theta^j) \frac{\partial f^l}{\partial u_i^j}, \qquad l = 1, \dots, m. \tag{3.9}$$

3.4. Nonlocal setting and shadows. Now let φ be a covering of (3.3) determined by nonlocal variables $w^1, w^2, \dots$ with the extended total derivatives of the form

$$\widetilde{D}_x = D_x + \sum_s X_s \frac{\partial}{\partial w_s}, \qquad \widetilde{D}_t = D_t + \sum_s T_s \frac{\partial}{\partial w_s}, \tag{3.10}$$

satisfying

$$[\widetilde{D}_x, \widetilde{D}_t] = 0. \tag{3.11}$$

Denote by $\mathcal{F}(\mathcal{E}_\varphi)$ the corresponding algebra of functions and by $\Lambda^*(\mathcal{E}_\varphi)$ and $D(\mathcal{E}_\varphi)$ differential forms and vector fields on $\mathcal{F}(\mathcal{E}_\varphi)$ respectively. Then the structural element for the covering object is

$$U_\varphi = U + \sum_s (d\omega_s - X_s\, dx - T_s\, dt) \otimes \frac{\partial}{\partial w_s}$$

and the identity $[\![U_\varphi, U_\varphi]\!] = 0$ is fulfilled due to (3.11). If

$$\Theta = \sum_{i=0}^{\infty} \sum_{j=1}^{m} \theta_i^j \otimes \frac{\partial}{\partial u_i^j} + \sum_s \rho_s \otimes \frac{\partial}{\partial w_s}$$

is an element of $\Lambda^p(\mathcal{E}_\varphi) \otimes D^v(\mathcal{E}_\varphi)$, then one can easily see that

(3.12)
$$\begin{aligned}
\partial_\varphi(\Theta) &= [\![U_\varphi , \Theta]\!] \\
&= \sum_{i=0}^{\infty} \sum_{j=1}^{m} \Big(dx \wedge \{\theta^j_{i+1} - \widetilde{D}_x(\theta^j_i)\} \\
&\qquad + dt \wedge \Big\{ \sum_{\beta=0}^{\infty} \sum_{\alpha=1}^{m} \theta^\alpha_\beta \frac{\partial D^i_x f^j}{\partial u^\alpha_\beta} - \widetilde{D}_t(\theta^j_i) \Big\} \Big) \otimes \frac{\partial}{\partial u^j_i} \\
&\quad + \sum_s \Big(dx \wedge \Big\{ \sum_{\beta=0}^{\infty} \sum_{\alpha=1}^{m} \theta^\alpha_\beta \frac{\partial X_s}{\partial u^\alpha_\beta} + \sum_\gamma \rho_\gamma \frac{\partial X_s}{\partial w_\gamma} - \widetilde{D}_x(\rho_s) \Big\} \\
&\qquad + dt \wedge \Big\{ \sum_{\beta=0}^{\infty} \sum_{\alpha=1}^{m} \theta^\alpha_\beta \frac{\partial X_s}{\partial u^\alpha_\beta} + \sum_\gamma \rho_\gamma \frac{\partial T_s}{\partial w_\gamma} - \widetilde{D}_t(\rho_s) \Big\} \Big) \otimes \frac{\partial}{\partial w_s} .
\end{aligned}$$

Again, confining ourselves to the case $\Theta \in \mathcal{C}^p\Lambda(\mathcal{E}_\varphi) \otimes D^v(\mathcal{E}_\varphi)$ we get the following

THEOREM 3.2. *Let $\mathcal{E}$ be an equation of the form* (3.3) *and φ be its covering with nonlocal variables $w_1, w_2, \dots$ and extended total derivatives given by* (3.10). *Then $H^{p,0}_{\mathcal{C}}(\mathcal{E}_\varphi)$ consists of the elements*

$$\partial_{\theta,\rho} = \sum_{i=0}^{\infty} \sum_{j=1}^{m} \widetilde{D}^i_x(\theta^j) \otimes \frac{\partial}{\partial u^j_i} + \sum_s \rho_s \otimes \frac{\partial}{\partial w_s} , \tag{3.13}$$

where $\theta = (\theta^1, \dots, \theta^m)$, $\rho = (\rho_1, \dots, \rho_s, \dots)$, $\theta^j, \rho_s \in \mathcal{C}^p\Lambda(\mathcal{E}_\varphi)$, are vector-valued forms satisfying the equations

$$\tilde{l}^{[p]}_{\mathcal{E}}(\theta) = 0 , \tag{3.14}$$

(3.15)
$$\begin{aligned}
\sum_{\beta=0}^{\infty} \sum_{\alpha=1}^{m} \widetilde{D}^\beta_x(\theta^\alpha) \frac{\partial X_s}{\partial u^\alpha_\beta} + \sum_j \rho_j \frac{\partial X_s}{\partial w_j} &= \widetilde{D}_x(\rho_s) , \\
\sum_{\beta=0}^{\infty} \sum_{\alpha=1}^{m} \widetilde{D}^\beta_x(\theta^\alpha) \frac{\partial T_s}{\partial u^\alpha_\beta} + \sum_j \rho_j \frac{\partial T_s}{\partial w_j} &= \widetilde{D}_t(\rho_s) ,
\end{aligned}$$

$s = 1, 2, \dots$, *$\tilde{l}^{[p]}_{\mathcal{E}}$ being the natural extension of $l^{[p]}_{\mathcal{E}}$ with D_x and D_t replaced by $\widetilde{D}_x$ and $\widetilde{D}_t$ in* (3.9).

Following [9, 10], we call (3.14) *shadow equations* and (3.15) *relation equations* for the element (θ, ρ); solutions of (3.14) are called *shadow solutions*, or simply *shadows*. Our main concern lies in reconstructing elements of $H^{p,0}_{\mathcal{C}}(\mathcal{E}_\varphi)$ from their shadows. Denote the set of such shadows by $SH^{p,0}_{\mathcal{C}}(\mathcal{E}_\varphi)$.

REMARKS. 1. Let φ be a covering. Consider horizontal 1-forms

$$\omega^s_\varphi = d_h w_s = X_s \, dx + T_s \, dt , \qquad s = 1, 2, \dots ,$$

where d_h is the horizontal de Rham differential associated to φ. Then (3.15) can be rewritten as

$$\ni_{\theta,p}(\omega_\varphi^s) = d_h \rho_s, \qquad s = 1, 2, \dots. \tag{3.16}$$

2. When X_s, T_s do not depend on nonlocal variables, the conditions of φ being a covering is equivalent to

$$d_h \omega_\varphi^s = 0, \qquad s = 1, 2, \dots,$$

d_h being the horizontal differential on $\mathcal{E}$. In particular, one-dimensional coverings are identified with $\ker(d_h)$. We say that a one-dimensional covering φ is *trivial* if the corresponding form ω_φ is exact (for motivations see [10]). Thus, the set of classes of nontrivial one-dimensional coverings φ with ω_φ independent of nonlocal variables is identified with the cohomology group $H_h^1(\mathcal{E})$, or with the group of nontrivial conservation laws for $\mathcal{E}$.

3.5. Functors K and T. Having in mind the reconstruction of recursion operators from their shadows, we introduce two functors in the category $\mathrm{GDE}(M)$. One of them is known from the classical (nongraded) theory (see [4, 10]), the other is specific to graded equations (though it was introduced in a somewhat different form in [16]).

Let $(\mathcal{F}, \nabla_{\mathcal{F}})$ be an object of $\mathrm{GDE}(M)$ and $H_h^1(\mathcal{F})$ be the $\mathbb{R}$-module of its first horizontal cohomology. Let $\{w_\alpha\}$ be the set of generators for $H_h^1(\mathcal{F})$, each w_α being the cohomology class of a form $\omega_\alpha \in \Lambda_h^1(\mathcal{F})$, $\omega_\alpha = \sum_{i=1}^m X_\alpha^i \, dx_i$. We define the functor $K\colon \mathrm{GDE}(M) \Rightarrow \mathrm{GDE}(M)$ killing $H_h^1(\mathcal{F})$ as follows. The algebra $K\mathcal{F}$ is a graded commutative algebra freely generated by $\{w_\alpha\}$ over $\mathcal{F}$ with $\mathrm{gr}(w_\alpha) = \mathrm{gr}(X_\alpha^i)$. The connection $\nabla_{K\mathcal{F}}$ has the form

$$\nabla_{K\mathcal{F}}\left(\frac{\partial}{\partial x_i}\right) = \nabla_{\mathcal{F}}\left(\frac{\partial}{\partial x_i}\right) + \sum_\alpha X_\alpha^i \frac{\partial}{\partial w_\alpha}.$$

From the fact that H_h^1 is a covariant functor from $\mathrm{GDE}(M)$ to the category of R-modules it easily follows that K is a functor as well.

To define the functor $T\colon \mathrm{GDE}(M) \Rightarrow \mathrm{GDE}(M)$, set $T\mathcal{F} = \mathcal{C}^*\Lambda(\mathcal{F})$, where $\mathcal{C}^*\Lambda^*(\mathcal{F}) = \sum_{p\geqslant 0} \mathcal{C}^p\Lambda(\mathcal{F})$ is the module of all Cartan forms on $\mathcal{F}$ (see 2.7). If $\mathcal{F}$ is n-graded, then $T\mathcal{F}$ carries an obvious structure of $(n+1)$-graded algebra. The action of vector fields $\nabla_{\mathcal{F}}(X)$, $X \in D(M)$, on $\Lambda^*(\mathcal{F})$ by Lie derivatives preserves the submodule $\mathcal{C}^*\Lambda(\mathcal{F})$. Since $\mathcal{C}^*\Lambda(\mathcal{F})$, as a graded algebra, is generated by the elements χ and $d_{\mathcal{C}}\psi$, $\chi, \psi \in \mathcal{F}$, this action can be written as

$$\begin{aligned} L_{\nabla_{\mathcal{F}}(X)}\chi &= \nabla_{\mathcal{F}}(X)\chi, \\ L_{\nabla_{\mathcal{F}}(X)} d_{\mathcal{C}}\psi &= d_{\mathcal{C}} \nabla_{\mathcal{F}}(X)(\psi), \\ L_{\nabla_{\mathcal{F}}(X)}(\chi d_{\mathcal{C}}\psi) &= L_{\nabla_{\mathcal{F}}}(\chi) \cdot d_{\mathcal{C}}\psi + \chi L_{\nabla_{\mathcal{F}}(X)} d_{\mathcal{C}}\psi. \end{aligned}$$

Moreover, for any $X \in D(M)$ and $\omega \in \mathcal{C}^*\Lambda(\mathcal{F})$ we have $\nabla_{\mathcal{F}}(X) \lrcorner \, \omega = 0$; hence, for any $\theta \in \mathcal{C}^*\Lambda(\mathcal{F})$,

$$\begin{aligned} (\theta \wedge L_{\nabla_{\mathcal{F}}(X)})(\omega) &= \theta \wedge L_{\nabla_{\mathcal{F}}(X)}(\omega) + (-1)^{\theta_1} d\theta \wedge (\nabla_{\mathcal{F}}(X) \lrcorner \, \omega) \\ &= \theta \wedge L_{\nabla_{\mathcal{F}}(X)}(\omega). \end{aligned}$$

This means that we have a natural extension of the connection $\nabla_{\mathcal{F}}$ in $\mathcal{F}$ to a connection $\nabla_{T\mathcal{F}}$ in $T\mathcal{F}$. It is easy to see that the correspondence $T\colon (\mathcal{F}, \nabla_{\mathcal{F}}) \Rightarrow (T\mathcal{F}, \nabla_{T\mathcal{F}})$ is functorial. We call $(T\mathcal{F}, \Delta_{T\mathcal{F}})$ the Cartan covering of $(\mathcal{F}, \Delta_{\mathcal{F}})$.

In the case when $(\mathcal{F}, \nabla_{\mathcal{F}})$ is an evolution equation $\mathcal{E}$ of the form (3.3), $T(\mathcal{F}, \nabla_{\mathcal{F}})$ is again an evolution equation $T\mathcal{E}$ with additional dependent variables $v^1, \dots, v^m$ and additional relations

$$v_t^1 = \sum_{i,j} \frac{\partial f^1}{\partial u_i^j} v_i^j, \dots, v_t^m = \sum_{i,j} \frac{\partial f^m}{\partial u_i^j} v_i^j. \tag{3.17}$$

3.6. Reconstructing shadows. Computer computations on nonlocal objects, such as symmetries and recursion operators, can be effectively realized for shadows of these objects (see the next section). Here we describe a setting that guarantees the existence of symmetries and, in general, elements of $H_C^{p,0}(\mathcal{E})$ corresponding to the computed shadows. Below we still consider only evolution equations.

PROPOSITION 3.3. *Let $\mathcal{E}$ be an evolution equation and φ be its covering. Let $\theta \in SH_C^p(\mathcal{E}_\varphi)$. Then, if the coefficients X_s, T_s for the extensions of total derivatives do not depend on nonlocal variables for all s,*

(i) *for any extension $\ni_{\theta,p}$ of θ to a vector field on $\mathcal{E}_\varphi$, the forms*

$$\ni_{\theta,p}(\omega_\varphi^s) \overset{\text{def}}{=} \Omega^s$$

(*see Remark* 1 *in* 3.4) *are d_h-closed on $\mathcal{E}_\varphi$;*

(ii) *θ is extendable to an element of $H_C^{p,0}(\mathcal{E}_\varphi)$ iff all Ω^s are d_h-exact.*

PROOF. To prove the first statement, note that Proposition 2.3(i) implies

$$\begin{aligned} 0 = \partial_\varphi^2(\ni_{\theta,p}) &= \partial_\varphi\left(\sum_s (\ni_{\theta,p}(\omega_\varphi^s) + d_h\rho_s) \otimes \frac{\partial}{\partial \omega_s}\right) \\ &= -\sum_s d_h\Omega^s \otimes \frac{\partial}{\partial \omega_s}. \end{aligned} \tag{3.18}$$

The second statement immediately follows from (3.15). □

REMARK. If X_s, T_s depend on $w_1, w_2, \dots$, then (3.18) transforms into

$$\sum_s \left(d_h\Omega^s - (\Omega^s + d_h\rho_s) \wedge \left(\frac{\partial X_s}{dw_s} dx + \frac{\partial T_s}{\partial w_s} dt\right)\right) \otimes \frac{\partial}{\partial w_s} = 0. \tag{3.19}$$

Now let $\theta \in SH_C^{p,0}(\mathcal{E}_\varphi)$ and $\Phi \in H_C^{q,0}(\mathcal{E}_\varphi)$. Then Proposition 2.3(iii) implies

$$[i_\Phi, \partial_\varphi]\theta = (-1)^q(\partial_\varphi\Phi) \lrcorner\, \theta = 0.$$

Since, by the definition of shadows, $\partial_\varphi\theta$ is a φ-vertical element, it follows that $i_\Phi\partial_\varphi\theta$ is vertical too. This means that $\partial_\varphi i_\Phi\theta$ is a φ-vertical element, i.e., $i_\varphi\theta \in SH_C^{p+q-1}(\mathcal{E}_\varphi)$. This proves the following

PROPOSITION 3.4. *For any* $\theta \in SH_{\mathcal{C}}^{p,0}(\mathcal{E}_\varphi)$ *and* $\Phi \in H_{\mathcal{C}}^{q,0}(\mathcal{E}_\varphi)$ *the element* $\Phi \lrcorner \theta$ *lies in* $SH_{\mathcal{C}}^{p+q-1}(\mathcal{E}_\varphi)$. *In particular, when applying a shadow of a recursion operator to a symmetry, one gets a shadow of a symmetry.*

The next result follows directly from the previous ones.

THEOREM 3.5. *Let* $\mathcal{E}$ *be an evolution equation of the form* (3.3) *and* $\mathcal{E}_\varphi$ *be its covering constructed by infinite application of the functor* $K\colon \mathcal{E}_\varphi = K^{(\infty)}\mathcal{E}$, *where*

$$K^{(\infty)}\mathcal{E} = \lim_{n\to\infty} (K^n\mathcal{E}), \qquad K^n\mathcal{E} = (K \circ \cdots \circ K)\mathcal{E},$$

n times. Then for any shadow $\mathcal{R}$ *of a recursion operator in* $\mathcal{E}_\varphi$ *and a symmetry* $\Phi \in \operatorname{sym}\mathcal{E}_\varphi$, *the shadow* $\mathcal{R}(\Phi)$ *can be extended to a symmetry of* $\mathcal{E}_\varphi$. *Thus, an action of* $SH_{\mathcal{C}}^{1,0}(\mathcal{E}_\varphi)$ *on* $\operatorname{sym}(\mathcal{E}_\varphi)$ *is defined modulo "shadowless" symmetries.*

To be sure that elements of $SH_{\mathcal{C}}^{1,0}(\mathcal{E}_\varphi)$ can be extended to recursion operators in an appropriate setting, we prove the following two results.

PROPOSITION 3.6. *Let* $\mathcal{E}$ *be an equation and* $\mathcal{E}_\varphi$ *be its covering by means of* $T\mathcal{E}$. *Then there exists a natural embedding*

$$T_{\mathrm{sym}}\colon H_{\mathcal{C}}^{*,0}(\mathcal{E}) \to \operatorname{sym}(T\mathcal{E})$$

of graded Lie algebras.

PROOF. Let $\Phi \in H_{\mathcal{C}}^{*,0}(\mathcal{E})$. Then L_Φ acts on $\Lambda^*(\mathcal{E})$ and this action preserves $\mathcal{C}^*\Lambda(\mathcal{E}) \subset \Lambda^*(\mathcal{E})$, since $[L_\Phi, d_{\mathcal{C}}] = L_{[\![\Phi, U_{\mathcal{E}}]\!]} = 0$.

Let $X \in \mathcal{C}D(\mathcal{E})$. Then, due to (1.11), $[\![X, \Phi]\!] \lrcorner U_{\mathcal{E}} = 0$. But, using (1.11) again, one can see that $[\![X, \Phi]\!]$ is a vertical element. Hence,

$$[\![X, \Phi]\!] \lrcorner U_{\mathcal{E}} = [\![X, \Phi]\!] = 0. \quad \square$$

Proposition 3.6 allows one to compute elements of $H_{\mathcal{C}}^{*,0}(\mathcal{E})$ as nonlocal symmetries in $\mathcal{E}_\varphi = T\mathcal{E}$. This is the basis of the computational technology used in §4.

The last result of this subsection follows from the previous ones.

THEOREM 3.7. *Let* $\mathcal{E}$ *be an evolution equation and* $\mathcal{E}_\varphi$ *be its covering constructed by infinite application of the functor* $K \circ T$. *Then any shadow* $\Phi \in SH_{\mathcal{C}}^{*,0}(\mathcal{E}_\varphi)$ *can be extended to an element of* $H_{\mathcal{C}}^{*,0}(\mathcal{E}_\varphi)$. *In particular, to any shadow* $SH_{\mathcal{C}}^{1,0}(\mathcal{E}_\varphi)$ *corresponds a recursion operator in* $\mathcal{E}_\varphi$.

REMARK. For "fine obstructions" to shadows reconstruction one must use corresponding term of A. Vinogradov's $\mathcal{C}$-spectral sequence ([17], cf. [9]).

3.7. Lie algebra structure. We finish this section by describing the Lie algebra structure of $\operatorname{sym}(\mathcal{E})$ for the equations $\mathcal{E}$ possessing a recursion operator (cf. [5]).

Let $\mathcal{E}$ be such an equation and $\mathcal{R}$ its recursion operator. We call $\mathcal{E}$ 2-trivial if $H_{\mathcal{C}}^{2,0}(\mathcal{E}) = 0$, and say that $\mathcal{R}$ is even if $\mathcal{R} \cdot \mathcal{R} \in 2\mathbb{Z}$. For any symmetry $X \in \operatorname{sym}(\mathcal{E})$, we set $X_n \overset{\text{def}}{=} \mathcal{R}^n X = (\cdots(X \lrcorner \mathcal{R}) \lrcorner \cdots) \lrcorner \mathcal{R}$, n times.

PROPOSITION. *Let $\mathcal{E}$ be a 2-trivial equation and $\mathcal{R}$ be an even recursion operator for $\mathcal{E}$. Then for any $X, Y \in \operatorname{sym}(\mathcal{E})$ and $m, n \geqslant 1$ one has*

$$[X_m, Y_n] = [X_m, Y]_n + (-1)^{mY\cdot\mathcal{R}}([X, Y_n]_m - [X, Y]_{m+n}). \tag{3.20}$$

PROOF. Induction with respect to $m+n$. The first step is given by (2.16):

$$[X_1, Y_1] = [X_1, Y]_1 + (-1)^{Y\cdot\mathcal{R}}([X, Y_1]_1 - [X, Y]_2).$$

Now assume (3.20) is proved for all $m+n \leqslant s$, $m, n \geqslant 1$. Then for $m+n = s+1$,

$$\begin{aligned}
[X_m, Y_n] &= [(X_{m-1})_1, (Y_{n-1})]_1 \\
&= [X_m, Y_{n-1}]_1 + (-1)^{Y\cdot\mathcal{R}}([X_{m-1}, Y_n]_1 - [X_{m-1}, Y_{n-1}]_2) \\
&= \{[X_m, Y]_{n-1} + (-1)^{mY\cdot\mathcal{R}}([X, Y_{n-1}]_m - [X, Y]_{m+n-1})\}_1 \\
&\quad + (-1)^{YR}\{[X_{m-1}, Y]_n \\
&\qquad\qquad + (-1)^{(m-1)Y\cdot\mathcal{R}}([X, Y_n]_{m-1} - [X, Y]_{m+n-1})\}_1 \\
&\quad - (-1)^{Y\cdot\mathcal{R}}\{[X_{m-1}, Y]_{n-1} \\
&\qquad\qquad + (-1)^{(m-1)Y\cdot\mathcal{R}}([X, Y_{n-1}]_{m-1} - [X, Y]_{m+n-2})\}_2 \\
&= [X_m, Y]_n + (-1)^{mY\cdot\mathcal{R}}([X, Y_n]_m - [X, Y]_{m+n}). \quad \square
\end{aligned}$$

LEMMA 3.9. *Let $(\mathcal{F}, \nabla_{\mathcal{F}})$ be an object of $\mathrm{GDE}(M)$, $X, Y \in D^v(\mathcal{F})$, $\mathcal{R} \in \Lambda^*(\mathcal{F}) \otimes D^v(\mathcal{F})$. Then*

$$[X, \mathcal{R}(Y)] = \mathcal{R}[X, Y] + (-1)^{X\cdot Y}[\![X, \mathcal{R}]\!](Y). \tag{3.21}$$

PROOF. This lemma is a direct consequence of (1.11). □

COROLLARY 3.10. *If $X \in \operatorname{sym}(\mathcal{F}) = H^0_{\mathcal{C}}(\mathcal{F})$ and $\mathcal{R} \in H^{1,0}_{\mathcal{C}}(\mathcal{F})$, then $\mathcal{R}_X \stackrel{\text{def}}{=} [\![X, \mathcal{R}]\!]$ lies in $H^{1,0}_{\mathcal{C}}(\mathcal{F})$ as well, i.e., $\operatorname{sym}(\mathcal{F})$ acts on the algebra of recursion operators of $(\mathcal{F}, \nabla_{\mathcal{F}})$ by the Frölicher–Nijenhuis bracket.*

PROOF. In (3.21) let $X \in \operatorname{sym}(\mathcal{F})$ and $Y \in \mathcal{C}D(\mathcal{F})$. Then, since $\mathcal{R}(Y) = 0$ and $[X, Y] = 0$, one has

$$[\![X, \mathcal{R}]\!](Y) = 0. \quad \square$$

PROPOSITION 3.11. *Let $\mathcal{E}$ and $\mathcal{R}$ satisfy the assumptions of Proposition 3.8. Then for any $X, Y \in \operatorname{sym}(\mathcal{E})$ and $m, n \geqslant 1$ we have*

$$[X, Y_n] = [X, Y]_n + (-1)^{X\cdot Y}\sum_{i=0}^{n-1}(-1)^{iX\cdot\mathcal{R}}(\mathcal{R}_X Y_i)_{n-i-1}, \tag{3.22}$$

$$[X_m, Y] = (-1)^{mY\cdot\mathcal{R}}\left([X, Y]_m - \sum_{i=0}^{m-1}(-1)^{iY\cdot\mathcal{R}}(\mathcal{R}_Y X_i)_{m-i-1}\right). \tag{3.23}$$

PROOF. This follows by a straightforward induction based on (3.20).

From the results obtained one gets the following

THEOREM 3.12. *Let $\mathcal{E}$ be a 2-trivial equation and $\mathcal{R}$ be an even recursion operator for $\mathcal{E}$. Then for any X, $Y \in sym(\mathcal{E})$ and $m, n \geqslant 1$ we have*

$$[X_m, Y_n] = (-1)^{mY\cdot\mathcal{R}}\Big\{[X, Y]_{m+n} + (-1)^{X\cdot Y}\sum_{i=0}^{n-1}(-1)^{iX\cdot\mathcal{R}}(\mathcal{R}_X Y_i)_{m+n-i-1} - \sum_{i=0}^{m-1}(-1)^{iY\cdot\mathcal{R}}(\mathcal{R}_Y X_i)_{m+n-i-1}\Big\}. \tag{3.24}$$

COROLLARY 3.13. *If X and Y are such that $\mathcal{R}_x$ and $\mathcal{R}_y$ commute with $\mathcal{R}$ with respect to the Richardson–Nijenhuis bracket, then*

$$[X_m, Y_n] = (-1)^{mY\cdot\mathcal{R}}\{[X, Y]_{m+n} + (-1)^{X\cdot Y}n(\mathcal{R}_X Y)_{m+n-1} - m(\mathcal{R}_Y X)_{m+n-1}\}. \tag{3.25}$$

COROLLARY 3.14. *If a recursion operator $\mathcal{R}$ is invariant with respect to an even symmetry X, then the hierarchy of symmetries generated by X and $\mathcal{R}$ is commutative: $[X_m, X_n] = 0$ for all m, n.*

PROOF. Taking $X = Y$ in (3.23), one gets

$$[X_m, X_n] = (-1)^{mX\cdot\mathcal{R}}[X, X]_{m+n} = 0$$

due to the evenness of X. □

From the above it follows that the Lie algebra structure of sym$(\mathcal{E})$ heavily depends on the triviality of $H_{\mathcal{C}}^{2,0}(\mathcal{F})$. We shall consider the problem of computations for $H_{\mathcal{C}}^{p,0}(\mathcal{F})$, $p \geqslant 2$, elsewhere.

§4. Applications

We illustrate the theory of deformations of supersymmetric equations and systems developed above by several examples.

First of all, we develop the theory for the supersymmetric extension of KdV equation [3, 11, 12, 14]. We construct the recursion operator for symmetries, which in this case is exactly the contraction of a symmetry and the deformation. Moreover, we construct a new hierarchy of conserved quantities and a hierarchy of (x, t)-dependent symmetries, which was lost in a previous paper [3].

As a second application, we consider the two supersymmetric extensions of nonlinear Schrödinger equation leading to the recursion operators for symmetries and new hierarchies of odd and even symmetries.

Finally, we construct a supersymmetric extension of the Boussinesq equation, again leading to the recursion operator and hierarchies of odd and even symmetries and conservation laws.

4.1. Supersymmetric KdV equation. We start with the supersymmetric extension of the KdV equation [11, 12] and restrict our considerations to the case $a = 3$ in

$$(4.1)\qquad \begin{aligned} u_t &= -u_3 + 6uu_1 - a\varphi\varphi_2 ,\\ \varphi_t &= -\varphi_3 + (6-a)\varphi_1 u + a\varphi u_1 . \end{aligned}$$

Basic properties of the equation were discussed in several papers cf. [3, 11, 13, 14]. In order to construct a deformation of (4.1), we must construct an appropriate covering by the introduction of several nonlocal variables. These nonlocal variables, which arise classically from conserved densities related to conservation laws, have been computed, and are

$$\begin{aligned} &q_{1/2} = D^{-1}(\varphi), \quad q_{3/2} = D^{-1}(p_1\varphi), \quad q_{5/2} = D^{-1}(\tfrac{1}{2}p_1^2\varphi - u\varphi),\\ (4.2)\qquad &q_{7/2} = D^{-1}(p_3\varphi + q_{1/2}[\tfrac{1}{2}p_1^2 u - p_1u_1 + u_2 - u^2]),\\ &q_{9/2} = D^{-1}(6p_3p_1\varphi + q_{1/2}[p_1^3 u - 3p_1^2u_1 + 6p_1u_2 - 6p_1u^2 + 36uu_1 - 6u_3]),\end{aligned}$$

$$(4.3)\qquad \begin{aligned} &p_1 = D^{-1}(u), \quad p_0 = D^{-1}(p_1), \quad \overline{p}_1 = D^{-1}(\varphi q_{1/2}),\\ &p_3 = D^{-1}(u^2 - \varphi\varphi_1), \quad \overline{p}_3 = D^{-1}(u^2 - 2u\varphi q_{1/2} + uq_{1/2}q_{3/2}),\end{aligned}$$

where $D = D_x$. Odd nonlocal variables will be denoted by q , while even nonlocal variables will be denoted by p , $\overline{p}$.

For hierarchies of symmetries of (4.1) $(Y_{n+1/2})$, (X_{2n+1}) $(n \in \mathbb{N})$ we refer to [3], where

$$\begin{aligned} Y_{1/2} &= \varphi_1\partial_u + u\partial_\varphi ,\\ Y_{3/2} &= (2q_{1/2}u_1 - p_1\varphi_1 + u\varphi - \varphi_2)\partial_u + (2q_{1/2}\varphi_1 - p_1u + u_1)\partial_\varphi ,\\ (4.4)\qquad X_1 &= u_1\partial_u + \varphi_1\partial_\varphi , \qquad X_3 = -u_t\partial_u - \varphi_t\partial_\varphi ,\\ X_5 &= -(u_5 - 10u_3u - 20u_2u_1 + 30u_1u^2\\ &\quad + 5\varphi\varphi_4 + 5\varphi_1\varphi_3 - 20u\varphi\varphi_2 - 20u_1\varphi\varphi_1)\partial_u\\ &\quad - (\varphi_5 - 5u\varphi_3 - 10u_1\varphi_2 - 10u_2\varphi_1 + 10u^2\varphi_1 + 20u_1u\varphi - 5u_3\varphi)\partial_\varphi .\end{aligned}$$

Moreover we found a supersymmetric analog of the (x, t)-dependent symmetry which acts as recursion on the even hierarchy (X_{2n+1}) $(n \in \mathbb{N})$, i.e.,

$$(4.5)\qquad V_2 = -6tX_5 - 2xX_3 + H_2 ,$$

where

(4.6)

$$\begin{aligned} H_2 = \{&-q_{1/2}(\varphi_2 + p_1\varphi_1 - \varphi u)\\ &+ 3q_{3/2}\varphi_1 - 13\varphi\varphi_1 + 4p_1u_1 - 2\overline{p}_1u_1 - 8u_2 + 16u^2\}\partial_u\\ &+ \{-q_{1/2}(p_1u - u_1) + 3q_{3/2}u + 2p_1\varphi_1 - 2\overline{p}_1\varphi_1 - 7\varphi_2 + 14\varphi u\}\partial_\varphi .\end{aligned}$$

It should be noted that the vector fields

$$(4.7)\qquad \begin{aligned} Y_{-1/2} &= \partial_{q_{1/2}} + \cdots = \partial_{q_{1/2}} - q_{1/2}\partial_{\overline{p}_1} + \overline{p}_1\partial_{q_{3/2}} ,\\ \overline{X}_{-1} &= \partial_{p_1} + \cdots = \partial_{p_1} , \qquad X_{-1} = \partial_{\overline{p}_1} + \cdots \end{aligned}$$

are symmetries of (4.1) in the covering defined by (4.2), (4.3). These are so-called symmetries vertical with respect to the covering (or shadowless symmetries); computation of graded Lie brackets leads to

$$(4.8)\quad [Y_{-1/2}, V_2] = Y_{3/2}, \quad [Y_{-1}, Y_2] = 2Z_1 + 4X_1, \quad [\overline{X}_{-1}, V_2] = -2X_1.$$

In order to construct a deformation of (4.1) we formally construct the infinite-dimensional Cartan covering (see 3.5) over the infinite covering of (4.1) by (4.2), (4.3), etc. We take the covering of (4.1) by (4.2), (4.3) and take the Cartan covering of that by introducing the Cartan forms

$$(4.9)\qquad \omega_{q_{1/2}}, \ \ldots, \ \omega_{q_{9/2}}, \ \omega_{p_0}, \ \omega_{p_1}, \ \omega_{\overline{p}_1}, \ \omega_{p_3}, \ \omega_{\overline{p}_3},$$

where $\omega_f = L_{U_\varphi}(f)$ denotes the Cartan form corresponding to the potential f (see 3.4). According to (3.8), (3.9), we look for a generalized vector *field* linear with respect to the Cartan forms. Applying the deformation condition to this vector field and taking into account the grading of (4.1)–(4.3), (4.9) we arrive at the following deformation

(4.10)

$$\begin{aligned} U_1 = \{&\omega_{u_2} + \omega_u(-4u) + \omega_{\varphi_1}(-2\varphi) + \omega_\varphi(\varphi_1) + \omega_{q_{1/2}}(q_{1/2}u_1 + p_1\varphi_1 + \varphi_2 - u\varphi) \\ &+ \omega_{q_{3/2}}(-\varphi_1) + \omega_{p_1}(-2u_1) + \omega_{\overline{p}_1}(u_1)\}\,\partial_u \\ &+ \{\omega_{\varphi_2} + \omega_\varphi(-2u) + \omega_u(-2\varphi) + \omega_{q_{1/2}}(-q_{1/2}\varphi_1 + p_1 u - u_1) \\ &+ \omega_{p_1}(-\varphi_1) + \omega_{\overline{p}_1}(\varphi_1) + \omega_{q_{3/2}}(-u)\}\,\partial_\varphi. \quad \square \end{aligned}$$

It can be proved that U_1 satisfies $[\![U_1, U_1]\!] = 0$ which means that it is a graded Nijenhuis operator [5].

We redefine our hierarchies in the following way. First we put

$$\begin{gathered} Y_{1/2} = \varphi_1\partial_u + u\partial_\varphi, \\ Y_{3/2} = (2q_{1/2}u_1 - p_1\varphi_1 + u\varphi - \varphi_2)\partial_u + (2q_{1/2}\varphi_1 - p_1 u + u_1)\partial_\varphi, \\ X_1 = u_1\partial_u + \varphi_1\partial_\varphi, \qquad \overline{X}_1 = (q_{1/2}\varphi_1)\partial_u + (q_{1/2}u - \varphi_1)\partial_\varphi = Z_1 \quad (\text{ref. [3]}), \\ V_0 = -x\partial_x - 3t\partial_t + 2u\partial_u + (3/2)\varphi\partial_\varphi \quad (\text{scaling}) \end{gathered}$$

and define

$$\begin{gathered} Y_{2n+1/2} = \underbrace{((((Y_{1/2} \lrcorner N) \lrcorner N)\cdots) \lrcorner N)}_{n \text{ times}} \stackrel{\text{def}}{=} Y_{1/2} \lrcorner N^n, \\ Y_{2n+3/2} = Y_{3/2} \lrcorner N^n, \qquad X_{2n+1} = X_1 \lrcorner N^n, \\ \overline{X}_{2n+1} = \overline{X}_1 \lrcorner N^n, \qquad V_{2n} = V_0 \lrcorner N^n. \end{gathered}$$

4.2. Passing from deformations to recursion operators. Here we rewrite the main result of the previous subsection in more conventional terms.

Let $X = (F, G)$ be a nonlocal symmetry of (4.1) in the covering (4.2), (4.3) with a 2-component generating function (F, G). This means that X is of the form

$$X = \sum_{i=0}^{\infty}(D^i(F)\partial_{u_i} + D^i(G)\partial_{\varphi_i}),$$

where F and G satisfy the shadow equation for the covering in question and D denotes the extension of the total derivative D_x to the covering. Then

$$i_X(\omega_{u_i}) = D^i(F), \qquad i_X(\omega_{\varphi_i}) = D^i(G)$$

for all $i = 0, 1, \ldots$.

From the definition of nonlocal variables (see (4.2, 4.3)), one also has

$$\begin{aligned} i_X(\omega_{p_1}) &= D^{-1}(F), \qquad i_x(\omega_{q_{1/2}}) = D^{-1}(G), \\ i_X(\omega_{\overline{p}_1}) &= D^{-1}(G_{q_{1/2}} + \varphi D^{-1}(G)), \\ i_X(\omega_{q_{3/2}}) &= D^{-1}(p_1 G + D^{-1}(F)\varphi). \end{aligned}$$

Then the recursion operator $\mathcal{R}$ corresponding to the deformation U_1 from (4.10) acts as $\mathcal{R}(X) = i_X(U_1)$ and is of the form

$$\mathcal{R}(F, G) = (F_1, G_1),$$

where

$$\begin{aligned} F_1 = {} & D^2(F) - 2Fu - 2D^{-1}(F)u_1 - D^{-1}(D^{-1}(F)\varphi)\varphi_1 \\ & - 2D(G)\varphi + G\varphi_1 + D^{-1}(G)(q_{1/2}u_1 + p_1\varphi_1 + \varphi_2 - u\varphi) \\ & - D^{-1}(p_1 G)\varphi_1 + D^{-1}(Gq_{1/2} + \varphi D^{-1}(G))u_1, \\ G_1 = {} & -2F\varphi - D^{-1}(F)\varphi_1 - D^{-1}(D^{-1}(F)\varphi)u \\ & + D^2(G) - 2Gu - D^{-1}(G)(q_{1/2}\varphi_1 - p_1 u + u_1) \\ & - D^{-1}(p_1 G)u + D^{-1}(Gq_{1/2} + \varphi D^{-1}(G))\varphi_1. \end{aligned}$$

In what follows we restrict ourselves to representations similar to (4.10).

4.3. Supersymmetric extensions of nonlinear Schrödinger equation. In this subsection we discuss deformations and recursion operators for the two supersymmetric extensions of the nonlinear Schrödinger equation [15]

$$\begin{aligned} u_t &= -v_2 + kv(u^2 + v^2) - u_1(c_1 - c_2)\varphi\psi - 4kv\psi\psi_1 \\ &\quad - u(c_1 + c_2 + 4k)\psi\varphi_1 + c_2 u\varphi\psi_1 + c_1 v\varphi\varphi_1, \\ v_t &= u_2 - ku(u^2 + v^2) - v_1(c_1 - c_2)\varphi\psi + 4ku\varphi\varphi_1 \\ &\quad + v(c_1 + c_2 + 4k)\varphi\psi_1 - c_1 u\psi\psi_1 - c_2 v\psi\varphi_1, \\ \varphi_t &= -\psi_2 + (\tfrac{1}{2}c_2 u^2 + ku^2 + kv^2)\psi - \tfrac{1}{2}c_2 uv\varphi - (c_1 - c_2)\varphi\psi\varphi_1, \\ \psi_t &= \varphi_2 - (\tfrac{1}{2}c_2 v^2 + ku^2 + kv^2)\varphi + \tfrac{1}{2}c_2 uv\psi - (c_1 - c_2)\varphi\psi\psi_1, \end{aligned} \tag{4.11}$$

where in

$$\begin{aligned} qc_1 &= -4k, \quad c_2 = 0, && \text{in Case A}, \\ c_1 &= c, \quad c_2 = 4k, && \text{in Case B}. \end{aligned} \tag{4.12}$$

The construction of deformations follows exactly along the same lines as for the supersymmetric KdV equation, presented in 4.1, so for the nonlinear Schrödinger equation we shall only state the results.

Case A. In order to work in the appropriate covering for the supersymmetric extension of the nonlinear Schrödinger equation, we construct the following set of nonlocal variables, associated to conserved quantities

$$p_0,\ p_1,\ p_2,\ \overline{p}_0,\ \overline{p}_1,\ \overline{p}_2,\ q_{1/2},\ \overline{q}_{1/2},\ q_{3/2},\ \overline{q}_{3/2},\ q_{5/2},\ \overline{q}_{5/2},$$

which are defined by

$$\begin{aligned}
p_0 &= D^{-1}(\varphi\psi), \qquad \overline{p}_0 = D^{-1}(p_1),\\
p_1 &= D^{-1}(u^2+v^2-2\varphi\varphi_1-2\psi\psi_1),\\
\overline{p}_1 &= D^{-1}(k(\psi v+\varphi u)\overline{q}_{1/2}+k(\psi u-\varphi v)q_{1/2}-2\psi\psi_1-2\varphi\varphi_1),\\
p_2 &= D^{-1}(uv_1+2\varphi_1\psi_1),\\
\overline{p}_2 &= D^{-1}[k(2\psi_1 v+2\varphi_1 u+k\psi v p_1+k\varphi u p_1)q_{1/2}\\
&\quad +k(-2\psi_1 u+2\varphi_1 v-k\psi u p_1+k\varphi v p_1)\overline{q}_{1/2}+2uv_1],\\
q_{1/2} &= D^{-1}(\psi u-\varphi v), \qquad \overline{q}_{1/2} = D^{-1}(\psi v+\varphi u),\\
q_{3/2} &= D^{-1}(k\psi u p_1-k\varphi v p_1+2\psi_1 u-2\varphi_1 v),\\
\overline{q}_{3/2} &= D^{-1}(k\psi v p_1+k\varphi u p_1+2\psi_1 v+2\varphi_1 u).
\end{aligned} \tag{4.13}$$

After introducing the associated Cartan forms, we find the Nijenhuis operator for this case to be

(4.14)

$$\begin{aligned}
U_1 = \{&\omega_{v_1}+\omega_{p_1}(-kv)-2\omega_{p_0}ku_1+\omega_u(-2k\varphi\psi)+\omega_\varphi(-ku\psi-kv\varphi)\\
&+\omega_\psi(-kv\psi+ku\varphi)-\omega_{q_{1/2}}(k\varphi_1)+\omega_{\overline{q}_{1/2}}(k\psi_1)\}\partial_u\\
+\{&-\omega_{u_1}+\omega_{p_1}(ku)-2\omega_{p_0}kv_1+\omega_v(-2k\varphi\psi)+\omega_\varphi(-kv\psi+ku\varphi)\\
&+\omega_\psi(ku\psi+kv\varphi)+\omega_{q_{1/2}}(-k\psi)+\omega_{\overline{q}_{1/2}}(-k\varphi_1)\}\partial_v\\
+\{&\omega_{\psi_1}+\omega_\varphi(k\varphi\psi)+\omega_{p_1}(-\tfrac{1}{2}k\psi)\\
&+\omega_{p_0}(-2k\varphi_1)+\omega_{q_{1/2}}(-\tfrac{1}{2}ku)+\omega_{\overline{q}_{1/2}}(-\tfrac{1}{2}kv)\}\partial_\varphi\\
+\{&-\omega_{\varphi_1}+\omega_\psi(-k\varphi\psi)+\omega_{p_1}(+\tfrac{1}{2}k\varphi)\\
&+\omega_{p_0}(-2k\psi_1)+\omega_{q_{1/2}}(-\tfrac{1}{2}kv)+\omega_{\overline{q}_{1/2}}(+\tfrac{1}{2}ku)\}\partial_\psi .
\end{aligned}$$

By starting with the symmetries [15]

$$\begin{aligned}
X_0 &= v\partial_u-u\partial_v, \qquad \overline{X}_0=\psi\partial_\varphi-\varphi\partial_\psi,\\
Y_{1/2} &= -\psi_1\partial_u+\varphi_1\partial_v+\tfrac{1}{2}v\partial_\varphi-\tfrac{1}{2}u\partial_\psi,\\
\overline{Y}_{1/2} &= \varphi_1\partial_u+\psi_1\partial_v+\tfrac{1}{2}u\partial_\varphi+\tfrac{1}{2}v\partial_\psi,
\end{aligned} \tag{4.15}$$

$$S_0 = -x\partial_x-2t\partial_t+u\partial_u+v\partial_v+\tfrac{1}{2}\varphi\partial_\varphi+\tfrac{1}{2}\psi\partial_\psi, \tag{4.16}$$

the recursion operator $U_1 = R$ generates five hierarchies of symmetries

$$\begin{aligned}
X_n &= R^n X_0, \quad Y_{n+1/2}=R^n Y_{1/2}, \quad \overline{S}_n = R^n S_0,\\
\overline{X}_n &= R^n X_0, \quad \overline{Y}_{n+1/2} = R^n\overline{Y}_{1/2},
\end{aligned} \tag{4.17}$$

where $R^n X_0, \ldots$ should be understood as

$$X_n = (\cdots(((X_0 \lrcorner \underbrace{U_1) \lrcorner U_1) \lrcorner \cdots) \lrcorner U_1)}_{n}. \tag{4.18}$$

Case B ($c_2 = 4k$). In this case the supersymmetric nonlinear Schrödinger equation is

$$\begin{aligned}
u_t &= -v_2 + kv(u^2+v^2) - (c_1-4k)u_1\varphi\psi - 4kv\psi\psi_1 \\
&\quad - (c_1+8k)u\psi\varphi_1 + 4ku\varphi\psi_1 + c_1 v\varphi\varphi_1, \\
v_t &= u_2 - ku(u^2+v^2) - (c_1-4k)v_1\varphi\psi + 4ku\varphi\varphi_1 \\
&\quad + (c_1+8k)v\varphi\psi_1 - c_1 u\psi\psi_1 - 4kv\psi\varphi_1, \\
\varphi_1 &= -\psi_2 + (3ku^2+kv^2)\psi - 2kuv\varphi - (c_1-4k)\varphi\psi\varphi_1, \\
\psi_1 &= \varphi_2 - (ku^2+3kv^2)\varphi + 2kuv\psi - (c_1-4k)\varphi\psi\psi_1.
\end{aligned} \tag{4.19}$$

We introduce the following nonlocal variables

$$\begin{aligned}
p_0 &= D^{-1}(\varphi\psi), \qquad \overline{p}_0 = D^{-1}(p_1 + (c_1+4k)\overline{p}_1), \\
p_1 &= D^{-1}(u^2+v^2+(2k)^{-1}(c_1+4k)(\varphi\varphi_1+\psi\psi_1)), \\
\overline{p}_1 &= D^{-1}((u\psi - v\varphi)q_{1/2} - (2k)^{-1}(\varphi\varphi_1+\psi\psi_1)), \\
q_{1/2} &= D^{-1}(u\psi - v\varphi), \qquad \overline{q}_{1/2} = D^{-1}(q_{3/2} + \tfrac{1}{2}(c_1+4k)\varphi\psi q_{1/2}), \\
q_{3/2} &= D^{-1}(v\psi_1 + u\varphi_1)
\end{aligned} \tag{4.20}$$

and additionally

$$\begin{aligned}
q_{-1/2} &= D^{-1}(q_{1/2}), \\
p_2 &= D^{-1}(-uv_1 + (4k)^{-1}(c_1+4k)\varphi_1\psi_1 + \tfrac{1}{4}(c_1+12k)(u^2+v^2)\varphi\psi), \\
\overline{p}_2 &= D^{-1}(-(v\psi_1+u\varphi_1)q_{1/2} + (2k)^{-1}\varphi_1\psi_1).
\end{aligned} \tag{4.21}$$

Within this covering we constructed a deformation of the form

$$\begin{aligned}
U_1 = \{&\omega_{v_1} + \omega_u\tfrac{1}{2}(c_1-4k)\varphi\psi + \omega_\varphi(-4ku\psi + \tfrac{1}{4}(c_1-4k)v\varphi) \\
&+ \omega_\psi(\tfrac{1}{4}(c_1-4k)v\psi + 4ku\varphi) + \omega_{p_0}\tfrac{1}{2}(c_1-4k)u_1 \\
&+ \omega_{p_1}(-kv) + \omega_{\overline{p}_1}(-\tfrac{1}{2}k(c_1+12k)v) \\
&+ \omega_{q_{3/2}}(\tfrac{1}{2}k(c_1+12k)vq_{1/2} + \tfrac{1}{2}(c_1+4k)\varphi_1) + \omega_{q_{3/2}}(-\tfrac{1}{2}(c_1+4k)\psi)\}\partial_u \\
+ \{&-\omega_{u_1} + \omega_v\tfrac{1}{2}(c_1-4k)\varphi\psi + \omega_\varphi(-4kv\psi - \tfrac{1}{4}(c_1-4k)u\varphi) \\
&+ \omega_\psi(-\tfrac{1}{4}(c_1-4k)u\psi + 4kv\varphi) + \omega_{p_0}\tfrac{1}{2}(c_1-4k)v_1 \\
&+ \omega_{p_1}(ku) + \omega_{\overline{p}_1}\tfrac{1}{2}k(c_1+12k)u \\
&+ \omega_{q_{1/2}}(-\tfrac{1}{2}k(c_1+12k)uq_{1/2} + \tfrac{1}{2}(c_1+4k)\psi_1) + \omega_{q_{3/2}}\tfrac{1}{2}(c_1+4k)\varphi\}\partial_v \\
+ \{&\omega_{\psi_1} + \tfrac{1}{4}\omega_\varphi(c_1-4k)\varphi\psi + \omega_{p_0}\tfrac{1}{2}(c_1-4k)\varphi_1 - \omega_{p_1}k\psi \\
&- \omega_{\overline{p}_1}\tfrac{1}{2}k(c_1+12k)\psi + \omega_{q_{1/2}}(-2ku - \tfrac{1}{2}k(c_1+12k)\psi q_{1/2})\}\partial_\varphi \\
+ \{&-\omega_{\varphi_1} + \omega_\psi\tfrac{1}{4}(c_1-4k)\varphi\psi + \omega_{p_0}\tfrac{1}{2}(c_1-4k)\psi_1 + \omega_{p_1}k\varphi \\
&+ \omega_{\overline{p}_1}\tfrac{1}{2}k(c_1+12k)\varphi + \omega_{q_{1/2}}(-2kv) + \tfrac{1}{2}k(c_1+12k)\varphi q_{1/2}\}\partial_\psi.
\end{aligned} \tag{4.22}$$

The action of U_1 on the symmetries

$$
\begin{gathered}
X_1 = u_1\partial_u + \varphi_1\partial_u + \varphi_1\partial_\varphi + \psi_1\partial_\psi\,, \qquad \overline{X}_1 = Z_1\,,\\
Y_{1/2} = \varphi_1\partial_u + \psi_1\partial_v + \tfrac{1}{2}u\partial_\varphi + \tfrac{1}{2}v\partial_\psi\,,\\
\overline{Y}_{1/2} = q_{1/2}v\partial_u - q_{1/2}u\partial_u + (+q_{1/2}\psi - (4k)^{-1}u)\partial_\varphi + (-q_{1/2}\varphi - (4k)^{-1}v)\partial_\psi\,,\\
X_0 = v\partial_u - u\partial_v + \psi\partial_\varphi - \varphi\partial_\psi\,,\\
\overline{X}_0 = -q_{1/2}\psi\partial_u + q_{1/2}\varphi\partial_v - (8k)^{-1}\varphi\partial_\varphi - (8k)^{-1}\psi\partial_\psi \quad (c_1 = 4k)\,,\\
Y_{-1/2} = \psi\partial_u - \varphi\partial_v
\end{gathered}
\tag{4.23}
$$

creates hierarchies of symmetries in a similar way as in the preceding subsection. The properties of the additional symmetry $\overline{X}_0$ in the case $c_1 = 4k$ will be discussed elsewhere. For explicit formula for Z_1 the reader is referred to [15].

4.4. Supersymmetric Boussinesq equation. We discuss the construction of a supersymmetric extension of the Boussinesq equation. Conservation laws, nonlocal variables, symmetries and recursion operators for this supersymmetric system will be discussed as well.

4.4.1. *The construction of the supersymmetric Boussinesq equation.* We start our discussion with the classical system [1, 13]

$$
\begin{aligned}
u_t &= -\tfrac{1}{2}u_{xx} + uu_x + v_x\,,\\
v_t &= +\tfrac{1}{2}v_{xx} + uv_x + u_xv\,.
\end{aligned}
\tag{4.24}
$$

We construct the so-called fermionic extension [12] by setting

$$
\Phi = \varphi + \theta u\,, \qquad \Psi = \psi + \theta v\,,
\tag{4.25}
$$

where φ, ψ, θ are odd variables. Due to the classical grading of (4.24), i.e.,

$$
\deg(u) = 1\,, \quad \deg(v) = 2\,, \quad \deg(x) = -1\,, \quad \deg(t) = -2\,,
\tag{4.26}
$$

and the grading of the odd variables

$$
\deg(\theta) = -1/2\,, \quad \deg(\varphi) = 1/2\,, \quad \deg(\psi) = 3/2\,,
\tag{4.27}
$$

the variables Φ, Ψ are graded by

$$
\deg(\Phi) = 1/2\,, \quad \deg(\Psi) = 3/2\,.
\tag{4.28}
$$

Now we construct a formal extension of (4.24) by setting

$$
\begin{aligned}
u_t &= f_1[u, v, \varphi, \psi]\,, \qquad & v_t &= f_2[u, v, \varphi, \psi]\,,\\
\varphi_t &= f_3[u, v, \varphi, \psi]\,, \qquad & \psi_t &= f_4[u, v, \varphi, \psi]\,,
\end{aligned}
\tag{4.29}
$$

where f_1, f_2, f_3, and f_4 are functions defined on the jet bundle $J(u, v, \varphi, \psi)$ of degrees 3, 4, $5/2$, and $7/2$, respectively.

The construction of f_1 and f_2 should be done in such a way that in the absence of odd variables f_1, f_2 reduce to the right-hand sides of (4.24).

We now impose on the following requirements on system (4.29) [15].

1. The existence of an odd symmetry of (4.29), i.e.,

$$
\begin{aligned}
Y_{1/2} &= \varphi_1\partial_u + \psi_1\partial_v + u\partial_\varphi + v\partial_\psi , \\
[Y_{1/2}, Y_{1/2}] &= 2(u_1\partial_u + v_1\partial_v + \varphi_1\partial_\varphi + \psi_1\partial_\psi) = -2\partial_x .
\end{aligned}
\tag{4.30}
$$

2. The existence of an even symmetry of appropriate degree of (4.29) which reduces to the classical first higher order symmetry of (4.24) in the absence of odd variables, i.e.,

$$
\begin{aligned}
X_3^{\text{clas}} &= (\tfrac{1}{3}u_3 - u_1^2 + 2uv_1 + 2vu_1 - uu_2 + u^2u_1)\partial_u \\
&\quad + (\tfrac{1}{3}v_3 + u_1v_1 + 2vv_1 + uv_2 + 2uu_1v + u^2v_1)\partial_v .
\end{aligned}
\tag{4.31}
$$

From the above requirements we obtain the following supersymmetric extension of (4.24)

$$
\begin{aligned}
u_t &= -\tfrac{1}{2}u_{xx} + uu_x + v_x , \\
v_t &= \tfrac{1}{2}v_{xx} + u_xv + uv_x + \varphi_x\psi_x + \varphi_{xx}\psi , \\
\varphi_t &= -\tfrac{1}{2}\varphi_{xx} + \psi_x + u\varphi_x , \\
\psi_t &= \tfrac{1}{2}\psi_{xx} + u\psi_x + u_x\psi
\end{aligned}
\tag{4.32}
$$

while the symmetry X_3 is given by

(4.33)

$$
\begin{aligned}
X_3 = {} & (\tfrac{1}{3}u_3 - u_1^2 + 2vu_1 - uu_2 + 2uv_1 + u^2u_1 + \varphi_x\psi_x + \varphi_{xx}\psi)\partial_u \\
& + (\tfrac{1}{3}v_3 + u_1v_1 + 2vv_1 + uv_2 + 2uvu_1 + u^2v_1 \\
& \qquad + \varphi_2\psi_1 + \varphi_1\psi_2 + 2u\varphi_1\psi_1 - \psi\psi_2 + 2\varphi_2\psi u + 2u_1\varphi_1\psi)\partial_v \\
& + (\tfrac{1}{3}\varphi_3 - u\varphi_2 + 2u\psi_1 + u^2\varphi_1 + v\varphi_1 - u_1\varphi_1 + u_1\psi)\partial_\varphi \\
& + (\tfrac{1}{3}\psi_3 + u\psi_2 + u^2\psi_1 + v\psi_1 + u_1\psi_1 + 2uu_1\psi + v_1\psi)\partial_\psi .
\end{aligned}
$$

4.4.2. *Construction of conserved quantities and nonlocal variables.* For the supersymmetric extension (4.32) of the Boussinesq equation we construct the following set of conserved densities (X), associated conserved quantities $(\int_{-\infty}^{\infty} X\,dx)$ and nonlocal variables $\mathbb{X} = D^{-1}(X)$, i.e., p_i of degree i, q_j of degree j.

$$
\begin{gathered}
p_0 = D^{-1}(u), \qquad p_1 = D^{-1}(v), \qquad p_2 = D^{-1}(uv + \varphi_1\psi), \\
p_3 = D^{-1}(v^2 + uv_1 + u^2 v + 2u\varphi_1\psi + \varphi_1\psi_1 - \psi\psi_1), \\
\overline{p}_0 = D^{-1}(p_1), \qquad \overline{p}_1 = D^{-1}(\psi q_{1/2} + \varphi\psi), \\
\overline{p}_2 = D^{-1}(p_1\varphi\psi - u\varphi\psi + \varphi_1\psi - u\psi q_{1/2} + p_1\varphi_1 q_{1/2}), \\
\overline{p}_3 = 2D^{-1}\{(p_2\varphi_1 - u^2\psi - 2v\psi + u_1\psi - p_1 v\varphi - \varphi v_1)q_{1/2} + (u\psi - p_1\psi)\overline{q}_{3/2} \\
+ (-u^2 + p_1 u - 2v + u_1 - p_1^2 + p_2)\varphi\psi - u^2 v - uv_1 - v^2\},
\end{gathered} \tag{4.34}
$$

$$
\begin{gathered}
q_{1/2} = D^{-1}(\psi), \quad q_{3/2} = D^{-1}(u\psi + v\varphi), \quad \overline{q}_{3/2} = D^{-1}(q_{1/2}v + p_1\varphi_1), \\
q_{5/2} = D^{-1}(-\varphi_1 p_1^2 + p_2(2\psi - 2\varphi_1) - 2(p_1 v + v_1)q_{1/2} - 2v\varphi_1), \\
\overline{q}_{5/2} = D^{-1}(\tfrac{1}{2}\varphi_1 p_1^2 + p_2(-2\psi + \varphi_1) + (uv - 2uu_1 + u_1 p_1 + u_2)q_{1/2} + v\varphi_1).
\end{gathered} \tag{4.35}
$$

Note the higher order nonlocalities in $\overline{p}_0, \overline{p}_1, \dots$. The nonlocal variables $p_0, \overline{p}_0, \dots, q_{1/2}, q_{3/2}, \overline{q}_{3/2}, \dots$ are essential in the construction of nonlocal symmetries, while the associated Cartan forms $\omega_{p_0}, \omega_{\overline{p}_0}, \dots, \omega_{q_{1/2}}, \dots$ play a significant role in the construction of deformations or recursion operators.

4.4.3. *Symmetries of supersymmetric Boussinesq equation.* For the supersymmetric extension of Boussinesq equation, we obtain the following symmetries

$$
\begin{aligned}
Y_{1/2} &= \varphi_1\partial_u + \psi_1\partial_v + u\partial_\varphi + v\partial_\psi, \\
\overline{Y}_{1/2} &= \psi\partial_u + \psi_1\partial_v + (u - p_1)\partial_\varphi, \\
X_1 &= u_1\partial_u + v_1\partial_v + \varphi_1\partial_\varphi + \psi_1\partial_\psi, \\
\overline{X}_1 &= (\varphi\psi + \varphi_1 q_{1/2})\partial_u + (\varphi\psi_1 + \varphi_1\psi + \psi_1 q_{1/2})\partial_v \\
&\quad + (-uq_{1/2} - q_{3/2} - \varphi_1 + u\varphi)\partial_\varphi + (-vq_{1/2} - \psi_1 - u\psi)\partial_\psi, \\
Y_{3/2} &= (-2q_{1/2}u_1 - \varphi_2 + u\varphi_1 + p_1\varphi_1 - 3u\psi + u_1\varphi)\partial_u \\
&\quad + (-2q_{1/2}v_1 - \psi_2 + 2u\psi_1 + p_1\psi_1 - v\varphi_1 - v\psi - 2u_1\psi + v_1\varphi)\partial_v \\
&\quad + (2\varphi_1 q_{1/2} + \varphi\varphi_1 - u^2 + up_1 + u_1 + 2p_2)\partial_\varphi \\
&\quad + (2\psi_1 q_{1/2} + 2\varphi_1\psi + \varphi\psi_1 + uv + vp_1 + v_1)\partial_\psi, \\
\overline{Y}_{3/2} &= (-q_{1/2}u_1 - \psi_1 - 2u\psi + p_1\psi)\partial_u \\
&\quad + (-q_{1/2}v_1 - \psi_2 - 2u\psi_1 + p_1\psi_1 - 2u_1\psi)\partial_v \\
&\quad + (\varphi_1 q_{1/2} - u^2 + up_1 - v + u_1 - \tfrac{1}{2}p_1^2 + p_2)\partial_\varphi\psi_1 q_{1/2} + \psi_1 q_{1/2}\partial_\psi.
\end{aligned}
$$

Note that $Y_{3/2} - 2\overline{Y}_{3/2}$ is a *local* vector field.

4.4.4. *Deformation and recursion operator for supersymmetric Boussinesq equations.*

Similarly to the previous applications, we construct a deformation of the equation structure U related to the supersymmetric Boussinesq equation, i.e.,

$$\begin{aligned}
U_1 = {} & (\omega_{u_1} - 2\omega_v - \omega_u u - \omega_{p_0} u_1 - \omega_\varphi \psi + \omega_{q_{1/2}}(2\psi - \varphi_1))\partial_u \\
& + (-\omega_{v_1} - \omega_v u - 2\omega_u v - 2\omega_{\varphi_1}\psi - \omega_\varphi \psi_1 \\
& \qquad + \omega_\psi(\varphi_1 + \psi) - \omega_{p_0} v_1 + \omega_{q_{1/2}}\psi_1)\partial_v \\
& + (\omega_{\varphi_1} - 2\omega_\psi + \omega_\varphi(2p_1 - u) - \omega_{p_0}\varphi_1 \\
& \qquad + \omega_{p_1}(2q_{1/2} + \varphi) - \omega_{q_{3/2}} - 2\omega_{\overline{q}_{3/2}} + \omega_{q_{1/2}} u)\partial_\varphi \\
& + (-\omega_{\psi_1} - \omega_\psi u - 2\omega_u \psi - \omega_{p_0}\psi_1 + \omega_{p_1}\psi - \omega_{q_{1/2}} v)\partial_\psi \,.
\end{aligned}$$

From the deformation U, we obtain four hierarchies of (x, t)-independent symmetries $(Y_{n+1/2})$, $(\overline{Y}_{n+1/2})$, (X_{n+1}), $(\overline{X}_{n+1})$ $(n \in \mathbb{N})$ by

$$\begin{aligned}
Y_{n+1/2} &= (\cdots(Y_{1/2} \lrcorner U_1)\cdots \lrcorner U_1), \\
\overline{Y}_{n+1/2} &= (\cdots(\overline{Y}_{1/2} \lrcorner U_1)\cdots \lrcorner U_1), \\
X_{n+1} &= (\cdots(X_1 \lrcorner U_1)\cdots \lrcorner U_1), \\
\overline{X}_{n+1} &= (\cdots(\overline{X}_1 \lrcorner U_1)\cdots \lrcorner U_1)
\end{aligned}$$

and an (x, t)-dependent hierarchy defined by

$$S_n = (\cdots(S_0 \lrcorner U_1)\cdots \lrcorner U_1),$$

where S_0 is defined by

$$S_0 = -x\partial_x - 2t\partial_t + u\partial_u + 2v\partial_v + \tfrac{1}{2}\varphi\partial_\varphi + \tfrac{3}{2}\psi\partial_\psi \,.$$

Actually the hierarchies $(\overline{Y}_{n+1/2})$ and $(\overline{X}_{n+1})$ start at

$$\overline{Y}_{-1/2} = \partial_\varphi \quad \text{and} \quad \overline{X}_0 = (2q_{1/2} - \varphi)\partial_\varphi + \psi\partial_\psi \,,$$

respectively.

Acknowledgements. The first author (P. H. M. K.) is grateful to his co-author (I. S. Krasil′shchik) for his warm cooperation during their joint work. The second author (I. S. K.) expresses his gratitude to the Faculty of Applied Mathematics, University of Twente, and to NWO for the financial support during his stay at the University of Twente in February–July 1993. He is also grateful to his co-author (P. H. M. Kersten) for his warm cooperation during their joint work.

References

1. L. J. F. Broer, *Approximate equations for long water waves*, Appl. Sci. Res. **31** (1975), 377–395.

2. M. Gerstenhaber and S. D. Schack, *Algebraic cohomology and deformation theory*, Deformation theory of algebras and structures and applications (M. Hazewinkel and M. Gerstenhaber, eds.), Kluwer, Dordrecht, 1988, pp. 11–264.

3. P. H. M. Kersten, *Higher order symmetries and fermionic conservation laws of the supersymmetric extension of the KdV equation*, Phys. Lett. A **134** (1988), no. 1, 25–30.

4. N. G. Khor′kova, *Conservation laws and nonlocal symmetries*, Mat. Zametki **44** (1988), 134–144; English transl. in Math. Notes **44** (1988).

5. Y. Kosmann-Schwarzbach and F. Magri, *Poisson–Nijenhuis structures*, Ann. Inst. H. Poincaré **53** (1990), no. 1, 35–81.

6. I. S. Krasil′shchik, *Some new cohomological invariants for nonlinear differential equations*, Diff. Geom. Appl. **2** (1992), 307–350.

7. ———, *Supercanonical algebras and Schouten brackets*, Mat. Zametki **49** (1991), 70–76; English transl. in Math. Notes **49** (1991).

8. I. S. Krasil′shchik, V. V. Lychagin, and A. M. Vinogradov, *Geometry of jet spaces and nonlinear partial differential equations*, Gordon & Breach, New York, 1986.

9. I. S. Krasil′shchik and P. H. M. Kersten, *Deformations of differential equations and recursion operators*, Geometry in partial differential equations (A. Prastaro and Th. M. Rassias, eds.), World Scientific, Singapore, 1993.

10. I. S. Krasil'shchik and A. M. Vinogradov, *Nonlocal trends in the geometry of differential equations: Symmetries, conservation laws, and Bäcklund transformations*, Acta Appl. Math. **15** (1989), 161–209.

11. Yu. I. Manin and A. O. Radul, *A supersymmetric extension of the Kadomtsev–Petviashvili hierarchy*, Comm. Math. Phys. **98** (1985), 65.

12. P. Mathieu, *Supersymmetric extension of the Korteweg–de Vries equation*, J. Math. Phys. **29** (1988), 2499.

13. P. J. Olver, *Applications of Lie groups to differential equations*, Springer-Verlag, New York, 1986.

14. G. H. M. Roelofs, *Prolongation structures of supersymmetric systems*, Doctoral Thesis. Dept. of Appl. Math. Univ. of Twente, Enschede (1993).

15. G. H. M. Roelofs and P. H. M. Kersten, *Supersymmetric extensions of the nonlinear Schrödinger equation: Symmetries and coverings*, J. Math. Phys. **33** (1992), 2185.

16. T. Tsujishita, *Homological method of computing invariants of differential equations*, Diff. Geom. Appl. **1** (1991), 3–34.

17. A. M. Vinogradov, *The C-spectral sequence, Lagrangian formalism, and conservation laws*, J. Math. Anal. Appl. **100** (1984), 2–129.

18. ———, *Category of nonlinear differential equations*, Global Analysis – Studies and Applications (Yu. G. Borisovich and Yu. E. Gliklikh, eds.), Lecture Notes in Math., vol. 1108, Springer-Verlag, Berlin, 1984, pp. 77–102.

19. ———, *The main functors of differential calculus in graded algebras*, Uspekhi Mat. Nauk **44** (1989), no. 3, 151–152; English transl. in Russian Math. Surveys **44** (1989).

1 Tverskoĭ-Yamskoĭ per., 14, Apt. 45, 125047, Moscow, Russia

University of Twente, Department of Applied Mathematics, P.O. Box 217, 7500 AE Enschede, The Netherlands

Amer. Math. Soc. Transl.
(2) Vol. **167**, 1995

Symplectic Geometry of Mixed Type Equations

A. KUSHNER

ABSTRACT. For a wide class of Monge–Ampère equations a complete system of invariants related to the symplectic group is established. The normal form of equations for the homogeneous case is constructed.

§1. Introduction

1.1. In 1979 V. Lychagin established a connection between differential forms on the manifold of 1-jets of smooth functions and differential Monge–Ampère equations [1], which allowed him to formulate and solve many classical problems of the theory of partial differential equations in geometrical terms. This was the case for the quasilinearization problem [2] and for the classification problem of Monge–Ampère equations that can be reduced to equations with constant coefficients [2]. Using this approach, new methods for solving the Cauchy problem for hyperbolic equations were constructed [4] and necessary and sufficient conditions for equivalence of mixed Monge–Ampère equations and linear mixed equations (Tricomi, Chaplygin, and Keldysh equations) were found [3].

The most substantial results occur in the case of Monge–Ampère equations on the plane. These equations have the form:

$$a\,\frac{\partial^2 u}{\partial q_1^2} + 2b\,\frac{\partial^2 u}{\partial q_1 \partial q_2} + c\,\frac{\partial^2 u}{\partial q_2^2} + e\,\mathrm{Hess}\,u + f = 0\,, \tag{1.1}$$

where a, b, c, e, f are smooth functions in the variables q_1, q_2, u, $\partial u/\partial q_1$, $\partial u/\partial q_2$, and

$$\mathrm{Hess}\,u = \frac{\partial^2 u}{\partial q_1^2}\,\frac{\partial^2 u}{\partial q_2^2} - \left(\frac{\partial^2 u}{\partial q_1 \partial q_2}\right)^2$$

is the Hessian of the function $u = u(q_1, q_2)$.

The present work is devoted to classification of mixed type equations. The problem of classifying Monge–Ampère equations with respect to the group of contact diffeomorphisms was formulated by Sophus Lie in [5]. In 1979 T. Morimoto obtained classification results for Monge–Ampère equations of a special kind by using

1991 *Mathematics Subject Classification*. Primary 58G37; Secondary, 35A30, 35M10.

G-structure theory [6]. Important results were obtained in [2]. In the paper [7], the author classified effective differential 2-forms of mixed type or, what is equivalent, mixed equations up to the multiplication by a smooth function.

1.2. Let M be a two-dimensional smooth manifold, J^1M the manifold of 1-jets of smooth functions on M, and U_1 the Cartan form on J^1M. In the special coordinate system q_1, q_2, u, p_1, p_2 on J^1M it has the form

$$U_1 = du - p_1 dq_1 - p_2 dq_2 .$$

The Cartan form determines a contact structure on J^1M. Following [1], for any differential 2-form ω on J^1M construct a nonlinear differential operator $\Delta_\omega : C^\infty(M) \to \Lambda^2(M)$ acting by the formula:

$$\Delta_\omega(h) = \sigma^*_{j_1(h)}(\omega),$$

where $\sigma_{j_1(h)} : M \ni x \mapsto j_1(h)|_x \in J_1M$ is the section corresponding to the 1-jet of a function $h \in C^\infty(M)$, $\Lambda^2(M)$ is the module of differential 2-forms on the manifold M.

Note that in general one operator can be assigned to different forms. In order to establish a one-to-one correspondence between forms and operators, we must restrict ourselves to effective forms.

Let $a \in J^1M$ and $C_a = \operatorname{Ker} U_{1,a}$ be the Cartan subspace of the space $T_a(J^1M)$. Consider the restriction $\Omega_a = dU_1|_{C_a}$ of the differential of the form U_1 to the subspace C_a. The form Ω_a is a symplectic structure on C_a.

Let X_1 be a contact vector field such that its generating function is 1 (see [8]). A differential 2-form $\omega \in \Lambda^2(J^1M)$ is called *effective* [1] if

1) $X_1 \lrcorner\, \omega = 0$,

2) $\omega \wedge \Omega = 0$.

Let ω be an effective differential 2-form and Δ_ω the corresponding differential operator. The operator Δ_ω is called the *Monge–Ampère operator*, and the corresponding equation $E_\omega = \{\Delta_\omega = 0\} \subset J^2M$ is called the *Monge–Ampère equation*. A *multivalued solution* (or *generalized solution* in the terminology of [1]) of the equation E_ω is a two-dimensional submanifold $L \subset J^1M$ such that

$$U_1|_L = 0 \quad \text{and} \quad \omega|_L = 0 .$$

It should be pointed out that two forms ω and $\lambda\omega$, where λ is a nonvanishing function, generate the same equation $E = E_\omega = E_{\lambda\omega}$.

Two equations E_{ω_1} and E_{ω_2} are contactly equivalent if and only if there exists a contact diffeomorphism φ of the manifold J^1M such that $\varphi^*(\omega_1) = \lambda\omega_2$ for some nonvanishing function $\lambda \in C^\infty(J^1M)$.

Sometimes one can consider the symplectic geometry of Monge–Ampère equations instead of their contact geometry. Let $m \in J^1M$. Let us assume that there exists a contact vector field X_f such that $f(m) \neq 0$. Suppose that the Lie derivative of the 2-form ω with respect to X_f vanishes, $L_{X_f}(\omega) = 0$. Then in a neighborhood of the point m there exists a contact diffeomorphism φ such that

$$\varphi_*(X_f) = X_1, \qquad \varphi(m) = m .$$

This means that we can assume $L_{X_1}(\omega) = 0$. We can therefore consider ω as a form on the cotangent bundle T^*M and classify equations with respect to the group of symplectic diffeomorphisms. The symplectic structure on T^*M is determined by the 2-form Ω. Let us note that in the appropriate coordinates the field X_1 has the form $X_1 = \partial/\partial u$ and the condition $L_{X_1}(\omega) = 0$ in terms of coordinates means that the coefficients a, b, c, e, f in equation (1.1) do not depend on u.

Let $\omega \in \Lambda^2(T^*M)$ be an effective differential 2-form, $m \in T^*M$. Recall [2] that the number $\operatorname{Pf}(\omega_m)$ is called the *Pfaffian* of the form ω at the point m if

$$\omega_m \wedge \omega_m = \operatorname{Pf}(\omega_m)\Omega_m \wedge \Omega_m .$$

A differential 2-form $\omega \in \Lambda^2(T^*M)$ at the point $m \in T^*M$ is called

1) *elliptic* if $\operatorname{Pf}(\omega_m) > 0$,
2) *hyperbolic* if $\operatorname{Pf}(\omega_m) < 0$,
3) *parabolic* if $\operatorname{Pf}(\omega_m) = 0$.

The corresponding equations E_ω are called *elliptic*, *hyperbolic* or *parabolic* at the point m, respectively. The equation E_ω is said to be a *mixed type equation* (or *mixed equation*) if the function $\operatorname{Pf}(\omega)$ changes sign at the point m.

1.3. Let $\omega \in \Lambda^2(T^*M)$ be an effective differential 2-form on the cotangent bundle T^*M. Using the symplectic structure ω, one can associate to the form ω_a the linear operator $A_{\omega,a}\colon T_a(T^*M) \to T_a(T^*M)$ defined by the formula

$$A_{\omega,a}X \lrcorner \Omega_a = X \lrcorner \omega_a ,$$

where $X \in T_a(T^*M)$, see [2]. So, we get a field of linear operators A_ω on the cotangent bundle T^*M. It is not hard to prove that A_ω^2 is a scalar matrix and $A_\omega^2 + \operatorname{Pf}(\omega) = 0$. Since the form ω is antisymmetric, the operator A_ω is symmetric with respect to the structure form Ω, i.e.,

$$\Omega(A_\omega X, Y) = \Omega(X, A_\omega Y)$$

for any vector fields X, Y.

Note that the vector fields X and $A_\omega X$ are skew orthogonal, i.e.,

$$\Omega(A_\omega X, X) = \omega(X, X) = 0, \qquad \omega(X, A_\omega X) = \Omega(A_\omega X, A_\omega X) = 0.$$

Let us additionally indicate the following three properties:

i) $\det A_\omega = (\operatorname{Pf}(\omega))^2$,
ii) if $\omega_0 = \lambda\omega$, then $A_{\omega_0} = \lambda A_\omega$,
iii) $\operatorname{Pf}(\omega_0) = \operatorname{Pf}(\lambda\omega) = \lambda^2 \operatorname{Pf}(\omega)$.

A Lagrangian manifold $L \subset T^*M$ is said to be an *ω-plane* if $\omega|_L = 0$ [4]. In terms of the cotangent bundle, a multivalued solution of the equation E_ω is an ω-plane Lagrangian manifold.

§2. Invariants of Monge–Ampère equations

2.1. Consider a Monge–Ampère equation $E = E_\omega$, where ω is an effective 2-form on the cotangent bundle T^*M. By definition, put $F = \operatorname{Pf}(\omega)$. Let $m \in T^*M$. Suppose the function F satisfies the conditions

$$F(m) = 0, \quad dF_m \neq 0. \tag{2.1}$$

Note that conditions (2.1) do not change when the form ω is multiplied by a smooth function $\lambda \in C^\infty(T^*M)$, $\lambda(m) \neq 0$. Hence, conditions (2.1) are determined by the equation E itself, and not by the form ω.

Let us construct four relative vector invariants Y^i_ω $(i = 1, 2, 3, 4)$ related to the multiplication of the last form by an arbitrary function $\lambda \in C^\infty(T^*M)$, $\lambda(m) \neq 0$. In the neighborhood of the point m define vector fields X_F, Z_ω, W_ω, V_ω [7]:

$$X_F \lrcorner \Omega = -dF, \quad Z_\omega = A_\omega X_F, \quad W_\omega \lrcorner \Omega^2 = 2\, d\omega, \quad V_\omega = A_\omega W_\omega,$$

where $\Omega^2 = \Omega \wedge \Omega$ is the volume form. It is not hard to check that

$$A_\omega Z_\omega = -FX_F, \qquad A_\omega V_\omega = -FW_\omega.$$

Let $\omega_0 = \lambda\omega$, $F_0 = \mathrm{Pf}(\omega_0)$. Let us calculate the vector fields X_{F_0}, Z_{ω_0}, W_{ω_0}, V_{ω_0}. Note that

$$d\lambda \wedge \omega = (A_\omega X_\lambda \lrcorner \Omega) \wedge \Omega, \quad \text{where } X_\lambda \lrcorner \Omega = -d\lambda. \tag{2.2}$$

Indeed, since the form ω is effective, we see that $\omega \wedge \Omega = 0$. Then

$$0 = X_\lambda \lrcorner (\omega \wedge \Omega) = (A_\omega X_\lambda \lrcorner \Omega) \wedge \Omega - d\lambda \wedge \omega.$$

Since the vector field X_f is Hamiltonian, it follows that $X_{F_0} = X_{\lambda^2 F} = \lambda^2 X_F + 2F\lambda X_\lambda$. Since $A_{\omega_0} = \lambda A_\omega$, we have $Z_{\omega_0} = A_{\omega_0} X_{F_0} = \lambda^3 Z_\omega + 2\lambda^2 F A_\omega X_\lambda$. For the vector field W_{ω_0}, we have (see (2.2))

$$W_{\omega_0} \lrcorner \Omega^2 = 2\, d\omega_0 = 2((A_\omega X_\lambda + \lambda W_\omega) \lrcorner \Omega) \wedge \Omega.$$

The form Ω is nondegenerate, therefore $W_{\omega_0} = \lambda W_\omega + A_\omega X_\lambda$. For the vector field V_{ω_0} we have $V_{\omega_0} = \lambda^2 V_\omega - F\lambda X_\lambda$. Thus,

$$\begin{cases} X_{F_0} = \lambda^2 X_F + 2F\lambda X_\lambda, \\ V_{\omega_0} = \lambda^2 V_\omega - F\lambda X_\lambda, \\ Z_{\omega_0} = \lambda^3 Z_\omega + 2\lambda^2 F A_\omega X_\lambda, \\ W_{\omega_0} = \lambda W_\omega + A_\omega X_\lambda. \end{cases} \tag{2.3}$$

Construct the vector fields $Y^1_\omega = X_F + 2V_\omega$ and $Y^2_\omega = Z_\omega - 2FW_\omega$. The relation (2.3) implies

$$Y^1_{\omega_0} = X_{F_0} + 2V_{\omega_0} = \lambda^2(X_F + 2V_\omega) = \lambda^2 Y^1_\omega,$$
$$Y^2_{\omega_0} = Z_{\omega_0} - 2F_0 W_{\omega_0} = \lambda^3(Z_\omega - 2FW_\omega) = \lambda^3 Y^2_\omega.$$

Note that $Y^2_\omega = A_\omega Y^1_\omega$. Let σ^i_ω $(i = 1, 2)$ be differential 1-forms conjugate to the fields Y^i_ω with respect to Ω: $\sigma^i_\omega = Y^i_\omega \lrcorner \Omega$ $(i = 1, 2)$. Construct the vector fields Y^3_ω and Y^4_ω:

$$Y^3_\omega \lrcorner \Omega^2 = 2\sigma^1_\omega \wedge d\sigma^1_\omega, \qquad Y^4_\omega = A_\omega Y^3_\omega$$

and differential 1-forms σ_ω^3 and σ_ω^4 by putting $\sigma_\omega^i = Y_\omega^i \lrcorner \Omega$ $(i = 3, 4)$. Note that

$$\sigma_{\omega_0}^1 \wedge d\sigma_{\omega_0}^1 = \lambda^2\sigma_\omega^1 \wedge d(\lambda^2\sigma_\omega^1) = \lambda^4\sigma_\omega^1 \wedge d\sigma_\omega^1 .$$

Therefore $Y_{\omega_0}^3 = \lambda^4 Y_\omega^3$ and $Y_{\omega_0}^4 = \lambda^5 Y_\omega^4$.

Hence, the vector fields Y_ω^i $(i = 1, 2, 3, 4)$ are relative invariants of the form ω related to the multiplication of the form ω and the function λ.

2.2. Let us calculate the values of the forms σ_ω^i on the vector fields Y_ω^j $(i, j = 1, 2, 3, 4)$. By definition, put

$$r_\omega = \Omega(Y_\omega^1, Y_\omega^3), \qquad s_\omega = \Omega(Y_\omega^1, Y_\omega^4).$$

Then

1) $\sigma_\omega^i(Y_\omega^i) = 0$ $\quad (i = 1, 2, 3, 4)$,

2) $\sigma_\omega^i(Y_\omega^{i+1}) = \Omega(Y_\omega^i, A_\omega Y_\omega^i) = 0$ $\quad (i = 1, 3)$,

3) $\sigma_\omega^{i+1}(Y_\omega^i) = \Omega(A_\omega Y_\omega^i, Y_\omega^i) = 0$ $\quad (i = 1, 3)$,

4) $\sigma_\omega^2(Y_\omega^3) = \Omega(A_\omega Y_\omega^1, Y_\omega^3) = s_\omega$,

5) $\sigma_\omega^2(Y_\omega^4) = \Omega(A\omega Y_\omega^1, A_\omega Y_\omega^3) = -Fr_\omega$.

The obtained results are tabulated in Table 2.1.

TABLE 2.1

	Y_ω^1	Y_ω^2	Y_ω^3	Y_ω^4
σ_ω^1	0	0	r_ω	s_ω
σ_ω^2	0	0	s_ω	$-Fr_\omega$
σ_ω^3	$-r_\omega$	$-s_\omega$	0	0
σ_ω^4	$-s_\omega$	Fr_ω	0	0

LEMMA 2.1. *We have*

(2.4) $\quad Y_\omega^i \lrcorner \omega = \sigma_\omega^{i+1}$ $\quad (i = 1, 3)$,

(2.5) $\quad Y_\omega^{i+1} \lrcorner \omega = -F\sigma_\omega^i$ $\quad (i = 1, 3)$,

(2.6) $\quad \sigma_\omega^3 \wedge \Omega = \sigma_\omega^1 \wedge d\sigma_\omega^1$,

(2.7) $\quad \sigma_\omega^i \wedge \omega = -\sigma_\omega^{i+1} \wedge \Omega$ $\quad (i = 1, 3)$,

(2.8) $\quad \sigma_\omega^{i+1} \wedge \omega = F\sigma_\omega^i \wedge \Omega$ $\quad (i = 1, 3)$,

(2.9) $\quad \sigma_\omega^2 \wedge \sigma_\omega^3 \wedge \Omega = \sigma_\omega^1 \wedge \sigma_\omega^4 \wedge \Omega$,

(2.10) $\quad \sigma_\omega^4 \wedge \sigma_\omega^2 \wedge \Omega = F\sigma_\omega^1 \wedge \sigma_\omega^3 \wedge \Omega$,

(2.11) $\quad \sigma_\omega^1 \wedge \sigma_\omega^2 \wedge \sigma_\omega^3 = s_\omega\sigma_\omega^1 \wedge \Omega - r_\omega\sigma_\omega^2 \wedge \Omega$.

PROOF. 1) The proofs of (2.4) and (2.5) are similar to each other. For instance, let us prove (2.5)

$$Y_\omega^{i+1} \lrcorner \omega = -A_\omega Y_\omega^{i+1} \lrcorner \Omega = A_\omega^2 Y_\omega^i \lrcorner \Omega = -FY_\omega^i \lrcorner \Omega = -F\sigma_\omega^i .$$

2) Let us prove (2.6):

$$2\sigma^1_\omega \wedge d\sigma^1_\omega = Y^3_\omega \lrcorner\, \Omega^2 = 2(Y^3_\omega \lrcorner\, \Omega) \wedge \Omega = 2\sigma^3_\omega \wedge \Omega .$$

3) The proofs of (2.7) and (2.8) are similar to each other. Let us prove (2.7). Since $\omega \wedge \Omega = 0$, we have

$$0 = Y^i_\omega \lrcorner\, (\omega \wedge \Omega) = \sigma^{i+1}_\omega \wedge \Omega + \sigma^i_\omega \wedge \omega .$$

4) Let us prove (2.9). From (2.7) and (2.8) it follows, that

$$\sigma^3_\omega \wedge \sigma^2_\omega \wedge \Omega = -\sigma^3_\omega \wedge \sigma^1_\omega \wedge \omega = \sigma^1_\omega \wedge \sigma^3_\omega \wedge \omega = -\sigma^1_\omega \wedge \sigma^4_\omega \wedge \Omega .$$

Identity (2.10) can be proved similarly.

5) Applying $\lrcorner\, Y^1_\omega$ to identity (2.9), we obtain (2.11). □

LEMMA 2.2. *We have* $2\sigma^1_\omega \wedge \sigma^2_\omega \wedge \sigma^3_\omega \wedge \sigma^4_\omega = (Fr^2_\omega + s^2_\omega)\Omega^2$.

PROOF. Applying $\lrcorner\, Y^4_\omega$ to identity (2.11), we obtain:

$$(Fr^2_\omega + s^2_\omega)\Omega = s_\omega \sigma^2_\omega \wedge \sigma^3_\omega + r_\omega F \sigma^1_\omega \wedge \sigma^3_\omega + r_\omega \sigma^4_\omega \wedge \sigma^2_\omega + s_\omega \sigma^1_\omega \wedge \sigma^4_\omega .$$

Taking the exterior product of the last identity by the form Ω and using (2.10) and (2.11), we obtain:

$$(2.12) \qquad (Fr^2_\omega + s^2_\omega)\Omega^2 = 2(s_\omega \sigma^2_\omega \wedge \sigma^3_\omega + r_\omega \sigma^4_\omega \wedge \sigma^2_\omega) \wedge \Omega .$$

Taking the exterior product of identity (2.11) and σ^4 and using (2.9) and (2.12), we obtain:

$$\sigma^1_\omega \wedge \sigma^2_\omega \wedge \sigma^3_\omega \wedge \sigma^4_\omega = s_\omega \sigma^2_\omega \wedge \sigma^3_\omega \wedge \Omega + r_\omega \sigma^4_\omega \wedge \sigma^2_\omega \wedge \Omega = (Fr^2_\omega + s^2_\omega)\Omega^2/2 . \quad \square$$

By definition, put

$$h_\omega = Fr^2_\omega + s^2_\omega .$$

From Lemma 2.2 it follows that the vectors $Y^1_{\omega,m}, \ldots, Y^4_{\omega,m}$ are linearly independent at the point $m \in \{F = 0\}$ if and only if

$$(2.13) \qquad s_\omega(m) \neq 0 .$$

In the following we shall assume that condition (2.13) is satisfied. Note that

$$(2.14) \qquad r_{\omega_0} = \lambda^6 r_\omega , \quad s_{\omega_0} = \lambda^7 s_\omega , \quad h_{\omega_0} = \lambda^{14} h_\omega .$$

The last identities enable us to construct absolute functional invariants with respect to multiplication of the form ω by smooth functions, i.e., functional invariants of the equation E:

$$I^0_E = Fs_\omega^{-2/7} , \quad I^1_E = r_\omega s_\omega^{-6/7} .$$

From (2.14) it follows that conditions (2.13) characterizes not only the form ω, but also the equation E itself.

2.3. In the neighborhood of the point $m \in \{F = 0\}$ construct a locally free *canonical basis* $\{X_E^i\}$ of the module $D(T^*M)$ (here $D(T^*M)$ is the module of vector fields on T^*M):

$$X_E^1 = h_\omega^{-1} s_\omega^{2/7}(Fr_\omega Y_\omega^3 + s_\omega Y_\omega^4), \qquad X_E^2 = -s_\omega^{-2/7} Y_\omega^1,$$
$$X_E^3 = h_\omega^{-1} s_\omega^{3/7}(s_\omega Y_\omega^3 - r_\omega Y_\omega^4), \qquad X_E^4 = -s_\omega^{-3/7} Y_\omega^2$$

and a *canonical basis* $\{\theta_E^i\}$ of the module $\Lambda^1(T^*M)$ of differential 1-forms on T^*M:

$$\theta_E^1 = s_\omega^{-2/7}\sigma_\omega^1, \qquad \theta_E^2 = h_\omega^{-1} s_\omega^{2/7}(Fr_\omega\sigma_\omega^3 + s_\omega\sigma_\omega^4),$$
$$\theta_E^3 = s_\omega^{-3/7}\sigma_\omega^2, \qquad \theta_E^4 = h_\omega^{-1} s_\omega^{3/7}(s_\omega\sigma_\omega^3 - r_\omega\sigma_\omega^4).$$

The vector fields X_E^i and differential 1-forms θ_E^i $(i = 1, 2, 3, 4)$ are absolute invariants of the form ω with respect to multiplication of the form ω by smooth functions $\lambda \in C^\infty(T^*M)$:

$$X_{E_{\lambda\omega}}^i = X_{E_\omega}^i, \qquad \theta_{E_{\lambda\omega}}^i = \theta_{E_\omega}^i \qquad (i = 1, 2, 3, 4).$$

Therefore, X_E^i and θ_E^i are invariants of the equation E with respect to symplectic diffeomorphisms. Note that the operator $A_E = s_\omega^{-1/7} A_\omega$ is invariant also.

REMARKS. 1) $\theta_E^1 = -X_E^2 \lrcorner\, \Omega$, $\theta_E^2 = X_E^1 \lrcorner\, \Omega$, $\theta_E^3 = -X_E^4 \lrcorner\, \Omega$, $\theta_E^4 = X_E^3 \lrcorner\, \Omega$;
2) $A_E X_E^2 = X_E^4$, $A_E X_E^3 = X_E^1$;
3) the bases $\{X_E^i\}$ and $\{\theta_E^i\}$ are dual, i.e., $\theta_E^i(X_E^j) = \delta_{ij}$, where δ_{ij} is the Kronecker delta.

In terms of the basis $\{\theta_E^i\}$, the following decompositions hold:

$$\Omega = \theta_E^1 \wedge \theta_E^2 + \theta_E^3 \wedge \theta_E^4, \qquad \omega = s_\omega^{1/7}\theta_E^3 \wedge \theta_E^2 + F s_\omega^{-1/7}\theta_E^4 \wedge \theta_E^1. \tag{2.15}$$

Instead of ω, consider the form $\widetilde{\omega}$, also corresponding to the equation E. Then

$$\widetilde{\omega} = \theta_E^3 \wedge \theta_E^2 + F_E \theta_E^4 \wedge \theta_E^1, \tag{2.16}$$

where $F_E = I^0$ is a functional invariant of equation E.

So, to the equation E we have assigned a uniquely determined 2-form $\widetilde{\omega}$. We shall say that the form $\widetilde{\omega}$ is *associated to the equation* E. Note that

$$s_{\widetilde{\omega}} = 1, \quad h_{\widetilde{\omega}} = 1 + I_E^0 I_E^1, \quad r_{\widetilde{\omega}} = I_E^1.$$

DEFINITION. An equation E is said to be a *binomial equation of kind* 1 *at the point* m if conditions (2.1) and (2.13) are satisfied.

§3. The normal form of equations

3.1. We can use of decomposition (2.15) and (2.16) to solve a problem of local equivalence of mixed type Monge–Ampère equations.

Equations $E_1 = E_{\omega_1}$ and $E_2 = E_{\omega_2}$ are called *locally symplectically equivalent* in a neighborhood of the point m if there exists a local diffeomorphism φ such that $\varphi(m) = m$ and

$$\varphi^*(\omega_2) = \lambda\omega\,,$$

for some nonvanishing function $\lambda \in C^\infty(T^*M)$ (see §1).

It is clear that binomial equations of kind 1 are locally symplectically equivalent if and only if there exists a local symplectic diffeomorphism φ such that $\varphi(m) = m$ and $\varphi^*(\widetilde{\omega}_2) = \widetilde{\omega}_1$, where $\widetilde{\omega}_1$ and $\widetilde{\omega}_2$ are associate forms.

Let E_1 and E_2 be binomial equations of kind 1 at the point m. Let us express the covariant derivatives f_1^i, f_2^i of the functions F_{E_1}, F_{E_2} in terms of bases $\{\theta^i_{E_1}\}$ and $\{\theta^i_{E_1}\}$, respectively; we get

$$dF_{E_j} = \sum_{i=1}^4 f_j^i \theta^i_{E_j}\,.$$

Then

$$f_j^i = df_{E_j}(X^i_{E_j}) = X^i_{E_j}(F_{E_j})\,.$$

We can similarly define kth order covariant derivatives of F_{E_j}:

$$df_j^{i_1\ldots i_{k-1}} = \sum_{l=1}^4 f_j^{i_1\ldots i_{k-1}l}\theta^l_{E_j}\,.$$

Suppose there exists a number $k > 1$ such that all covariant derivatives of order $\geqslant k$ of the function F_{E_1} are equal to zero. From decompositions (2.15) and (2.16) it follows that equations E_1 and E_2 are locally symplectically equivalent in a neighborhood of the point m if and only if

1) there exists a local diffeomorphism φ of the cotangent bundle T^*M such that

$$\varphi(m) = m \quad \text{and} \quad \varphi^*(\theta^i_{E_2}) = \theta^i_{E_1} \qquad (i = 1, 2, 3, 4);$$

2) $f_1^{i_1\ldots i_n}(m) = f_2^{i_1\ldots i_n}(m)$ for any $n < k$ and for any $i_l = 1, 2, 3, 4$ $(l = 1, \ldots, n)$.

Thus, the problem of equivalence for binomial equations of kind 1 can be reduced to the problem of equivalence of $\{e\}$-structures (absolute parallelisms), which can be solved by already known methods.

3.2. Suppose that for binomial equations E_1 and E_2 of kind 1 at the point m the following *homogeneity conditions* are satisfied:

1) the first covariant derivatives f_j^i of the functions F_{E_j} are constant $(i = 1, 2, 3, 4,\ j = 1, 2)$,

2) $\mathfrak{g}_{E_j} = \bigoplus_{i=1}^4 \mathbb{R}X^i_{E_j}$ is a Lie algebra $(j = 1, 2)$.

THEOREM. *Two equations E_1 and E_2 are locally symplectically equivalent in a neighborhood of the point m if and only if the corresponding structure constants of the Lie algebras $\mathfrak{g}_{E_1}$ and $\mathfrak{g}_{E_2}$ coincide in terms of the bases $\{X^i_{E_1}\}$ and $\{X^i_{E_2}\}$ respectively.*

PROOF. 1) The *necessity* is evident.

2) *Sufficiency.* Suppose

$$[X^i_{E_l}, X^j_{E_l}] = \sum_{k=1}^{4} c^k_{ij} X^k_{E_l} \qquad (l = 1, 2),$$

i.e., the c^k_{ij} are structure constants of the algebras $\mathfrak{g}_{E_1}$ and $\mathfrak{g}_{E_2}$. The bases $\{X^i_{E_l}\}$ and $\{\theta^i_{E_l}\}$ are dual, therefore

$$d\theta^k_{E_l}(X^i_{E_l}, X^j_{E_l}) = -\theta^k_{E_l}([X^i_{E_l}, X^j_{E_l}]) = -c^k_{ij}.$$

Consequently, the Maurer–Cartan identities

$$d\theta^k_{E_l} = -\sum_{1\leqslant i<j\leqslant 4} c^k_{ij}\theta^i_{E_l} \wedge \theta^j_{E_l}$$

are satisfied. From the Maurer–Cartan identities it follows that there exists a local diffeomorphism φ such that $\varphi(m) = m$ and $\varphi^*(\theta^i_{E_2}) = \theta^i_{E_1}$, $i = 1, 2, 3, 4$. Below it will be shown that the covariant derivatives of the functions F_{E_j} can be written in terms of the structure constants of the Lie algebra $\mathfrak{g}_{E_j}$ $(j = 1, 2)$ (see (3.6)).

3.3. It turns out that under our assumptions there exists only one Lie algebra $\mathfrak{g}_E$ and, therefore, only one normal form of the equation. In order to find this normal form, let us first specify the conditions on the structure constants of $\mathfrak{g}_E$.

1) Since Ω is a closed form, we see that

$$\begin{aligned} &c^1_{13} + c^2_{23} + c^4_{12} = 0, \qquad c^1_{14} + c^2_{24} - c^3_{12} = 0, \\ &c^2_{34} - c^3_{13} - c^4_{14} = 0, \qquad c^1_{34} + c^3_{23} - c^4_{24} = 0. \end{aligned} \tag{3.1}$$

2) Consider the associated form $\widetilde{\omega}$:

$$\widetilde{\omega} = \theta^3_E \wedge \theta^2_E + F_E\theta^4_E \wedge \theta^1_E.$$

We have $\sigma^1_{\widetilde{\omega}} = \theta^1_E$, $\sigma^2_{\widetilde{\omega}} = \theta^3_E$, $\sigma^3_{\widetilde{\omega}} = \theta^4_E + r_{\widetilde{\omega}}\theta^2_{\omega}$, $\sigma^4_{\widetilde{\omega}} = \theta^2_E - F_E r_{\widetilde{\omega}}\theta^4_E$. From (2.6) it follows that

$$c^1_{23} = 0, \quad c^1_{34} = 0, \quad c^1_{24} = -1, \tag{3.2}$$

and $r_{\widetilde{\omega}} = 0$.

3) Note that

$$\begin{aligned} X_{F_E} &= -f^2X^1_E + f^1X^2_E - f^4X^3_E + f^3X^4_E, \\ V_{\widetilde{\omega}} &= \alpha_3X^1_E - F_E\alpha_4X^2_E - F_E\alpha_1X^3_E + \alpha_2X^4_E, \end{aligned}$$

where f^i $(i = 1, 2, 3, 4)$ are covariant derivatives of the function F_E in terms of the basis $\{\theta_E^i\}$,

$$\tag{3.3} \begin{aligned} \alpha_1 &= -c_{34}^3 - c_{24}^2, & \alpha_2 &= c_{14}^2 + (c_{34}^4 - c_{13}^1) F_E - f^3, \\ \alpha_3 &= c_{14}^3 + (c_{12}^1 - c_{24}^4) F_E + f^2, & \alpha_4 &= -c_{12}^2 - c_{13}^3 + c_{23}^4 F_E. \end{aligned}$$

Then

$$\begin{aligned} Y_{\tilde{\omega}}^1 = (2\alpha_3 - f^2) X_E^1 + (f^1 - 2\alpha_4 F_E) X_E^2 \\ + (-f^4 - 2\alpha_1 F_E) X_E^3 + (f^3 + 2\alpha_2) X_E^4. \end{aligned}$$

We have $Y_{\tilde{\omega}}^1 = -X_E^2$, therefore

$$\tag{3.4} \begin{aligned} 2\alpha_3 - f^2 = 0, \; f^1 - 2\alpha_4 F_E &= -1, \\ f^4 + 2\alpha_1 F_E = 0, \; f^3 + 2\alpha_2 \quad &= 0. \end{aligned}$$

Relations (3.3) and (3.4) imply

$$\begin{aligned} f^1 &= -1 - 2(c_{12}^2 + c_{13}^3) F_E + 2c_{23}^4 F_E^2, & f^2 &= -2c_{14}^3 + 2(c_{24}^4 - c_{12}^1) F_E, \\ f^3 &= 2c_{14}^2 + 2(c_{34}^4 - c_{13}^1) F_E, & f^4 &= 2(c_{24}^2 + c_{34}^3) F_E. \end{aligned}$$

By the homogeneity conditions, we obtain

$$\tag{3.5} \begin{aligned} c_{23}^4 = 0, \quad c_{12}^2 + c_{13}^3 = 0, \quad c_{13}^1 - c_{34}^4 = 0, \\ c_{12}^1 - c_{24}^4 = 0, \quad c_{34}^3 + c_{24}^2 = 0, \end{aligned}$$

$$\tag{3.6} f^1 = -1, \quad f^2 = -2c_{14}^3, \quad f^3 = 2c_{14}^2, \quad f^4 = 0.$$

4) From the Jacobi identity it follows that

$$\tag{3.7} c_{13}^1 - c_{23}^2 + c_{34}^4 = 0.$$

5) We have

$$f^{ji} - f^{ij} = [X_E^i, X_E^j](F_E) = \sum_{k=1}^{4} c_{ij}^k f^k,$$

where f^{ij} $(i, j = 1, 2, 3, 4)$ are second order covariant derivatives of F_E. But $f^{ij} = 0$ for any $i, j = 1, 2, 3, 4$ and $f_4 = 0$. Therefore,

$$\sum_{k=1}^{3} c_{ij}^k f^k = 0$$

for any $i, j = 1, 2, 3, 4$. So, we obtain (see (3.2) and (3.6)):

$$\tag{3.8} \begin{aligned} c_{14}^1 &= 0, & 2(c_{24}^2 c_{14}^3 - c_{24}^3 c_{14}^2) - 1 &= 0, \\ c_{14}^2 c_{23}^3 - c_{23}^2 c_{14}^3 &= 0, & c_{34}^2 c_{14}^3 - c_{34}^3 c_{14}^2 &= 0, \\ c_{12}^1 + 2(c_{12}^2 c_{14}^3 - c_{12}^3 c_{14}^2) &= 0, & c_{13}^1 + 2(c_{13}^2 c_{14}^3 - c_{13}^3 c_{14}^2) &= 0. \end{aligned}$$

TABLE 3.1

	1	2	3	4
12	0	0	0	0
13	0	0	0	0
14	0	b	a	0
23	0	0	0	0
24	-1	0	$-(2b)^{-1}$	0
34	0	0	0	0

There exists only one Lie algebra $\mathfrak{g}_E$ satisfying the conditions (3.1), (3.2), (3.5), (3.7), and (3.8). Let us write out the structure constants of $\mathfrak{g}_E$ in Table 3.1.

The element c_{ij}^k appears at the intersection of the kth column and the (ij)th line $(i, j, k = 1, 2, 3, 4)$. In Table 3.1, $a \neq 0$, $b \neq 0$ are arbitrary constants.

Let us construct a representation of the Lie algebra $\mathfrak{g}_E$ by vector fields on T^*M. Let q_1, q_2, p_1, p_2 be coordinates on the cotangent bundle T^*M. Then

$$X_E^1 = \frac{\partial}{\partial q_1} - bq_1\frac{\partial}{\partial q_2}, \quad X_E^2 = \frac{\partial}{\partial p_1} - p_1\frac{\partial}{\partial q_2}, \quad X_E^3 = \frac{\partial}{\partial q_2},$$

$$X_E^4 = \frac{\partial}{\partial p_2} + bq_1\frac{\partial}{\partial p_1} - p_1\frac{\partial}{\partial q_1} + \left(aq_1 - \frac{1}{2b}p_1\right)\frac{\partial}{\partial q_2}.$$

Therefore,

$$\theta_E^1 = dq_1 + p_1 dp_2, \qquad \theta_E^2 = dp_1 - bq_1 dp_2,$$

$$\theta_E^3 = dq_2 + bq_1 dq_1 + p_1 dp_1 + ((2b)^{-1}p_1 - aq_1)\,dp_2, \qquad \theta_E^4 = dp_2,$$

and

$$F_E = -q_1 + b^2q_1^2 + 2bq_2 - 2ap_1 + bp_1^2.$$

Construct the effective differential 2-form ω (see (2.15))

$$\omega = bq_1 dq_1 \wedge dp_1 + (-F_E - b^2q_1^2)\,dq_1 \wedge dp_2 + dq_2 \wedge dp_1 - bq_1 dq_2 \wedge dp_2 + (aq_1 - (2b)^{-1}p_1 - bp_1q_1)\,dp_1 \wedge dp_2.$$

The symplectic structure Ω has the form (see (2.14))

$$\Omega = dq_1 \wedge dp_1 + dq_2 \wedge dp_2.$$

Let us write out the equation $E = E_\omega$:

$$\frac{\partial^2 u}{\partial x^2} - 2bx\frac{\partial^2 u}{\partial x \partial y} + \left(2by - x - 2a\frac{\partial u}{\partial x} - b\left(\frac{\partial u}{\partial x}\right)^2\right)\frac{\partial^2 u}{\partial y^2} + \left(\left(\frac{1}{2b} + bx\right)\frac{\partial u}{\partial x} - ax\right)\operatorname{Hess} u = 0, \tag{3.9}$$

where $x = q_1$, $y = q_2$.

Equation (3.9) is the normal form of binomial equations of kind 1 if the homogeneity conditions are satisfied.

REMARK. For equation (3.9), the two-dimensional distribution $\mathcal{F}(X_E^1, X_E^3)$ is a complete integrable distribution and

$$\Omega(X_E^1, X_E^3) = \omega(X_E^1, X_E^3) = 0.$$

Therefore the maximal integral manifolds of $\mathcal{F}(X_E^1, X_E^3)$ are multivalued solutions of equation (3.9).

References

1. V. Lychagin, *Contact geometry and nonlinear second order differential equations*, Uspekhi Mat. Nauk **34** (1979), no. 1, 101–171; English transl., Russian Math. Surveys **34** (1979), 149–180.
2. V. Lychagin, V. Rubtsov, and I. Chekalov, *A local classification of Monge–Ampère equations*, Ann. Sci. École Norm. Sup. (4) **26** (1993), no. 1, 281–308.
3. A. Kushner, *Chaplygin's and Keldysh's normal forms of mixed type Monge–Ampère equations*, Mat. Zametki **52** (1992), no. 5, 63–67; English transl. in Math. Notes **52** (1992).
4. V. Lychagin, *Lectures on geometry of differential equation, part* I, Ciclo di conferenze tenute presso il Dipartimento di Metodi e Modelli Matematici per le Scienze Applicate Universita "La Sapienza" (1992), Rome.
5. S. Lie, *Gesamette Ahandlungen*, vol. 1–7, Leipzig–Oslo, 1922–1935.
6. T. Morimoto, *La problem d'equivalence des equationes de Monge–Ampère*, C. R. Acad. Sci. Paris **289** (1979), 1–2.
7. A. Kushner, *Classification of mixed type Monge–Ampère equations*, Geometry in Nonlinear Equations (A. Prastaro and Ph. Rassias, eds.), 1993 (to appear).
8. V. Lychagin, *The local classification of nonlinear differential equations in first order partial derivatives*, Uspekhi Mat. Nauk **30** (1975), no. 1, 101–171; English transl., Russian Math. Surveys **30** (1975), 105–175.

Translated by THE AUTHOR

INTERNATIONAL SOPHUS LIE CENTER, UL. MECHNIKOVA, 2 APT. #2, 414000 ASTRAKHAN′, RUSSIA

Amer. Math. Soc. Transl.
(2) Vol. **167**, 1995

Homogeneous Geometric Structures and Homogeneous Differential Equations

V. LYCHAGIN

Introduction

In this paper we introduce the notion of *homogeneous geometric structure* and show that these structures arise naturally both in the geometric theory of differential equations and in differential geometry.

Homogeneous geometric structures (under certain additional conditions) may be endowed with a connection of special type which we call the *Cartan connection*, because the construction of such a connection is based on the *Cartan distribution* on a jet manifold. In the case when the *Weyl tensor* of the Cartan connection is trivial, the latter becoming a *Bott connection*.

We would like to make three remarks concerning Cartan connections and homogeneous geometric structures.

First, most of the connections used in differential geometry are Cartan connections (*Levi–Civita connections* on Riemannian manifolds, *normal connections* in Projective Geometry, etc.).

Second, our definition of homogeneous geometric structures differs from the definition of G-*structures*; in our approach a homogeneous geometrical structure is, actually, a differential equation. It allows us to use the geometry of differential equations and the main constructions of differential geometry (connections, curvatures, covariant derivations) directly to answer the standard (but most important) question in the theory of differential equations: whether or not a given differential equation has a solution?

Third, under this approach the curvature tensor of the Cartan connection contains the torsion tensor as a part.

The paper consists of three main sections. In §1 we recall in a suitable form the main properties of the metasymplectic structure on jet spaces, define Cartan connections, and show that the restriction of the metasymplectic structure to the horizontal subspaces of the Cartan connection provides a curvature tensor.

In §2 the definition of homogeneous geometric structures and their generalizations is given. We also introduce the notion of a nonsingular map with respect

1991 *Mathematics Subject Classification*. Primary 53B15, 58A20.

Key words and phrases. Geometric structure, jet spaces, Cartan distribution, Cartan connection, characteristics, Lie equations.

to a given geometrical structure. These nonsingular mappings are morphisms in category of geometrical structures.

The main technical result of §2 is Theorem 2.6, which provides a criterion of the existence of the Cartan connection for generalized geometric structures.

In §3 the Lie algebras of symmetries for differential equations are studied. The main results are collected in Theorems in 3.4; they describe conditions for a symmetry algebra to be elliptic or of finite type. Note that these structures are intimately connected with the structure of characteristic manifold of the initial system. The proofs of these theorems are based on a new approach to characteristics regarded as straight lines on Grassmannians.

We also introduce the notion of homogeneous model and give conditions for a given system to be equivalent to the corresponding model. These conditions are formulated in terms of the Weyl tensor of the Cartan connections for the appropriate homogeneous geometric structure.

This paper was prepared when the author was visiting a professor at the Mathematics Department of Rome University "La Sapienza" and was completed at IHES. I would like to thank many colleagues who helped me in this very profitable period for my work. In particular my thanks go to Professor A. Prastaro for interesting discussions. Further, I would also like to thank Professors P. Benvenuti and M. Berger for their kind hospitality.

§1. Differential equations and connections

1.1. Jet spaces. We recall some facts from the geometry of jet spaces [3, 5, 11]. Let M be a smooth manifold of dimension $m+n$, where $m>0$ is a fixed number. For any $k \geqslant 0$ and each submanifold $N \subset M$ of codimension m, we denote by $[N]_a^k$ the k-jet of the submanifold at the point $a \in M$.

Let $J_n^k(M)_a$ be the space of all k-jets of submanifolds (of codimension m) at the fixed point $a \in M$ and

$$J_n^k(M) = \bigcup_{a\in M} J_n^k(M)_a$$

be the space of all k-jets. Reduction of k-jets $[N]_a^k$ to s-jets $[N]_a^s$ for $k>s$ generates a projection $\pi_{k,s}\colon J_n^k(M) \to J_n^s(M)$. For any submanifold $N \subset M$, $\operatorname{codim} N = m$, one has the natural embedding:

$$j_k\colon N \hookrightarrow J_n^k(M), \quad \text{where } j_k\colon N \ni x \mapsto [N]_x^k \in J_n^k(M).$$

The submanifold $N^{(k)} = j_k(N) \subset J_n^k(M)$ will be called the *k-jet extension of N*.

EXAMPLES. (1) $J_n^0(M) = M$.

(2) $J_n^1(M)_a$ is identified with the *Grassmannians* $\operatorname{Grass}_n(T_aM)$ of all n-dimensional subspaces in the tangent space T_aM. Hence

$$J_n^1(M) = \bigcup_{a\in M} \operatorname{Grass}_n(T_aM).$$

(3) If $n=k=1$, then $J_1^1(M)$ is the *projectivization of the tangent bundle* $\tau_M\colon TM \to M$.

(4) If $m=k=1$, then $J_{n-1}^1(M)$ is the *projectivization of the cotangent bundle* $\tau_M^*\colon T^*M \to M$.

1.2. Affine structures. Keeping in mind the description of the structure of jet spaces $J_n^k(M)$ in the general case ($k \geqslant 2$), we consider the action of the pseudogroup $G(M)$ of local diffeomorphisms of the manifold M on the space $J_n^k(M)$.

Any local diffeomorphism $\varphi\colon M \to M$ defines (local) diffeomorphisms

$$\varphi^{(k)}\colon J_n^k(M) \to J_n^k(M),$$

where

$$\varphi^{(k)}([N]_a^k) = [\varphi(N)]_{\varphi(a)}^k. \tag{1.2.1}$$

The diffeomorphisms $\varphi^{(k)}$ will be called the *kth prolongation of* φ. Obviously,

$$\pi_{k,s} \circ \varphi^{(k)} = \varphi^{(s)} \circ \pi_{k,s}, \qquad k > s,$$

and

$$(\varphi \circ \psi)^{(k)} = \varphi^{(k)} \circ \psi^{(k)}, \qquad (\mathrm{id})^{(k)} = \mathrm{id},$$

for any (local) diffeomorphisms φ and ψ.

Denote by $G_{a,b}^k$, $a, b \in M$, a set of all k-jets $[\varphi]_a^k$ of local diffeomorphisms $\varphi\colon M \to M$, taking the point $a \in M$ to the point $b = \varphi(a)$.

The composition of local diffeomorphisms generates a pairing:

$$G_{b,c}^k \times G_{a,b}^k \to G_{a,c}^k, \qquad [\psi]_b^k \circ [\varphi]_a^k \mapsto [\psi \circ \varphi]_a^k,$$

which defines a Lie group structure on the space $G_a^k = G_{a,a}^k$.

EXAMPLE. $G_a^0 = \{\mathrm{id}\}$, $G_a^1 = GL(T_aM)$.

It is obvious that the natural projections $\pi_{k,k-1}\colon G_a^k \to G_a^{k-1}$, $k \geqslant 1$, are epimorphisms of Lie groups.

If $k \geqslant 2$, then the kernels H_a^k of the projections are abelian groups isomorphic to the tensor products $S^k(T_a^*M) \otimes T_aM$, where $S^k(V)$ denotes the kth symmetric degree of the vector space V.

The group G_a^k acts on the space $J_n^k(M)_a$ in the natural way:

$$[\varphi]_a^k([N]_a^k) = [\varphi(N)]_a^k, \quad \text{where } [\varphi]_a^k \in G_a^k,\ [N]_a^k \in J_n^k(M)_a.$$

Under this action, the kernel H_a^k acts transitively on the fibers $F(a_{k-1}) \subset J_n^k(M)$ of the projection $\pi_{k,k-1}\colon J_n^k(M) \to J_n^{k-1}(M)$ over the point $a_{k-1} = [N]_a^{k-1}$. For $k \geqslant 2$, the stationary subgroup of the element $a_k \in F(a_{k-1})$ in H_a^k under this action is

$$(\operatorname{Ann} T_aN) \circ S^{k-1}(T_a^*M) \otimes T_aM + S^k(T_a^*M) \otimes T_aN,$$

where $(\operatorname{Ann} T_aN) \circ S^{k-1}(T_a^*M)$ denotes the symmetric product of the annihilator $\operatorname{Ann} T_aN \subset T_a^*M$ and the space $S^{k-1}(T_a^*M)$.

Therefore, the fibers $F(a_{k-1})$, $k \geqslant 2$, are homogeneous spaces with respect to the action of the connected abelian group H_a^k and thus carry an affine structure associated with the vector space $S^k(\tau_a^*) \otimes \nu_a$, where $\tau_a = T_aN$ and $\nu_a = T_aM/T_aN$ are the tangent and the normal subspaces of the submanifold $N \subset M$ at the point $a \in N$.

Obviously, the prolongations $\varphi^{(k)}$ of any local diffeomorphism $\varphi\colon M \to M$ for $k \geqslant 2$ preserve the affine structure described above.

So, we get the following result.

THEOREM [3, 11]. (1) *The jet bundles* $\pi_{k,k-1}\colon J_n^k(M) \to J_n^{k-1}(M)$ *are affine for* $k \geqslant 2$. *Here the fiber* $F(a_{k-1})$ *over the point* $a_{k-1} = [N]_a^{k-1}$ *is an affine space having* $S^k\tau_a^* \otimes \nu_a$ *as associated vector space.*

(2) *The prolongations* $\varphi^{(k)}\colon J_n^k(M) \to J_n^k(M)$ *of the local diffeomorphisms* $\varphi\colon M \to M$ *are affine automorphisms for* $k \geqslant 2$ *and are linear collineations of the Grassmannians* Grass(TM) *for* $k = 1$.

1.3. Cartan distributions. Any point $a_{k+1} = [N]_a^{k+1} \in J_n^{k+1}(M)_a$ may be also regarded as a subspace:

$$L(a_{k+1}) = T_{a_k}(N^{(k)}) \subset T_{a_k}(J_n^k(M)),$$

i.e., is the tangent space to the submanifold $N^{(k)} = j_k(N) \subset J_n^k(M)$.

Let $C(a_k) \subset T_{a_k}(J_n^k(M))$ be the linear space generated by all subspaces $L(a'_{k+1})$, $\forall a'_{k+1} \in F(a_k)$.

The space $C(a_k)$ will be called the *Cartanian*. The distribution

$$C\colon J_n^k(M) \ni a_k \mapsto C(a_k) \subset T_{a_k}(J_n^k(M))$$

obtained in this way will be called the *Cartan distribution* on the k-jet manifold $J_n^k(M)$.

The definition of the Cartan subspace $C(a_k)$ implies that the differential of the projection $\pi_{k,k-1}$ maps this subspace onto $L(a_k)$. Furthermore, there is another way of defining the Cartan subspaces:

$$C(a_k) = (\pi_{k,k-1})_*^{-1}(L(a_k)). \tag{1.3.1}$$

In other words, the Cartan subspaces $C(a_k)$ can be represented as a direct sum of subspaces

$$C(a_k) = T_{a_k}(F(a_{k-1})) \oplus L(a_{k+1}), \tag{1.3.2}$$

for any elements $a_{k+1} \in F(a_k)$, or, using the affine structure,

$$C(a_k) = S^k\tau_a^* \otimes \nu_a \oplus L(a_{k+1}). \tag{1.3.3}$$

By definition of the Cartan distribution, submanifolds of the form $N^{(k)} = j_k(N) \subset J_n^k(M)$ are integral manifolds of the Cartan distribution projected without singularities on the manifold M. The converse assertion is also true. Any integral manifold $L \subset J_n^k(M)$ of the Cartan distribution of dimension n for which the mapping $\pi_{k,0}\colon L \to M$ is an embedding has the form $L = N^{(k)}$ for some submanifold $N \subset M$.

1.4. Cartan forms. In order to obtain a dual description of the Cartan distribution, we consider the following construction. Denote by U_{k,a_k} the composition

$$\begin{array}{ccc} T_{a_k}(J_n^k(M)) & \xrightarrow{(\pi_{k,k-1})_*} & T_{a_{k-1}}(J_n^{k-1}(M)) \\ & \searrow^{U_{k,a_k}} & \downarrow \\ & & T_{a_{k-1}}(J_n^{k-1}(M))/L(a_k), \end{array}$$

where $a_k \in J_n^k(M)$, $a_{k-1} = \pi_{k,k-1}(a_k)$.

By definition of the Cartan space, we have $C(a_k) = \operatorname{Ker} U_{k,a_k}$. Moreover, one can define a vector bundle v_k over $J_n^k(M)$ by putting

$$v_k(a_k) = T_{a_{k-1}}(J_n^{k-1}(M))/L(a_k)$$

and consider the operator U_k as a differential v_k-valued 1-form on the manifold $J_n^k(M)$:

$$U_k \in \Omega^1(J_n^k(M)) \otimes v_k .$$

For any smooth section $f \in C^\infty(v_k^*)$ of the dual bundle v_k^* one gets the ordinary differential 1-form

$$\omega_f = \langle f , U_k \rangle \in \Omega^1(J_n^k(M)),$$

which will be called the *Cartan form*.

THEOREM. *The distribution on the manifold $J_n^k(M)$ defined by the zeros of the Cartan form ω_f, $f \in C^\infty(v_k^*)$, coincides with the Cartan distribution.*

1.5. Metasymplectic structure. Keeping in mind the description of tangent spaces to integral manifolds of the Cartan distribution, we consider the restrictions of the differentials of Cartan forms to subspaces $C(a_k)$. We fix an element $a_k \in J_n^k(M)$ and define the operator

$$\text{(1.5.1)} \quad \Omega_{a_k} : C^\infty(v_k^*) \to \Lambda^2(C^*(a_k)), \quad \text{where } \Omega_{a_k}(f)(X, Y) = d\omega_f(X, Y),$$

for all $f \in C^\infty(v_k^*)$, $X, Y \in C(a_k)$. We get $\Omega_{a_k}(\lambda f) = \lambda(a_k)\Omega_{a_k}(f)$, for any smooth function $\lambda \in C^\infty(J_n^k(M))$. Therefore, the operator Ω_{a_k} defines the operator

$$\widetilde{\Omega}_{a_k} : v^*(a_k) \to \Lambda^2(C^*(a_k)).$$

Moreover, one has the following short exact sequences:

$$0 \longrightarrow S^{k-1}(\tau_a^*) \otimes v_a \longrightarrow v_k(a_k) \xrightarrow{(\pi_{k-1,k-2})_*} v_{k-1}(a_{k-1}) \longrightarrow 0$$

and the dual sequences

$$0 \longrightarrow v_{k-1}^*(a_{k-1}) \longrightarrow v_k^*(a_k) \longrightarrow S^{k-1}(\tau_a) \otimes v_a^* \longrightarrow 0.$$

It is obvious that $\Omega_{a_k}(f) = 0$ for any section $f \in C^\infty(v_{k-1}^*) \subset C^\infty(v_k^*)$. Hence, the mapping $\widetilde{\Omega}_{a_k}$ defines the operator

$$\text{(1.5.2)} \quad \Omega_k : S^{k-1}(\tau_a) \otimes v_a^* \to \Lambda^2(C^*(a_k)),$$

which we call the *metasymplectic structure* on the Cartan space $C(a_k)$.

Moreover, one can define the vector bundles τ and v over the jet manifold $J_n^k(M)$ by taking $\tau(a_k) = \tau_a$, $v(a_k) = v_a$, where $a = \pi_{k,0}(a_k)$. Then

$$\Omega_k \in \Lambda^2(C^*) \otimes S^{k-1}(\tau^*) \otimes v$$

may be regarded as $S^{k-1}(\tau^*) \otimes \nu$-valued 2-form on the Cartan distribution.

Let $\varphi\colon M \to M$ be a local diffeomorphism, and $\varphi(a) = a'$, $\varphi^{(k)}(a_k) = a'_k$. Then the differential $\varphi_*\colon T_aM \to T_{a'}M$ induces mappings $\varphi_*\colon \tau_a \to \tau_{a'}$ and $\varphi_*\colon \nu_a \to \nu_{a'}$. It immediately follows from the construction of the metasymplectic structure that the diagram

$$\begin{array}{ccc} S^{k-1}\tau_a \otimes \nu_a^* & \xrightarrow{\Omega_k} & \Lambda^2(C^*(a_k)) \\ {\scriptstyle S^{k-1}(\varphi_*^{-1})\otimes\varphi^*}\uparrow & & \uparrow{\scriptstyle \Lambda^2(\varphi^{(k)})^*} \\ S^{k-1}\tau'_a \otimes \nu_{a'}^* & \xrightarrow{\Omega_k} & \Lambda^2(C^*(a'_k)) \end{array}$$

is commutative.

Before we calculate the value of Ω_k, we note that for each element $\lambda \in S^{k-1}(\tau_a)\otimes\nu_a^*$ the exterior 2-form $\Omega_k(\lambda) \in \Lambda^2(C^*(a_k))$ vanishes on the subspaces $L(a_{k+1})$, $a_{k+1} \in F(a_k)$, and $T_{a_k}(F(a_{k-1})) = S^k\tau_a^* \otimes \nu_a$. Therefore, by virtue of decomposition (1.3.2), it suffices to determine the value of the exterior 2-form $\Omega_k(\lambda)$ on a pair vectors $X \in L(a_{k+1}) \simeq \tau_a$, $\theta \in T_{a_k}(F(a_{k-1})) \simeq S^k\tau_a^* \otimes \nu_a$.

One gets [5, 11]:

$$\Omega_k(\lambda)(X, \theta) = \langle \lambda, X \lrcorner \delta\theta \rangle, \tag{1.5.3}$$

where by $\delta S^k\tau_a^* \otimes \nu_a \to \tau_a^* \otimes S^{k-1}\tau_a^* \otimes \nu_a$ we denote the Spencer δ-operator.

Using this formula, we can describe the degeneration subspace of the exterior 2-form $\Omega_k(\lambda)$:

$$\operatorname{Ker}\Omega_k(\lambda) = \{\theta \in S^k\tau_a^* \otimes \nu_a \mid \delta\theta \in \tau_a^* \otimes g_\lambda\},$$

where

$$g_\lambda = \operatorname{Ker}\lambda = \{\gamma \in S^{k-1}\tau_a^* \otimes \nu_a \mid \langle\lambda, \gamma\rangle = 0\}.$$

In other words, the degeneration space $\operatorname{Ker}\Omega_k(\lambda)$ coincides with the first prolongation of the symbol space g_λ.

Definition. (1) We say that the vectors $X, Y \in C(a_k)$ are *in involution* if $\Omega_k(\lambda)(X, Y) = 0$ for all tensors $\lambda \in S^{k-1}\tau_a \otimes \nu_a^*$.

(2) A subspace $L \subset C(a_k)$ is called *isotropic* if any vectors $X, Y \in L$ are in involution.

(3) We say that a subspace $L \subset C(a_k)$ is a *maximal isotropic subspace* if L is not a proper subspace of any other isotropic subspace.

Keeping in mind the description of all maximal isotropic subspaces, we consider the following construction. Let us fix a subspace $U \subset \tau_a^*$. Using decomposition (1.3.3) and the identification $(\pi_{k,0})_*\colon L(a_{k+1}) \xrightarrow{\sim} \tau_a$ we define the subspace $E(a_{k+1}, U) \subset C(a_k)$ generated by vectors of the form $X \oplus \theta$, where $X \in \operatorname{Ann} U \subset \tau_a$, $\theta \in S^kU \otimes \nu_a \subset S^k\tau_a^* \otimes \nu_a$.

Theorem [11]. *Any maximal isotropic subspace $E \subset C(a_k)$ has the form $E = E(a_{k+1}, U)$ for some subspace $U \subset \tau_a^*$, and $a_{k+1} \in F(a_k)$. Here the subspace U is uniquely determined by the subspace E: $U = \operatorname{Ann}(\pi_{k,0})_*(E)$.*

COROLLARY. *Let* $E \subset C(a_k)$ *be an isotropic subspace for which the mapping* $(\pi_{k,0})_*: E \to \tau_a$ *is an isomorphism. Then* E *is a maximal isotropic subspace and* $E = L(a_{k+1})$ *for some element* $a_{k+1} \in F(a_k)$.

REMARK. Let $N \subset M$ be a submanifold, $\operatorname{codim} N = m$, and let $N_0 \subset N$ be a submanifold in N. Denote by $N_0^{(k-1)} \subset J_n^{k-1}(M)$ the submanifold formed by all $(k-1)$-jets $[N]_x^{k-1}$ where the point x runs over the entire submanifold N_0:

$$N_0^{(k-1)} = \{[N]_x^{k-1} \mid x \in N_0\}.$$

Now suppose that $N_0^{(k)}(N) \subset J_n^k(M)$ is the set of k-jets $x_k \in J_n^k(M)$ for which $\pi_{k,k-1}(x_k) \in N_0^{(k-1)}$ and the subspaces $L(x_{k+1})$ contain the tangent planes to the submanifold $N_0^{(k-1)}$ at the points $x_{k-1} = \pi_{k,k-1}(x_k)$.

Then the tangent planes to the submanifold $N_0^{(k)}(N)$ coincide with the maximal involutive subspaces described in the previous theorem. Therefore, $N_0^{(k-1)}(N)$ is the maximal integral manifold of the Cartan distribution.

1.6. Specializations. Now we consider various specializations of the general constructions. Let M be the total space of a smooth fiber bundle:

$$\alpha: E(\alpha) = M \to B.$$

Considering local sections of α as submanifolds of codimension $m = \dim \alpha$ in M transversal to the fibers of α, we obtain k-jet spaces of local sections of α:

$$J^k(E(\alpha)) = \bigcup_{b \in B} J_b^k(\alpha),$$

where $J_b^k(\alpha)$ is the space of k-jet local sections at the point $b \in B$, and also get the natural embedding, $\mathcal{H}_k: J^k(E(\alpha)) \hookrightarrow J_n^k(M)$.

Since the transversality conditions are conditions on the 1-jets, the fibers of the projections

$$\alpha_{k,k-1} = \pi_{k,k-1} \circ \mathcal{H}_k: J^k(E(\alpha)) \to J^{k-1}(E(\alpha))$$

coincide for $k \geqslant 2$ with the fibers of the general projection $\pi_{k,k-1}: J_n^k(M) \to J_n^{k-1}(M)$ and thus inherit the affine structure. Prolongations of local automorphisms of the bundle α are affine automorphisms.

If α is a smooth vector bundle, then $\alpha_k = \alpha \circ \alpha_{k,0}: J^k(E(\alpha)) \to B$ are vector bundles too, and $\alpha_{k,s}$, $k > s$, are morphisms of these vector bundles. The affine structure coincides with the structure defined by the vector structure.

Let $v\tau: vTM \to M$ be a vertical subbundle of the tangent bundle TM:

$$vT_aM = \{X \in T_aM \mid \alpha_*(X) = 0\}, \qquad a \in M.$$

Then the bundle v_k can be identified with the pullback $\pi_{k,0}^*(v\tau)$. Hence, one gets

$$\Omega_{k,a_k} \in \Lambda^2(C^*(a_k)) \otimes S^{k-1}(\tau_{B,b}^*) \otimes vT_aM$$

for all points $a_k \in J^k(E(\alpha))$, $a = \pi_{k,0}(a_k)$, $b = \alpha(a)$.

Moreover, if $\alpha\colon M \to B$ is a trivial bundle

$$M = B \times Q, \quad \alpha(b, q) = b, \qquad b \in B, \ q \in Q,$$

for some manifold Q, then the manifold $J^k(E(\alpha))$ may be identified with the k-jet space $J^k(B, Q)$ of all local mappings $f\colon B \to Q$. In this case the bundle $\nu\tau$ may be identified with the pullback of the tangent bundle $\tau_Q\colon TQ \to Q$ and

$$\Omega_{k, a_k} \in \Lambda^2(C^*(a_k)) \otimes S^{k-1}(\tau^*_{B,b}) \otimes \tau_{Q,q},$$

for all points $a_k = [f]^k_b$, where $a = (b, q)$, $q = f(b)$.

EXAMPLES. (1) Let $k = m = 1$; then the Cartan distribution on $J^1_{n-1}(M) = \mathbb{P}(T^*M)$ is the classical contact structure and $\Omega_1 \in \Lambda^2(C^*) \otimes \nu$ is a *conformal symplectic structure* on the contact distribution.

(2) Let $k = m = 1$ and $\alpha\colon B \times \mathbb{R} \to B$ be a trivial line bundle. Then the Cartan distribution on $J^1(E(\alpha)) = T^*B \times \mathbb{R}$ is a contact structure defining by the differential 1-form $U_1 \in \Lambda^1(J^1(E(\alpha)))$ and $\Omega_1 = dU_1$ is the classical symplectic structure.

(3) Let $k = n = 1$ and $\alpha\colon Q \times \mathbb{R} \to \mathbb{R}$ be a trivial bundle. Then $J^1(E(\alpha)) = TQ \times \mathbb{R}$ and $U_1 \in \Lambda^1(J^1(E(\alpha))) \otimes \nu = \Lambda^1(J^1(E(\alpha))) \otimes TQ$ is the *classical tangent structure*.

Now we return to the general situation and consider some element $a_k = [N]^k_a \in J^k_n(M)$. Let $\mathcal{O}(N) \supset N$ be a tubular neighborhood of the submanifold $N \subset M$ and $\alpha\colon \mathcal{O}(N) \to N$ be a normal bundle. The image of the natural embedding

$$\mathcal{H}_k\colon J^k(\mathcal{V}(N)) \hookrightarrow J^k_n(M)$$

covers some neighborhood of the element a_k and will be called an *affine map* on $J^k_n(M)$.

The construction described above is similar to the construction of affine maps on a projective space. We must note also that the affine structure on $J^k_n(M)$ is induced by the linear one in an affine map.

1.7. Differential equations. By a *system of differential equations* (d.e.) of kth order on submanifolds of codimension m of manifold M (or on sections of a fiber bundle α) we mean a smooth submanifold $E_k \subset J^k_n(M)$ (respectively, $E_k \subset J^k(E(\alpha))$).

The (*regular*) *solution* of such d.e. system is a smooth submanifold $N \subset M$ (respectively, section $h\colon B \to M$) whose k-jet extension lies in the manifold E_k: $N^{(k)} \subset E_k$ (respectively, $j_k(h)(B) \subset E_k$). Equivalently, such an integral manifold of the Cartan distribution lies in E_k and can be "nicely" (= without singularities) projected to M.

A *prolongation of order* $l \geqslant 0$ of a d.e. system $E_k \subset J^k_n(M)$ is defined to be a subset $E^{(l)}_k \subset J^{k+l}_n(M)$ formed by elements $[N]^{k+l}_a$ of the k-jet extensions $N^{(k)}$ that are tangent to the submanifold E_k at the point $[N]^k_a \in E_k$ with order $\geqslant l$.

Generally speaking, $E^{(l)}_k$ is not a smooth submanifold of $J^{k+l}_n(M)$.

The d.e. system E_k will be called *formally integrable* if all prolongations $E_k^{(l)}$, $l \geqslant 0$, are smooth submanifolds and the projections

$$\pi_{k+l+1,k+l}\colon E_k^{(l+1)} \to E_k^{(l)}, \qquad l = 0, 1, \ldots$$

are smooth bundles. The conditions for formal integrability are usually stated in terms of symbols of d.e. systems [3] (below we give other conditions in terms of curvature tensors). The intersection

$$g(a_k) = T_{a_k}(E_k) \cap T_{a_k}(F(a_{k-1})) \subset S^k\tau_a^* \otimes \nu_a$$

will be called the *symbol of d.e. system* E_k at the point $a_k = [N]_a^k \in E_k$.

If all prolongations $E_k^{(l)}$, $0 \leqslant l \leqslant l_0$, are smooth submanifolds, then their symbols

$$g^{(l)}(a_{k+l}) = g^{(l)}(a_k) \subset S^{k+l}\tau_a^* \otimes \nu_a$$

at the points $a_{k+l} \in E_k^{(l)}$, $\pi_{k+l,k}(a_{k+l}) = a_k$, are the *lth prolongations of the symbol* $g(a_k)$ at the point $a_k \in E_k$. Hence, if we denote by

$$\delta\colon \Lambda^r(\tau_a^*) \otimes S^t(\tau_a^*) \otimes \nu_a \to \Lambda^{r+1}(\tau_a^*) \otimes S^{t-1}(\tau_a^*) \otimes \nu_a$$

the Spencer δ-operator, we can define the *lth prolongation of the symbol* $g(a_k)$

$$g^{(l)}(a_k) \subset S^{k+l}(\tau_a^*) \otimes \nu_a$$

by induction from the condition

$$\delta(g^{(l)}(a_k)) \subset \tau_a^* \otimes g^{(l-1)}(a_k), \qquad l = 1, 2, \ldots.$$

Then

$$g^{(l)}(a_k) = T_{a_{k+l}}(E_k^{(l)}) \cap T_{a_{k+l}}(F(a_{k+l-1})), \qquad l = 1, 2, \ldots$$

if $E_k^{(l)}$ are smooth submanifolds in $J_n^{k+l}(M)$.

Therefore, at each point $a_k \in E_k$ the *Spencer δ-complex*:

$$\begin{aligned} 0 \longrightarrow g^{(l)}(a_k) &\xrightarrow{\delta} \tau_a^* \otimes g^{(l-1)}(a_k) \xrightarrow{\delta} \cdots \xrightarrow{\delta} \Lambda^r(\tau_a^*) \otimes g^{(l-r)}(a_k) \\ &\xrightarrow{\delta} \Lambda^{r+1}(\tau_a^*) \otimes g^{(l-r-1)}(a_k) \xrightarrow{\delta} \cdots \xrightarrow{\delta} \Lambda^n(\tau_a^*) \otimes g^{(l-n)}(a_k) \longrightarrow 0, \end{aligned}$$

where $g^{(t)}(a_k) = S^{k+t}(\tau_*^a) \otimes \nu_a$ if $t < 0$, is defined. The cohomology of this complex at the term $\Lambda^i(\tau_a^*) \otimes g^{(j)}(a_k)$ is denoted by $H^{j,i}(E_k, a_k)$. It is called the Spencer δ-cohomology of d.e. system E_k at the point $a_k \in E_k$. We shall say that a d.e. system E_k is *r-acyclic* if $H^{j,i}(E_k, a_k) = 0$ for all $0 \leqslant i \leqslant r$, $j \geqslant 0$, $a_k \in E_k$.

THEOREM [3, 11]. *Let $E_k \subset J_n^k(M)$ be a 2-acyclic d.e. system for which projections*

$$\pi_{k+1,k}\colon E_k^{(1)} \to E_k, \qquad \pi_{k,0}\colon E_k \to M$$

are smooth bundles. Then E_k is a formally integrable d.e. system.

1.8. Connections. Denote by C_E the restriction of the Cartan distribution to the submanifold $E_k \subset J_n^k(M)$:

$$C_E\colon E_k \ni a_k \mapsto C(a_k) \cap T_{a_k}(E_k) = C_E(a_k).$$

We shall say that the point $a_k \in E_k$ is a *regular point* of the d.e. system E_k if in some neighborhood of this point the function $x_k \in E_k \mapsto \dim C_E(x_k)$ is constant. Otherwise, the point $a_k \in E_k$ is called *singular*.

If we suppose that all points $a_k \in E_k$ are regular, then the family of subspaces $C_E(x_k)$, $x_k \in E_k$, defines a distribution C_E on the submanifold E_k. The solutions of the d.e. system E_k are integral manifolds of the distribution C_E of dimension n which are projected without singularities to the manifold M.

DEFINITION 1. A smooth field H of n-dimensional planes

$$H\colon a_k \in E_k \mapsto H(a_k) \subset T_{a_k}(E_k)$$

on the submanifold $E_k \subset J_n^k(M)$ will be called a *Cartan connection* if the following conditions hold:

(i) $H(a_k) \subset C_E(a_k)$,
(ii) $(\pi_{k,k-1})_* H(a_k) = L(a_k)$,

for all points $a_k \in E_k$.

Now suppose that the first prolongation $E_k^{(1)} \subset J_n^{k+1}(M)$ is a smooth submanifold and $\pi_{k+1,k}\colon E_k^{(1)} \to E_k$ is a smooth bundle.

DEFINITION 2. A *Bott connection* is any smooth section $H\colon E_k \to E_k^{(1)}$ of the bundle $\pi_{k+1,k}\colon E_k^{(1)} \to E_k$.

1.9. Curvature tensors. Let H be a Cartan connection on d.e. system $E_k \subset J_n^k(M)$. We consider the restriction of the metasymplectic form Ω_k to the horizontal planes $H(a_k)$. We get

$$\Omega_H = \Omega_k|_{H(a_k)}\colon S^{k-1}\tau_a \otimes \nu_a^* \to \Lambda^2(\tau_a^*), \tag{1.9.1}$$

if we identify the space $H(a_k)$ with the tangent space τ_a:

$$(\pi_{k,0})_*\colon H(a_k) \xrightarrow{\sim} \tau_a.$$

Suppose that $E_{k-1} = \pi_{k,k-1}(E_k) \subset J_n^{k-1}(M)$ is a smooth submanifold. Then $\Omega_H(\lambda) = 0$, if $\lambda \in S^{k-1}\tau_a \otimes \nu_a^*$ is a tensor such that $\langle \lambda, \theta\rangle = 0$ for all tensors θ in the symbol space $\hat{g}(a_{k-1})$ of the d.e. system E_{k-1}:

$$\theta \in \hat{g}(a_{k-1}) = T_{a_{k-1}}(E_{k-1}) \cap T_{a_{k-1}}(F(a_{k-2})),$$

where $a_{k-1} = \pi_{k,k-1}(a_k)$, $a_{k-2} = \pi_{k,k-2}(a_k)$. Hence, we obtain the mapping

$$\Omega_H\colon S^{k-1}\tau_a \otimes \nu_a^* / \operatorname{Ann} \hat{g}(a_{k-1}) \to \Lambda^2(\tau_a^*),$$

for all points $a_k \in E_k$.

DEFINITION. The tensor Ω_H will be called a *curvature tensor of the Cartan connection* H.

PROPOSITION. *Cartan connection* H *is a Bott connection if and only if* $\Omega_H = 0$.

PROOF. The subspace $H(a_k) \subset C(a_k)$ has the form $H(a_k) = L(a_{k+1})$ for some point $a_{k+1} \in J_n^{k+1}(M)$ if and only if the metasymplectic forms $\Omega_k(\lambda)$ vanish on $H(a_k)$. In this case $a_{k+1} = a_{k+1}^H \in E_k^{(1)}$ by definition.

1.10. Weyl tensors. Let H and H' be Cartan connections of the d.e. system E_k. Since we have

$$C_E(a_k) = g(a_k) \oplus H(a_k), \tag{1.10.1}$$

the connection H' defines the tensor $\lambda = \lambda_{H,H'} \in \tau_a^* \otimes g(a_k)$, which we call the *soldering form*.

By the definition of the metasymplectic structure and the curvature tensor of Cartan connection we obtain

$$\Omega_{H'} = \Omega_H + \delta\lambda, \tag{1.10.2}$$

for any point $a_k \in E_k$.

PROPOSITION 1 (Bianchi identity). $\delta\Omega_H = 0$.

PROOF. The Spencer δ-operator is the symbol of the exterior differential. The form Ω_H is defined by the restrictions of the differentials $d\omega_f$ of the Cartan form. Hence, Ω_H is δ-closed.

DEFINITION. The δ-cohomology class

$$W_k(a_k) = \Omega_H \bmod \delta(\tau_a^* \otimes g(a_k)) \in H^{k-1,2}(E_k, a_k)$$

is called the *Weyl tensor* of the d.e. system E_k at the point $a_k \in E_k$.

Proposition 1.9 and formula (2) imply the following result.

PROPOSITION 2. *There exists a point* $a_{k+1} \in E_k^{(1)}$ *over the point* $a_k \in E_k$ *if and only if* $W_k(a_k) = 0$.

REMARKS. (1) This proposition gives an alternative approach to the integrability problem. Suppose that

$$g^{(l)} \colon E_k \ni a_k \mapsto g^{(l)}(a_k) \subset S^{k+l}\tau_a^* \otimes v_a \tag{1.10.3}$$

are vector bundles over the submanifold $E_k \subset J_n^k(M)$. Then, if the Weyl tensor W_k vanishes, the projection $\pi_{k+1,k} \colon E_k^{(1)} \to E_k$ is a smooth affine bundle. Hence, we can consider the next Weyl tensor W_{k+1} on $E_k^{(1)}$, and so on. Thus, the conditions

$$W_k = 0, \quad W_{k+1} = 0, \quad \ldots$$

together with (1.10.3), ensure the formal integrability of the d.e. system E_k (cf. [3, 5]).

(2) The d.e. system E_k is said to be of *finite type* if $g^{(l)}(a_k) = 0$ for all $a_k \in E_k$, $l \geqslant l_0$. In this case a unique point $a_{k+l_0} \in F(a_{k+l_0-1})$ over $a_{k+l_0-1} \in E_k^{(l_0-1)}$ defines the subspace $L(a_{k+l_0}) \subset T_{a_{k+l_0-1}}(E_k^{(l_0-1)})$ and hence we get the distribution $L \colon a_{k+l_0} \mapsto L(a_{k+l_0+1})$ on the submanifold $E_k^{(l_0-1)}$. This distribution is integrable if and only if $W_{k+l_0} = 0$ (Frobenius theorem). Thus, the vanishing of $W_k, \ldots, W_{k+l_0}$ is a sufficient condition for integrability.

1.11. Covariant differentials and parallel transport. Let $N \subset M$ be an arbitrary solution of the d.e. system $E_k \subset J_n^k(M)$. Then we have two subspaces $H(a_k)$ and $T_{a_k}(N^{(k)})$ and the soldering form

$$\nabla_{H, a_k} \in \tau_a^* \otimes g(a_k) \tag{1.11.1}$$

at each point $a_k = [N]_a^k$. Now we consider two vector bundles τ^* and g over $E_k^{(1)}$ with τ_a^* and $g(a_k)$ as fibers over the point $a_{k+1} = [N]_a^{k+1}$.

DEFINITION. The section (1.11.1) $\nabla_H \in C^\infty(\tau \otimes g)$ will be called the *covariant differential of the Cartan connection* H.

For any vector field X tangent to the submanifold $N \subset M$ and any solution $N^{(k)} \subset E_k$ we have $\nabla_H(X) \in C^\infty(g|_{N^{(k)}})$, where $\nabla_H(X)(a) = \langle \nabla_{H, a_k}, X_a \rangle \in g(a_k)$ and $a_k = [N]_a^k$, $\forall a \in N$.

Let us point out the geometrical meaning of this construction. Let $N \subset M$ be a solution of the d.e. system E_k and let $\gamma \subset N$ be a smooth curve. Denote by $N_\gamma^k \subset J_n^k(M)$ the set of all k-jets $x_k \in J_n^k(M)$ such that $L(x_k) \supset T_{x_{k-1}}(\gamma_{k-1})$, where $\gamma_{k-1} \subset N^{(k-1)}$ is the image of the curve γ with respect to the mapping $j_{k-1}\colon N \to J_n^{k-1}(M)$. Then N_γ^k is the integral manifold of the Cartan distribution on $J_n^k(M)$.

Let $N_\gamma^k(E_k) = N_\gamma^k \cap E_k$ and let H_γ be the restriction of the horizontal distribution H on $N_\gamma^k(E_k)$:

$$H_\gamma\colon a_k \in N_\gamma^k(E_k) \mapsto H(a_k) \cap T_{a_k}(N_\gamma^k(E_k)).$$

Since $\dim H_\gamma = 1$, we get a fibration of the submanifold $N_\gamma^k(E_k)$ by means of integral curves of the distribution H_γ. Let $L_\gamma(a_k)$ be the integral curve passing through a point $a_k \in E_k$. Any element $a_k' \in L_\gamma(a_k)$ will be called a *parallel transport* of the k-jet solution $a_k \in E_k$ along the curve γ.

Note that the k-jet $a_k \in E_k$ can be transported only along curves γ such that $\dot\gamma \in \tau_a$. The value of the covariant differential on the vector $v = \dot\gamma$ is the difference between the two tangent vectors $\dot L_\gamma(a_k)$ and $\dot\gamma(a_k)$.

§2. Homogeneous geometric structures

2.1. Klein models. Let M_0 be a smooth manifold, and let G be a Lie group with Lie algebra $\mathfrak{G}$, the later being identified with the algebra of left-invariant vector fields on G. We fix a left transitive and effective action of G on M_0: $G \times M_0 \to M_0$. Such a pair (M_0, G) will be called a *Klein model.*

We denote by $G_x = \{g \in G \mid g(x) = x\}$ the *isotropy subgroup* of x for any point $x \in M_0$.

The Lie algebra $\mathfrak{G}$ can be identified with a subalgebra $\mathfrak{G} \subset \mathcal{D}(M_0)$ of the Lie algebra $\mathcal{D}(M_0)$ of all smooth vector fields on M_0. Then the Lie algebra $\mathfrak{G}_x$ corresponding to the Lie group G_x can be described as follows:

$$\mathfrak{G}_x = \{X \in \mathfrak{G} \mid X_x = 0\}.$$

The pair $(\mathfrak{G}, \mathfrak{G}_x)$ defines a natural filtration $\{\mathfrak{G}_i\}$ of the Lie algebra $\mathfrak{G}$. Namely, let $\mathfrak{G}_0 = \mathfrak{G}$, $\mathfrak{G}_1 = \mathfrak{G}_x$, and

$$\mathfrak{G}_{i+1} = \{X \in \mathfrak{G}_i \mid [X, \mathfrak{G}] \subset \mathfrak{G}_i\}. \tag{2.1.1}$$

We get the filtration:

$$\mathfrak{G}_0 \supset \mathfrak{G}_1 \supset \cdots \supset \mathfrak{G}_i \supset \mathfrak{G}_{i+1} \supset \cdots \supset \mathfrak{G}_r = 0. \tag{2.1.2}$$

An easy computation gives

$$[\mathfrak{G}_i, \mathfrak{G}_j] \subset \mathfrak{G}_{i+j-1}. \tag{2.1.3}$$

Thus, $\mathfrak{G}_{r-1}$ is an abelian subalgebra if $r \geqslant 3$.

By putting $V = \mathfrak{G}/\mathfrak{G}_1$ one obtains a representation of $\mathfrak{G}_1$ in the vector space V induced by the adjoint representation of $\mathfrak{G}_1$:

$$X \in \mathfrak{G}_1 \mapsto \mathrm{ad}_X \colon V \to V.$$

Denote by $\overline{\mathfrak{G}}_1$ the image of $\mathfrak{G}_1$ under the representation $\overline{\mathfrak{G}}_1 \subset V^* \otimes V$.

Any element $\overline{\theta} = \theta \mod \mathfrak{G}_{i+1}$, $\theta \in \mathfrak{G}_i$, can be regarded as a tensor $\overline{\theta} \in S^i V^* \otimes V$, where

$$\begin{gathered} \overline{\theta}(\overline{X}_1, \dots, \overline{X}_i) = [X_1, [X_2, \dots, [X_i, \theta]]\dots] \mod \mathfrak{G}_1, \\ \overline{X}_j = X_j \mod \mathfrak{G}_1, \qquad X_j \in \mathfrak{G}, \ j = 1, \dots, i. \end{gathered} \tag{2.1.4}$$

Denote by $\overline{\mathfrak{G}}_i$ the image of $\mathfrak{G}_i$ in $S^i V^* \otimes V$. One gets

$$\delta(\mathfrak{G}_i) \subset V^* \otimes \overline{\mathfrak{G}}_{i-1}, \qquad i = 2, \dots, r,$$

where δ is Spencer δ-operator.

Thus the first prolongation $(\overline{\mathfrak{G}}_{i-1})^{(1)}$ contains $\overline{\mathfrak{G}}_i$ and

$$(\overline{\mathfrak{G}}_i)^{(j)} \supset \overline{\mathfrak{G}}_{i+j}. \tag{2.1.5}$$

DEFINITION. (1) The smallest number r such that $\mathfrak{G}_r = 0$, and $\mathfrak{G}_{r-1} \neq 0$ will be called the *order* $\mathrm{ord}(\mathfrak{G}, \mathfrak{G}_1)$ of the pair $(\mathfrak{G}, \mathfrak{G}_1)$.

(2) The smallest number s such that $(\overline{\mathfrak{G}}_{i-1})^{(s)} = 0$, and $(\overline{\mathfrak{G}}_{i-1})^{(s-1)} \neq 0$ will be called the *δ-order* $\mathrm{ord}_\delta(\mathfrak{G}, \mathfrak{G}_1)$ of the pair $(\mathfrak{G}, \mathfrak{G}_1)$.

From (5) we easily get the following

PROPOSITION. $\mathrm{ord}(\mathfrak{G}, \mathfrak{G}_1) \leqslant \mathrm{ord}_\delta(\mathfrak{G}, \mathfrak{G}_1) + 1$.

2.2. Homogeneous structures. Let (M_0, G) a Klein model and M a smooth manifold of dimension n, $\dim M_0 = \dim M$.

DEFINITION 1. The k-jet $[\varphi]_a^k$ of a local diffeomorphism $\varphi: M \to M_0$ will be called a *free k-coframe* at the point $a \in M$. Denote by $R_a^k(M, M_0)$, or simply $R_a^k(M)$ when no confusion is possible, the set of all free k-coframes at the point $a \in M$.

Let

$$R^k(M) = \bigcup_{a \in M} R_a^k(M)$$

be the manifold of all free k-coframes.

We have an embedding $R^k(M) \hookrightarrow J_n^k(M \times M_0)$ and natural projections $\pi_{k,s}$ determining the bundles: $\pi_{k,s}: R^k(M) \to R^s(M)$, $k > s$.

Since $R^k(M) \subset J_n^k(M \times M_0)$ is an open set, the bundles

$$\pi_{k,k-1}: R^k(M) \to R^{k-1}(M), \qquad k \geqslant 1 \tag{2.2.1}$$

are affine bundles.

The vector spaces associated with the fibers $F(a_{k-1})$, $a_{k-1} = [\varphi]_a^{k-1}$, are tensor products

$$S^k(T_a^*M) \otimes T_b M_0 \simeq S^k(T_a^*M) \otimes \mathfrak{G}/\mathfrak{G}_1, \tag{2.2.2}$$

where $b = \varphi(a) \in M_0$.

The pseudogroup of local diffeomorphisms of M acts on $R^k(M)$ in the natural way. There is also a natural left action (via gauge transformations) of the Lie group G on $R^k(M)$:

$$g([\varphi]_a^k) = [g \circ \varphi]_a^k, \tag{2.2.3}$$

where all $g \in \Gamma$, $[\varphi]_a^k \in R^k(M)$.

DEFINITION 2. The smallest number k such that action (2.2.3) is a free action will be called the *order* $\operatorname{ord}(M_0, G)$ *of the homogeneous space* (M_0, G).

DEFINITION 3. A subbundle $E_k \subset R^k(M)$ of a free k-coframe bundle $\pi_k: R^k(M) \to M$ will be called a *homogeneous* (M_0, G)*-structure* on the manifold M if the following conditions hold:

(i) the Lie group G acts on the manifold E_k freely;
(ii) the Lie group G acts on fibers of the bundle $\pi_k: E_k \to M$ transitively.

In other words, the subbundle $\pi_k: E_k \to M$ is a principal G-bundle.

EXAMPLES. (1) *G_0-structures.* Let V be a vector space, $\dim V = n$ and let $G_0 \subset \operatorname{End} V$ be a Lie subgroup. Then the semidirect product $G = G_0 \ltimes V$ acts on V in a natural (and affine) way. The order of homogeneous space (V, G) is equal to 1. Let $E_1 \subset J^1(M, V)$ be a homogeneous (V, G)-structure. Thus we have the following bundle: $\pi_{1,0}: E_1 \to M \times V$. The fibers $F(a, 0)$ of this bundle over the points $(a, 0)$, where $a \in M$, $0 \in V$, may be identified with coframes $T_aM \to V$ in the usual sense. Therefore $\widetilde{E} = \pi_{1,0}^{-1}(M \times \{0\})$ is a (co) G_0-structure in the classical sense.

Note that $\widetilde{E} \subset E_1$ is not a "good" d.e. system .

(2) *Projective structures.* Let $P_2 \subset J^2(\mathbb{R}, M)$ be an ordinary second order d.e. system. Denote by $P_2^0 \subset J^2(\mathbb{R}, \mathbb{R}P^n)$ the ordinary d.e. system describing the set of all straight lines in the projective space $\mathbb{R}P^n$. The system $P_2 \subset J^2(\mathbb{R}, M)$ will be called a *projective structure* on M if there exist local diffeomorphisms $\varphi_a\colon M \to \mathbb{R}P^n$ such that $[\varphi_a]_a^2(P_{2,a}) = P^0_{2,\varphi_a(a)}$ for any point $a \in M$. Here we denote by $P_{2,a}$ (respectively, $P^0_{2,b}$) the fiber of the projection $\pi_2\colon P_2 \to M$ at the point $a \in M$.

The set $E_2 \subset J^2(M, \mathbb{R}P^n)$ of such diffeomorphisms defines a homogeneous $(\mathbb{R}P^n, SL(n+1, \mathbb{R}))$-structure which will be called a *projective structure.*

2.3. Generalized homogeneous structures.

Definition. Let X be an arbitrary manifold. A *generalized homogeneous* (M_0, G)*-structure* on the manifold X is a d.e. system $E_k \subset J^k(X, M_0)$ such that conditions (i), (ii) of the previous definition hold.

Note that in the above definition we do not require that $\dim X = \dim M_0$ and also we do not assume any regularity conditions (e.g., the local diffeomorphism condition) with respect to k-jets $[\varphi]_x^k \in E_k$.

Examples. (1) Let $M_0 = \mathbb{R}$ and let $G = \mathrm{Aff}(1)$ be the group of affine transformations of the straight line. Since $\mathrm{ord}(M_0, G) = 1$, we consider the first order generalized $(\mathbb{R}, \mathrm{Aff}(1))$-structures $E_1 \subset J^1(X, \mathbb{R}) = T^*X \oplus \mathbb{R}$. It is obvious that E_1 is defined by a line bundle $l \subset T^*X$ or a distribution of codimension 1 on X.

(2) Let $M_0 = \mathbb{R}^s$ and let $G = \mathrm{Aff}(s)$ be the group of affine transformations of $\mathbb{R}^s$. In this case, $E_1 \subset J^1(X, \mathbb{R}^s) = T^*X \otimes \mathbb{R}^s \oplus \mathbb{R}^s$ and the $(\mathbb{R}^s, \mathrm{Aff}(s))$-homogeneous structure is defined by a distribution of codimension s on manifold X.

(3) Let $M_0 = \mathbb{R}P^1$ and let $G = SL(2, \mathbb{R})$ be the group of projective transformations. Since $\mathrm{ord}(\mathbb{R}P^1, SL(2, \mathbb{R})) = 2$, we consider second order generalized $(\mathbb{R}P^1, SL(2, \mathbb{R}))$-structures. Let $E_2 \subset J^2(X, \mathbb{R}P^1)$ be such a structure. Then $E_1 = \pi_{2,1}(E_2) \subset J^1(X, \mathbb{R}P^1)$ be an affine structure and hence it is defined by some distribution Σ on the manifold X, $\operatorname{codim}\Sigma = 1$. Let $\mathbb{P}(T^*X)$ be the projectivization of the cotangent bundle $T^*X \to X$. Then the $SL(2, \mathbb{R})$-orbits are completely determined by the Lagrangian planes $L_x \subset C(x_\Sigma) \subset T_{x_\Sigma}(\mathbb{P}T^*X)$ that are projected on the plane $\Sigma(x) \subset T_xX$ without singularities. Here $x_\Sigma \subset \mathbb{P}T^*X$ is a point defined by a hyperplane $\Sigma(x)$ and $C(x_\Sigma)$ is a Cartan space at the point $x_\Sigma \in \mathbb{P}(T^*X) = J^1_{n-1}(X)$. Therefore, the generalized $(\mathbb{R}P^1, SL(2, \mathbb{R}))$-structure is defined by a distribution Σ of codimension 1 on X and the field of Lagrangian planes $X \ni x \mapsto L_x \subset C(x_\Sigma)$ "nicely" projected on the distribution Σ.

2.4. Singularities and morphisms. Let (M_0, G) be a Klein model.

Definition. (1) A point $a_k = [\varphi]_a^k \in J^k(M_0, G)$ will be called *singular* if the isotropy subgroup $G_{a_k} \subset G$ is nontrivial.

(2) Let $E_k \subset J^k(M_0, G)$ be a generalized (M_0, G)-structure and $F\colon Y \to X$ a smooth mapping. A point $y \in Y$ will be called a *singular point of* F *with respect*

to E_k if $[\varphi \circ F]^k_y \in J^k(Y, M_0)$ is a singular point for each point $[\varphi]^k_y \in E^k$, $a = F(y)$.

(3) We say that $F\colon Y \to X$ is a *regular mapping with respect to* E_k if there are no singular points on Y.

For a regular mapping $F\colon Y \to X$ the pullback $F^*(E_k) \subset J^k(Y, M_0)$ is a well-defined homogeneous structure; here

$$F^*(E_k) = \{[F \circ \varphi]^k_y \mid \forall y \in Y,\ [\varphi]^k_{F(y)} \in E_k\}.$$

EXAMPLE. Let $E_1 \subset J^1(X, \mathbb{R})$ be a $(\mathbb{R}, \mathrm{Aff}(1))$-homogeneous structure, i.e., a distribution Σ of codimension 1 on X. Then $F\colon Y \to X$ is a regular mapping with respect to this structure if $F^*(\theta) \neq 0$ for any form $\theta \in \Omega^1(X)$ such that (locally) $\Sigma = \operatorname{Ker}\theta$. In this case, $\Sigma_F = F_*^{-1}(\Sigma)$ is a well-defined distribution of codimension 1 on Y.

Let $E_k \subset J^k(X, M_0)$, $E'_k \subset J^k(X, M_0)$ be generalized (M_0, G)-homogeneous structures, and let $F\colon Y \to X$ be a mapping that is regular with respect to E_k. We shall say that F is a *morphism* of homogeneous (M_0, G)-structures if $F^*(E_k) = E'_k$. It is obvious that generalized homogeneous (M_0, G)-structures (where (M_0, G) is fixed) and their morphisms form a category.

We consider only two examples.

(1) Let $F\colon Y \hookrightarrow X$ be an embedding. We say that Y with a (M_0, G)-homogeneous structure is a (M_0, G)-*submanifold* of X if and only if F is a (M_0, G)-morphism.

(2) Let $F\colon Y \to X$ be a smooth bundle. We say that F is a (M_0, G)-*homogeneous bundle* if F is a (M_0, G)-morphism.

2.5. Connections. Let $E_k \subset J^k(M, M_0)$ be a generalized (M_0, G)-homogeneous structure. We introduce the notion of general connection and some of its specifications.

(i) *General case.* A *general connection* H on E_k is a G-invariant connection on the principle G-bundle $\pi_k\colon E_k \to M$.

We shall denote by $H(a_k)$ the horizontal plane at the point $a_k \in E_k$. Thus $H\colon E_k \ni a_k \mapsto H(a_k)$ is a G-invariant horizontal distribution on E_k.

We denote a *connection form* on E_k by $\theta \in \Omega^1(E_k) \otimes \mathfrak{G}$. This form satisfies the following conditions:

(1) $\theta(V_X) = X$;

(2) $L_g^*(\theta) = \operatorname{Ad} g \circ \theta$;

(3) $\operatorname{Ker}\theta_{a_k} = H(a_k)$, $a_k \in E_k$,

where V_X is a vertical vector field corresponding to $X \in \mathfrak{G}$ and $L_g\colon E_k \to E_k$ is the left action of elements $g \in G$.

(ii) *Cartan connections.* A general connection H on E_k is called a *Cartan connection* if

$$H(a_k) \subset C_E(a_k), \tag{2.5.1}$$

for all elements $a_k \in E_k$.

(iii) *Bott connections.* A G-invariant smooth section $H\colon E_k \to E_k^{(1)}$ is called a *Bott connection.*

In this definition we assume the existence of the first prolongation $E_k^{(1)}$ and require that $\pi_{k+1,k}\colon E_k^{(1)} \to E_k$ be a smooth affine bundle.

It follows directly from the definition that the curvature form Ω_H of a general connection is an equivariant horizontal 2-form $\Omega_H \in \Omega^2(E_k)\otimes\mathfrak{G}$. If H is a Cartan connection, then Ω_H is the restriction of the metasymplectic structure Ω_k on the distribution H.

Let $[\varphi]_a^k \in E_k$ and $\mathfrak{G} \supset \mathfrak{G}_{1,b} \supset \mathfrak{G}_{2,b} \supset \cdots \supset \mathfrak{G}_{k,b} \supset 0$ be the filtration (2.1.2) defined by the point $b = \varphi(a) \in M_0$. It follows from (1.9.2) that

$$\Omega_{H,a_k} \in \Lambda^2(T_a^*M)\otimes(\mathfrak{G}_{k-1,b}/\mathfrak{G}_{k,b}). \tag{2.5.2}$$

Denote by

$$\eta_{k-1}(b) = \mathfrak{G}_{k-1,b}/\mathfrak{G}_{k,b}. \tag{2.5.3}$$

Then Ω_H is a G-equivariant horizontal 2-form on the trivial bundle $\pi_0\colon M\times M_0 \to M$ with values in the vector bundle η_{k-1}:

$$\Omega_H \in \Lambda^2(T^*M)\otimes\eta_{k-1}. \tag{2.5.4}$$

It follows from the description of involutive subspaces that a Cartan connection is a Bott connection if and only if $\Omega_H = 0$.

2.6. The existence of Cartan connections.

THEOREM. *Let $E_k \subset J^k(M, M_0)$ be a generalized homogeneous structure. Then E_k admits a Cartan connection if and only if the Lie subalgebra $\mathfrak{G}_b^{k-1} \subset \mathfrak{G}$ has a $G_{a_{k-1}} \subset G$-invariant summand for any elements $a_k = [\varphi]_a^k$. Here $G_{a_{k-1}} = \{g \in G \mid g(a_{k-1}) = a_{k-1}\}$ is the isotropy subgroup.*

PROOF. Denote by vT_{a_k} and $vT_{a_{k-1}}$ the tangent spaces to the fibers of $\pi_k\colon E_k \to M$ and $\pi_{k-1}\colon E_{k-1} \to M$ at the points $a_k = [\varphi]_a^k \in E_k$ and $a_{k-1} = \pi_{k,k-1}(a_k)$, $b = \varphi(a)$. Recall that $E_{k-1} = \pi_{k,k-1}(E_k)$.

Let H be an arbitrary connection on the principal G-bundle $\pi_k\colon E_k \to M$ and let θ be the connection form, $\theta\colon T_{a_k}E_k \to vT_{a_k}E_k$, for any point $a_k \in E_k$.

Consider also the restriction of the Cartan form U_k on E. One gets

$$U_k\colon T_{a_k}E_k \to vT_{a_{k-1}}E_{k-1}.$$

By definition, the restriction of U_k to the subspace $vT_{a_k}E_k$ coincides with the epimorphism $(\pi_{k,k-1})_*\colon vT_{a_k}E_k \to vT_{a_{k-1}}E_{k-1}$. Therefore, the mappings

$$\gamma = U_k - (\pi_{k,k-1})_*\circ\theta\colon T_{a_k}E_k \to vT_{a_{k-1}}E_{k-1}$$

have the following properties:

(i) γ is G-invariant,

(ii) γ is horizontal, i.e., $\gamma(v) = 0$ for any $v \in T_{a_k}E_k$.

Now, if one has a G-invariant 1-form λ on E_k, $\lambda\colon T_{a_k}E_k \to vT_{a_k}E_k$, such that

(1) λ is horizontal,

(2) $(\pi_{k,k-1})_* \circ \lambda = \gamma$,

then the 1-form $\theta + \lambda$ is a connection form, and

$$(\pi_{k,k-1})_*(\operatorname{Ker}(\theta+\lambda)_{a_x}) \subset \operatorname{Ker}((\pi_{k,k-1})_* \circ (\theta+\lambda)_{a_x}) = \operatorname{Ker} U_{k,a_k} = C(a_k),$$

hence, $\theta + \lambda$ defines a Cartan connection.

To define λ, we consider the following diagram

$$\begin{array}{ccccccccc} 0 & \longrightarrow & \mathfrak{G}_b^{k-1} & \overset{s}{\longleftrightarrow} & vT_{a_k}E_k & \xrightarrow{(\pi_{k,k-1})_*} & vT_{a_{k-1}}E_{k-1} & \longrightarrow & 0 \\ & & & & \uparrow{\scriptstyle\lambda} & \nearrow{\scriptstyle\gamma} & & & \\ & & & & T_{a_k}E_k & & & & \end{array}$$

Therefore, the construction of λ is possible if and only if the Lie subalgebra $\mathfrak{G}_b^{k-1}$ admits a $G_{a_{k-1}}$-invariant summand $\mathfrak{G} = \mathfrak{G}_b^{k-1} \oplus \mathfrak{M}^{k-1}$ and $\operatorname{Ad}_g(\mathfrak{M}^{k-1}) \subset \mathfrak{M}^{k-1}$ for all elements of the isotropy subgroup $G_{a_{k-1}}$.

REMARK. If $G_{a_{k-1}}$ is a connected Lie group, then it is sufficient that $\mathfrak{G}_b^{k-1} \subset \mathfrak{G}$ be a reductive subalgebra.

COROLLARY 1. *If* $k > \operatorname{ord}(M_0, G)$, *then any generalized homogeneous structure has a unique Cartan connection.*

COROLLARY 2. *If the mapping* $E_k^{(1)} \to E_k$ *is a diffeomorphism, then* E_k *admits a Cartan connection.*

EXAMPLES. 1) The Levi–Civita connection is the Cartan connection defined by Corollary 2.

2) The *normal connection* (introduced by E. Cartan) for a projective structure is the Cartan connection defined by Corollary 2.

§3. Homogeneous geometric structures

3.1. Symmetries and Lie equations. (A) Let $E_k \subset J_n^k(M)$ be an arbitrary d.e. system.

DEFINITION. (1) A local diffeomorphisms $\varphi\colon M \to M$ is called a (*finite*) *symmetry* of E_k if $\varphi^{(k)}(E_k) \subset E_k$.

(2) A vector field $X \in \mathcal{D}(M)$ is called an (*infinitesimal*) *symmetry* of E_k if the vector field $X^{(k)}$ is tangent to E_k.

Here $X^{(k)}$ is a *prolongation of the vector field* X, i.e., the vector field corresponding to the one-parameter group $\varphi_t^{(k)}\colon J_n^k(M) \to J_n^k(M)$, where $\varphi_t\colon M \to M$ is the flow of X.

(b) Let $E_k \subset J^k(E(\alpha))$ be a d.e. system on the bundle $\alpha\colon M = E(\alpha) \to B$.

DEFINITION. A local automorphism $(\varphi, \overline{\varphi}) \in \operatorname{Aut}\alpha$

$$\begin{array}{ccc} M & \xrightarrow{\varphi} & M \\ {\scriptstyle\alpha}\downarrow & & \downarrow{\scriptstyle\alpha} \\ B & \xrightarrow{\overline{\varphi}} & B \end{array}$$

is said to be a *(finite) symmetry* of E_k if $\varphi^{(k)}(E_k) \subset E_k$.

Respectively, a vector field $(X, \overline{X}) \in \operatorname{aut}(\alpha)$ of infinitesimal automorphism of α will be called an *infinitesimal symmetry* of E_k if $X^{(k)}$ is tangent to E_k.

Let $\tau: TM \to M$ be the tangent bundle of M and let $\operatorname{Lie}(E_k) \subset J^k(TM)$ be the differential equation of infinitesimal symmetries

$$\operatorname{Lie}(E_k) = \{[X]^k_a, \ X \in \mathcal{D}(M) \mid X^{(k)}_{a_k} \text{ is tangent to } E_k \text{ at any point } a_k \in E_k\}.$$

The differential equation $\operatorname{Lie}(E_k)$ will be called the *Lie equation of* E_k.

Note that $\operatorname{Lie}(E_k)$ is a linear d.e.

Let $\alpha: M = E(\alpha) \to B$ be a smooth bundle. We denote by $J^k(\operatorname{aut}(\alpha))$ the space of all k-jets of the infinitesimal automorphism of α. If $E_k \subset J^k(E(\alpha))$, then we denote by $\operatorname{Lie}(E_k) \subset J^k(\operatorname{aut}(\alpha))$ the differential equation of infinitesimal symmetries of E_k.

Note that by definition the set of solutions of Lie equations $\operatorname{Lie}(E_k)$ is a Lie algebra.

3.2. Straight lines on Grassmannians. Let V be a vector over $\mathbb{R}$, $\dim V = n+m$, and let $\operatorname{Grass}_n(V)$ be the Grassmannian of all n-dimensional subspaces of V. In order to regard $\operatorname{Grass}_n(V)$ as a projective manifold lying in the projective space $\mathbb{P}(\Lambda^n(V))$ (via the Plucker mapping), we give the following

DEFINITION. A projective line $l \subset \mathbb{P}(\Lambda^n(V))$ will be called a *straight line on the Grassmannian* $\operatorname{Grass}_n(V)$ if $l \subset \operatorname{Grass}_n(V) \subset \mathbb{P}(\Lambda^n(V))$.

PROPOSITION. *Let $p \in \operatorname{Grass}_n(V)$ be an n-dimensional subspace, $p = X_1 \wedge \cdots \wedge X_n$, where $X_1, \dots, X_n$ is some basis in p. Then any straight line l passing through the point p has the form*

$$p([t:s]) = (sX_1 + tv) \wedge X_2 \wedge \cdots \wedge X_n,$$

for some vector $v \in V$, $v \notin p$, $[s:t] \in \mathbb{R}P^1$.

Hence any straight line l passing through $p \in \operatorname{Grass}_n(V)$ is uniquely determined by a subspace $q \subset p$, $\operatorname{codim}_p q = 1$ (in our case, $q = X_2 \wedge \cdots \wedge X_n$) and a vector $v \notin p$, $v \notin p \cap q$.

Denote by $S(p) \subset \operatorname{Grass}_n(V)$ the cone of all straight lines passing through p. Then one has $\dim S(p) = m + n$. Now we give another description of straight lines on Grassmannians.

DEFINITION. The transformations

$$A^t_{r,\lambda}: X \mapsto X + t\langle\lambda, X\rangle v, \qquad \lambda \in V^*, \ v \in V \tag{3.2.1}$$

are called *transvections*.

PROPOSITION. *Let $p \in \mathrm{Grass}_n(V)$ be an n-plane such that $v \notin p$, $\lambda|_p \neq 0$. Then $p(t) = A^t_{r,\lambda}(p) \subset \mathrm{Grass}_n(V)$ is a straight line.*

Hence, the tensors

$$\overline{v} \otimes \overline{\lambda} \in (V/p) \otimes (V^*/\,\mathrm{Ann}\, p), \tag{3.2.2}$$

where $\overline{v} = v \mod p$, $\overline{\lambda} = \lambda \mod (\mathrm{Ann}\, p)$, provide a representation of the straight line $p(t)$.

3.3. Characteristics and straight lines. Let $E_1 \subset J^1_n(M)$ be a d.e. system of the first order. Denote by $FE(a) \subset F(a)$ the intersection of E_1 and the fiber $F(a)$.

DEFINITION. A straight line $l \subset F(a) = \mathrm{Grass}_n(T_a M)$ is said to be a *characteristic line at the point* $a_1 \in E_1$ if l tangent to $FE(a)$ at the point a_1.

The line l is defined by the tensor $\overline{v} \otimes \overline{\lambda}$ (see (3.2.2)), $\overline{v} \neq 0$, $\overline{\lambda} \neq 0$, then l is a characteristic line if and only if $\overline{v} \otimes \overline{\lambda} \in g(a_1)$, where $g(a_1) = T_{a_1}(FE(a))$ is a symbol of E_1 at the point $a_1 \in E_1$. Hence, the definition of characteristic lines is equivalent to the usual notion of characteristics.

3.4. Symmetries and characteristics.

THEOREM 1. *Let $E_1 \subset J^1_n(M)$ be a d.e. system such that the following conditions hold:*

(i) *the submanifolds $FE(a) \subset \mathrm{Grass}(T_a M)$, $a \in M$ do not contain straight lines;*

(ii) *$FE(a)$ are nondegenerate submanifolds, i.e., submanifolds that do not lie in hyperplane sections of $\mathrm{Grass}_n(T_a M)$;*

(iii) *for any point $a \in M$ and any vector $v \in T_a M$ there exists point $a_1 \in E_1$ such that $v \notin L(a_1)$.*

Then $\mathrm{Lie}(E_1) \subset J^1(TM)$ is an elliptic d.e. system.

PROOF. Let $v \otimes \lambda \in T_a M \otimes T^*_a M$ be a characteristic tensor for the Lie equation $\mathrm{Lie}(E_1)$ at the point $a \in M$. Then the straight line $a_1(t) = A^t_{r,\lambda}(a_1)$ lies on $FE(a)$ for any point $a_1 \in FE(a)$ if $\overline{v} \neq 0$, $\overline{\lambda} \neq 0$. From (i) it follows that $A^t_{r,\lambda}(a_1)$ has to "vanishing line" for all points $a_1 \in FE(a)$; i.e., $A^t_{r,\lambda}(a_1) = a_1$. Hence,

(1) $\lambda = 0$ on all n-planes $L(a_1)$ for any $a_1 \in FE(a)$, or

(2) $v \in L(a_1)$ for any $a_1 \in FE(a)$.

The proof is concluded by comparing with (ii), (iii).

In a similar way one obtains the following

THEOREM 2. *Let $E_1 \subset J^1_n(M)$ be a d.e. system such that the following conditions hold:*

(i) *$FE(a)$ is a projective manifold for all points $a \in M$;*

(ii) *for the complexification $FE(a)^{\mathbb{C}} \subset \mathrm{Grass}_n(T^{\mathbb{C}}_a M)$ conditions* (i)–(iii) *of Theorem* 1 *hold.*

Then $\mathrm{Lie}(E_1)$ is a finite type d.e. system.

For a d.e. system $E_1 \subset J^1(E(\alpha))$ on the bundle $\alpha\colon E(\alpha) = M \to B$ Theorems 1 and 2 take the following form.

THEOREM 1′. *Let* $E_1 \subset J^1(E(\alpha))$ *be a d.e. system such that the following conditions hold*:

(i) $FE(a)$, $a \in M$, *does not contains straight lines*;
(ii) *the projective characteristic manifolds* $\mathrm{Char}(E_1, a_1) \subset \mathbb{P}(T_b^* B)$, $a_1 = [h]_b^1$, *do not lie in a hyperplane*;
(iii) *the subspace* $\bigcup_\lambda K_\lambda$, *where* $K_\lambda \subset vT_a(\alpha)$ *is defined by the relations* $K_\lambda \otimes \lambda \subset g(a_1)$, $\lambda \neq 0$, *generates the vertical tangent space* $vT_a(\alpha)$, $a = h(b)$.

Then $\mathrm{Lie}(E_1) \subset J^1(\mathrm{aut}(\alpha))$ *is an elliptic d.e. system.*

THEOREM 2′. *Let* $E_1 \subset J^1(E(\alpha))$ *be a d.e. system such that the following conditions hold*:

> $FE(a)$ *is an analytic submanifold of* $F(a)$, $a \in M$, *and for the complexification* $(FE(a))^{\mathbb{C}}$ *conditions* (i)–(iii) *of Theorem* 1′ *hold.*

Then $\mathrm{Lie}(E_1) \subset J^1(\mathrm{aut}(\alpha))$ *is a finite type d.e. system.*

REMARK. The same results hold for a d.e. system of the kth order $k \geqslant 2$. But in this case we must use an integral Grassmannians (see [12]) and straight lines on such manifolds.

3.5. Homogeneous differential equations.

DEFINITION. A d.e. system $E_k \subset J_n^k(M)$ (or $E_k \subset J^k(E(\alpha))$) is called a *homogeneous system* if there exists a Lie group G of symmetries E_k acting transitively on E_k.

We shall consider only d.e. systems E_k that are projected to the manifold M, i.e., $\pi_{k,0}\colon E_k \to M$ is a smooth bundle. Now let us consider models of homogeneous d.e. system. First of all, M must be a homogeneous space. Fix a homogeneous space (M_0, G) with a given structure group G. Let $k \geqslant \mathrm{ord}(M_0, G)$, and we have a free action of G on $J_n^k(M_0)$. Consider any orbit $E_k^0 \subset J_n^k(M_0)$. We say that E_k^0 is a *homogeneous model* if G is the group of all symmetries of E_k^0.

REMARK. Let a d.e. system E_k^0 satisfy conditions (ii) of Theorem 2 in 3.4. Then the group $\overline{G}$ of all symmetries of E_k^0 is a Lie group and $\overline{G} \supset G$ by definition. Hence, we can consider the new Klein model $(M_0, \overline{\Gamma})$ and the same model system E_k^0.

3.6. Classification problems. Let $E_k^0 \subset J_n^k(M_0)$ be a model homogeneous d.e. system and let $E_k \subset J_n^k(M)$ be an arbitrary d.e. system, $\dim M = \dim M_0$.

DEFINITION. We say that E_k has E_k^0*-type* if the d.e. system

$$Q_k = \{[\varphi]_a^k \in R^k(M, M_0) \mid \varphi^{(k)}(E_{k,a}) = E^0_{k,\varphi(a)}\},$$

where $E_{k,a} = E_k \cap J_n^k(M)_a$, defines a homogeneous (M, M_0)-structure on the manifold M.

REMARK. Since the Lie group G is a symmetric group on E_k^0, Q_k defines a homogeneous structure if and only if the following condition holds:

(i) for any point $a \in M$ there exists a local diffeomorphism $\varphi\colon M \to M_0$ such that $\varphi^{(k)}(E_{k,a}) = E^0_{k,\varphi(a)}$.

THEOREM 1. *A d.e. system* $E_k \subset J_n^k(M)$ *of* E_k^0*-type is (locally) equivalent to the model system* E_k^0 *if and only if the d.e. system* Q_k *has a (local) solution.*

Hence we can use the results of §1 to obtain classification theorems. Here we formulate one of them.

THEOREM 2. *Let* $E_k \subset J_n^k(M)$ *be a d.e. system of* E_k^0*-type such that the conditions of Theorem* 3.4.2 *hold for the d.e. system* Q_k. *Then the d.e. system* E_k *is locally equivalent to its model* E_k^0 *if and only if all Weyl tensors of the d.e. system* Q_k *vanish.*

By using some simple topological arguments and the above theorem we get the following result.

THEOREM 3. *Let* $E_k \subset J_n^k(M)$ *be a d.e. system of* E_k^0*-type such that conditions of Theorem* 3.6.2 *hold. If* E_k *and* E_k^0 *are compact simply-connected manifolds, then the d.e. system* E_k *is globally equivalent to* E_k^0.

EXAMPLES. (1) Any Monge–Ampère equation can be identified with an effective n-form θ on $(2n+1)$-dimensional contact manifold M (see [8]). In general ($n \geqslant 3$) conditions (ii) of Theorem 2 in 3.4 hold (see [9, 10]). Hence we can use Theorem 3.6.2 and a description of effective n-forms on contact homogeneous manifolds to obtain a conclusive classification of Monge–Ampère equations.

(2) Any quasilinear 2×2 d.e. system of first order may be regarded as a quasicomplex (= elliptic case) or semiproduct (= hyperbolic case) structure on 4-dimensional manifolds. Theorem 3.6.2 gives models for such systems. For the elliptic case, see also [6, 7].

References

1. E. Cartan, *Les groupes d'holonomie des espaces généralisés*, Acta Math. **48** (1926), 1–42.
2. C. Ehresmann, *Introduction à la théorie des structures infinitésimales et des pseudogroupes de Lie*, Colloc. Internat. Centre Nat. Rech. Sci. Strasbourg (1953), 97–111.
3. H. Goldshmidt, *Integrability criteria for systems of nonlinear partial differential equations*, J. Differential Geom. **1** (1967), 269–307.
4. V. Guillemin, *The integrability problem for G-structures*, Trans. Amer. Math. Soc. **116** (1965), 544-560.
5. I. S. Krasil′shchik, V. V. Lychagin, and A. M. Vinogradov, *Geometry of jet spaces and nonlinear partial differential equations*, Gordon and Breach, New York, 1986.
6. P. Libermann, *Sur le problème d'équivalence de certaines structures infinitesimales*, Ann. Math. Pura Appl. **36** (1954), 27–120.
7. ———, *Sur les prolongements des fibrés principaux et groupoides différentiables*, Sem. Anal. Glob. Montréal (1969), 7–108.
8. V. Lychagin, *Contact geometry and nonlinear second order differential equations*, Uspekhi Mat. Nauk **34** (1979), no. 1, 101–171; English transl., Russian Math. Surveys **34** (1979), 149–180.
9. V. Lychagin and V. Rubtsov, *Local classification of Monge–Ampère differential equations*, Dokl. Akad. Nauk SSSR **272** (1983), 34–38; English transl. in Soviet Math. Dokl. **28** (1983).
10. V. Lychagin, V. Rubtsov, and I. Chekalov, *A classification of Monge–Ampère equations*, Reports Dept. of Math. Univ. of Stockholm **28** (1983), 16–58.
11. V. Lychagin, *Geometric theory of singularities of solutions of nonlinear differential equations*, J. Soviet Math. **51** (1990), 2735–2757.
12. ———, *Homogeneous structures on manifolds: differential geometry from PDEs viewpoint*, Mat. Zametki **51** (1992), no. 4, 54–68; English transl. in Math. Notes **51** (1992).

Translated by THE AUTHOR

S. LIE CENTER, P.O.BOX 546, 119618, MOSCOW, RUSSIA

Amer. Math. Soc. Transl.
(2) Vol. **167**, 1995

Geometry of Quantized Super PDE's

AGOSTINO PRÁSTARO

ABSTRACT. In this paper we announce some results on the geometrization of super PDE's, i.e., PDE's defined in the category of supermanifolds [46, 47]. These results generalize previous ones for PDE's [45].

§1. Geometry of super PDE's

(see also [3–5, 9, 16, 20, 24. 25, 41, 46, 47, 49, 50, 54, 56])

Let $\mathbb{K}$ be the field of real numbers $\mathbb{R}$ or of complex numbers $\mathbb{C}$. Let A be a superalgebra. (In this paper "superalgebra" means "Banach–Grassmann algebra" in the sense of [16].) Set

$$A_0, \quad A_1, \quad A^{m,n} \equiv (A_0)^m \times (A_1)^n, \quad A^{\overline{m},\overline{n}} \equiv (A_1)^m \times (A_0)^n,$$
$$A^{m+n} \equiv (A^{m+n})_0 \oplus (A^{m+n})_1 \equiv A^{m,n} \oplus A^{\overline{m},\overline{n}}.$$

Let A be a superalgebra and U an open subset of $A^{m,n}$. A *superdifferentiable function* $f: U \to A$ of class G^s_w, $s \geqslant 0$, on $U \subset A^{m,n}$ is a weakly s-times differentiable map (in the sense of [18]) (or of class C^s_w) such that

$$Df(x) \in L_{A_0}(A^{m,n}, A) \equiv (A^{m,n})^+ \cong A^{m+n} \quad \text{for all } x \in U.$$

Supersmooth functions (or functions of class G^∞_w) are (weak) superdifferentiable functions belonging to the classes G^s_w, $s \geqslant 0$. The set of all such functions is denoted by $G^\infty_w(U; A)$. Similarly we can define *superanalytic functions* (or of class G^ω_w), functions $U \to A$ of class G^ω_w such that their derivatives are A_0-linear. The set of all such functions is denoted by $G^\omega_w(U; A)$. A function $f: U \to A^{m',n'}$ is called *superdifferentiable* (or of class G^s_w, $s \geqslant 0$) if all its components $f^A: U \to A$, are of class G^s_w. The set of all such functions is denoted by $G^s_w(U; A^{m',n'})$ and is a $G^s_w(U; A)_0$-module. Let $f: U \to A$ be a superdifferentiable function defined on an open subset $U \subset A^{m,n}$. Then, for each $p \in \mathbb{N}$, the p-derivative of f is a

---

1991 *Mathematics Subject Classification*. Primary 58A50, 58G99; Secondary 35Q40.
Work partially supported by grants CNR/GNFM and MURST 40

map

$$D^p f : U \to L_{A_0}(A^{m,n} \odot_{A_0} \cdots p \cdots \odot_{A_0} A^{m,n}; A).$$

So we have the maps (*pth partial derivative* of f):

$$((\partial_{A_p \dots A_1} f)) : U \to ((A_1 \odot_{A_0} \cdots r \cdots \odot_{A_0} A_1)^{\oplus})^{\binom{p}{r} m^{p-r} n^r}, \qquad 1 \leqslant r \leqslant p,$$

if $|A_1| = \cdots = |A_{p-r}| = 0$, $|A_{p-r+1}| = \cdots = |A_p| = 1$;

$$((\partial_{A_p \dots A_1} f)) : U \to A^{m^p} \quad \text{if } |A_1| = \cdots = |A_p| = 0,$$

such that

$$D^p f(x)(t_1, \dots, t_p) = t_p^{A_p} \cdots t_1^{A_1} (\partial_{A_1 \dots A_p} f)(x)$$
$$\text{for all } t_i = (t_i^{A_i}) \in A^{m,n} \quad (i = 1, \dots, p),$$
$$(\partial_{A_p \dots A_1} f) = (\partial_{A_p}(\partial_{A_{p-1} \dots A_1} f)),$$
$$(\partial_{A_p \dots A_{k+1} A_k \dots A_1} f) = (-1)^{|A_k||A_{k+1}|} (\partial_{A_p \dots A_k A_{k+1} \dots A_1} f).$$

A *supermanifold of dimension* (m, n) (over a superalgebra A) (of class G_w^k, $0 \leqslant k \leqslant \infty$, ω) is a locally convex manifold M, modeled on $A^{m,n}$ and with a G_w^k atlas Ξ of local coordinate mappings. So for each open coordinate set $U \subset M$ we have a set of $m + n$ coordinate functions $(x^A): U \to A^{m,n}$ such that $x^A: U \to A_0$, $1 \leqslant A \leqslant m$, $x^A: U \to A_1$, $m + 1 \leqslant A \leqslant m + n$. We denote by $\mathcal{A}_p(M)$ the graded algebra of germs of A-valued weak supersmooth functions at $p \in M$. Let M be a supermanifold of dimension (m, n). The *tangent space* at $p \in M$ is the (m, n)-dimensional vector superspace T_pM defined in the following way: $T_pM \equiv$ space of equivalence classes $v = [f]$ of C_w^1 (or equivalently C^1) curves $f: I \to M$, $I =$ open neighborhood of $0 \in \mathbb{R}$, $f(0) = p$; two curves f, f' are equivalent if for each (equivalently, for some) coordinate system μ around p, the functions $\mu \circ f$, $\mu \circ f': I \to A^{m,n}$, have the same derivative at $0 \in \mathbb{R}$. T_pM turns out to be either a $\mathbb{K}$-vector space or a topological locally convex $\mathbb{K}$-vector space, so there exists, accordingly, the *dual*: $(T_pM)^{\cdot} \equiv L_{\mathbb{K}}(T_pM; \mathbb{K})$, of T_pM, or the *topological dual*: $(T_pM)' \equiv \mathcal{L}_{\mathbb{K}}(T_pM; \mathbb{K})$. On the other hand T_pM is also a (m, n)-dimensional vector superspace, hence an A_0-module. So we can give the following definitions.

DEFINITION 1. Let M be a supermanifold of dimension (m, n). For any $p \in M$ put

$$\widetilde{T}_pM \equiv AO_{A_0}T_pM \equiv \mathrm{Hom}_{A_0}(A, T_pM) \quad (\textit{supertangent space}),$$
$$(T_pM)^+ \equiv T_pMO_{A_0}A \equiv \mathrm{Hom}_{A_0}(T_pM, A) \quad (\textit{superdual of } T_pM).$$

NOTATION. We shall denote by $\overline{T}M$ the odd part of $\widetilde{T}M$: $\widetilde{T}M = TM \oplus \overline{T}M$. Similar notations will be used for the corresponding tensor extensions. Furthermore, we shall denote by $\dot{T}_s^r M$ the tensor bundle of type (r, s) if each tangent space T_pM is regarded as an A_0-module. The corresponding symmetric and skewsymmetric bundles will be denoted by $\dot{S}^r_s M$ and $\dot{\Lambda}^r_s M$ respectively. Furthermore, we shall denote by $G^s_w(W)$ the set of sections of class G^s_w of a fiber bundle $\pi: W \to M$. A *regular supermanifold* is a supermanifold M of dimension (m, n) such that it admits a projection on a m-dimensional manifold M_B. We will call M_B the *body* of M. Below we only consider regular supermanifolds.

By using standard results of algebra [6, 14, 31], we can prove the following theorems.

THEOREM 1. 1) *One has the canonical isomorphisms*:

$$(T_pM)^+ \cong AO_{A_0}(T_pM)^+_0 \cong AO_{A_0}(T_pM)^+_1,$$
$$T_pM \cong (T_pM)^+_0 \cong \mathrm{Hom}_{A_0}(T_pM; A_0)^+.$$

2) *Let M be a supermanifold of dimension (m, n). For any section i of $b: M \to M_B$ we have a canonical A-homomorphism*: $i_\bullet: (T_{i(p)}M)^+ \to A \oplus_{\mathbb{K}} (T_pM_B)^\bullet$.

Thus we can calculate the pullback of any graded p-differential form $\alpha: M \to \widetilde{\Lambda}^0_p M$ on M:

$$i^\bullet\alpha \equiv i_\bullet \circ \alpha \circ i: M_B \to A \oplus_{\mathbb{K}} \Lambda^0_p M_B.$$

DEFINITION 2. The *superderivative space* $\underline{D}(M, N)$ for supermanifolds M and N of dimension (m, n) and (m', n') respectively is defined by

$$\underline{D}(M, N) \equiv TMO_{A_0}TN \equiv \bigcup_{(p,q)\in M\times N} T_pMO_{A_0}T_qN,$$

where $T_pMO_{A_0}T_qN$ is the space of superlinear mappings $T_pM \to T_qN$.

2) If $\pi: W \to M$ is a fiber bundle between supermanifolds, the *superderivative space of sections* of π, $\underline{D}(W)$ is defined by $\underline{D}(W) \equiv d^\bullet\underline{D}(M, W)$, where d is the diagonal mapping $d: M \to M \times M$.

3) *Superderivative spaces of higher orders* between supermanifolds can be defined by putting $\underline{D}^k(M, N) \equiv \underline{D}(\underline{D}^{k-1}(M, N))$.

REMARK. One has a natural injection over $M \times N$: $\underline{D}(M, N) \to D(M, N)$, where $D(M, N)$ denotes the derivative space with considered M and N as locally convex manifolds [18, 51]. $\underline{D}(M, N)$ turns out to be a supermanifold of dimension $(m + m' + mm' + nn', n + n' + mn' + nm')$. If (x^A) is a coordinate system on M and (y^B) is a coordinate system on N, then (x^A, y^B, y^B_A), $1 \leqslant A \leqslant m + n$, $1 \leqslant B \leqslant m' + n'$ is a coordinate system on $\underline{D}(M, N)$. The first *derivative* of a superdifferentiable mapping $f: M \to N$ is a section $Df: M \to \underline{D}(M, N)$ of the fiber bundle $\pi_1: \underline{D}(M, N) \to M$. The local expression of Df is as follows: $x^A \circ Df = x^A$, $y^B \circ Df = f^B \equiv y^B \circ f$, $y^B_A \circ Df = (\partial x_A \cdot f^B)$. The *velocity* of a superdifferentiable curve in M, $f: A \to M$ is the map $\dot{f} \equiv \mathrm{pr}_2 \circ Df: A \to \widetilde{T}M$, where pr_2 is the canonical mapping $\mathrm{pr}_2: \underline{D}(A, M) \cong A \times \widetilde{T}M \to \widetilde{T}M$. The *differential* of a superdifferentiable map $f: M \to A$ is a map $df \equiv \mathrm{pr}_2 \circ Df: M \to (TM)^+$, where pr_2 is the canonical mapping $\mathrm{pr}_2: \underline{D}(M, A) \cong A \times (TM)^+ \to (TM)^+$. For a superdifferentiable map $f: A \times M \to N$ we can define its *infinitesimal variation* $\partial f: M \to f_0^\bullet\widetilde{T}M$ as $\partial f =$

$D_1 f \circ j$ being j the inclusion $j\colon M \to A \times M$ given by $p \mapsto (0, p)$. So, if f is a superdeformation of a superdifferentiable section s of the fiber bundle $p\colon W \to M$, such that $f_0 = s$, then $\partial f\colon M \to s^{\bullet}\mathrm{v}\widetilde{T}W$, where the symbol v denotes "vertical". From the above considerations we see that $\widetilde{T}M$ can be identified with the space of vectors tangent to superdifferentiable curves $A \to M$ of M, and $(TM)^{+})$ can be identified with the space of differentials of superdifferentiable A-valued functions on M.

REMARK. 1) We get natural structures of fiber bundles on $\underline{D}^k(M, N)$ as in the case of case of l. c. manifolds: $\pi_{k,h}\colon \underline{D}^k(M, N) \to \underline{D}^h(M, N)$, $0 \leqslant h \leqslant k$, $\pi_k\colon \underline{D}^k(M, N) \to M$, $\overline{\pi}_k\colon \underline{D}^k(M, N) \to N$. If $\pi\colon W \to M$ is a fiber bundle between supermanifolds, then the *k-superderivative space of sections* of π is a subbundle $\underline{D}^k(W)$ of $\underline{D}^k(M, N)$ over M, and in $\underline{D}^k(W)$ we can distinguish some further subbundles homonymous to classical ones. For example the *holonomic k-jet superderivative space*:

$$J\underline{D}^k(W) \equiv \{u \in \underline{D}^k(W) \mid \text{there exist superdifferential sections } s \text{ of } \pi \text{ such that } D^k s(p) = u\,,\ p = \pi_k(u)\}.$$

A system of coordinates on a k-jet superderivative space can be written as

$$\{x^A, y^B, y^B_A, y^B_{A_1A_2}, \dots, y^B_{A_1\dots A_k}\}, \quad 1 \leqslant A \leqslant m+n\,,\ 1 \leqslant B \leqslant m'+n',$$

where

$$y^B_{A_1\dots A_s A_{s+1}\dots A_q} = (-1)^{|A_s||A_{s+1}|} y^b_{A_1\dots A_{s+1}A_s\dots A_q}\,.$$

So $J\underline{D}^k(W)$ is a supermanifold of dimension $(m' + (m'-m)P + (n'-n)Q,\ n' + (m'-m)Q + (n'-n)P)$, with

$$P = \sum_{0 \leqslant 2j \leqslant \min(n,k)} \binom{n}{2j}\left[\binom{m+k-2j}{k-2j} - 1\right],$$

$$Q = \sum_{0 \leqslant 2j+1 \leqslant \min(n,k)} \binom{n}{2j+1}\left[\binom{m+k-2j-1}{k-2j} - 1\right].$$

One has:

$$|y^B_{A_1\dots A_p}| = |B| + |A_1| + \dots + |A_p|\,.$$

2) $J\underline{D}^k(W)$ is an affine bundle over $J\underline{D}^{k-1}(W)$ with associated vector bundle $\pi^{\bullet}_{k,0}(\dot{S}^k_0 MO_{A_0}\mathrm{v}TW)$. So we have the following exact sequence of vector bundles over $J\underline{D}^k(W)$:

$$0 \to \pi^{\bullet}_{k,0}(\dot{S}^k_0 MO_{A_0}\mathrm{v}TW) \to \mathrm{v}TJ\underline{D}^k(W) \to \pi^{\bullet}_{k,k-1}\mathrm{v}TJ\underline{D}^{k-1}(W) \to 0\,.$$

3) The relation with k-jet derivative spaces is given by the following exact commutative diagram:

$$\begin{array}{ccccccccc} & & 0 & & 0 & & 0 & & \\ & & \downarrow & & \downarrow & & \downarrow & & \\ 0 & \longrightarrow & \pi^{\bullet}_{k,0}(\dot{S}^k_0 MO_{A_0}\mathrm{v}TW) & \longrightarrow & \mathrm{v}TJ\underline{D}^k(W) & \longrightarrow & \pi^{\bullet}_{k,k-1}\mathrm{v}TJ\underline{D}^{k-1}(W) & \longrightarrow & 0 \\ & & \downarrow & & \downarrow & & \downarrow & & \\ 0 & \longrightarrow & \pi^{\bullet}_{k,0}(S^k_0 MO_{\mathbb{K}}\mathrm{v}TW) & \longrightarrow & \mathrm{v}TJD^k(W) & \longrightarrow & \pi^{\bullet}_{k,k-1}\mathrm{v}TJD^{k-1}(W) & \longrightarrow & 0 \end{array}$$

In order to develop a formal theory of super PDE's we shall use a (co)homology theory for supermanifolds. So the aim of this chapter is to explicitly characterize the (co)homological structures of supermanifolds. Set

$$\widetilde{\Omega}^{\bullet}(M) \equiv \bigoplus_{p\geqslant 0} \widetilde{\Omega}^p(M), \quad \text{where } \widetilde{\Omega}^p(M) \equiv G^{\infty}_w(\Lambda^p T^+M) = G^{\infty}_w(\widetilde{\Lambda}^0_p M).$$

$\widetilde{\Omega}^{\bullet}(M)$ has a natural A-module structure. Then we can define a cochain complex (the *de Rham supercomplex* of M), $\{\widetilde{\Omega}^{\bullet}(M), \tilde{d}\}$. (By abuse of notation we shall write d for $\tilde{d}$.) The *de Rham supercohomology* of the supermanifold M is the homology $\widetilde{H}^{\bullet}(M)$ of the de Rham supercomplex on M.

DEFINITION 3. The *p-supercochain group* is the superdual $\widetilde{C}^p(X; A)$ of $C_p(X; A)$:

$$\widetilde{C}^p(X; A) \equiv \mathrm{Hom}_A(C_p(X; A), A),$$

where $C_p(X; A) \equiv A \otimes_{\mathbb{K}} C_p(X; \mathbb{K})$ and $C_p(X; \mathbb{K})$ is the group of p-chains of X. ($\widetilde{C}^p(X; A)$ is a $\mathbb{Z}_2$-graded A-module:

$$\widetilde{C}^p(X; A)_r = \{\alpha \in \widetilde{C}^p(X; A) \mid \langle \alpha, c_s\rangle \in A_{r+s},\ c_s \in C_p(X; A)_s\},$$

$\langle\ ,\ \rangle$ denotes the following pairing: $\langle\ ,\ \rangle\colon \widetilde{C}^p(X; A) \times C_p(X; A) \to A$). The corresponding cochain complex $\{\widetilde{C}^p(X; A), \delta\}$ is defined by considering the *coboundary* of a p-supercochain $\alpha \in \widetilde{C}^p(X; A)$ as the p-supercochain $\delta\alpha \in \widetilde{C}^{p+1}(X; A)$ such that

$$\langle \delta\alpha, c\rangle + (-1)^p \langle \alpha, \partial c\rangle = 0.$$

The *singular supercohomology of* X with coefficients in A, is the cohomology $\widetilde{H}^{\bullet}(X; A)$ of the cochain complex $\{\widetilde{C}^p(X; A), \delta\}$.

In the Table 1 we also define some useful cochain complexes and the corresponding cohomology spaces that will be used later.

NOTE. $\dot{\Lambda}^0_p M = (\Lambda^p T^+M)^+_0$, $\overline{\Lambda}^0_p M = (\Lambda^p T^+M)^+_1$, $\Omega^p(M) \equiv C^{\infty}_w(\Lambda^0_p M)$, $\dot{\Omega}^p(M) \equiv G^{\infty}_w(\dot{\Lambda}^0_p M)$, $\overline{\Omega}^p(M) \equiv G^{\infty}_w(\overline{\Lambda}^0_p M)$, $\widetilde{\Omega}^p(M) \equiv G^{\infty}_w(\widetilde{\Lambda}^0_p M)$, $\widetilde{\Lambda}^0_0 M \equiv$

TABLE 1A

Cochain complex	Cohomology spaces	Name
$\{\widetilde{\Omega}^{\cdot}(M), \tilde{d}\}$	$\widetilde{H}^{\cdot}(M)$	de Rham supercohomology
$\{\dot{\Omega}^{\cdot}(M), \dot{d}\}$	$\dot{H}^{\cdot}(M)$	de Rham even supercohomology
$\{A_1 \circ_{A_0} \dot{\Omega}^{\cdot}(M), \overline{d}\}$	$\overline{H}^{\cdot}(M)$	de Rham odd supercohomology
$\{\Omega^{\cdot}(M), d\}$	$H^{\cdot}(M)$	de Rham cohomology
$\{A \otimes_{\mathbb{K}} \Omega^{\cdot}(M), d\}$	$H^{\cdot}(M; A)$	de Rham cohomology with coefficients in A
$\{A_0 \otimes_{\mathbb{K}} \Omega^{\cdot}(M), d\}$	$H^{\cdot}(M; A_0)$	de Rham cohomology with coefficients in A_0
$\{A_1 \otimes_{\mathbb{K}} \Omega^{\cdot}(M), d\}$	$H^{\cdot}(M; A_1)$	de Rham cohomology with coefficients in A_1
$\{\widetilde{C}^{\cdot}(X, A), \delta\}$	$\widetilde{H}^{\cdot}(X; A)$	Singular supercohomology
$\{C^p(X, A) \equiv \mathrm{Hom}_{\mathbb{K}}(C_p(X; \mathbb{K}); A), \delta\}$	$H^{\cdot}_S(X; A)$	Singular supercohomology with coefficients in A

TABLE 1B. Important isomorphic (super)cohomology spaces.

Name	(Super)cohomology spaces
de Rham cohomology with coefficients in A	$H^{\cdot}(X; A) \cong A \otimes_{\mathbb{K}} H^{\cdot}(X) \cong A \otimes_{\mathbb{K}} H^{\cdot}(X_B)$
Singular supercohomology	$\widetilde{H}^{\cdot}(X; A) \cong \mathrm{Hom}_A(H_{\cdot}(X; A); A)$
Singular cohomology with coefficients in A	$H^{\cdot}_S(X; A) \cong \mathrm{Hom}_{\mathbb{K}}(H_{\cdot}(X; A); \mathbb{K})$

$M \times A$, $\dot{\Lambda}^0_0 M \equiv M \times A_0$, $\overline{\Lambda}^0_0 M \equiv M \times A_1$, $\widetilde{\Omega}^0(M) = G^\infty_w(M, A)$, $\dot{\Omega}^0(M) = G^\infty_w(M, A_0)$, $\overline{\Omega}^0(M) = G^\infty_w(M, A_1)$. Furthermore, we put $\bigoplus^0_{A_0} T_pM \equiv A_0$.

DEFINITION 4. 1) A *super* PDE (SPDE) of order k on the fiber bundle $\pi: W \to M$, defined in the category of supermanifolds, is a subbundle $E_k \subset J\underline{D}^k(W)$ of the jet superderivative space $J\underline{D}^k(W)$ over M.

2) The *r-superprolongation* of $\underline{E}_k \subset J\underline{D}^k(W)$ is

$$\underline{E}_{k+r} \equiv J\underline{D}^r(E_k) \cap J\underline{D}^k(W) \subset J\underline{D}^{k+r}(W).$$

3) The *supersymbol* of $\underline{E}_{k+r}$, $r \geqslant 0$, is a family of A_0-modules over $\underline{E}_k$ given by

$$\dot{g}_{k+r} = \mathrm{v}T\underline{E}_{k+r} \cap \pi^{\cdot}_{k+r,0}(\dot{S}^{k+r}_0 MO_{A_0}\mathrm{v}TW).$$

4) A SPDE $\underline{E}_k \subset J\underline{D}^k(W)$ is *formally superintegrable* if for any $r \geqslant 0$ the map $\dot{g}_{k+r+1} \to \underline{E}_k$ is a bundle of A_0-modules and $\underline{E}_{k+r+1} \to \underline{E}_{k+r}$ is surjective.

REMARK. If $\underline{E}_k$ is a formally superintegrable SPDE, then one has the following sequence of A_0-bundles over $\underline{E}_k$

$$0 \to \pi^{\cdot}_{k+r,k}\dot{g}_{k+r} \to \mathrm{v}T\underline{E}_{k+r} \to \pi^{\cdot}_{k+r,k+r-1}\mathrm{v}T\underline{E}_{k+r-1},$$

which means that $\underline{E}_{k+r}$ is an affine bundle over $\underline{E}_{k+r-1}$ with associated vector bundle

$$\pi^{*}_{r+k-1,k}\dot{g}_{k+r} \rightarrow \underline{E}_{k+r-1}\,.$$

Associated to $\underline{E}_k$ one has the following complexes of A_0-modules over $\underline{E}_k$ (*δ-supersequence*):

$$0 \rightarrow \dot{g}_m \rightarrow TMO_{A_0}\dot{g}_{m-1} \rightarrow \dot{\Lambda}^2_0 MO_{A_0}\dot{g}_{m-2} \rightarrow \cdots$$
$$\rightarrow \dot{\Lambda}^{m-k}_0 MO_{A_0}\dot{g}_k \rightarrow \dot{\Lambda}^{m-k+1}_0 MO_{A_0}(\dot{S}^{k-1}_0 MO_{A_0}\mathrm{v}TW) \rightarrow \cdots.$$

DEFINITION 5. We call the corresponding cohomology $\{H^{m-j,j}_q\}_{q\in\underline{E}_k}$ at $(\dot{\Lambda}^j_0 MO_{A_0}\dot{g}_{m-j})_q$ the *Spencer supercohomology* of $\underline{E}_k$.

We say that $\underline{E}_k$ is *r-superacyclic* if $H^{m,j}=0$, $m\geqslant k$, $0\leqslant j\leqslant r$. We say that E_k is *superinvolutive* if $H^{m,j}=0$, $m\geqslant k$, $j\geqslant 0$ (e.g., the δ-supersequences are exact for $m\geqslant k$).

Then we have the following important criterion of formal superintegrability that extends the previous one given by H. Goldschmidt for PDE's [12].

THEOREM 2 [46]. 1) (Formal superintegrability criterion.) *Let $\underline{E}_k\subset J\underline{D}^k(W)$ be a SPDE in the category of supermanifolds. Then, there is an integer $k_0=k+h\geqslant k$, depending only on k, the dimension of M and the dimension of W such that $\underline{E}_k$ is superinvolutive and such that if $\dot{g}_{k+r+1}$ is a bundle of A_0-modules over $\underline{E}_k$, and $\underline{E}_{k+r+1}\rightarrow\underline{E}_{k+r}$ is surjective for $0\leqslant r\leqslant h$, then $\underline{E}_k$ is formally superintegrable.*

2) *In the superanalytic case, under the hypotheses of item 1, we have that, for any initial condition $q\in\underline{E}_k\subset J\underline{D}^k(W)$, there exists a G^{ω}_w solution of $\underline{E}_k$. Actually, we can choice points $q_{k+t}\in\underline{E}_{k+t}$, $t\geqslant -k$, where for $t<0$, $\underline{E}_{k+t}=\pi_{k,k+t}(\underline{E}_k)$, such that $\pi_{k+t,k+t-1}(q_{k+t})=q_{k+t-1}$. Then a local solution can be represented in a suitable neighborhood U of $p\equiv\pi_k(q)$, by the following functions:*

$$s^C(x,\theta)=\sum_{\substack{0\leqslant r+s\leqslant\infty\\ 0\leqslant s\leqslant n}}\frac{1}{r!}(x-b)^{A_1}\cdots(x-b)^{A_r}(\theta-\beta)^{B_1}\cdots(\theta-\beta)^{B_s}\times(\partial_{A_1\ldots A_r B_1\ldots B_s}s^C)(b,\theta),$$

$1\leqslant c\leqslant m''+n''$ (fiber dimension of $W=(m'',n'')$), where $(x,\theta)\in U$, $(b,\beta)=p$, and

$$(\partial_{A_1\ldots A_r B_1\ldots B_s}s^C)(b,\theta)=y^C_{A_1\ldots A_r B_1\ldots B_s}(q_{r+s}),$$

where $y^C_{A_1\ldots A_r B_1\ldots B_s}$ are vertical coordinates on $J\underline{D}^{r+s}(W)$.

REMARK. It is a simple corollary of Theorem 2 that in the superanalytic case, any superintegrable SPDE is also *completely superintegrable*: it admits a local solution of class G^{ω}_w for any initial condition.

REMARK. Similarly to the ordinary case [19, 21–23, 45, 57–59], we can develop a geometric theory of singular supersolutions for SPDE's. In fact, one recognizes a canonical distribution on $J\underline{D}^k(W)$, called the *Cartan superdistribution*, $\mathbb{E}_k\subset TJ\underline{D}^k(W)$, i.e., the distribution on $J\underline{D}^k(W)$ spanned by tangent spaces to graphs of k-superderivatives of sections of the bundle $\pi: W \to M$. If $\dim M = (m, n)$, we call *Vinogradov–supermanifolds* (*V-supermanifolds*) the (m, n)-dimensional integral supermanifolds of $\mathbb{E}_k$. We call also *Ehresmann–Vinogradov supermanifolds* (*EV-supermanifolds*) the (m, n)-dimensional integral supermanifolds of $\mathbb{E}_k$ that locally can be represented as the image of the k-superderivative of some section of π. Then a *singular supersolution* of $\underline{E}_k \subset J\underline{D}^k(W)$ is a closed subsupermanifold $V \subset J\underline{D}^k(W)$ with the property that for each $q \in V$ there exists a supersmooth V-supermanifold $\overline{V} \subset J\underline{D}^{k+r}(W)$, $r \geqslant 0$, such that V coincides with $\pi_{k+r,k}(\overline{V})$ in some neighborhood of q and $\overline{V} \cap U \subset \underline{E}_{k+r}$, for some neighborhood $\overline{U}$ of some point $\overline{q} \in \pi^{-1}_{k+r,k}(q) \cap \overline{V}$. By using the "blow up" $\overline{V}$ at each point, it is possible to construct a smooth supermanifold $\overline{V}$, called the *blow-up* of V. The points $q \in V$ for which $q \in \Sigma(\overline{V}) \equiv$ set of singular points of the map $\pi_{k+r}|\overline{V}: \overline{V} \to M$, will be called the *geometric singularities* of the supersolution V. A *nonsingular* supersolution of $\underline{E}_k$ is a EV-supermanifold $V \subset J\underline{D}^k(W)$ also contained in $\underline{E}_k$. We say that a super PDE $\underline{E}_k \subset J\underline{D}^k(W)$ is *V-formally superintegrable at $q \in \underline{E}_k$* if there exists a V-supermanifold L tangent to $\underline{E}_k$ at this point with infinite order. $\underline{E}_k$ is *formally superintegrable at $q \in \underline{E}_k$* if it is V-formally superintegrable at q and L is diffeomorphically projected onto M. A super PDE $\underline{E}_k \subset J\underline{D}^k(W)$ is formally superintegrable iff the projections $\pi_{k+s+1,k+s}(\underline{E}_k)_{+(s+1)} \to (\underline{E}_k)_{+s}$ are supersmooth bundles, $s \geqslant 0$. An equation $\underline{E}_s \subset J\underline{D}^s(W)$ is called an *intermediate superintegral* of the super PDE $\underline{E}_k$ if its $(k-s)$-superprolongation is contained in $\underline{E}_k$: $(\underline{E}_s)_{+(k-s)} \subset \underline{E}_k$. The number s is the *order* of the intermediate superintegral. An equation $\underline{E}_k \subset J\underline{D}^k(W)$ is called *V-superinvolutive* if: (a) its first superprolongation $\underline{E}_k$ is *superregular*, i.e., at each point $q \in (\underline{E}_k)_{+1}$ the subspaces $T_q((\underline{E}_k)_{+1})$ and $\dot{S}_0^{k+1}(T_xM)\,O_{A_0}\mathrm{v}T_pW$, $p \equiv \pi_{k+1,0}(q)$, intersect transversally; (b) the map $\pi_{k+1,k}: (\underline{E}_k)_{+1} \to \underline{E}_k$ is surjective; (c) $\underline{E}_k$ is superacyclic. Then one can directly obtain results that generalize previous ones obtained for PDE's on ordinary manifolds. In particular, we have the following important theorems. (Note that the hypothesis that the superalgebra A is the inductive limit of Noetherian rings is essential in these theorems. In fact, this allows us to use well-known results of algebraic geometry [13] and overcome the difficulties due to the fact that, in general, A is not a Noetherian ring.)

THEOREM 3. 1) *Let $\underline{E}_k \subset J\underline{D}^k(W)$ be a superregular system of order k on $\pi: W \to M$. If $\dot{g}_{k+1}$ is a bundle of A_0-modules over $\underline{E}_k$ and if $\dot{g}_k$ is 2-superacyclic, then*

$$\underline{E}^{(1)}_{k+r} = (\underline{E}^{(1)}_k)_{+k}, \quad \textit{where } \underline{E}^{(1)}_{k+r} \equiv \pi_{k+r+1,k+r}(\underline{E}_{k+r+1}) \subset J\underline{D}^{k+r}(W).$$

2) *Let $\underline{E}_k \subset J\underline{D}^k(W)$ be a sufficiently superregular SPDE of order k on $\pi: W \to M$, i.e., $\underline{E}_k$ is a superregular system such that $\pi_{k+r+s,k+r}: \underline{E}_{k+r+s} \to \underline{E}_{k+r}$ are the morphisms of constant superrank, for all $r, s \geqslant 0$. If A is the inductive limit of Noetherian rings A_λ,*

$$A = \lim_{\lambda \to \infty} A_\lambda,$$

then there exist integer $s \geqslant 0$ and $k_s \geqslant k$ such that $\underline{E}^{(s)}_{k_s} \subset J\underline{D}^{k_s}(W)$ is a superinvolutive formally superintegrable system with the same supersolutions as $\underline{E}_k$

and such that

$$(\underline{E}_{k_s}^{(s)})_{+r} = \underline{E}_{k_s+r}^{(s)} .$$

3) *Let* $\underline{E}_k \subset J\underline{D}^k(W)$ *be a SPDE of order* k *on* $\pi\colon W \to M$, $\dim M = (m, n)$, $\dim W = (m+m', n+n')$, *such that* $\underline{E}_k = \ker K$, *where* $K\colon J\underline{D}^k(W) \to F$ *is a differential operator of class* G^∞, F *being a bundle of vector superspaces of dimension* (p, q) *over* M. *Let* $L \subset \underline{E}_k$ *be a supersolution. Let us denote by* $\Sigma^{(i,j)}(L)$ *the set of singular points of type* (i, j), *i.e.,*

$$\Sigma^{(i,j)}(L) \equiv \{q \in L \mid \dim\ker(T\pi_k \mid T_qL) = (i, j)\}, \qquad 1 \leqslant j \leqslant m,\ 1 \leqslant i \leqslant n.$$

Let Ξ *be the subspace* $\Xi \equiv T(\pi_k)(T_qL) \subset T_xM$, $x \equiv \pi_k(q)$. *Let*

$$\widetilde{\Xi} \equiv \mathrm{Hom}_{A_0}(A, \Xi) \subset \widetilde{T}(\pi_k)(\widetilde{T}_qL) \subset \widetilde{T}_xM$$

be the graded extension of Ξ. *Set*

$$\tilde{g}_{k+s,q}(\widetilde{\Xi}) = \tilde{g}_{k+s,q} \cap \widetilde{\Xi}^{(k+s)}(W)_p \subset S_0^{k+s}(\widetilde{T}_xM)\, O_A \mathrm{v}\widetilde{T}_pW ,$$

where $\widetilde{\Xi}^{(k+s)}(W)_p \subset S_0^{k+s}(\widetilde{T}_xM)\, O_A \mathrm{v}\widetilde{T}_pW$ *is the subspace consisting of all* A-*linear* $(k+s)$-*symmetric applications* $\widetilde{T}_xM \times \dots_{k+s} \dots \times \widetilde{T}_xM \to \mathrm{v}\widetilde{T}_pW$ *that annihilate on* $\widetilde{\Xi}$, $p = \pi_{k,0}(q)$. *Then there exists a number* s_0 *such that* $(\underline{E}_k)_{+s}$, $s \geqslant s_0$, *has no more singular supersolutions of type* $\Sigma^{(i,j)}(L)$ *for any point* $\tilde{q} \in (\underline{E}_k)_{+s}$, *such that* $\pi_{k+s,k}(\overline{q}) = q$, *if the following conditions are satisfied*: (a) A *is the inductive limit of Noetherian rings*: $A = \lim_{\lambda\to\infty} A_\lambda$; (b) *the complexified projective spaces* $P(\widetilde{\Xi}^{\mathbb{C}})$ *and* $P(\widetilde{\mathrm{Char}}_{(i,j)}(\underline{E}_k)_q^{\mathbb{C}})$ *intersect in a finite number of points,* $\lambda_1, \dots, \lambda_r$, *and*

$$\sum_{1\leqslant j\leqslant r} \dim_{A^{\mathbb{C}}} \tilde{g}_{k,q}(\lambda_j)^{\mathbb{C}} < \operatorname{codim} \widetilde{\Xi} = i + j .$$

Furthermore, if the above conditions are satisfied for all subspaces $\Xi \subset T_xM$, $\operatorname{codim} \Xi = (i, j)$, *then the system* $\underline{E}_{k+s} \subset J\underline{D}^{k+s}(W)$, $s \geqslant s_0$, *does not have any solutions passing through points at which there is a singularity of type* $\Sigma^{(i,j)}$.

Let us now characterize singularities by means of characteristic superclasses.

Let $\pi\colon W \to M$ be a fiber bundle over an (m, n)-dimensional supermanifold M. Set:

$$\underline{I}_k(W) = \bigcup_{u\in J\underline{D}^k(W)} \underline{I}_k(W)_u ,$$

where $\underline{I}_k(W)_u \equiv$ Grassmannian of integral superplanes at q. An *integral superplane* at a point $u \in J\underline{D}^k(W)$ is defined to be an (m, n)-dimensional subspace of $(\mathbb{E}_k)_u$ tangent to some integral supermanifold of the Cartan distribution $\mathbb{E}_k \subset TJ\underline{D}^k(W)$. It is also useful to consider the graded counterpart. More precisely we call *integral graded superplanes* at $u \in J\underline{D}^k(W)$ the $(m+n)$-A-dimensional subsuperspaces L lying in the supertangent to some integral supermanifold passing through u. Let us denote by $\widetilde{\underline{I}}_k(W)_u$ Grassmannian of integral graded superplanes at u and by $\widetilde{\underline{I}}_k(W)$ the corresponding bundle over $J\underline{D}^k(W)$. Note that the graded counterpart of the superstructures considered allow us to work with free A-modules of finite type and so we can easily reproduce well-known results obtained for ordinary manifolds [21, 22, 45, 57–59]. Let $\widetilde{\underline{I}}(\underline{E}_{k+s})$ be the fiber bundle of Grassmannian $(m+n)$-dimensional integral superplanes of the Cartan graded superdistribution $\mathbb{E}(\underline{E}_{k+s})$ on $\underline{E}_{k+s}$, where $\underline{E}_k \subset J\underline{D}^k(W)$ is a SPDE. If $\underline{E}_{k+s} = J\underline{D}^{k+s}(W)$ one has

$$\widetilde{\underline{I}}(J\underline{D}^{k+s}(W)) = \widetilde{\underline{I}}_{k+s}(W).$$

DEFINITION 6. An (m, n)-dimensional integral supermanifold $V \subset J\underline{D}^k(W)$ determines a *tangential map* $i_V : V \to \widetilde{\underline{I}}_k(W)$ given by $i_V(q) = \widetilde{T}_q V \in \widetilde{\underline{I}}_k(W)_q$. Then any integral supermanifold $V \subset J\underline{D}^k(W)$ determines a map

$$\widetilde{H}^i(\widetilde{\underline{I}}_k(W); A) \to \widetilde{H}^i(V; A)$$

given by

$$\omega \mapsto i_V^*\omega \equiv \textit{characteristic superclass} \text{ on } V$$
$$\text{corresponding to } \omega \in \widetilde{H}^i(\widetilde{\underline{I}}_k(W); A).$$

Then we have the following important theorem that extends the previous one obtained by Lychagin for ordinary manifolds [21, 22].

THEOREM 4. 1) *Let $\underline{E}_k \subset J\underline{D}^k(W)$ be a SPDE. Each cohomology superclass*

$$\omega \in \widetilde{H}^i(\widetilde{\underline{I}}(\underline{E}_{k+s}); A)$$

defines characteristic superclass $\omega_V \equiv i_V^*\omega$ on the singular solutions $V \subset \underline{E}_{k+s}$ of $\underline{E}_k$.

2) *$\widetilde{H}^\bullet(\widetilde{\underline{I}}(\underline{E}_{k+s}); A)$ is an algebra over $\widetilde{H}^\bullet(\underline{E}_{k+s}; A)$.*

3) *Let $\underline{E}_k$ be a formally superintegrable SPDE; then the fiber bundles*

$$\pi_{k+s,k+s-1} : \underline{E}_{k+s} \to \underline{E}_{k+s-1}, \qquad s \geqslant 1$$

are affine subbundles in $\pi_{k+s,k+s-1} : J\underline{D}^{k+s}(W) \to J\underline{D}^{k+s-1}(W)$ and, hence

$$\widetilde{H}^\bullet(\underline{E}_{k+s}; A) \cong \widetilde{H}^\bullet(\underline{E}_k; A).$$

4) *If V is a nonsingular integral supermanifold, then all its supercharacteristics classes are zero in dimension $\geqslant 1$.*

5) (Cauchy problem for SPDE and characteristic supernumbers). *The Cauchy problem for (singular) supersolutions of SPDE can be formulated in the following way. Cauchy data for the system $\underline{E}_k \subset J\underline{D}^k(W)$ is defined by a $(m-1, n-1)$-dimensional integral supermanifold $\sigma \subset \underline{E}_{k+r}$ of the Cartan superdistribution $\mathbb{E}_{k+r}$, beside a section h of the bundle $\widetilde{\underline{I}}(\underline{E}_{k+r})|\sigma \to \sigma$, such that $h(\overline{q}) \supset T_{\overline{q}}\sigma$, for all $\overline{q} \in \sigma$. A supersolution of the Cauchy problem with initial data, (σ, h), is defined as an (m, n)-dimensional integral supermanifold V, with boundary such that*

$V \subset \underline{E}_{k+r}$, $\partial V = \sigma$ and $T_{\overline{q}}V = h(\overline{q})$, for all $\overline{q} \in \sigma$. With each supercohomology class $\omega \in \widetilde{H}^{m-1}(\underline{\widetilde{I}}(\underline{E}_{k+r}); A)$ and each Cauchy data (σ, h) we associate the characteristic supernumber $\chi_\omega(\sigma, h) = \langle h^*\omega, z_\sigma\rangle$, where $Z_\sigma \in \widetilde{H}_{m-1}(\sigma; A)$ is a fundamental cycle of σ. For the solvability of the Cauchy problem (σ, h) for SPDE's, it is necessary that all the characteristic supernumbers $\chi_\omega(\sigma, h)$ be equal to zero.

REMARK. Spectral sequences can be associated to super PDE's and are useful to characterize the space of conservation laws of super PDE's. The purely cohomological character of super PDE's can be used and a general criterion to recognize conservation laws for super PDE's can be given in such a way to generalize previous results formulated for PDE's. (For a detailed study of this subject see [26, 27, 34, 35, 38, 45, 57–59] for the ordinary case, and [46, 47] for super PDE's).

§2. Quantization of super PDE's

In this section we formulate the superanalog of the quantization of PDE's previously given in [37–40, 42, 43, 45]. So, to a super PDE $\underline{E}_k \subset J\underline{D}^k(W)$, we associate its *quantum supersitus* $\Omega(\underline{E}_k)$, i.e., the set of *quantum supersolutions* of $\underline{E}_k$, a quantum supersolution being an integral supermanifold in $J\underline{D}^k(W)$, that cobords two Cauchy hypersurface data belonging to $\underline{E}_k$. The *classical limit* $\Omega(\underline{E}_k)_c$ of $\Omega(\underline{E}_k)$ is the set of solutions of $\underline{E}_k$, and the *canonical quantization* of $\underline{E}_k$ can be formulated by associating a (quantum) noncommutative superalgebra to the set of *physical superobservables* of $\underline{E}_k$, which are functions on $\Omega(\underline{E}_k)_c$, by means of a method of deformation of $\underline{E}_k$ induced by each physical superobservable. Then we shall show that singular supersolutions describe either quantum tunneling effects or canonical quantization.

DEFINITION 1. The *category of superdifferential equations* $\underline{\mathcal{PDE}}$ is defined by the following:

1) $X \in \mathrm{Ob}(\underline{\mathcal{PDE}})$ if and only if X is a supermanifold equipped with a finite dimensional Frobenius graded superdistribution $\widetilde{C}(X) \subset \widetilde{T}X$, which is locally the same as $\underline{\mathbb{E}}_\infty \equiv$ Cartan distribution of $\underline{E}_\infty$ for some SPDE $\underline{E}_k \subset J\underline{D}^k(W)$. We set:

$$\mathrm{sdim}\, X \equiv \dim \widetilde{C}(X) = m + n \equiv \textit{Cartan superdimension of } X \in \mathrm{Ob}(\underline{\mathcal{PDE}}).$$

2) $f \in \mathrm{Hom}(\underline{\mathcal{PDE}})$ if and only if it is a smooth map $f: X \to Y$, $X, Y \in \mathrm{Ob}(\underline{\mathcal{PDE}})$, which conserves the corresponding Frobenius graded superdistributions:

$$\begin{gathered} \widetilde{T}(f): \widetilde{C}(X) \to \widetilde{C}(Y), \qquad f \in \mathrm{Hom}_{\underline{\mathcal{PDE}}}(X, Y), \\ \mathrm{sdim}\, X = m + n, \quad \mathrm{sdim}\, Y = m' + n', \\ \mathrm{srank}\, f = r = \dim(\widetilde{T}(f)_x(\widetilde{C}(X)_x)), \qquad x \in X. \end{gathered}$$

Then the fibers $f^{-1}(y)$, $y \in \mathrm{Im}\, f \subset Y$, are $(m + n - r)$-superdimensional objects of $\underline{\mathcal{PDE}}$.

Isomorphisms of $\underline{\mathcal{PDE}}$: supermorphisms with fibers consisting of separate points.

Covering maps of $\underline{\mathcal{PDE}}$: supermorphisms with zero-superdimensional fibers.

Proposition 1 (relation between solutions of X, $\overline{X} \in \mathrm{Ob}(\underline{\mathcal{PDE}})$, if $\tau\colon \overline{X} \to X$ is a covering). *Let $X \equiv \underline{E}_\infty$ for some $\underline{E}_k \subset J\underline{D}^k(W)$, $\pi\colon W \to M$,* $\dim M = (m, n)$. *One has:*

1) $\widetilde{T}(\tau)(\widetilde{C}(\widetilde{X})) = \widetilde{C}(X) \subset \widetilde{T}X$.

2) *If $\widetilde{V} \subset \widetilde{X}$ is an integral manifold of $\widetilde{C}(\widetilde{X})$, then $\tau(\widetilde{V}) \equiv V$ is an integral manifold of $\widetilde{C}(X)$, which is a supersolution of $\underline{E}_k$.*

3) *Conversely, if $V \subset \underline{E}_\infty$ is a supersolution of $\underline{E}_k$, then $\widetilde{C}(\widetilde{X})|\tau^{-1}(V)$ is an integrable $(m+n)$-superdimensional graded superdistribution.*

4) *If fiber* $\mathrm{sdim}\,\tau = N$, *then $\tau^{-1}(V)$ is foliated by an N-parameter family of integrable supermanifolds of $\widetilde{C}(\widetilde{X})|\tau^{-1}(V)$. Furthermore, $\widetilde{\underline{E}}_\infty$ can be locally represented as $\widetilde{\underline{E}}_\infty = \underline{E}_\infty \times W$, where $W \subset A^N$.*

Definition 2. For any $X \in \mathrm{Ob}(\underline{\mathcal{PDE}})$ set:

1) $\widetilde{\Omega}^i \equiv G_w^\infty(\Lambda_i^0 X)$;

2) $\mathcal{C}\widetilde{\Omega}^1 \subset \Omega^1$: annulator of the graded superdistribution $\widetilde{C}(X)$;

3) $\mathcal{C}\widetilde{\Omega}^i \equiv \mathcal{C}\widetilde{\Omega}^1 \wedge \widetilde{\Omega}^{i-1}$;

4) $\mathcal{C}\widetilde{\Omega} \equiv \bigoplus_{i\geqslant 0} \mathcal{C}\widetilde{\Omega}^i$;

5) $\overline{\widetilde{\Omega}}^i \equiv \widetilde{\Omega}^i/\mathcal{C}\widetilde{\Omega}^i$.

$\mathcal{C}\widetilde{\Omega}$ is a differential ideal of the Grassmann algebra $\widetilde{\Omega} \equiv \bigoplus_{i\geqslant 0} \widetilde{\Omega}^i$, i.e., $\mathcal{C}\widetilde{\Omega}$ is an ideal of $\widetilde{\Omega}$ such that $d\mathcal{C}\widetilde{\Omega} \subset \mathcal{C}\widetilde{\Omega}$.

The operator d induces the factor-operator $\overline{d}$: $\overline{d}_i\colon \overline{\widetilde{\Omega}}^i \to \overline{\widetilde{\Omega}}^{i+1}$.

In the case $X = J\underline{D}^\infty(W)$, elements $\omega \in \overline{\widetilde{\Omega}}^{m+n}$ are horizontal superdifferential forms locally written as follows:

$$\omega = L\,dx^1 \wedge \cdots \wedge dx^{m+n}, \qquad L\colon J\underline{D}^\infty(W) \to A.$$

Theorem 1 (category of SPDE's and superspectral sequences). *For any $X \in \mathrm{Ob}(\underline{\mathcal{PDE}})$ there exists a spectral sequence $\{\widetilde{E}_r^{p,q}, d_r\}$ associated to it such that*

$$\widetilde{E}_0^{0,i} \equiv \overline{\widetilde{\Omega}}^i, \quad d_0^{0,i} = \overline{d}_i, \quad \widetilde{E}_1^{0,i} = \overline{\widetilde{H}}^i \equiv \ker \overline{d}_i / \operatorname{Im} \overline{d}_{i-1}.$$

Remark. If $X = \underline{E}_\infty \subset J\underline{D}^\infty(W)$ one has the following:

1) $d_1^{0,m}\colon \overline{\widetilde{H}}^m \to \widetilde{E}_1^{1,m}$ is the Euler–Lagrange superoperator for variational problems constrained by the SPDE $\underline{E}_k$.

2) $\widetilde{E}_1^{0,m-1} = \overline{\widetilde{H}}^{m-1}$ may be interpreted as the group of all higher conservation superlaws of $\underline{E}_k$.

3) Every finite-dimensional supermanifold M, when equipped with the zero-superdimensional superdistribution $\widetilde{C}(M) \equiv 0 \subset \widetilde{T}M$, may be regarded as a zero-superdimensional object of $\underline{\mathcal{PDE}}$. In this case one has the following:

1) $\mathcal{C}\widetilde{\Omega} \equiv \bigoplus_{i\geqslant 1} \widetilde{\Omega}^i$, $\widetilde{\Omega}^i \equiv G_w^\infty(\Lambda_i^0 M)$;

2) $\mathcal{C}^k\widetilde{\Omega} = \bigoplus_{i\geqslant k} \widetilde{\Omega}^i$;

3) $\widetilde{E}_0^{p,0} \equiv \widetilde{\Omega}^p$;

4) $\widetilde{E}_0^{p,q} = 0$ if $q > 0$;

5) Since $d_0^{p,q}\colon \widetilde{E}_0^{p,q} \to \widetilde{E}_0^{p,q+1}$ it follows that $d_0^{p,q} = 0$ for every p, q;
6) $\widetilde{E}_0^{p,q} = \widetilde{E}_1^{p,q}$;
7) $d_1^{p,0}\colon \widetilde{E}_1^{p,0} = \widetilde{\Omega}^p \to \widetilde{E}_1^{p+1,0} = \widetilde{\Omega}^{p+1} \implies d_1^{p,0} = d$.
8) $d_2^{p,0}\colon \widetilde{E}_2^{p,0} = \widetilde{H}^p(M) \to \widetilde{E}_2^{p+1,0} = \widetilde{H}^{p+1}(M)$.

DEFINITION 3. Let M be a (m, n)-dimensional supermanifold. Let $\pi\colon E \to M$ be a fiber bundle of (m', n')-dimensional vector superspaces; E can be considered either as a fiber bundle of A_0-modules or a fiber bundle of $\mathbb{K}$-vector spaces over M. So we can define the structures presented in Table 1 for any section $i\colon M_B \to M$, where M_B is the body of M.

THEOREM 2. *Let $K\colon C^\infty(E) \to C^\infty(F)$ be a k-order A_0-linear differential operator. Let $\underline{E}_k \equiv \ker_f K \subset J\underline{D}^k(E)$ be an affine superequation, where $f \in C^\infty(F)$. Suppose that $L \subset J\underline{D}^{k+s}(E)$ is a (m, n)-dimensional V-supermanifold such that $\pi_{k+s}|L\colon L \to M$ is a proper map, and $\omega_1^{(k+s)}(L_B) = 0$. Then, for any section i of $b\colon M \to M_B$, $F[L|i] \in C_0^\infty(E^\phi)^+$ is a superdistribution-solution of $\underline{E}_k$ iff L is a multivalued supersolution. Furthermore, for any section i of b we have that the superdistribution solution of $\underline{E}_k$ can be written as follows: $v = G[i]_f$, where $G[i]$ is Green's kernel of K at the section i, i.e., $G[i]$ is the superdistribution kernel*

$$G[i] \in C_0^\infty(E^\phi \boxtimes F)^+ \cong C_0^\infty(F \boxtimes E^\phi)^+$$

such that

$$(\widetilde{K}[i] \otimes \mathbf{1})(G[i]) = (\mathbf{1} \otimes \widetilde{K}[i])(G[i]) = \widetilde{\mathbb{D}}[i],$$

where $\widetilde{\mathbb{D}}[i] \in C_0^\infty(F \boxtimes F^\phi)^+$ is the Dirac superkernel of F, at the section i, i.e.,

$$\widetilde{\mathbb{D}}[i](f \otimes \alpha) = \int_{M_B} i^\cdot\langle f, \alpha\rangle, \qquad \forall f \in C_0^\infty(F),\ \alpha \in C_0^\infty(F^\phi).$$

2) *Any singular supersolution L of an affine SPDE $\underline{E}_k \equiv \ker_f K \subset J\underline{D}^k(E)$, that satisfies some boundary conditions, i.e., $\partial L = X_1 \dot\cup X_2$, where X_1 and X_2 are some fixed hypersurfaces of Cauchy data, and such that:*

(i) $\pi_{k+s}|_L\colon L \to M$ *is a proper map, and* (ii) $\omega_1^{(k+s)}(L_B) = 0$,

determines, for any section $i\colon M_B \to M$, a superdistribution kernel

$$G[L|i] \in C_0^\infty(E^\phi \boxtimes F)^+.$$

REMARK. We call $G[L|i]$ the *generalized Green kernel* of the singular solution $L \subset \underline{E}_{k+s}$ satisfying the boundary condition $\partial L = X_1 \dot\cup X_2$, where $\dot\cup$ denotes disjoint union. In particular, if L is the singular solution corresponding to the Green kernel $G[i]$ of K, i.e., $F[L|i] = G[i]_f$, then the corresponding generalized Green kernel is defined by

$$G[L|i](\alpha \otimes \phi) = G[i]_f(\alpha) \int_{M_B} i^\cdot\langle \phi, f\rangle.$$

TABLE 2. Definitions associated to $\pi: E \to M$.

(a)	$C_0^\infty(i^\bullet E) \equiv \mathbb{K}$-vector space of C^∞ sections of $\pi\|i: i^\bullet E \to M_B$ with support compact. (It is a nuclear LF and Montel, Lindelöf locally convex topological vector space, hence paracompact and normal. It is also reflexive.)
(b)	$(i^\bullet E)' \equiv (i^\bullet E)^* \otimes_{\mathbb{K}} \Lambda_m^0 M_B \equiv$ *formal adjoint* of $i^\bullet E$. $((i^\bullet E)'' \cong i^\bullet E)$.
(c)	$(C_0^\infty(i^\bullet E))' \equiv \mathrm{Hom}_{\mathbb{K}}(i^\bullet E; \mathbb{K}) \equiv$ *space of distribution sections* of $i^\bullet E$ (it is nuclear).
(d)	$(C_0^\infty((i^\bullet E)'))' \equiv$ *space of distribution-sections* of $(i^\bullet E)'$.
(e)	$E' \equiv E^* \otimes_{\mathbb{K}} \Lambda_m^0 M \equiv$ *formal adjoint* of E.
(f)	$E^\phi \equiv E^+ \otimes_A \widetilde{\Lambda}_m^0 M \equiv$ *formal superadjoint* of E.
(g)	$C_0^\infty(E)^\bullet \equiv L_{\mathbb{K}}(C_0^\infty(E); \mathbb{K}) \equiv$ *algebraic dual* of $C_0^\infty(E)$.
(h)	$C_0^\infty(E)^+ \equiv L_{A_0}(C_0^\infty(E); A) \equiv$ *space of superdistributions* on E.

In this case, we call L the *Green singular solution of the boundary value problem* $\partial L = X_1 \dot{\cup} X_2$.

DEFINITION 4. 1) Let $\mathcal{M}$ be the *category of measurable spaces*, i.e., $\mathrm{Ob}(\mathcal{M})$ are all abstract measurable spaces $(X, \mathcal{T})$, where X is a set and $\mathcal{T}$ is a σ-algebra. Morphisms $\mathrm{Hom}(\mathcal{M})$ are measurable transformations.

2) Let $\mathcal{PDE}$ be the subcategory of $\underline{\mathcal{PDE}}$ such that $X \in \mathrm{Ob}(\mathcal{PDE})$ iff $X = \underline{E}_\infty = \varinjlim \underline{E}_{k+s}$, where $\underline{E}_k \subset J\underline{D}^k(W)$ is a SPDE.

THEOREM 3. *There exists a canonical covariant functor (formal superquantization)* $\Omega: \mathcal{DE} \to \mathcal{M}$, *such that for any* $\underline{E}_\infty \in \mathrm{Ob}(\mathcal{PDE})$, $\underline{\Omega}(\underline{E}_\infty)$ *is the quantum supersitus of* $\underline{E}_\infty$, *that, is the set of supersolutions of* $J\underline{D}^\infty(W)$ *that defines a cobordism of two Cauchy hypersurfaces data on* $\underline{E}_\infty$.

REMARK. Let us consider formal superquantization at finite-order. The *formal superquantization* of a SPDE $\underline{E}_k \subset J\underline{D}^k(W)$ of order k on a fiber bundle $\pi: W \to M$ built in the category of supermanifolds, is the couple $Q(E_k) = (\underline{SP}_k, \underline{C}_\alpha(\underline{E}_k))$, where: (i) $\underline{SP}_k \subset J\underline{D}(J\underline{D}^k(W))$ is the *nonlinear Spencer superequation* on $J\underline{D}^k(W)$ and (ii) $\underline{C}_\alpha(E_k)$ is the set of G^∞-Cauchy data superhypersurfaces on $\underline{E}_k$. Here $\underline{SP}_k$ is obtained by direct generalization of the ordinary case. $J\underline{D}^{k+1}(W)$ is the formally superintegrable SPDE that has the same set of solutions as $\underline{SP}_k$ (that is not formally superintegrable). The set $\underline{\Omega}_s(\underline{E}_k)$ of singular solutions of $\underline{SP}_k$ with boundary conditions $(\sigma_1, \sigma_2) \in \underline{C}_\alpha(\underline{E}_k) \times \underline{C}_\alpha(\underline{E}_k)$ is the *singular quantum supersitus* of $\underline{E}_k$. Any element of $\underline{\Omega}_s(\underline{E}_k)$ is called a *quantum supercobord.* In $\underline{\Omega}_s(\underline{E}_k)$ there are also singular supersolutions of $\underline{E}_k$. These constitute a subset $\underline{\Omega}_s(E_k)_c$ of $\underline{\Omega}_s(E_k)$ called the *classical superlimit* of the singular quantum supersitus of $\underline{E}_k$. The subset $\underline{\Omega}_s \equiv \underline{\Omega}_s(\sigma_1, \sigma_2)$ of $\underline{\Omega}_s(\underline{E}_k)$ for a fixed couple $(\sigma_1, \sigma_2) \in \underline{C}_\alpha(E_k) \times \underline{C}_\alpha(E_k)$ is called the *statistical set* of the scattering process $\sigma_1 \to \sigma_2$. The set $\underline{\Omega}_s(\underline{E}_k)$ has a natural structure of manifold (of infinite dimension modeled on locally convex spaces) and each statistical set is a closed submanifold of $\underline{\Omega}_s(\underline{E}_k)$.

REMARK. The formal superquantization becomes effective if on the quantum supersitus we recognize a (pre)spectral measure. In fact, in this way we can represent physical superobservables (represented as random variables on $\Omega_s(\underline{E}_k)$) as linear operators. To show this, we redefine the notion of super-Hilbert space.

DEFINITION 5. 1) A *super-Hilbert space* is a $\mathbb{Z}_2$-graded commutative A-module $\mathcal{H}$ such that: (A) A is endowed with a map (complex conjugation) that satisfies the following properties: (a) $(\lambda+\lambda')^- = (\overline{\lambda}+\overline{\lambda'})$; (b) $(\lambda\lambda')^- = (\overline{\lambda'}\overline{\lambda})$; (c) if $\lambda \in \mathbb{C}$, $\overline{\lambda}$ is the ordinary complex conjugation ($\lambda \in A$ is called *real* if $\overline{\lambda} = \lambda$ and *imaginary* if $\overline{\lambda} = -\lambda$). (B) $\mathcal{H}$ is endowed with an *inner product*, i.e., with a one-to-one mapping $^+ \colon \mathcal{H} \to \mathcal{H}^+ \equiv L_A(\mathcal{H}; A)$. (We shall denote vectors of $\mathcal{H}$ by the Dirac-symbolism $|\phi\rangle \in \mathcal{H}$, (*ket*) and the vectors of the dual $\mathcal{H}^+$ by $\langle\phi|$, (*bra*). So we can write $|\phi\rangle^+ = \langle\phi|$ and the inner product of $|\phi\rangle, |\chi\rangle \in \mathcal{H}$ will be denoted by $\langle\chi|\phi\rangle$.) We note that the mapping satisfies the following axioms: (a) $||\phi\rangle^+| = ||\phi\rangle|$; (b) $(\lambda|\phi\rangle)^+ = \langle\phi|\overline{\lambda}$ for any $\lambda \in A$; (c) $(|\phi\rangle + |\chi\rangle)^+ = \langle\phi| + \langle\chi|$; (d) $(\langle\chi|\phi\rangle)^+ = \langle\phi|\chi\rangle$.

2) An element $|\phi\rangle \in \mathcal{H}$ is called *physical* if it has the nonvanishing body. Physical elements of $\mathcal{H}$ are also called *state vectors*.

REMARK. 1) If $B \in L_A(\mathcal{H})$, we can assume, by abuse of notation (and by using the $+$-isomorphism) that $B \in L_A(\mathcal{H}^+)$. So we write indifferently $B(|\phi\rangle) = B|\phi\rangle$, $B(\langle\chi|) = \langle\chi|B$. One has the following properties: (a) $(\lambda\langle\chi|)B|\phi\rangle = \lambda\langle\chi|B|\phi\rangle$; (b) $(\langle\chi| + \langle\phi|)B|\alpha\rangle = \langle\chi|B|\alpha\rangle + \langle\phi|B|\alpha\rangle$; (c) $(\lambda\langle\chi|)B = \lambda\langle\chi|B$; (d) $(B+C)|\phi\rangle = B|\phi\rangle + C|\phi\rangle$, $(\lambda B)|\phi\rangle$, $(B\lambda)|\phi\rangle = B\lambda|\phi\rangle$.

2) Since $L_A(\mathcal{H}) \cong L_A(\mathcal{H}^+)$ are $\mathbb{Z}_2$-graded commutative A-modules, we have

$$\begin{aligned}\langle\chi|\lambda_B|\phi\rangle &= (-1)^{|\lambda||\chi|}\lambda\langle\chi|B|\phi\rangle \\ &= (-1)^{|\lambda|(|B|+||\phi\rangle|)}\langle\chi|B|\phi\rangle\lambda = (-1)^{|\lambda||B|}\langle\chi|B\lambda|\phi\rangle \Longrightarrow \lambda_B = (-1)^{|\lambda||B|}B\lambda.\end{aligned}$$

DEFINITION 6. 1) The *adjoint* B^+ of $B \in L_A(\mathcal{H})$ is defined by $B^+|\phi\rangle = (\langle\phi|B)^+$. The operator B is said to be *selfadjoint* iff $B^+ = B$.

2) A linear operator $B \in L_A(\mathcal{H})$ is said to be a *physical observable* iff: (i) B is selfadjoint; (ii) all its eigenvalues belong to B_0 (i.e., are *c-numbers*); (iii) for every eigenvalue there is at least one corresponding *physical eigenvector* (i.e., the corresponding body is nonvanishing); (iv) the set of physical eigenvectors that correspond to *soulless* eigenvalues (i.e., having no component in $B' \equiv B'_0 \oplus B_1$), contains a complete basis. The soulless eigenvalues will be called *physical eigenvalues*.

REMARK. We can prove the following statements. 1) All the eigenvalues of a physical observable are real.

2) Eigenvectors corresponding to different physical eigenvalues of physical observable are orthogonal.

DEFINITION 7. 1) Let $\mathcal{L}_{\mathcal{H}}$ be the category defined by the following conditions: (i) $\mathrm{Ob}(\mathcal{L}_{\mathcal{H}})$ is the set of topological vector spaces like $L_A(\mathcal{H})$, where $\mathcal{H}$ is defined as in the above theorem. (ii) $\mathrm{Hom}_{\mathcal{L}_{\mathcal{H}}}(L_A(\mathcal{H}_1), L_A(\mathcal{H}_2)) = L(\mathrm{Hom}(\mathcal{H}_1, \mathcal{H}_2))$, where $\mathrm{Hom}(\mathcal{H}_1, \mathcal{H}_2)$ is the group of isomorphisms between $\mathcal{H}_1$ and $\mathcal{H}_2$.

2) Let $[[\underline{E}_\infty]]$ be the subcategory of $\mathcal{PDE}$ consisting of SPDE's equivalent to $\underline{E}_\infty \subset J\underline{D}^\infty(W)$. We call *superquantum (pre-)spectral measure* on $\underline{E}_\infty$ a covariant functor

$$\widehat{E} \colon [[\underline{E}_\infty]] \to \mathcal{L}_{\mathcal{H}}$$

that can be factorized as follows:

$$\begin{array}{ccc} [[\underline{E}_\infty]] & \xrightarrow{\widehat{E}} & \mathcal{L}_{\mathcal{H}} \\ \Omega \downarrow & & \uparrow E \\ \mathcal{M} & = & \mathcal{M} \end{array}$$

where E is a functor super(pre)spectral measure. The definition of super(pre)-spectral measure is the natural extension of the ordinary one. (See [17, 52, 53].)

THEOREM 4. *A superquantum (pre)spectral measure of $\underline{E}_\infty \subset J\underline{D}^\infty(W)$ determines a representation of the algebra $\mathcal{F}_A(\underline{E}_\infty)$ of supernumerical functions $\underline{\Omega}(\underline{E}_\infty) \to A$:*

$$\widehat{\ }: \mathcal{F}_A(\underline{E}_\infty) \to L(\mathcal{H}), \quad f \mapsto \hat{f}$$

such that

$$\langle \psi' | \hat{f} | \psi \rangle = \int_{\Omega(\underline{E}_\infty)} f \, dE_{\psi, \psi'}, \qquad \textit{for any } \psi \in \mathcal{H}, \ \psi' \in \mathcal{H}^+.$$

REMARK. In practice the road that links superquantum spectral measures of SPDE's with quantized physical superobservables is reversed. This is obtained by means of canonical superquantization (see below). Note also that Dirac quantization is related to the so-called "geometric quantization" [17] since it realizes a representation of physical superobservables by linear operators in suitable super-Hilbert spaces.

DEFINITION 8. We define a *physical superobservable* for $\underline{E}_k \subset J\underline{D}^k(W)$ as a random variable

$$f[i]: \underline{\Omega}(\underline{E}_\infty)_c \to A \quad \text{such that } f[i](s) = \int_{M_B} i^\cdot(\underline{f} \circ D^p s) \, d\mu,$$

where μ is a suitable measure on M_B, $\underline{f}: J\underline{D}^p(W) \to A$ is a numerical function on $J\underline{D}^p(W)$, $p \geqslant 0$, and i is a section of $b: M \to M_B$. Let us denote by $\mathcal{A}(\underline{E}_k)$ the set of physical superobservables of $\underline{E}_k$.

REMARK. We shall assume also that on the body manifold M_B a "space–time" foliation is defined. So, on each even manifold $i(M_B) \subset M$ corresponding to a section i of $b: M \to M_B$, there is a space–time foliation too. Then the canonical Dirac-superquantization of physical superobservables can be directly obtained by using the natural Lie algebra just induced by $\underline{E}_k$ on the set $\mathcal{A}(\underline{E}_k)$ of physical superobservables on $\underline{E}_k$. The process is similar to the one developed in [42, 45] for ordinary manifolds. This allows us to associate to a geometric situation (super PDE) a quantum algebra in the sense of A. Connes [8].

THEOREM 5. 1) *Let $\underline{E}_k = \ker_\chi K \subset J\underline{D}^k(W)$ be a SPDE obtained as the kernel of a k-order superdifferential operator $K: J\underline{D}^k(W) \to K$, with respect to a G^∞_w section $\chi: M \to K$ of the fiber bundle $\underline{\pi}: K \to M$, defined in the category of supermanifolds. Then for any section $s \in G^\infty_w(W)$, any solution of $\underline{E}_k$, and*

$j[\chi]=\partial\widetilde{\chi}$, where $\widetilde{\chi}$ is a superdeformation of χ, we can associate to $\underline{E}_k$ an affine superdifferential equation

$$\underline{E}_k[s]=\ker_{j[\chi]}J[s]\subset J\underline{D}^k(s^{\bullet}\mathrm{v}TW),$$

where $J[s]\colon C^\infty(s^{\bullet}\mathrm{v}TW)\to C^\infty(\chi^{\bullet}\mathrm{v}TK)$ is a A_0-linear map. We call $\underline{E}_k[s]$ the Jacobi superequation of $\underline{E}_k$.

2) *For any section i of $b\colon M\to M_B$, we have the following natural extension of $J[s]$:*

$$\begin{array}{ccc}(C_0^\infty(s^{\bullet}\mathrm{v}TW)^\phi)^+ & \xrightarrow{J[s|i]} & (C_0^\infty(\chi^{\bullet}\mathrm{v}TK)^\phi)^+\\ \uparrow & & \uparrow\\ G^\infty(s^{\bullet}\mathrm{v}TW) & \xrightarrow{J[s]} & G^\infty(\chi^{\bullet}\mathrm{v}TK)\end{array}$$

So, the distribution solution of $\underline{E}_k[s]$ can be written as follows: $v=G[s|i]_{j[s]}$, where $G[s|i]$ is Green's kernel of $J[s]$ (associated to the section i of b).

3) *We have the following structure of $\mathbb{Z}_2$-graded commutative Lie superalgebra (for any section $i\colon M_B\to M$):*

$$\begin{aligned}(f_1,f_2)[i](s)&=G^+[s|i](jf_1(s)\otimes jf_2(s)\otimes\eta)-G^-[s|i](jf_1(s)\otimes jf_2(s)\otimes\eta)\\&\equiv\widetilde{G}[s|i](jf_1(s)\otimes jf_2(s)\otimes\eta),\end{aligned}$$

where

$$jf\colon C^\infty(W)\to\bigcup_{s\in C^\infty(W)}C_0^\infty(s^{\bullet}\mathrm{v}T^+W)$$

is the current associated to f as follows: $jf(s)=(\mathrm{v}df)\circ s$, where $\mathrm{v}df$ is the vertical differential of the super differentiable function $f\colon W\to A$. $G^\pm[s|i]$ are advanced and retarded Green kernels respectively with respect to the space–time foliation on $i(M_B)$.

DEFINITION 9. We call $\widetilde{G}[s|i]$ the *superpropagator* at the sections s and i.

REMARK (transition from the superclassical PDE to the quantum SPDE).

The canonical Dirac-superquantization for physical superobservables of $\underline{E}_k$ is given by the following quantum bracket:

$$[\hat{f}_1,\hat{f}_2][i](s)=i\hbar\widetilde{G}[s|i](df_1(s)\otimes df_2(s)\otimes\eta)\otimes\mathrm{id}_{\mathcal{H}},$$

where $\mathcal{H}$ is a suitable super-Hilbert space, $(f_j)[i]\in\mathcal{A}(\underline{E}_k)$, $j=1,2$.

Each physical superobservable $A^j\colon\underline{\Omega}(\underline{E}_\infty)_c\to A$ corresponding to field coordinates ($\equiv$ vertical coordinates on W) (*dynamical variables*) is replaced by a selfadjoint linear operator $\widehat{A^j}$.

Each superquantized random supervariable determines a quantum superspectral measure

$$E\colon(\underline{\Omega}(\underline{E}_k)_c,\Sigma)_0\to L_A(\mathcal{H})\subset L_{\mathbb{K}}(\mathcal{H}).$$

REMARK. Note that our approach to canonical quantization is a geometrical generalization of the previous one made for Lagrangian super PDE's (see, e.g., [9]).

EXAMPLE. *An harmonic superoscillator.* Let $A \equiv A_0 \oplus A_1$ be a BG-algebra. Let us consider the fiber bundle $\pi: W \equiv \mathbb{R} \times A_0 \times A_1^2 \to \mathbb{R}$ with fibered coordinates (t, x^i), $i = 1, 2, 3$. A section s of π is identifiable with a triplet of functions $f \equiv (f_i)$, $f_1: \mathbb{R} \to A_0$, $f_2, f_3: \mathbb{R} \to A_1$. One has the following representation in coordinates:

$$J\underline{D}^2(W): \quad (t, x^i, \dot{x}^i, \ddot{x}^i),$$
$$E_2 \subset J\underline{D}^2(W): \quad \ddot{x}^i + p^i(x^1, x^2, x^3) = 0, \qquad i = 1, 2, 3,$$

where p^i are polynomial functions with values p^1 in A_0 and p^2, p^3 in A_1 respectively. One has also $K \equiv \mathbb{R} \times A_0 \times A_1^2$, $\chi^{\bullet}\mathrm{v}TK = 0^{\bullet}\mathrm{v}TK \cong W$, $s^{\bullet}\mathrm{v}TW \cong W$. The Jacobi operator at the solution s is a map $J[s]: C_0^{\infty}(W) \to C_0^{\infty}(W)$ given by $J[s]\cdot v = (\partial t \cdot \partial t \cdot v^j) + \Phi_i^j v^i$, where $\Phi_i^j \equiv (\partial x_i \cdot p^j) \circ f$ is a function-matrix on $\mathbb{R}$ with the following entries: $|\Phi_i^j| = |i| + |j|$. Of course one also has $|v^i| = |j|$ and $|(\partial t \cdot \partial t \cdot v^j)| = |j|$. The corresponding Green kernel $G^{ij}[f](t, t')$ allows us to represent the distribution solution of the affine Jacobi equation $J[s] \cdot v = j[0]$ as $v^j = G[f]^j(t, t')_{j[0]}$. Then we have the following quantum bracket

$$[\hat{x}^i(t), \hat{x}^j(t')]_{\pm} = i\hbar \widetilde{G}[f]^{ij}(t, t') \,\mathrm{id}_{\mathcal{H}},$$

where $+$ is taken if $|i| = |j| = 1$, and $-$ otherwise. Now, taking the derivative with respect to t' for $t' = t$, we get

$$[\hat{x}^i(t), \dot{\hat{x}}^j(t')]_{\pm} = i\hbar(\partial t' \cdot \widetilde{G}[f]^{ij}(t, t'))|_{t=t'} \,\mathrm{id}_{\mathcal{H}}.$$

Finally, taking into account the fact that $(\partial t' \cdot \widetilde{G}[f]^{ij}(t, t'))|_{t=t'} = \delta^{ij}$ we get $[\hat{x}^i(t), \hat{x}^j(t')]_{\pm} = i\hbar\delta^{ij} \,\mathrm{id}_{\mathcal{H}}$,

REMARK (Quantum supercobordism and cobordism). Generalizations of previous theorems on cobordism in PDE's to super PDE's can be also given [47]. These allow to characterize quantum tunneling effects in super PDE's, by generalizing classical results of differential topology [10, 15]. In particular, let us state the following result.

THEOREM 6. 1) *Let $\underline{E}_k \subset J\underline{D}^k(W)$ be k-order super PDE on the fiber bundle $\pi: W \to M$, $\dim M = (m, n)$. Let (N_i, h_i), $i = 0, 1$, be a couple of Cauchy data for the system $\underline{E}_k$, defined on $\underline{E}_{k+r}$. Then one has a canonical mapping*

$$\phi_{(N, h)}: \widetilde{H}^{m-1}(\widetilde{\underline{I}}_{k+r}(W); A) \to A,$$

where $N \equiv N_0 \times 0 \cup N_1 \times 1$, and h is the natural extension of h_i, $i = 0, 1$, to N. We call $\widetilde{H}^{m-1}(\widetilde{\underline{I}}_{k+r}(W); A)$ the $(m-1)$-Leray–Serre supercohomology of $J\underline{D}^{k+r}(W)$. Then, the necessary condition for the existence of a quantum supercobord V cobording N_0 and N_1 is that the mapping $\phi_{(N, h)}$ be zero.

2) *Furthermore, in order that the topology does not change passing from N_0 to N_1 it is necessary that the supercharacteristic classes $i_V^*\omega$ be zero for all $\omega \in \widetilde{H}^i(\widetilde{\underline{I}}_{k+r}(W); A)$, $i \geqslant 1$.*

REMARK. The above theorem can be regarded as a generalization of a recent result by Gibbons and Hawking [11] proposing some selection rules on the change of topology in Lorentzian-spin cobordism for compact 3-dimensional manifolds, based on the Kervaire semicharacteristic:

$$u(N) = \dim_{\mathbb{Z}_2}(H_0(N; \mathbb{Z}_2) \oplus H_1(N; \mathbb{Z}_2)) \bmod 2.$$

THEOREM 7 (relation between quantum cobordism and superpropagators). *Let us assume that* $\underline{E}_k \equiv \ker_f K \subset J\underline{D}^k(W)$ *is a SPDE as given in Theorem 5. Then the superpropagator* $\widetilde{G}[s|i]$ *defines a quantum cobord* $L \subset \underline{\Omega}(\underline{E}_k)$ *such that if* $\widetilde{G}[s|i]$ *satisfies the boundary conditions* $X_1 \dot{\cup} X_2$, *then*

$$\partial L = X_1 \dot{\cup} X_2 .$$

The following theorem relates quantum supercobordism to the axioms of "topological quantum field theory" as introduced by Atiyah [1, 2]. (See also papers by Witten [60–62].)

THEOREM 8. 1) *Let* $\underline{E}_k \subset J\underline{D}^k(W)$ *be a formally superintegrable SPDE on a fiber bundle* $\pi: W \to M$ *in the category of supermanifolds,* $\dim M = (m, n)$. *Let* $\pi_H: H \to J\underline{D}^k(W)$ *be a fiber bundle of* $\mathbb{Z}_2$*-graded commutative* A*-modules over* $J\underline{D}^k(W)$ *such that on the fibers a* $+$*-operation is defined and satisfies the axioms for super-Hilbert space,* (H, W, Q) *being a super-bundle of geometric objects over* M, *for some functor* $Q: \mathcal{C}(W) \to \mathcal{C}(H)$ [32–35]. *We denote* $\mathcal{H} \equiv C_0^\infty(H)$. *Let us assume that* $L \subset J\underline{D}^k(W)$ *is a* V*-supermanifold such that* $\omega_1^{(k)}(L_B) = 0$. *Then, for any section* $i: L_B \to L$ *and* $\eta: J\underline{D}^k(W) \to \widetilde{\Lambda}_m^0 J\underline{D}^k(W)$, *the fiber bundle* $H \to J\underline{D}^k(W)$ *determines a super-Hilbert space, that we denote by* $\mathcal{H}(L)$.

2) *If* $N \subset \underline{E}_k$ *is a Cauchy superhypersurface data on* $\underline{E}_k$ *such that* $N \subset L$, *then one also has a super-Hilbert space* $\mathcal{H}(N) \subset \mathcal{H}(L)$. *Furthermore, let* $N^\bullet$ *denote* N *with the opposite orientation, and we set:*

$$\text{(a)} \quad \mathcal{H}(N^\bullet) = \mathcal{H}(N)^+ ; \qquad \text{(b)} \quad \mathcal{H}(N_1 \dot{\cup} N_2) = \mathrm{Hom}_A(\mathcal{H}(N_1), \mathcal{H}(N_2)) .$$

Then there exists a vector $U(L) \in \mathcal{H}(N)$ *for each compact oriented* V*-supermanifold with boundary* N. *Furthermore, if*

$$L = L_1 \cup_{N_2} L_2 , \quad \partial L_1 = N_1^\bullet \dot{\cup} N_2 , \quad \partial L_2 = N_2^\bullet \dot{\cup} N_3 ,$$

the vectors $U(L_1)$, $U(L_2)$, $U(L_3)$ *are related by*

$$U(L) = j(U(L_1), U(L_2)) = U(L_2) \circ U(L_1) \in \mathrm{Hom}_A(\mathcal{H}(N_1), \mathcal{H}(N_2)) ,$$

where j *is the canonical mapping:*

$$j: \mathrm{Hom}_A(\mathcal{H}(N_1), \mathcal{H}(N_2)) \times \mathrm{Hom}_A(\mathcal{H}(N_2), \mathcal{H}(N_3)) \to \mathrm{Hom}_A(\mathcal{H}(N_1), \mathcal{H}(N_3)) ,$$

given by $j(A, B) = A \circ B$.

The linear mappings $U(L)$ *and* $U(L^\bullet)$ *are adjoint to each other. We also define*

$$\text{(c)} \quad U(\varnothing) = A \quad (\varnothing \equiv \textit{empty superhypersurface});$$

$$\text{(d)} \quad U(N \times I) \in \mathrm{Hom}_A(\mathcal{H}(N_0), \mathcal{H}(N_1)) ,$$

where $I \equiv [0, 1] \subset \mathbb{R}$, $N_0 \equiv N \times 0$, $N_1 \equiv N \times 1$. *Moreover, the correspondence that associates to each Cauchy superhypersurface* N *the space* $\mathcal{H}(N)$ *is of functorial nature and homotopy invariant, i.e., the group,* $\mathrm{SDiff}^+(N)$ *of orientation preserving superdiffeomorphisms of* N *acts on* $\mathcal{H}(N)$ *via its group of connected components* $\Gamma(N)$.

3) *To any closed* V*-supermanifold contained in* $J\underline{D}^k(W)$ *there corresponds a supernumerical invariant belonging to* A. *This numerical invariant can be computed from any decomposition* $L = L_1 \cup_N L_2$. *One has* $U(L^\bullet) = \overline{U(L)} \in A$.

DEFINITION 10. 1) Let L be a V-supermanifold such that $\partial L = N_1^{\bullet} \dot{\cup} N_2$. Let $|\psi_1\rangle \in \mathcal{H}(N_1)$, $|\psi_2\rangle \in \mathcal{H}(N_2)$ be state vectors. Then the real part of the body of the following supernumber $\langle\psi_2|U(L)|\psi_1\rangle \in A$ is called the *transition amplitude between the two state vectors* $|\psi_1\rangle$ and $|\psi_2\rangle$: $\mathrm{Re}\langle\psi_2|U(L)|\psi_1\rangle_B \in \mathbb{R}$.

2) In particular, if $N_2 = \varnothing$, $U(L) \in \mathcal{H}(N_1)^+$, then $U(L)|\psi_1\rangle \in A$, and $\mathrm{Re}(U(L)|\psi_1\rangle_B)$ is the *transition amplitude between the state* $|\psi_1\rangle$ *and vacuum.*

3) Conversely, if $N_1 = \varnothing$, $U(L) \in \mathcal{H}(N_2)$, then $\langle\psi_2|U(L)\rangle \in A$ and $\mathrm{Re}(\langle\psi_2|U(L)\rangle) \in \mathbb{R}$ is the *transition amplitude between the vacuum and the state* $|\psi_2\rangle$.

Sometimes it is possible to give a direct definition of the vector $U(L)$ (see [47]).

§3. Quantized supergravity

Supergauge theories have been formulated by many authors from different point of views. The first aim of the present section is to give a unified intrinsic description of such theories. We shall refer to our previous geometrical setting for gauge theory [34] and group model gauge theory [48]. Then, by using the theory of formal quantization of PDE's, we shall find a direct way to quantize a supergauge theory.

REMARK (geometric structures and superconnections on principal fiber bundles). Before we define a supergauge system, let us point out some geometrical structures related to superconnections on principal fiber bundles.

Any principal superconnection $k\colon M \to C(P) \cong J\underline{D}(P)/G$ ($=$ fiber bundle of principal superconnections) on the G-principal fiber bundle $(P, M, \pi; G)$ (in the category of supermanifolds), determines a section $\underline{k}$ of $\pi^{\bullet}C(P) \to P$ such that the following diagram is commutative:

$$\begin{array}{ccccc}
TPO_{A_0}A(G) & \xleftarrow{\;j\;} & \pi^{\bullet}C(P) & \longrightarrow & C(P) \\
{\scriptstyle \omega_k}\uparrow & & \updownarrow{\scriptstyle \underline{k}} & & \updownarrow{\scriptstyle k} \\
P & =\!= & P & \xrightarrow[\pi]{} & M
\end{array}$$

More precisely one has $\underline{k} = (\mathrm{id}_P, \check{k})$, where $\check{k} = k \circ \pi$, and ω_k is the Ehresmann superconnection associated to k; j is the canonical embedding over P given by $j(p, k(x)) = \omega_k(p)$, with $x = \pi(p)$. Set

$$\overline{\pi^{\bullet}C(P)} \equiv j(\pi^{\bullet}C(P)) \subset TPO_{A_0}A(G).$$

Of course, one has the following isomorphism: $\overline{\pi^{\bullet}C(P)} \cong \pi^{\bullet}C(P)$ and the following exact sequences of supervector bundles over P: $0 \to \pi^{\bullet}C(P) \to TPO_{A_0}A(G)$. The morphism j induces a morphism $j^k\colon \pi^{\bullet}J\underline{D}^k(C(P)) \to J\underline{D}^k(TPO_{A_0}A(G))$ over P given by $j^k(p, u = D^k k(x)) = D^k\omega_k(p)$, $x = \pi(p)$. Set

$$\overline{C(P)}^k \equiv j^k(\pi^{\bullet}J\underline{D}^k(C(P))) \subset J\underline{D}^k(TPO_{A_0}A(G)).$$

THEOREM 1. *Let $(P, M, \pi; G)$ be a principal fiber bundle in the category of supermanifolds. Let $\pi_E: E \to P$ be a vector fiber bundle over P. Let C denote the fiber bundle $\dot{\Lambda}^{\bullet} PO_{A_0}E$ over M with projection π_C and let $\underline{E}$ denote the fiber bundle $\dot{\Lambda}^{\bullet} PO_{A_0}E$ over P with projection $\pi_{\underline{E}}$. One has a canonical epimorphism of fiber bundles over $\dot{\Lambda}^{\bullet} PO_{A_0}E$:*

$$\text{(0)} \qquad \mathbf{n}\colon \pi^{\bullet}_{k,0} J\underline{D}^k(\underline{E}) \to J\underline{D}^k(C),$$

given by

$$\mathbf{n}(D^k s(x), D^k{}_{\urcorner}\mu(s(x))) = D^k({}_{\urcorner}\mu \circ s)(x) = D^k c(x),$$

where $\pi_{k,0}\colon J\underline{D}^k(P) \to P$ is the canonical epimorphism.

DEFINITION 1. A *k-order supergauge system with blow-up structure* is a k-order gauge continuum system $G(M) \equiv (\mathcal{B}, E_k)$ [34] in the category of supermanifolds, such that:

1) $\mathcal{B}$ is a superbundle of geometric objects $\mathcal{B} \equiv (P, C; B)$ (see [34, 35] and below), where (a) P is a principal fiber bundle $(P, M, \pi; G)$, G being a Lie supergroup of dimension (m'', n'') called *gauge structure supergroup* and M being a (m, n)-dimensional supermanifold called *base supermanifold.* The body M_B of M is called the (m-dimensional) *space–time*; (b) C is a fiber bundle over M (called the *configuration bundle*) that coincides with the fiber bundle of $\mathbb{Z}$-graded superconnections over a supervector fiber bundle $\pi_E: E \to P$ over P; (c) B is the natural functor $B\colon \mathcal{C}(P) \to \mathcal{C}(C)$, where $\mathcal{C}(P)$ (respectively, $\mathcal{C}(C)$) is the category whose objects are open subbundles of P (respectively, C), and whose morphisms are the local fiber bundle automorphisms between these objects. (We shall denote C by $\underline{E}$ when it is considered as a fiber bundle over P).

2) The *blow-up structure* is given by an embedding morphism $j\colon \pi^* C(P) \to \underline{E}$ over P given by means of the following commutative diagram with exact middle row:

$$\begin{array}{ccccc}
 & & \underline{E} & = & \dot{\Lambda}^{\bullet} PO_{A_0}E \\
 & & \uparrow & & \uparrow \\
0 & \xrightarrow{\ j\ } & \pi^{\bullet} C(P) & \longrightarrow & TPO_{A_0}E \\
 & & & & \uparrow \\
 & & & & 0
\end{array}$$

3) E_k is a k-order formally superintegrable SPDE on C, $E_k \subset J\underline{D}^k(C)$, given by

$$E_k = \mathbf{n}(\pi^{\bullet}_{k,0} \underline{E}_k),$$

where $\underline{E}_k$ is a k-order formally superintegrable SPDE on $\pi_{\underline{E}}: \underline{E} \to P$, $\underline{E}_k \subset J\underline{D}^k(\underline{E})$, such that $\mathbf{b}^k(\underline{E}_k) \subset \overline{C(P)^k}$, $\mathbf{b}^k\colon J\underline{D}^k(\underline{E}) \to J\underline{D}^k(TPO_{A_0}E)$ being the canonical epimorphism over P induced by the canonical epimorphism $b\colon \underline{E} \to TPO_{A_0}E$ over P.

REMARK. The set of G^∞-solutions of E_k, $\underline{\mathrm{Sol}}(E_k)$ is characterized by

$$\begin{aligned}\underline{\mathrm{Sol}}(E_k) = \{c \in G^\infty(C) \mid\ & c = {}_{\rceil}\mu \circ s\,,\ s = \pi_{\underline{E}} \circ c\,,\ {}_{\rceil}\mu : \mathbb{Z}\text{-graded} \\ & \text{superpseudoconnection on } P \text{ such that } D^k{}_{\rceil}\mu(s(x)) \in \underline{E}_k\,, \\ & s \in G^\infty(P)\text{; the corresponding superpseudoconnection} \\ & b \circ {}_{\rceil}\mu \colon P \to TPO_{A_0}E \text{ is a Ehresmann superconnection on } P\}.\end{aligned}$$

DEFINITION 2. A *k-order* FDA-*supergauge system* (FDA = free differential algebra) is a k-order supergauge system with blow-up structure, $G(M) \equiv (\mathcal{B}, E_k)$, such that:

1) the configuration bundle C in $\mathcal{B} = (P, C; B)$ is the fiber bundle of graded superpseudoconnections on the trivial vector bundle $\pi_E \colon E \equiv P \times V \to V$, where V is a vector superspace containing $A(G)$: $V = A(G) \times V_1 \times \cdots \times V_s$, such that there is a representation of $A(G)$ on V of the type

$$r = (\mathbf{1}, r_1, \ldots, r_s) \colon A(G) \to L(A(G) \times V_1 \times \cdots \times V_s),$$

where $\mathbf{1}$ is the identity representation on $A(G)$ and $r_i \colon A(G) \to L(V_i)$, $i = 1, \ldots, s$, are irreducible representations of $A(G)$ on V_i. (Superpseudoconnections are pseudoconnections [48] in the super sense. For details see [47].)

2) The blow-up structure

$$j \colon \pi^{\cdot}C(P) \to \underline{E} \equiv \dot{\Lambda}^{\cdot}PO_{A_0}V$$

is induced by the morphism of fiber bundles over P defined by means of the following commutative diagram of embedding mappings:

$$\begin{array}{ccc} \pi^{\cdot}C(P) & \xrightarrow{\ j\ } & \dot{\Lambda}^{\cdot}PO_{A_0}V \equiv \underline{E} \\ \downarrow & & \cup \\ TPO_{A_0}A(G) & \longrightarrow & TPO_{A_0}V \end{array}$$

3) The SPDE $\underline{E}_k \subset J\underline{D}^k(\underline{E})$ is defined by the variational principle by using a Lagrangian superdensity $J\underline{D}^h(E) \to \widetilde{\Lambda}^0_m J\underline{D}^h(\underline{E})$ (m = space–time dimension).

DEFINITION 3. A *supersymmetry* of a supergauge system with blow-up structure, $G(M) \equiv (B, E_k)$, is a diffeomorphism $f \colon P \to P$ (that does not necesseraly come from the action of the group structure of G on P) such that the natural diffeomorphism $B(f)$ induced on C determines a diffeomorphism of $J\underline{D}^k(C)$ that preserves E_k.

REMARK. One can see [48] that a supersymmetry f transforms any solution $c = {}_{\rceil}\mu \circ s$ of E_k into a new solution $c' = {}_{\rceil}\mu' \circ s'$ such that f is a symmetry of $\underline{E}_k$ too.

DEFINITION 4. The *curvature map* of a supergauge system with blow-up structure is a differential operator $R\colon J\underline{D}(\underline{E}) \to \dot{\Lambda}^{\bullet} PO_{A_0}E$ such that

$$ {}_{\rceil}R \equiv R \circ D_{\rceil}\mu = {}_{\rceil}d_{\rceil}\mu + P({}_{\rceil}\mu), $$

where $P({}_{\rceil}\mu)$ is an element of the polynomial algebra $A[G^{\infty}_{w}(\underline{E})]$ such that on the canonical constraint $TPO_{A_0}A(G) \subset \underline{E}$, R reduces to the corresponding curvature for superpseudoconnections:

$$ R \circ D_{\rceil}\mu = d_{\rceil}\mu + [{}_{\rceil}\mu, {}_{\rceil}\mu]/2. $$

REMARK. Note that the $\mathbb{Z}$-graded curvature

$$ {}_{\rceil}R = R \circ D_{\rceil}\mu = {}_{\rceil}d_{\rceil}\mu $$

is a particular curvature map. Note also that in order to define ${}_{\rceil}d$ we assume the existence of a bilinear morphism $[\ ,\ ]\colon E \times E \to E$ on the vector bundle E.

Below we give an example of supergravity by using our unified geometric model of supergauge theory. In this way we shall also be able to use the theory of formal superquantization to describe quantum supergravity. We shall specialize to a model for $N = 2$, $d = 4$, supergravity.

$A =$ BG algebra; $M = (4, 0)$-dimensional supermanifold; $(P, M, \pi; G) =$ principal bundle in the category of supermanifolds with structure group G equal to the $U(1)$-central extension of the Poincaré supergroup; $A(G)$ has the generators $\{J_{ab}, P_c, Q_{i\beta}, Z\}$, $0 \leqslant a, b, c, \beta \leqslant 3$, $i = 1, 2$, with the commutation relations as in Table 3, all other commutators being zero.

TABLE 3. Commutation relations for the $U(1)$-central extension of the Poincaré supergroup.

$[J_{ab}, J_{cd}] = \eta_{bc}J_{ad} + \eta_{ad}J_{bc} - \eta_{ac}J_{bd} - \eta_{bd}J_{ac}$	
$[J_{ab}, P_c] = \eta_{bc}P_a - \eta_{ac}P_b$	
$[Q_{\beta i}, Q_{\mu j}] = (C\gamma^{\alpha})_{\beta\mu}\delta_{ij}P_a + C_{\beta\mu}\varepsilon_{ij}Z$,	$[J_{ab}, Q_{\beta i}] = (\sigma_{ab})^{\mu j}_{\beta i} Q_{\mu j}$

Here we also use the following notations: $\eta_{\beta\mu} = \mathrm{diag}(-1, 1, 1, 1)$, $\sigma_{\beta\mu} = [\gamma_{\beta}, \gamma_{\mu}]/4$, γ^{μ} are the Dirac matrices, C is charge conjugation matrix,

$$ \pi_E\colon \underline{E} \equiv TPO_{A_0}A(G) \to P: $$

is the fiber bundle of superpseudoconnections on the trivial vector bundle $E \equiv P \times A(G) \to P$, $\pi_C\colon C \equiv TPO_{A_0}A(G) \to M$ is the configuration bundle; a configuration is a section c of π_C such that $c = {}_{\rceil}\mu \circ s$, where $s \in G^{\infty}(P)$ and ${}_{\rceil}\mu \in G^{\infty}(\underline{E})$ is locally given by the formulas

$$ {}_{\rceil}\mu_{\beta} = \omega^{a}_{b\beta}J^{b}_{a} + \theta^{\mu}_{\beta}P_{\mu} + \psi^{\mu j}_{\beta}Q_{\mu j} + A_{\beta}Z. $$

TABLE 4. Local expression of $\underline{E}_2$.

$$
\begin{array}{|l|}
\hline
(\partial x_{[\varepsilon} \cdot R^{\alpha\beta}_{\mu\nu]}) - \omega^{\alpha}_{\delta[\varepsilon} R^{\alpha\beta}_{\mu\nu]} + \omega^{\beta}_{\delta[\varepsilon} R^{\alpha\delta}_{\mu\nu]} = 0 \\
(\partial x_{[\varepsilon} \cdot R^{\alpha}_{\mu\nu]}) + \omega^{\alpha\delta}_{[\varepsilon} R_{\mu\nu]\delta} + \psi^{i}_{[\varepsilon} \gamma^{\alpha} \rho_{\mu\nu]i} - R^{\alpha\beta}_{[\mu\nu} \theta_{\varepsilon]\beta} = 0 \\
(\partial x_{[\varepsilon} \cdot \rho^{i}_{\mu\nu]}) + \tfrac{1}{2} \omega^{\alpha\beta}_{[\varepsilon} (\rho_{\mu\nu]} \sigma_{\alpha\beta})^{i} - \tfrac{1}{2} R^{\alpha\beta}_{[\mu\nu} (\psi_{\varepsilon]} \sigma_{\alpha\beta})^{i} = 0 \\
(\partial x_{[\varepsilon} \cdot F_{\mu\nu]}) + \psi^{j}_{[\varepsilon} \rho^{i}_{\mu\nu]} c \varepsilon_{ij} = 0 \\
\cdots\cdots\cdots\cdots\cdots\cdots\cdots\cdots\cdots\cdots\cdots\cdots \\
R^{\alpha\beta}_{\mu\nu} = (\partial x_{\mu} \cdot \omega^{\alpha\beta}_{\nu}) - (\partial x_{\nu} \cdot \omega^{\alpha\beta}_{\mu}) + \omega^{\alpha\gamma}_{\mu} \omega^{\beta}_{\gamma\nu} - \omega^{\alpha\gamma}_{\nu} \omega^{\beta}_{\gamma\mu} \quad \text{(curvature)} \\
R^{\alpha}_{\mu\nu} = (\partial x_{\mu} \cdot \theta^{\alpha}_{\nu}) - (\partial x_{\nu} \cdot \theta^{\alpha}_{\mu}) + \omega^{\alpha}_{\mu\gamma} \theta^{\gamma}_{\nu} - \omega^{\alpha}_{\nu\gamma} \theta^{\gamma}_{\mu} + \psi^{i}_{\mu} c^{\gamma\alpha} \psi_{\nu i} \quad \text{(torsion)} \\
\rho^{i}_{\mu\nu} = (\partial x_{\mu} \cdot \psi^{i}_{\nu}) - (\partial x_{\nu} \cdot \psi^{i}_{\mu}) + \tfrac{1}{2} \omega^{\gamma\delta}_{\mu} (\sigma_{\gamma\delta} \psi_{\nu})^{i} - \tfrac{1}{2} \omega^{\gamma\delta}_{\nu} (\sigma_{\gamma\delta} \psi_{\mu})^{i} \\
F_{\mu\nu} = (\partial x_{\mu} \cdot A_{\nu}) - (\partial x_{\nu} \cdot A_{\mu}) + \psi^{i}_{[\mu} c \psi^{j}_{\nu]} \varepsilon_{ij} \\
\cdots\cdots\cdots\cdots\cdots\cdots\cdots\cdots\cdots\cdots\cdots\cdots \\
R^{\alpha}_{\mu\nu} + 4 f^{ab} A_{[\mu} \theta_{\nu]b} = 0 \quad \text{(torsion equation)} \\
F_{\mu\nu} - f_{ab} \theta^{\alpha}_{[\mu} \theta^{b}_{\nu]} = 0 \quad \text{(matter field equation)} \\
\rho^{\alpha i}_{\mu\nu} - \rho^{\alpha i}_{ab} \theta^{a}_{[\mu} \theta^{b}_{\nu]} - \varepsilon^{ij} (\gamma_{\alpha} \psi^{\alpha}_{j[\mu} \theta_{\nu]b} f^{ab} + \tfrac{1}{2} \gamma_5 \gamma_a \psi^{\alpha i}_{[\mu} \theta_{\nu]b} f_{cd} \varepsilon^{abcd}) = 0 \\
\hfill \text{(gravitino equation)} \\
(\partial x_{\mu} \cdot f^{ab}) + 2 \omega^{ac}_{\mu} f^{b}_{c} + \tfrac{1}{2} \varepsilon^{abcd} \psi^{\alpha i}_{\mu} (c \gamma_5)_{\alpha\beta} \rho^{\beta j}_{cd} \varepsilon_{ij} = 0 \quad \text{(Maxwell's equation)} \\
\hline
\end{array}
$$

Blow-up structure is given by $\pi^{\bullet} C(P) \to \underline{E}$. The local expression of the dynamic equation $\underline{E}_2 \subset J\underline{D}^2(C)$ is given in Table 4.

The variational constraints are obtained by using the following supersymmetric Lagrangian superdensity:

$$
\begin{aligned}
\Omega \cdot \rceil \mu = {} & \theta^a \wedge \theta^b \wedge R^{cd} \varepsilon_{abcd} - \tfrac{1}{3} f_{mn} f^{mn} \theta^a \wedge \theta^b \wedge \theta^c \wedge \theta^d \varepsilon_{abcd} \\
& - 4 \psi^i \wedge C \gamma_5 \gamma_b \rho_i \wedge \theta^b - 4 (dA + \psi^i \wedge C \psi_i) \wedge \psi^k \wedge C \gamma_5 \psi_k \\
& + 4 (f^{cd} \wedge \psi^i \wedge C \gamma^a \psi_i - \omega^d f^{cd} \wedge \theta^a) \wedge \theta^b \wedge A \\
& + \tfrac{1}{2} f^{cd} \psi^i \wedge C \psi_i \wedge \theta^a \wedge \theta^b \varepsilon_{abcd} .
\end{aligned}
$$

Equation $\underline{E}_2$ can be quantized by means of the following commutative diagram

$$
\begin{array}{ccc}
\underline{E}_2 & \xrightarrow{Q|\underline{E}_2} & H \\
\cap & & \cap \\
j\underline{D}^2(C) & \xrightarrow[Q]{} & J\underline{D}(H) O_{A,C} H
\end{array}
$$

where:

1) H is a fiber bundle over C, $\pi_H : H \to C$ defined by

$$
H \equiv \bigcup_{q \in C} C(\widetilde{T}_p M, g(p))^{\mathbb{C}},
$$

where $p = \pi_C(q)$, $C(\widetilde{T}_p M, g(p))^{\mathbb{C}} \equiv$ Clifford algebra built on the complexificated supertangent space $\widetilde{T}_p M$ with respect to the metric

$$
g(p) = g_{\beta\mu}(p)\, dx^{\beta} \otimes dx^{\mu}, \qquad g_{\beta\mu}(p) = \eta_{\gamma\delta} \theta^{\gamma}_{\beta} \theta^{\delta}_{\mu} \quad \text{on } T_p M .
$$

Let $\mathcal{H}$ be the locally convex space of C^∞-sections of π_H with compact support. The couple (C, H) defines a superbundle of geometric objects (in the sense of [33, 34]) with respect to a functor $Q: \mathcal{C}(C) \to \mathcal{C}(H)$. Furthermore, there is a map $\mu: G^\infty(C) \to C^\infty(\Lambda_4^0 M)$ such that $\mu(c): M \to \Lambda_4^0 M$ is the volume form on the base manifold M associated to the metric g corresponding to c. Let $\mathcal{H}(c)$ be the completion of $C_0^\infty(c^\bullet H)$ corresponding to a configuration c, associated to the inner product

$$\langle \psi | \psi' \rangle_c = \int_M h(c)(\psi, \psi') \mu(c),$$

where $h(c)$ is the scalar product induced by the super-Hermitian structure of $c^\bullet H$.

2) In the above diagram $\mathbf{H}$ is the subbundle of $JD(H) O_{A,C} H$ characterized by means of sections of $J\underline{D}(H) OH$ over C that represent super(anti-)Hermitian superlinear operators belonging to $L_A(\mathcal{H})$, with domain of definition in super-Hilbert subspaces of $\mathcal{H}$.

3) The differential operator Q is defined by

$$\text{(Q)} \qquad Q \cdot c = \nabla_{X(c)} - f(c),$$

where $X(c)$ is a vector field on M defined by the following equation:

$$-4R(c) = \delta \underline{X}(c) + \phi(c),$$

where $\underline{X}(c)$ is the dual form of $X(c)$, $\phi(c)$ is the harmonic function associated to c, and $R(c)$ is the scalar curvature on the space–time (M, g), where g is the metric associated to c. The function $f(c): M \to A$ is defined in such a way that

$$(*) \qquad \delta \underline{X}(c) - 2f(c) = 0$$

on the set of sections of the dynamic equation E_2. More precisely, one has:

$$f(c) = -\tfrac{1}{2} R(c) - \tfrac{1}{2}\{(\rceil d^2 \rceil \mu)^H_{\mu\nu\varepsilon} g^{\mu\nu} \rceil \mu^\varepsilon_H + \Pi_{\mu\nu} F^{\mu\nu} + \Sigma^a_{\mu\nu} R^{\mu\nu}_a + \Delta^{\alpha i}_{\mu\nu} \rho^{\mu\nu}_{\alpha i} + \Theta^{ab}_\mu R^\mu_{ab}\} - \tfrac{1}{2}\phi(c),$$

where Σ, Π, Δ, and Θ are the first terms of the torsion, matter field, gravitino, and Maxwell equation respectively. Then, by using simple calculations, we see that $Q \cdot c$ is superantihermitian iff

$$\begin{aligned} 0 &= \int_M h(Q \cdot c(\psi), \psi') \cdot \mu(c) + \int_M h(\psi, Q \cdot c(\psi')) \cdot \mu(c) \\ &= \int_M [h(\nabla_{X(c)}\psi, \psi') + h(\psi, \nabla_{X(c)}\psi') - 2f(c) h(\psi, \psi')] \cdot \mu(c). \end{aligned}$$

So, $Q \cdot c$ is superantihermitian iff equation $(*)$ holds. Taking into account the expression of $X(c)$ and $f(c)$, we see that $Q \cdot c$ is surely superantihermitian on $\underline{E}_2$.

THEOREM 2. *The dynamic equation $\underline{E}_2$ for supergravity $N = 2$, $d = 4$, is μ-superquantizable, i.e., there exists a prespectral measure associated to any measure μ*

on the quantum supersitus $\underline{\Omega}(\underline{E}_2)$ of $\underline{E}_2$. More precisely, the μ-superquantization is obtained by the Dirac-superquantization (Q).

REMARK. The smeared quantum fields are obtained by derivation of the map $\widehat{\ }: C^\infty(C) \to L_A(\mathcal{H})$ associated to the differential operator (Q).

The canonical quantization of $\underline{E}_2$ is given by the following quantum bracket:

$$[\hat{c}^I(x), \hat{c}^J(x')] = i\hbar\widetilde{G}[c]^{IJ}(x, x'),$$

where $(c^I) \equiv (\lrcorner\mu^I \circ s)$, and $\widetilde{G}[c]$ is the propagator of $\underline{E}_2$ solution of the following Cauchy problem:

$$\begin{cases} J[c] \cdot \widetilde{G}[c] = 0, \\ \widetilde{G}[c]^{IJ}(x^0, \overline{x}, x^0, \overline{x}') = 0, \\ (\partial x_0 \cdot \widetilde{G}[c]^{IJ}(x, x'))|_{x^0 = x^{0\prime}} = \delta^{IJ}\delta(x, x')|_{x^0 = x^{0\prime}}, \end{cases}$$

$J[c]$ being the Jacobi operator of $\underline{E}_2$. As a consequence, we get the following commutation relations:

$$[\hat{c}^i(x^0, \overline{x}), \hat{c}^j(x^0, \overline{x})] = 0, \qquad [\hat{c}^i(x^0, \overline{x}), \dot{\hat{c}}^i(x^0, \overline{x}')] = i\hbar\delta^{ij}\delta(x, x')|_{x^0 = x^{0\prime}}.$$

References

1. M. Atiyah, *Topological quantum field theories*, Inst. Hautes Études Sci. Publ. Math. **68** (1988), 175–186.
2. ———, *The geometry and physics of knots*, Cambridge Univ. Press, Cambridge, 1990.
3. M. Bachelor, *The structure of supermanifolds*, Trans. Amer. Math. Soc. **253** (1979), 329–338; *Two approaches to supermanifolds* **258** (1980), no. 1, 257–270.
4. F. A. Berezin, *Introduction to superanalysis*, Reidel, Dordrecht, 1987.
5. F. A. Berezin and D. Leites, *Supermanifolds*, Dokl. Akad. Nauk SSSR **224** (1975), no. 3, 505–508; English transl. in Soviet. Math. Dokl. **16** (1975).
6. N. Bourbaki, *Algebre*, Hermann, Paris, 1970.
7. ———, *Theorie spectrales, Chap.* 1 *et Chap.* 2, Hermann, Paris, 1969.
8. A. Connes, *Noncommutative differential geometry*, Inst. Hautes Études Sci. Publ. Math. **62** (1986), 44–144.
9. B. de Witt, *Supermanifolds*, Cambridge Univ. Press, Cambridge, 1984.
10. A. T. Fomenko, *Differential geometry and topology*, Plenum, New York, 1987.
11. G. W. Gibbons and S. W. Hawking, *Selection rules for topology change*, Comm. Math. Phys. **148** (1992), 345–352.
12. H. Goldschmidt, *Integrability criteria for systems of nonlinear partial differential equations*, J. Differential Geom. **1** (1967), 269–307.
13. A. Grothendieck and J. Dieudonné, *Elements de géometrie algébrique.* I: *Le language des schémes*, Inst. Hautes Études Sci. Publ. Math. **4** (1960), 1–228; II: *Étude globale élémentaire de quelques classes de morphismes* **8** (1961), 1–222; III: *Étude cohomologique des faisceaux cohérents* **17** (1963), 137–223; IV: *Étude locale des schémeas et des morphismes de schémas* **32** (1967), 1–361.
14. J. Hilton and V. Stanbach, *A course in homological algebra*, Springer-Verlag, Berlin, 1971.
15. M. W. Hirsch, *Differential topology*, Springer-Verlag, Berlin, 1976.
16. A. Jadczyk and K. Pilch, *Superspaces and supersymmetries*, Comm. Math. Phys. **78** (1981), 373–390.
17. M. V. Karasev and V. P. Maslov, *Asymptotic and geometric quantization*, Uspekhi Mat. Nauk **39** (1984), no. 6, 133–205; English transl. in Russian Math. Surveys **39** (1984).
18. H. H. Keller, *Differential calculus in locally convex spaces*, Lecture Notes in Math., vol. 417, Springer-Verlag, Berlin, 1974, pp. 1–142.
19. I. S. Krasil′shchik, V. V. Lychagin, and A. M. Vinogradov, *Geometry of jet spaces and nonlinear partial differential equations*, Gordon and Breach, New York, 1986.

20. D. A. Leites, *Introduction to the theory of supermanifolds*, Uspekhi Mat. Nauk; English transl. in Russian Math. Surveys **35** (1980), no. 1, 1–64; English transl. in Russian Math. Surveys **35** (1980).
21. V. Lychagin, *Geometric singularities of solutions of nonlinear differential equations*, Soviet Math. Dokl. **24** (1981), no. 3, 680–685; *The geometry and topology of shock waves* **25** (1982), no. 3, 685–689; *Characteristic classes of solutions of differential equations* **28** (1983), no. 1, 275–279.
22. ______, *Geometric theory of singularities of solutions of nonlinear differential equations*, J. Soviet Math. **51** (1990), no. 6, 2735–2757.
23. V. Lychagin and A. Prástaro, *Singularities of Cauchy data, characteristics, cocharacteristics and integral cobordism*, J. Diff. Geom. Appl. (to appear).
24. Yu. I. Manin, *Holomorphic supergeometry and Yang–Mills superfields*, Sovremennye Problemy Matematiki. Itogi Nauki i Tekhniki, vol. 24, VINITI, Moscow, 1984, pp. 3–80; English transl., J. Soviet Math. **30** (1985), no. 2, 1927–1975.
25. ______, *New exact solutions and cohomology analysis of ordinary and supersymmetric Yang–Mills equations*, Trudy Mat. Inst. Steklov. **165** (1984), 98–114; English transl. in Proc. Steklov Inst. Math. **1985**, no. 3; *Some applications of algebraic geometry*, Trudy Mat. Inst. Steklov. **168** (1984), 110–132; English transl. in Proc. Steklov Inst. Math. **1986**, no. 3.
26. V. Marino and A. Prástaro, *On the geometric generalization of the Noether theorem*, Lecture Notes in Math., vol. 1209, Springer-Verlag, Berlin, 1986, pp. 222–234.
27. ______, *On the conservation laws of PDE's*, Rep. Math. Phys. **26** (1987–88), no. 2, 211–225.
28. J. McCleary, *User's guide to spectral sequences*, Publish or Perish in., Wilmington, DE, 1985.
29. J. Milnor, *Morse theory*, Princeton Univ. Press, Princeton, NJ, 1973.
30. J. Milnor and J. D. Stasheff, *Characteristic classes*, Ann. of Math. Stud., vol. 76, Princeton Univ. Press, Princeton, NJ, 1974.
31. D. G. Northcott, *An introduction to homological algebra*, Cambridge Univ. Press, Cambridge, 1962.
32. A. Prástaro, *On the general structure of continuum physics.* I: *Derivative spaces*, Boll. Un. Mat. Ital. **5** (1980), no. 17-B, 704–726; II: *Differential operators* **5** (1981), no. FM-S, 69–106; III: The physical picture **5** (1981), no. FM-S, 107–129.
33. ______, *Spinor super bundles of geometric objects on* spin^G *space-time structures*, Boll. Un. Mat. Ital. **6** (1982), no. 1-B, 1015–1028.
34. ______, *Gauge geometrodynamics*, Riv. Nuovo Cimento **5** (1982), no. 4, 1–122.
35. ______, *Geometrodynamics of nonrelativistic continuous media.* I: *Space–time structures*, Rend. Sem. Mat. Univ. Politec. Torino **40** (1982), no. 2, 89–117; II: *Dynamic and constitutive structures* **43** (1985), no. 1, 99–116.
36. ______, *Geometry and existence theorems for incompressible fluids*, Geometrodynamics Proceedings, 1983 (A. Prástaro, ed.), Pitagora Ed., Bologna, 1984, pp. 65–89.
37. ______, *A geometric point of view for the quantization of nonlinear field theories*, Atti VI Conv. Naz. Relativitá. Firenze 1984, Pitagora Ed., Bologna, 1986, pp. 289–292.
38. ______, *Dynamic conservation laws*, Geometrodynamics Proceedings (A. Prástaro, ed.), World Scientific, Singapore, 1985, pp. 283–420.
39. ______, *Quantum gravity and group-model gauge theory*, Journées Relativistes de Toulouse (1986), Univ. P. Sabatier, Toulouse, 213–222.
40. ______, *On the quantization of Newton equation*, Atti IX Congresso AIMETA, Bari 1988, AIMETA, 1988, pp. 13–16.
41. ______, *Wholly cohomological PDE's*, International Conference on Differential Geometry and Applications, Dubrovnik 1988, Univ. Beograd & Univ. Novi Sad, 1989, pp. 305–314.
42. ______, *Geometry of quantized PDE's*, Differential Geometry and Applications (J. Janyska and D. Krupka, eds.), World Scientific, Singapore, 1990, pp. 392–404.
43. ______, *On the singular solutions of PDE's*, Atti X Congresso AIMETA, Pisa 1990, AIMETA, 1990, pp. 17–20.
44. ______, *Cobordism of PDE's*, Boll. Un. Mat. Ital. **7** (1991), no. 5-B, 977–1001.
45. ______, *Quantum geometry of PDE's*, Rep. Math. Phys. **30** (1991), no. 3, 273–352.
46. ______, *Geometry of super PDE's*, Geometry in Partial Differential Equations (A. Prástaro and Th. Rassias, eds.), World Scientific, Singapore (to appear).
47. ______, *Quantum geometry of super PDE's* (to appear).
48. A. Prástaro and T. Regge, *The group structure of supergravity*, Ann. Inst. H. Poincaré **44** (1986), no. 1, 39–89.

49. A. Rogers, *A global theory of supermanifolds*, J. Math. Phys. **21** (1980), no. 6, 1352–1365; *Some examples of compact supermanifolds with non-Abelian fundamental group* **22** (1981), no. 3, 443–444.
50. M. Rothshtein, *The axioms of supermanifolds and a new structure arising from them*, Trans. Amer. Math. Soc. **297** (1986), no. 1, 159–180.
51. A. Roux, *Jets et connections*, Publ. Dipt. Math. Lyon **7** (1970), no. 4, 1–43.
52. H. Schaefer, *Topological vector spaces*, Springer-Verlag, New York, 1971.
53. M. Schreiber, *A functional calculus for general operators in Hilbert space*, Trans. Amer. Math. Soc. **87** (1958), 108–118.
54. A. S. Schwarz, *Supergravity complex geometry and G-structures*, Comm. Math. Phys. **87** (1982), 37–63; *Geometry of $N = 1$ supergravity* **95** (1984), 161–184; *Geometry of $N = 1$ supergravity* (II) **96** (1984), 161–184.
55. R. Switzer, *Algebraic topology–homotopy and homology*, Springer-Verlag, Berlin, 1975.
56. P. Van Nieuvwenhuizen, *Supergravity*, Phys. Rep. **68** (1981), no. 4, 189–398.
57. A. M. Vinogradov, *The geometry of nonlinear differential equations*, Problemy Geometrii, vol. 11, VINITI, Moscow, 1980, pp. 89–134; English transl., J. Soviet Math. **17** (1981), no. 1, 1624–1649.
58. ———, *Category of nonlinear differential equations*, Global Analysis-Studies and Applications, Lect. Notes Math. (Yu. G. Borisovich and Yu. E. Gliklikh, eds.), Lecture Notes in Math., vol. 1108, Springer-Verlag, Berlin, 1984, pp. 77–102.
59. ———, *The C-spectral sequence, Lagrangian formalism, and conservation laws.* I; II, J. Math. Anal. Appl. **100** (1981), no. 1, 1–129.
60. E. Witten, *Constraints and supersymmetry breaking*, Nucl. Phys. B **202** (1982), 253–316; *Dimensional gravity as an exactly solvable system* **311** (1988/89), 46–78.
61. ———, *Supersymmetry and Morse theory*, J. Differential Geom. **17** (1982), no. 4, 661–692.
62. ———, *Elliptic geners of quantum field theory*, Comm. Math. Phys. **109** (1987), 525–536; *Topological quantum field theory* **117** (1988), 353–386; *Topological sigma model* **118** (1988), 411–449; *Quantum field theory and the Jones polynomial* **121** (1989), 351–399.

UNIVERSITÁ DEGLI STUDI DI ROMA "LA SAPIENZA" DIPARTIMENTO DI METODI E MODELLI MATEMATICI PER IE SCIENZE APPLICATE, VIA A. SCARPA, 10–00161 ROMA–ITALY

Amer. Math. Soc. Transl.
(2) Vol. **167**, 1995

Symmetries of Linear Ordinary Differential Equations

ALEXEY V. SAMOKHIN

ABSTRACT. The description of the full symmetry algebra of the general nth order system of linear ordinary differential equations is obtained: it is isomorphic to the space of all vector fields on the linear space of solutions. The isomorphism is given by explicit formulas as well as in terms of generators of the natural sl and gl Lie subalgebras. The dimension of the point symmetry algebra is assessed and its structure studied.

§1. Introduction

This paper is concerned mainly with symmetries of systems of the form

$$L\mathbf{f}(x) \equiv \left[\left(\frac{d}{dx}\right)^n + \sum_0^{n-1} A_i(x)\left(\frac{d}{dx}\right)^i\right]\mathbf{f}(x) = \mathbf{g}(x), \tag{1.1}$$

where the A_i's are $(m \times m)$ matrices and $\mathbf{f}$, $\mathbf{g}$ are m-vectors. Throughout this paper "symmetry" means higher infinitesimal symmetry, see [1]. This concept can be briefly described as follows. The right-hand side of the evolution equation

$$\partial\mathbf{f}/\partial\tau = \mathcal{G}(x, \mathbf{f}, \mathbf{f}', \dots, \mathbf{f}^{(N)}) \tag{1.2}$$

defines a symmetry if

$$L\mathcal{G}|_{L\mathbf{f}=\mathbf{g}} = 0,$$

where d/dx acts on $\mathcal{G}$ as the total derivative.

Now, a point symmetry is a function $\mathcal{G}$ of the form

$$\mathcal{G} = \mathbf{e}(x, \mathbf{f}) + \mathbf{f}'\xi(x, \mathbf{f}). \tag{1.3}$$

Point symmetries were introduced by S. Lie, who obtained the first results for second order equations in the scalar ($m = 1$) case. He established in [2] that the dimension of the point symmetry algebra in this case is not greater than 8, and computed its generators for the equation $y'' = 0$. It was shown in [3] that this algebra is exactly 8-dimensional. In [4] the generators of the algebra where found for $y'' + ky = 0$,

1991 *Mathematics Subject Classification*. 58F37, 34A30.

Key words and phrases. Ordinary differential equations, systems of ordinary differential equations, point symmetries, higher symmetries.

$k \in \mathbb{R}$, and it was shown in [5] that they generate $\mathrm{sl}(3, \mathbb{R})$. For the general second order ODE $y'' + a(x)y' + b(x)y = 0$ it was found in [6] that its point symmetry algebra is $\mathrm{sl}(3, \mathbb{R})$ too, and its generators were explicitly presented. More recent papers [7–11] deal with different aspects of point symmetries of ordinary equations. The most complete results were obtained in [11] for nth order $(n > 2)$ scalar differential equations: they include an evaluation of the dimension of the symmetry algebra (it can be $n+1$, $n+2$, or $n+4$), a description of some subalgebras, and important applications concerning the possibility of linearization.

Section 2 of this paper contains an explicit description of the full symmetry algebra of an $(m \times m)$ system of ODE's (1.1) (which is isomorphic to the space of vector fields on the space of solutions of (1.1)) and its linear sl and gl subalgebras. These results are extended to the more general case of the operator

$$L = \sum_{i=0}^{n} B_i(x)\left(\frac{d}{dx}\right)^i, \qquad \operatorname{rank} B_n < m, \tag{1.4}$$

in (1.1).

There is one simple reason why point symmetries are studied separately and more extensively. One may try to construct a flow on solutions using (1.2), starting with some known solution $\mathbf{f}(x, \tau)|_{\tau=0}$. Yet it is not *a priori* a well-posed problem except for point symmetries (i.e., for $\mathcal{G}$ of the form (1.3)). Indeed the problem

$$\mathbf{f}_\tau = \mathcal{G}(\mathbf{f}, \mathbf{f}', \dots, \mathbf{f}^{(N)}), \qquad \mathbf{f}|_{\tau=0} = \mathbf{f}_0$$

generally does not possess a unique solution (take, for instance, the heat equation). However, in the particular case of linear ODE's there is no reason to distinguish between point and higher symmetries (since recent results of Chetverikov [12] imply that both types generate flows on the finite-dimensional linear space of solutions). Still there exists a historically motivated interest in point symmetries as shown by the references cited above.

In §3, the dimension of the point symmetry algebra is evaluated and its algebraic structure studied. In particular, this dimension is proved to be at least $mn + 1$ and at most $(m+n)m + 3$. The algebra itself is

$$\underbrace{\mathcal{B} \oplus \cdots \oplus \mathcal{B}}_{mn} \oplus_s (\mathcal{A} \oplus Z\{A_0, \dots, A_{n-1}\}),$$

where $\mathcal{B}$ is a one-dimensional abelian algebra, $\dim \mathcal{A} \leqslant 3$, and $Z \subset \mathrm{gl}(mn, \mathbb{R})$ is the centralizer of the coefficient matrices $\{A_i\}$; $\oplus_s$ stands for a semidirect sum. The maximal symmetry algebra for a nondegenerate equation (1.1) is

$$\underbrace{\mathcal{B} \oplus \cdots \oplus \mathcal{B}}_{mn} \oplus_s (\mathrm{sl}(2, \mathbb{R}) \oplus \mathrm{gl}(m, \mathbb{R})).$$

For the operator (1.4), the dimension of the point symmetry algebra of the equation $L\mathbf{f} = 0$ is at least $d + z$, where $d = \dim \operatorname{Ker} L = \sum \operatorname{rank} B_i$ and $z = \dim Z\{B_0, \dots, B_n\} \subset \mathrm{gl}(d, \mathbb{R})$.

§2. Full symmetry algebra

The equation (1.1), $L\mathbf{f} = \mathbf{g}$, is equivalent to the equation

$$L\widetilde{\mathbf{f}}(x) = \left[\left(\frac{d}{dx}\right)^n + \sum_0^{n-1} A_i(x)\left(\frac{d}{dx}\right)^i\right]\widetilde{\mathbf{f}}(x) = 0. \tag{2.1}$$

This equivalence is realized by the invertible pointwise transformation

$$\widetilde{\mathbf{f}}(x) = \mathbf{f}(x) - \mathbf{f}^*(x),$$

where $\mathbf{f}^*(x)$ is a solution of (1.1). This transformation may be interpreted as a transformation of the jet space $J^n(\mathbb{R}, \mathbb{R}^m)$. We denote coordinates in $J^n(\mathbb{R}, \mathbb{R}^m)$ by x, $\mathbf{p}^0, \dots, \mathbf{p}^n$. For any function $\mathbf{f}(x) = (f_1(x), \dots, f_m(x))$ with m-vector values the formulas

$$\mathbf{p}^i = \mathbf{f}^{(i)}(x), \qquad i = 1, \dots, n$$

define the jet section $j_n(f)$ of the natural projection

$$\pi\colon J^n(\mathbb{R}, \mathbb{R}^m) \to \mathbb{R}, \qquad \pi(x, \mathbf{p}^0, \dots, \mathbf{p}^n) = x.$$

The invertible transformation from (1.1) to (2.1) in these coordinates is

$$x = x, \quad \widetilde{\mathbf{p}}^i = \mathbf{p}^i - (\mathbf{f}^*)^{(i)}, \qquad i = 1, \dots, n.$$

(However, for simplicity we omit tildes for the solutions of (2.1).) The equation (2.1) defines a submanifold $\mathcal{E}$ in $J^n(\mathbb{R}, \mathbb{R}^m)$ by the formula

$$\mathbf{p}^n + \sum_{i=0}^{n-1} A_i(x)\mathbf{p}^i = 0.$$

Prolongations of $\mathcal{E}$, denoted by $\mathcal{E}^{(N)} \subset J^{n+N}(\mathbb{R}, \mathbb{R}^m)$, are defined by

$$\mathbf{p}^{n+k} + D^k\left(\sum_{i=0}^{n-1} A_i(x)\mathbf{p}^i\right) = 0, \qquad 0 \leqslant k \leqslant N \leqslant \infty,$$

where

$$D = \frac{\partial}{\partial x} + \sum_{i=0}^{\infty}\sum_{k=1}^{m} p_k^{i+1}\frac{\partial}{\partial p_k^i}$$

is the total differentiation with respect to x.

The space of solutions of (2.1) is isomorphic to $\mathbb{R}^d$, $d = mn$, since n initial conditions on m-vectors $\mathbf{f}^{(0)}, \dots, \mathbf{f}^{(n-1)}$ define a unique solution. Let us fix some basis $\{\mathbf{R}_i \mid 1 \leqslant i \leqslant mn\}$ in the space $\operatorname{Ker} L$ of solutions of (2.1). Then for an arbitrary solution $\mathbf{f}$ of (2.1) we have

$$\mathbf{f} = \sum_{i=1}^{d} c_i\mathbf{R}_i, \tag{2.2}$$

where $c_i = c_i(\mathbf{f}) \in \mathbb{R}$ are constants. The dependence of c on $\mathbf{f}$ is explicitly given by Wronskians:

$$c_i = W_i(\mathbf{f})/W. \tag{2.3}$$

Here $W = W(\mathbf{R}_1, \ldots, \mathbf{R}_i, \ldots, \mathbf{R}_d)$,

$$W = \det \begin{pmatrix} \mathbf{R}_1 & \ldots & \mathbf{R}_i & \ldots & \mathbf{R}_d \\ \mathbf{R}_1' & \ldots & \mathbf{R}_i' & \ldots & \mathbf{R}_d' \\ \vdots & \ddots & \vdots & \ddots & \vdots \\ \mathbf{R}_1^{(n-1)} & \ldots & \mathbf{R}_i^{(n-1)} & \ldots & \mathbf{R}_d^{(n-1)} \end{pmatrix} \tag{2.4}$$

and $W_i(f) = W(\mathbf{R}_1, \ldots, \mathbf{f}(x), \ldots, \mathbf{R}_d)$,

$$W_i(\mathbf{f}) = \det \begin{pmatrix} \mathbf{R}_1 & \ldots & \mathbf{f}(x) & \ldots & \mathbf{R}_d \\ \mathbf{R}_1' & \ldots & \mathbf{f}'(x) & \ldots & \mathbf{R}_d' \\ \vdots & \ddots & \vdots & \ddots & \vdots \\ \mathbf{R}_1^{(n-1)} & \ldots & \mathbf{f}^{(n-1)}(x) & \ldots & \mathbf{R}_d^{(n-1)} \end{pmatrix}. \tag{2.5}$$

The correspondences $\mathbf{f} \mapsto W_i(\mathbf{f})$ or $\mathbf{f} \mapsto c_i(\mathbf{f})$ are linear ordinary operators of order $n-1$. Let us interpret them as functions on the jet space $J^{n-1}(\mathbb{R}, \mathbb{R}^m)$. Define

$$\widetilde{W}_i = \det \begin{pmatrix} \mathbf{R}_1 & \ldots & \mathbf{p}^0 & \ldots & \mathbf{R}_d \\ \mathbf{R}_1' & \ldots & \mathbf{p}^1 & \ldots & \mathbf{R}_d' \\ \vdots & \ddots & \vdots & \ddots & \vdots \\ \mathbf{R}_1^{(n-1)} & \ldots & \mathbf{p}^{(n-1)} & \ldots & \mathbf{R}_d^{(n-1)} \end{pmatrix}. \tag{2.6}$$

Then $W_i(\mathbf{f}) = \widetilde{W}|_{j_{n-1}(\mathbf{f})}$.

Now the accurate definition of a symmetry in terms of jet spaces in this situation is as follows.

Definition (cf. [1]). A function $\mathcal{G} \in C^\infty(J^N(\mathbb{R}, \mathbb{R}^m))$ is an N-order symmetry of (2.1) if

$$\left[D^n + \sum_0^{n-1} A_i(x) D^i\right] \mathcal{G}\Big|_{\mathcal{E}^{(\infty)}} = 0. \tag{2.7}$$

Note that elements of the differential ideal $\mathcal{J} = \{\mathbf{p}^n + \sum A_i \mathbf{p}^i\,;\, D\} \subset C^\infty(J^\infty)$ generated by $\mathbf{p}^n + \sum A_i \mathbf{p}^i$ and the differentiation D satisfy (2.7). Elements of $\mathcal{J}$ define trivial symmetries (i.e., trivial flows on $\operatorname{Ker} L$). We denote by $\operatorname{Sym} L$ the algebra of all symmetries and by $\operatorname{sym} L = \operatorname{Sym} L/\mathcal{J}$ the algebra of nontrivial symmetries. The latter includes the point symmetry algebra $\operatorname{sym}_0 L$. Thus any point symmetry must be of the form

$$\mathbf{e}(x, \mathbf{p}^0) + \xi(x, \mathbf{p}^0)\mathbf{p}^1 \tag{2.8}$$

and must satisfy (2.7). Here x, ξ are scalars, $\mathbf{p}^0$, $\mathbf{p}^1$, and $\mathbf{e}$ are m-vectors.

The main result of this section is

THEOREM 1. *The full symmetry algebra* $\operatorname{sym} L$ *is isomorphic to the algebra of vector fields on the space* $\operatorname{Ker} L$ *of* (2.1) *solutions. The isomorphism is given by the formula*

$$\sum_{i=1}^{d} F_i(\xi_1, \dots, \xi_d)\frac{\partial}{\partial \xi_i} \mapsto \sum_{i=1}^{d} F_i\left(\frac{W_1}{W}, \dots, \frac{W_d}{W}\right)\mathbf{R}_i .$$

PROOF. Any function $\varphi \in C^\infty(J^N)$, $N \geqslant n$, is equivalent modulo $\mathcal{J}$ to some function $\widetilde{\varphi} \in C^\infty(J^{n-1})$, since $\mathbf{p}^{n+k} = -D^k(\sum A_i \mathbf{p}^i) \bmod \mathcal{J}$, $k \geqslant 0$. We have $\varphi(x, \mathbf{p}^0, \dots, \mathbf{p}^N) = \widetilde{\varphi}(x, \mathbf{p}^0, \dots, \mathbf{p}^{n-1}) \bmod \mathcal{J}$. Therefore, internal coordinates on $\mathcal{E}$ (see [1]) are x, $\mathbf{p}^0, \dots, \mathbf{p}^{n-1}$ and all nontrivial symmetries may be interpreted as functions on $J^{n-1}(\mathbb{R}, \mathbb{R}^m)$.

Let $s \in \operatorname{sym} L$, $s \in C^\infty(J^{n-1})$. Then by definition

$$\left(D^n + \sum_0^{n-1} A_i D^i\right)s\Big|_{\mathcal{E}^{(\infty)}} = \left[\left(D^n + \sum_0^{n-1} A_i D^i\right)s\right] \bmod \mathcal{J} = 0.$$

Thus, if $j_\infty(\mathbf{f})(\mathbb{R}) \subset \mathcal{E}^\infty$,

$$\begin{aligned}\left[\left(D^n + \sum A_i D^i\right)s\right]\Big|_{j_\infty(\mathbf{f})} &= \left[\left(\frac{d}{dx}\right)^n + \sum A_i \left(\frac{d}{dx}\right)^i\right](s|_{j_\infty(\mathbf{f})}) \\ &= L(s|_{j_{n-1}(\mathbf{f})}) = 0.\end{aligned} \tag{2.9}$$

Note that $s|_{j_{n-1}(\mathbf{f})} = s|_{j_\infty(\mathbf{f})}$ since $s \in C^\infty(J^{n-1})$. It follows from (2.9) that $s|_{j_{n-1}(\mathbf{f})}$ is a solution of (2.1). Therefore, $\mathbf{f} \mapsto s|_{j_{n-1}(\mathbf{f})}$ is a transformation of $\operatorname{Ker} L$. Using the basis $\{\mathbf{R}_i\}$ and formulas (2.2), (2.3), any such transformation can be written as follows

$$\left(\frac{W_1(\mathbf{f})}{W}, \dots, \frac{W_d(\mathbf{f})}{W}\right) \\ \to \left(F_1\left(\frac{W_1(\mathbf{f})}{W}, \dots, \frac{W_d(\mathbf{f})}{W}\right), \dots, F_d\left(\frac{W_1(\mathbf{f})}{W}, \dots, \frac{W_d(\mathbf{f})}{W}\right)\right)$$

for some F_i's. So

$$s|_{j_{n-1}(\mathbf{f})} = \left[\sum_{i=1}^{d} F_i\left(\frac{\widetilde{W}_1}{W}, \dots, \frac{\widetilde{W}_d}{W}\right)\mathbf{R}_i\right]\Big|_{j_{n-1}(\mathbf{f})}.$$

Here it is possible to omit $j_{n-1}(\mathbf{f})$. For any $y \in J^{n-1}(\mathbb{R}, \mathbb{R}^m)$ there exists a solution $\mathbf{f}(x)$ of (2.1) such that $j_{n-1}(\mathbf{f}) \circ \pi(y) = y$. (Recall that $y = (x_0, \mathbf{p}^0, \dots, \mathbf{p}^{n-1})$, $\pi(y) = x_0$, so that the usual initial conditions are $\mathbf{f}(x_0) = \mathbf{p}^0$, $\mathbf{f}'(x_0) = \mathbf{p}^1, \dots, \mathbf{f}^{(n-1)}(x_0) = \mathbf{p}^{n-1}$.) Thus

$$s = \sum_{i=1}^{d} F_i\left(\frac{\widetilde{W}_1}{W}, \dots, \frac{\widetilde{W}_d}{W}\right)\mathbf{R}_i \tag{2.10}$$

holds anywhere on $J^{n-1}(\mathbb{R}, \mathbb{R}^m) \approx \mathcal{E}$.

The converse, i.e., the fact that any s given by (2.10) is a symmetry of (2.1) (i.e., a solution of (2.7)), can be verified at any point $y \in J^{n-1}$ by using (2.9).

Now all nontrivial symmetries are of the form (2.10). Any such symmetry generates a flow

$$\partial \mathbf{f}/\partial t = \sum F_i(\mathbf{f})\mathbf{R}_i$$

on the space $\operatorname{Ker} L$ of solutions of (2.1). Since $\mathbf{f} = \sum c_i(\mathbf{f})\mathbf{R}_i$, we have

$$\partial c_i/\partial t = F_i(c_1, \dots, c_d), \qquad i = 1, \dots, d \tag{2.11}$$

which is the coordinate representation of an arbitrary vector field $\mathcal{V}$ on $\operatorname{Ker} L$ with respect to the basis $\{\mathbf{R}_i\}$,

$$\mathcal{V} = \sum_1^d F_i \frac{\partial}{\partial c_i}. \tag{2.12}$$

The Lie bracket for the pair of vector fields $\mathcal{V}_k = \sum_{i=1}^d F_{i,k}\partial/\partial c_i$, $k = 1, 2$,

$$\begin{aligned}[\mathcal{V}_1, \mathcal{V}_2] &= \sum_i (\mathcal{V}_1(F_{i,2}) - \mathcal{V}_2(F_{i,1}))\frac{\partial}{\partial c_i} \\ &= \sum_i \sum_j \left(F_{j,1}(c)\frac{\partial F_{i,2}(c)}{\partial c_j} - F_{j,2}(c)\frac{\partial F_{i,1}(c)}{\partial c_j}\right)\frac{\partial}{\partial c_i}\end{aligned}$$

corresponds to the commutator of the flows generated by the symmetries $s_k = \sum F_{i,k}\mathbf{R}_i$, $k = 1, 2$. Since this commutator defines the Lie algebra structure in the set of symmetries, the correspondence

$$\operatorname{sym} L \to \operatorname{Vect} \operatorname{Ker} L, \qquad s \mapsto \sum F_i \partial/\partial c_i$$

is an isomorphism of Lie algebras. □

It is easy to construct representations of various linear subalgebras of $\operatorname{Ker} L$ in $\operatorname{sym} L$ using the isomorphism from Theorem 1. The most important examples are given by

THEOREM 2. *The bases of the representation of* $\mathrm{sl}(d+1, \mathbb{R})$ *and* $\mathrm{gl}(d, \mathbb{R})$ *in* $\operatorname{sym} L$ *consist of*

$$\mathbf{p}^0, \ \mathbf{R}_k, \ (\widetilde{W}_l/W)\mathbf{R}_k, \ (\widetilde{W}_l/W)\mathbf{p}^0, \qquad l, k = 1, \dots, d$$

and

$$(\widetilde{W}_l/W)\mathbf{R}_k, \qquad i, k = 1, \dots, d$$

respectively; $d = \dim \operatorname{Ker} L$.

PROOF. Any solution $\mathbf{f}$ of (2.1) corresponds to a point

$$(c_1, \dots, c_d) = (W_1(\mathbf{f})/W, \dots, W_d(\mathbf{f})/W) \in \mathbb{R}^d \subset \mathbb{R}P^d.$$

The last inclusion is given by

$$(W_1/W, \dots, W_d/W) \to (W, W_1, \dots, W_d).$$

The standard action of $\mathrm{sl}(d+1, \mathbb{R})$ on $\mathbb{R}P^d$ in the affine chart $W \neq 0$ is described as follows. Let $h \in \mathrm{sl}(d, \mathbb{R})$. Then

$$\begin{gathered} h(\mathbf{f}) = \frac{d}{dt}\bigg|_{t=0} \left(\frac{W_1(\mathbf{f}, t)}{W(t)}, \dots, \frac{W_d(\mathbf{f}, t)}{W(t)} \right), \\ (W(t), W_1(\mathbf{f}, t), \dots, W_d(\mathbf{f}, t)) = e^{th}(W, W_1, \dots, W_d). \end{gathered} \tag{2.13}$$

Consider $\mathrm{sl}(d+1, \mathbb{R})$ as the space of null-trace $(d+1) \times (d_1)$ matrices. Its basis consists of $H_{k,l}$, $k \neq l$ and $Q_l = H_{0,0} - H_{l,l}$, where $k, l = 1, \dots, d$ and $H_{i,j}$ is a matrix such that $(H_{i,j})_{p,q} = \delta_i^p \delta_j^q$, δ being the Kronecker symbol. It is readily verifiable that by (2.13) we have

$$\begin{gathered} H_{k,l}(\mathbf{f}) = \frac{W_l(\mathbf{f})}{W} \mathbf{R}_k, \qquad k, l > 0, \\ H_{0,l}(\mathbf{f}) = -\frac{W_l(\mathbf{f})}{W} \sum \frac{W_i(\mathbf{f})}{W} \mathbf{R}_i, \qquad H_{k,0}(\mathbf{f}) = \mathbf{R}_k, \\ Q_l(\mathbf{f}) = -\sum \frac{W_i(\mathbf{f})}{W} \mathbf{R}_i - \frac{W_l(\mathbf{f})}{W} \mathbf{R}_l. \end{gathered}$$

Note that $\sum (W_i(\mathbf{f})/W)\mathbf{R}_i \equiv \mathbf{p}^0$. Indeed, for any $\mathbf{f} \in \operatorname{Ker} L$, $\mathbf{f} = \sum c_i \mathbf{R}_i$ one has

$$\mathbf{p}^0|_{j_{n-1}(\mathbf{f})} = \mathbf{f} \quad \text{and} \quad \sum \frac{W_i(\mathbf{f})}{W} \mathbf{R}_i \equiv \sum c_i \mathbf{R}_i = \mathbf{f}.$$

To prove the statement about the algebra $\mathrm{gl}(d, \mathbb{R})$ it is sufficient to notice that only $(W_l/W)\mathbf{R}_k$ result in linear transformations of $\operatorname{Ker} L$. Indeed, only these elements of $\mathrm{sl}(d+1, \mathbb{R})$ basis are linear in $\mathbf{p}^i$'s. □

REMARK. Theorems 1 and 2 are true in the more general case given by (1.4), that is for

$$L = \sum_{i=0}^{n} B_i(x) \left(\frac{d}{dx} \right)^i, \qquad \operatorname{rank} B_n < m. \tag{2.14}$$

Using invertible transformations of the dependent variables $\mathbf{f}$, one can get the canonical form for the coefficients

$$B_i(x) = \begin{pmatrix} \mathbf{b}_i(x) & 0 & 0 \\ 0 & E_i & 0 \\ 0 & 0 & 0 \end{pmatrix}.$$

Here $\mathbf{b}_i$ is a $(k_i \times k_i)$ matrix, E_i is the identity matrix of order $(k_{i-1} - k_i)$ and $0 = k_n \leqslant k_{n-1} \leqslant \cdots \leqslant k_0 = m$. The space of solutions of (1.4), (2.14) has the dimension $d = \sum k_i$.

Now the statements and the proofs of Theorems 1 and 2 are valid if W_i's are substituted with ω_i's,

$$\omega_i = \det \begin{pmatrix} \mathbf{R}_1 & \dots & \mathbf{R}_i & \dots & \mathbf{R}_d \\ \widetilde{\mathbf{R}}_1' & \dots & \widetilde{\mathbf{R}}_i' & \dots & \widetilde{\mathbf{R}}_d' \\ \vdots & \ddots & \vdots & \ddots & \vdots \\ \widetilde{\mathbf{R}}_1^{(n-1)} & \dots & \widetilde{\mathbf{R}}_i^{(n-1)} & \dots & \widetilde{\mathbf{R}}_d^{(n-1)} \end{pmatrix}, \tag{2.15}$$

where $\{\mathbf{R}_i = (R_{i,1}, \dots, R_{i,m})^t,\ i = 1, \dots, d\}$ is some basis of $\operatorname{Ker} L$ and

$$\widetilde{\mathbf{R}}_i^{(j)} = (R_{i,1}^j, \dots, R_{i,k_j}^{(j)})^t.$$

Note that ω_i has all the properties of a Wronskian. In particular

$$\omega' = -\left(\sum_{i=0}^{n-1} \sum_{s=k_{i-1}+1}^{k_i} (\mathbf{b}_i)_{s,s}\right)\omega,$$

$(\mathbf{b}_i)_{s,s}$ being diagonal elements of $\mathbf{b}_i$. We skip all subsequent tedious verifications.

Theorems 1 and 2 are, of course, of little practical importance since they do not help in solving (2.1) or (2.14) explicitly: to write down symmetries one has to have solutions in advance. Yet it is always illuminating to know the exact relation between an equation and its symmetries.

§3. Point symmetry algebra

What part of the full symmetry algebra of (2.1) is formed by point symmetries? There are $mn + 1$ of them in any case, namely $\mathbf{R}_k$, $k = 1, \dots, mn$ and $\mathbf{p}^0 = \sum (W_i/W)\mathbf{R}_i$. Both types are related to the linearity of (2.1). The action of the $\mathbf{R}_k$ symmetry reduces to adding a multiple of $\mathbf{R}_k$ to solutions of (2.1), while $\mathbf{p}^0$ acts on them as a homothetic transformation. Note that $\mathbf{R}_k$'s form a commutative algebra

$$\bigoplus_1^{mn} \mathcal{B} = \underbrace{\mathcal{B} \oplus \cdots \oplus \mathcal{B}}_{mn}, \qquad \dim_{\mathbb{R}} \mathcal{B} = 1.$$

Adding $\mathbf{p}^0$, we get the semidirect sum $\bigoplus_1^{mn} \mathcal{B} \oplus_s \mathcal{B}$.

In the scalar ($m = 1$) case it was discovered by S. Lie that the maximal dimension of the point symmetry algebra is $n+4$, exceeding $n\cdot 1+1$ by 3. The vector equation $(d/dx)^n \mathbf{f} = 0$, $\mathbf{f} = (f_1, \dots, f_m)$ is demonstrated here to have an $(mn+m^2+3)$-dimensional point symmetry algebra, which is the upper limit for (2.1).

The main result of this section is

THEOREM 3. *Any point symmetry of* (2.1) *is of the form*

$$\mathcal{G} = (-(n-1)\xi'(x)/2 + M)\mathbf{p}^0 + \xi(x)\mathbf{p}^1 + \mathbf{b}(x), \tag{3.1}$$

where ξ is a scalar function, $\mathbf{b}$ is some solution of (2.1) *and M is a constant $(m\times m)$ matrix commuting with all coefficient matrices $A_i(x)$ of* (2.1).

PROOF. It is straightforward enough. We must solve the equation

$$\left[D^n + \sum_0^{n-1} A_i(x) D^i\right](\mathbf{e}(x, \mathbf{p}^0) + \xi(x, \mathbf{p}^0)\mathbf{p}^1)|_{\mathcal{E}} = 0. \tag{3.2}$$

Yet it is not convenient to proceed this way. To simplify further calculations, we put (2.1) in the equivalent form

$$\left[\left(\frac{d}{dx}\right)^n + \sum_0^{n-2} B_i(x)\left(\frac{d}{dx}\right)^i\right]\mathbf{g}(x) = 0, \tag{3.3}$$

where $\mathbf{g} = P\mathbf{f}$ and the matrix P is any nondegenerate solution of the equation $(d/dx)P = -(1/n)A_{n-1}P$, say

$$P = \exp\left(-\frac{1}{n}\int A_{n-1}(x)\,dx\right).$$

Since this transformation is pointwise and invertible, the point symmetry algebras of (2.1) and (3.3) coincide. Now instead of (3.2) one has

$$\left[D^n + \sum_0^{n-2} B_i(x)D^i\right](\mathbf{e}(x, \mathbf{p}^0) + \xi(x, \mathbf{p}^0)\mathbf{p}^1)|_{\widetilde{\mathcal{E}}} = 0. \tag{3.4}$$

Here $\widetilde{\mathcal{E}} \subset J^n$ is the transformed equation $\mathbf{p}^n = -\sum B_i\mathbf{p}^i$.

Using the last equation and the relation $D^r\mathbf{p}^s = \mathbf{p}^{r+s}$, one may conclude that the symbol of the maximal order derivative appearing in (3.4) is $\mathbf{p}^{n-1}$. We start by determining its coefficient.

Rewrite (3.4) will two summands

$$\left\{\left[D^n + \sum_0^{n-2} B_i(x)D^i\right]\mathbf{e}(x, \mathbf{p}^0) + \left[D^n + \sum_0^{n-2} B_i(x)D^i\right]\xi(x, \mathbf{p}^0)\mathbf{p}^1\right\}\Big|_{\widetilde{\mathcal{E}}} = 0.$$

Since $D^i\mathbf{e}(x, \mathbf{p}^0)$ contains $\mathbf{p}^j$ with $j \le i$ only, in the first summand the term $\mathbf{p}^{n-1}$ appears only in $D^n\mathbf{e}(x, \mathbf{p}^0)$. Thus

$$\begin{aligned} D^n\mathbf{e} &= D^{n-1}(\mathbf{e}_x + \mathbf{e}_{\mathbf{p}^0}\mathbf{p}^1) \\ &= \left(\mathbf{e}_{x\mathbf{p}^0}\mathbf{p}^{n-1} + \binom{n-1}{n-2}D\mathbf{e}_{\mathbf{p}^0}\mathbf{p}^{n-1} - D^{n-1}\mathbf{e}\mathbf{p}^1 + \mathbf{e}_{\mathbf{p}^0}\mathbf{p}^{n-1}\right) \bmod J^{n-2}. \end{aligned}$$

Here $\bmod J^{n-2}$ means "*modulo* summands not containing $\mathbf{p}^{n-1}$" (such are the functions on J^{n-2}); $\binom{m}{l}$ is the binomial coefficient. We also point out that $\mathbf{e}_x$ is an m-vector, $\mathbf{e}_x = (e^i_x)$, and $\mathbf{e}_{\mathbf{p}^0}$ is an $(m\times m)$ matrix, $\mathbf{e}_{\mathbf{p}^0} = (e^i_{p^0_j})$, i being the line and j the column number. Using $\mathbf{p}^n = -\sum B_i\mathbf{p}^i$, we get $(\mathbf{e}_{\mathbf{p}^0}\mathbf{p}^n) \bmod J^{n-2} = 0$. So

$$D^n\mathbf{e} = \{[\mathbf{e}_{x\mathbf{p}^0} + (n-1)\mathbf{e}_{\mathbf{p}^0}]\mathbf{p}^{n-1} + \widetilde{\mathbf{e}}_{\mathbf{p}^0\mathbf{p}^0}\mathbf{p}^1\} \bmod J^{n-2}.$$

Here $\mathbf{e}^{x\mathbf{p}^0} = (e^i_{xp^0_j})$ and $\widetilde{\mathbf{e}}_{\mathbf{p}^0\mathbf{p}^0} = (\sum_k e^i_{p^0_j p^0_k} p^{n-1}_k)$ are $(m \times m)$ matrices. Changing the order of summation, we get $\widetilde{\mathbf{e}}_{\mathbf{p}^0\mathbf{p}^0}\mathbf{p}^1 = \widetilde{\widetilde{\mathbf{e}}}_{\mathbf{p}^0\mathbf{p}^0}\mathbf{p}^{n-1}$, where $\widetilde{\widetilde{\mathbf{e}}}_{\mathbf{p}^0\mathbf{p}^0} = (\sum_j e^i_{p^0_j p^0_k} p^1_j)$. Thus the summands containing $\mathbf{p}^{n-1}$ in $[D^n + \sum_0^{n-2} B_i(x)D^i] \times \mathbf{e}(x, \mathbf{p}^0)$ are

$$[\mathbf{e}^{x\mathbf{p}^0} + (n-1)\mathbf{e}^{\mathbf{p}^0} + \widetilde{\widetilde{\mathbf{e}}}_{\mathbf{p}^0\mathbf{p}^0}]\mathbf{p}^{n-1}. \tag{3.5}$$

Now we compute the second summand in (3.4)

$$\text{(3.6)}\qquad \left[D^n + \sum_0^{n-2} B_i(x) D^i\right](\xi(x\,,\,\mathbf{p}^0)\mathbf{p}^1) \bmod J^{n-2}$$
$$= \left\{D^n \xi \mathbf{p}^1 + nD^{n-1}\xi\mathbf{p}^2 + \xi\mathbf{p}^{n+1}\right.$$
$$\left. + nD\xi\mathbf{p}^n + \xi B_{n-2}\mathbf{p}^{n-1} + \binom{n}{2} D^2\xi\mathbf{p}^{n-1}\right\} \bmod J^{n-2}$$
$$= \left\{D^n \xi \mathbf{p}^1 + nD^{n-1}\xi\mathbf{p}^2 + \binom{n}{2} D^2\xi\mathbf{p}^{n-1}\right\} \bmod J^{n-2}.$$

To obtain the last expression, it is sufficient to notice that $\mathbf{p}^n \bmod J^{n-2} = 0$ and $\xi\mathbf{p}^{n+1} = [-\xi B_{n-2}\mathbf{p}^{n-1}] \bmod J^{n-2}$, since $\mathbf{p}^n = -\sum_0^{n-2} B_i\mathbf{p}^i$ and $\mathbf{p}^{n+1} = -D(\sum_0^{n-2} B_i\mathbf{p}^i)$ on $\mathcal{E}^{(\infty)}$.

We proceed with each summand in (3.6) separately:

$$D^n\xi\mathbf{p}^1 = D^{n-1}(\xi_x + \xi_{\mathbf{p}^0}\mathbf{p}^1)\mathbf{p}^1$$
$$= \{(\xi_{x\mathbf{p}^0}\mathbf{p}^{n-1})\mathbf{p}^1 + (n-1)(\mathbf{p}^1\xi_{\mathbf{p}^0\mathbf{p}^0}\mathbf{p}^{n-1})\mathbf{p}^1\} \bmod J^{n-2}.$$

Here $\xi_{x\mathbf{p}^0}\mathbf{p}^{n-1}$ is the scalar product and $\mathbf{p}^1\xi_{\mathbf{p}^0\mathbf{p}^0}\mathbf{p}^{n-1}$ is the value of the bilinear form $\xi_{\mathbf{p}^0\mathbf{p}^0} = (\xi_{\mathbf{p}_i^0\mathbf{p}_j^0})$ on the vectors $\mathbf{p}^1$ and $\mathbf{p}^{n-1}$.

Further

$$nD^{n-1}\xi\mathbf{p}^2 = [(n\xi_{\mathbf{p}^0}\mathbf{p}^{n-1})\mathbf{p}^2] \bmod J^{n-2},$$

where $\xi_{\mathbf{p}^0}\mathbf{p}^{n-1}$ is the scalar product, and

$$\binom{n}{2} D^2\xi\mathbf{p}^{n-1} = \left[\binom{n}{2}(\xi_{xx} + 2\xi_{x\mathbf{p}^0}\mathbf{p}^1 + \xi_{\mathbf{p}^0}\mathbf{p}^2 + \mathbf{p}^1\xi_{\mathbf{p}^0\mathbf{p}^0}\mathbf{p}^1)\mathbf{p}^{n-1}\right] \bmod J^{n-2}.$$

Finally for all summands in (3.6) we obtain

$$\text{(3.7)}\qquad \left[D^n + \sum_0^{n-2} B_i(x) D^i\right][\xi(x\,,\,\mathbf{p}^0)\mathbf{p}^1] \bmod J^{n-2}$$
$$= (\xi_{x\mathbf{p}^0}\mathbf{p}^{n-1})\mathbf{p}^1 + (n-1)(\mathbf{p}^1\xi_{\mathbf{p}^0\mathbf{p}^0}\mathbf{p}^{n-1})\mathbf{p}^1 + (n\xi_{\mathbf{p}^0}\mathbf{p}^{n-1})\mathbf{p}^2$$
$$+ \binom{n}{2}(\xi_{xx} + 2\xi_{x\mathbf{p}^0}\mathbf{p}^1 + \xi_{\mathbf{p}^0}\mathbf{p}^2 + \mathbf{p}^1\xi_{\mathbf{p}^0\mathbf{p}^0}\mathbf{p}^1)\mathbf{p}^{n-1}.$$

Putting together (3.5) and (3.7), we conclude that the left-hand side of (3.4) is an m-vector, its coordinates depending on p_k^i, $1 \leqslant i \leqslant n-1$ polynomially. The $\mathbf{p}^2$ components appear only in (3.7). The kth coordinate of (3.7) is

$$\sum_i \xi_{xp_i^0} p_i^{n-1} p_k^1 + (n-1)(\mathbf{p}^1\xi_{\mathbf{p}^0\mathbf{p}^0}\mathbf{p}^{n-1})\, p_k^1$$
$$+ n\left(\sum_i \xi_{p_i^0} p_i^{n-1}\right) p_k^2 + \binom{n}{2}\left(\xi_{xx} + 2\xi_{x\mathbf{p}^0}\mathbf{p}^1 + \xi_{\mathbf{p}^0}\mathbf{p}^2 + \mathbf{p}^1\xi_{\mathbf{p}^0\mathbf{p}^0}\mathbf{p}^1\right).$$

The last expression is a polynomial of the first order in p_k^2 and it must vanish identically. Hence the coefficients at p_k^2 are zero too, i.e.,

$$n\sum_i \xi_{p_i^0} p_i^{n-1} + p_k^{n-1}\binom{n}{2}\xi_{p_k^0} \equiv 0. \tag{3.8}$$

The left-hand side of (3.8) is linear in p_s^{n-1}, $s = 1, \dots, m$, so

$$n\xi_{p_s^0} + \binom{n}{2}\xi_{p_k^0}\delta_k^s \equiv 0,$$

which, in turn, means that $\xi_{p_s^0} = 0$ for all s. Thus $\xi = \xi(x)$. This implies that all nonzero terms of (3.4) containing $\mathbf{p}^{n-1}$ reduce to the sum

$$\left(\mathbf{e}_{x\mathbf{p}^0} + (n-1)D\mathbf{e}_{\mathbf{p}^0} + \widetilde{\widetilde{\mathbf{e}}}_{\mathbf{p}^0\mathbf{p}^0} + \binom{n}{2}\xi_{xx}\right)\mathbf{p}^{n-1}, \tag{3.9}$$

which must be identically zero. The ith coordinate of the m-vector (3.9) is

$$n\sum_j e^i_{xp_j^0} p_j^{n-1} + n\sum_k\sum_j e^i_{p_j^0 p_k^0} p_k^1 p_j^{n-1} + \binom{n}{2}\xi_{xx} p_i^{n-1}. \tag{3.10}$$

The last expression is linear in p_j^{n-1} for any j and its coefficient in (3.10) is

$$ne^i_{xp_j^0} + n\sum_k e^i_{p_j^0 p_k^0} p_k^1 + \binom{n}{2}\xi_{xx}\delta_j^i. \tag{3.11}$$

This coefficient must also be identically zero. Since (3.11) is linear in p_k^1, this implies

$$e^i_{p_j^0 p_k^0} = 0, \qquad ne^i_{xp_j^0} = -\binom{n}{2}\xi_{xx}\delta_j^i. \tag{3.12}$$

Thus

$$\mathbf{e} = \begin{pmatrix} e^1 \\ \vdots \\ e^m \end{pmatrix} = \begin{pmatrix} \sum_k a_k^1 p_k^0 - \frac{1}{2}(n-1)\xi_x(x)p_1^0 + b^1(x) \\ \vdots \\ \sum_k a_k^m p_k^0 - \frac{1}{2}(n-1)\xi_x(x)p_m^0 + b^m(x) \end{pmatrix}$$

or

$$\mathbf{e} = \mathbf{b}(x) + M\mathbf{p}^0 - \tfrac{1}{2}(n-1)\xi_x(x)\mathbf{p}^0, \tag{3.13}$$

where $M = (a_s^r)$ is a constant $(m \times m)$ matrix, $\xi(x)$ is some scalar function and $\mathbf{b}(x)$ is a vector function. We conclude that any point symmetry in our case must be of the form

$$s = \mathbf{b}(x) + M\mathbf{p}^0 - \tfrac{1}{2}(n-1)\xi_x(x)\mathbf{p}^0 + \xi(x)\mathbf{p}^1. \tag{3.14}$$

Equation (3.14) only states a necessary condition on s. To obtain the finite form for a symmetry, we must substitute $M\mathbf{p}^0$, $\mathbf{b}(x)$, and $\frac{1}{2}(-n+1)\xi_x(x)\mathbf{p}^0+\xi(x)\mathbf{p}^1$ into (3.4) for $\mathbf{e}+\xi\mathbf{p}^1$. We obtain

1. $$\left[D^n+\sum_0^{n-2}B_i(x)D^i\right]M\mathbf{p}^0\Big|_{\mathcal{E}} = \left[M\mathbf{p}^n+\sum_0^{n-2}B_i(x)M\mathbf{p}^i\right]\Bigg|_{\{\mathbf{p}^n=-\sum B_i\mathbf{p}^i\}}$$
$$= -\sum_0^{n-2}MB_i(x)\mathbf{p}^i+\sum_0^{n-2}B_i(x)M\mathbf{p}^i = \sum_0^{n-2}[B_i(x),\, M]\mathbf{p}^i \equiv 0,$$

which is only possible if for all i we have

$$[B_i(x),\, M]=0. \tag{3.15}$$

2. $$\left[D^n+\sum_0^{n-2}B_i(x)D^i\right]\mathbf{b}(x) = \frac{d^n\mathbf{b}(x)}{dx^n}+\sum_0^{n-2}B_i\frac{d^i\mathbf{b}(x)}{dx^n}=0,$$

which means

$$\mathbf{b}(x)\in \operatorname{Ker} L. \tag{3.16}$$

3. $$\left[D^n+\sum_0^{n-2}B_i(x)D^i\right]\left(-\frac{n-1}{2}\,\xi_x(x)\mathbf{p}^0+\xi(x)\mathbf{p}^1\right)\Bigg|_{\{\mathbf{p}^n=-\sum B_i\mathbf{p}^i\}}$$
$$=\sum_{i=2}^{n}K_i\mathbf{p}^{n-i}\equiv 0.$$

The matrix coefficients K_i must be identically zero:

$$\begin{aligned} K_i = {}&\left(\binom{n}{n-i}\frac{n-1}{2}-\binom{n}{n-i-1}\right)\xi^{(i+1)}\\ &+\sum_{j=2}^{i-1}B_{n-j}\xi^{(i-j+1)}\left(\binom{n-i}{n-j}\frac{n-1}{2}-\binom{n-j}{n-i-1}\right)\\ &-B_{n-i}\binom{n-i}{n-i-1}\xi'+n\xi'B_{n-i}+\xi B'_{n-i}=0, \end{aligned} \tag{3.17$_i$}$$

just as in [11], except that the B_i's are matrices instead of scalars.

All $(3.17)_i$ are linear ordinary differential equations of order $i+1$ for the same scalar function $\xi(x)$. Hence the system (3.17) is highly overdetermined. It has at most three independent solutions. (This upper limit is attained in the case when all $(3.17)_{i>2}$ are differential prolongations of the lowest order equation in $(3.17)_2$

$$n(n^2-1)\xi'''+24B_{n-2}\xi'+12B'_{n-2}\xi=0.$$

In this way the B_{n-i}, $i > 2$ are expressed in terms of B_{n-2}; for a complete argument, see [11].) Yet there is a new feature for systems of ODE's of type (3.4). In this case, B_{n-2} is an $(m \times m)$ matrix and $(3.17)_2$ is overdetermined itself (it is not in the scalar case). □

Now we can evaluate the dimension N of the point symmetry algebra $\operatorname{sym}_0 L$. The number of independent constant matrices M in (3.15) is the dimension z of the centralizer $Z\{B_0(x), \dots, B_{n-2}(x)\} \subset \mathbb{R}^{m^2}$ of the matrices $B_0(x), \dots, B_{n-2}(x)$. Obviously, $1 \leqslant z \leqslant m^2$: there are always scalar matrices ($z \geqslant 1$) and the upper limit is attained for instance for scalar B_i's. The dimension of the space of solutions of (3.17) is no more than 3, which is attained in particular for the system of equations $\mathbf{y}^{(n)} = 0$ or in the case of some delicate recursions connecting B_i's as described in [11]. The number of independent **b**'s is the dimension of $\operatorname{Ker} L$, which equals mn. Therefore, we have

COROLLARY 1. *The dimension of the point symmetry algebra of equation* (1.1) *or* (3.3) *satisfies*

$$mn + 1 \leqslant \dim \operatorname{sym}_0 L \leqslant (m+n)m + 3.$$

Three-dimensional algebra generated by solutions of (3.17) is $\mathrm{sl}(2, \mathbb{R})$, see [11]. Therefore we have

COROLLARY 2. *The maximal symmetry algebra for the nondegenerate equation* (1.1) *or, equivalently,* (3.3) *is*

$$\underbrace{\mathcal{B} \oplus \cdots \oplus \mathcal{B}}_{mn} \oplus_s (\mathrm{sl}(2, \mathbb{R}) \oplus \mathrm{gl}(m, \mathbb{R})).$$

Generally, it is

$$\underbrace{\mathcal{B} \oplus \cdots \oplus \mathcal{B}}_{mn} \oplus_s (\mathcal{A} \oplus Z\{B_0, \dots, B_{n-2}\}),$$

where $\mathcal{A}$ *is generated by solutions of* (3.17).

In the case of a degenerate equation (2.14), one can obtain a weaker result in a similar way. Namely, the dimension of the point symmetry algebra for (2.14) is no less than $d + z$, where $d = \dim \operatorname{Ker} L = \sum \operatorname{rank} B_i$ and $z = \dim(Z\{B_0, \dots, B_n\}) \subset \mathrm{gl}(d, \mathbb{R})$.

References

1. A. M. Vinogradov, *Symmetries and conservation laws of partial differential equations: Basic notions and results*, Acta Appl. Math. **15** (1989), 3–21.
2. S. Lie, *Über die Integration durch bestimmte Integrale von einer Klasse linearer partiellen Differentialgleichungen*, Arch. Math. og Naturvid. **6** (1881), 328–359.
3. L. V. Ovsyannikov, *Group analysis of differential equations*, "Nauka", Moscow, 1978; English transl., Academic Press, New York, 1982.
4. R. L. Anderson and S. M. Davidson, *A generalization of Lie's "Counting" theorem for second-order ordinary differential equations*, J. Math. Anal. Appl. **48** (1974), 301–315.
5. C. E. Wulfman and B. G. Wyborne, *The Lie group of Newton's and Lagrange's equations for harmonic oscillator*, J. Phys. A **9** (1976), 507–518.
6. A. V. Samokhin, *Symmetries of Sturm–Liouville equations and the Korteweg–de Vries equation*, Dokl. Akad. Nauk SSSR; English transl., Soviet Math. Dokl. **21** (1980), 488–492.

7. M. Aguirre and J. Krause, $SL(2, \mathbb{R})$ *as a group of symmetry transformations for all one-dimensional linear systems*, J. Math. Phys. **29** (1988), 9–15.
8. F. Gonzales-Gascon and A. Gonzales-Lopez, *Symmetries of differential equations,* IV, J. Math. Phys. **24** (1983), 2006–2021.
9. J. Krause and L. Michel, *Équations différentielles linéares d'ordre* $n > 2$ *ayant une algébre de Lie de symétrie de dimension* $n+4$, C. R. Acad. Sci. Paris Sér. I **307** (1988), 905–910.
10. F. M. Mahomed and P. G. L. Leach, *Lie algebras associated with second order ordinary differential equations*, J. Math. Phys. **30** (1989), 2770–2777.
11. ———, *Symmetry Lie algebras of* n*th order ordinary differential equations*, J. Math. Anal. Appl. **15** (1990), 80–107.
12. V. N. Chetverikov, *On the structure of integrable* C*-fields*, Diff. Geometry Appl. **1** (1991), 309–325.

Translated by THE AUTHOR

MOSCOW INSTITUTE OF CIVIL AVIATION ENGINEERS, DEPT. OF MATH., 6A PULKOVSKAYA ST., MOSCOW, 125493, RUSSIA

Amer. Math. Soc. Transl.
(2) Vol. **167**, 1995

Foliations of Manifolds and Weighting of Derivatives

N. A. SHANANIN

ABSTRACT. The purpose of this paper is to give a definition of the weighting of derivatives in differential operators via a foliation of the given manifold. Relationships between the approach presented here and the coordinate approach are described. Connected notions are adapted appropriately.

1. Foliations of manifolds. Let M be a C^∞-manifold of dimension n, $\mathcal{E}(M)$ the ring of real C^∞-functions on M, $\Lambda(M)$ the algebra of differential forms, $\Lambda^j(M)$ the $\mathcal{E}(M)$-module of j-forms on M. A foliation of the manifold

$$M = M_1 \to M_2 \to \cdots \to M_k \tag{1}$$

consisting of a family of differentiable manifolds M_j and their mappings $M_j \to M_{j+1}$ generates a grading preserving filtration of the algebra $\Lambda(M)$. In particular, the ring $\mathcal{E}(M)$ has the filtration

$$\mathcal{E}(M) = \mathcal{E}_1(M) \supset \mathcal{E}_2(M) \supset \cdots \supset \mathcal{E}_k(M) \supset \mathbb{R}, \tag{2}$$

where the subring $\mathcal{E}_j(M)$ is the image of $\mathcal{E}(M_j)$ induced by the appropriate composition of mappings, and the $\mathcal{E}(M)$-module of 1-forms $\Lambda^1(M)$ has the filtration

$$\Lambda^1(M) = \Lambda^1_1(M) \supset \Lambda^1_2(M) \supset \cdots \supset \Lambda^1_k(M) \supset 0, \tag{3}$$

where $\Lambda^1_j(M)$ is the $\mathcal{E}(M)$-module generated by the image of $\Lambda^1(M_j)$. The filtration (3) generates a filtration of the $\mathcal{E}(M)$-module $\mathcal{T}(M)$of vector fields on M:

$$0 \in \mathrm{Ann}_1(M) \subset \mathrm{Ann}_2(M) \subset \cdots \subset \mathrm{Ann}_k(M) = \mathcal{T}(M),$$

where $\mathrm{Ann}_j(M)$ is the submodule in $\mathcal{T}(M)$ consisting of vector fields that vanish on $\Lambda^1_{j+1}(M)$. Let

$$\widehat{\Lambda}^1(M) = \bigoplus_{j=1}^{k} \Lambda^1_j(M)/\Lambda^1_{j+1}(M) \tag{4}$$

be the associated $\mathcal{E}(M)$-module.

1991 *Mathematics Subject Classification*. Primary 58G99; Secondary 57R30.

2. Regular points of foliations. Let $f: M \to N$ be a differentiable mapping. The rank of a mapping is a lower semicontinuous function with integer values, so the variety of points of the maximal rank is open and everywhere dense in M. We say that a point x is a *regular point* of a mapping if it has a neighborhood in which the rank of the mapping at every point is equal to the rank at the point x. If the point x is regular, then it has a neighborhood whose image is a submanifold in the manifold N. In the case of the foliation (1), we say that a point x is *regular*, if it is regular for all mappings $M \to M_j$. Furthermore, by restricting the neighborhood of a regular point and replacing the manifold M by the proper image of the neighborhood, we can assume (for local purposes) that every mapping $M_j \to M_{j+1}$ is an epimorphism. The following propositions hold.

PROPOSITION 1. *Let every mapping of the foliation* (1) *be an epimorphism. Then the inclusions in* (2) *and* (3) *are imbeddings.*

PROPOSITION 2. *Let the mappings of the foliation* (1) *be regular at all points. Then for every* $j = 1, \dots, k$, *the dimension of the quotient space*

$$d_x\mathcal{E}_j / d_x\mathcal{E}_{j+1} \tag{5}$$

is constant and the vector bundle

$$\widehat{T}^*(M) = \bigcup_x \bigoplus_{j=1}^{k} d_x\mathcal{E}_j / d_x\mathcal{E}_{j+1} \tag{6}$$

is the total space for the $\mathcal{E}(M)$*-module* $\widehat{\Lambda}^1(M)$.

So elements of $\widehat{\Lambda}^1(M)$ are sections of the bundle $\widehat{T}^*(M)$. We denote by n_j the dimension of the quotient space (5).

PROPOSITION 3. *Let* y *be a regular point. Then there exist local coordinates* $(U; x^1, \dots, x^k)$, *where* U *is a neighborhood of the point* y, x^j *is a mapping of the neighborhood* U *into* $\mathbb{R}^{n_j}$, *such that the ring* $\mathcal{E}_j|_U$ *of restrictions of functions is generated by functions of type* $f(x^j, \dots, x^k)$.

If $(U; x^1, \dots, x^k)$ are the local coordinates from Proposition 3, then for every point $y_1 \in U$ and every $j = 2, \dots, k$ the submanifold given by

$$\begin{cases} x^1(y) = x^1(y_1), \\ \dots\dots\dots\dots\dots \\ x^{j-1}(y) = x^{j-1}(y_1), \end{cases} \qquad y \in U,$$

in the manifold M is taken by the foliation mappings onto some neighborhood of the image of a point y_1 in the manifold M_j. The system of local coordinates from Proposition 3 is said to be a system of local coordinates *associated with the foliation*. It is evident that the associated system of coordinates is not defined uniquely by the foliation. All points in the neighborhood U from Proposition 3 are regular. Associated local coordinates for a point exist if and only if this point is regular.

3. Anisotropic differential operators. Let $\mathcal{R}_j$ be a locally trivial vector bundle with base M whose fiber is a real vector space $(j = 1, 2)$. Let $\mathbb{R}^\infty(M, \mathcal{R}_j)$ be the $\mathcal{E}(M)$-module of $C^\infty(M)$ sections. Every function $a(x) \in \mathcal{E}(M)$ defines a differentiation of $\mathcal{E}(M)$-bimodule of $\mathbb{R}$-linear homomorphisms

$$L\colon \mathbb{R}^\infty(M, \mathcal{R}_1) \to \mathbb{R}^\infty(M, \mathcal{R}_2)$$

in the following way

$$(\partial_a L)u = a\,Lu - L(au),$$

where $u \in \mathbb{R}^\infty(M, \mathcal{R}_1)$. We say that an $\mathbb{R}$-homomorphism L is a *differential operator* in the domain $V \subset M$ of order no greater than m if for every family of functions $\{a_1, \dots, a_{m+1}\} \subset \mathcal{E}(M)$ the mapping

$$\partial_{a_1} \cdots \partial_{a_{m+1}} L$$

takes every section $u \in \mathbb{R}^\infty(M, \mathcal{R}_1)$ to a section that vanishes on V.

Now we introduce the weights. For every $j = 1, \dots, k$ we denote by ρ_j the weight corresponding to M_j. We assume that $\rho_j < \rho_{j+1}$ for $j = 1, \dots, k-1$.

DEFINITION 1. An $\mathbb{R}$-homomorphism L is said to be a *differential operator of weighted order of no more than* m in the domain U if for every family of functions $\{a_1^j, \dots, a_{p(j)}^j\} \subset \mathcal{E}_j$ $(j = 1, \dots, k)$ the $\mathbb{R}$-homomorphism $\partial_{a_1^1} \cdots \partial_{a_{p(k)}^k} L$ maps $\mathbb{R}^\infty(M, \mathcal{R}_1)$ to sections of $\mathbb{R}^\infty(M, \mathcal{R}_2)$ that vanishes on U, and

$$\rho_1 p(1) + \cdots + \rho_k p(k) > m.$$

Weighted order differential operators have a specific form in coordinate systems associated with the foliation. Let $(U; x^1, \dots, x^k)$ be a local coordinate system associated with the foliation. The weighted order $\langle \alpha, \rho \rangle$ with multi-indices $(\alpha^1, \dots, \alpha^k)$ is defined by

$$\langle \alpha, \rho \rangle = |\alpha^1|\rho_1 + \cdots + |\alpha^k|\rho_k,$$

where $\alpha^j \in (\mathbb{Z}_+)^{n_j}$, $|\alpha^j| = \alpha_1^j + \cdots + \alpha_{n_j}^j$. Let $U \times \mathbb{R}^l$ and $U \times \mathbb{R}^r$ be trivializations of the bundles $\mathcal{R}_1$ and $\mathcal{R}_2$ on U. We assume that U is sufficiently small.

THEOREM 1. *Let the weighted order of an* $\mathbb{R}$*-homomorphism* L *be no more than* m *in a neighborhood* U *of a regular point* x_0. *Then there exists neighborhood* $V \subset U$ *of a point* x_0 *such that*

(1) *the relation* $(\operatorname{supp}(u)) \cap V = \varnothing$ *implies* $(\operatorname{supp}(Lu)) \cap V = \varnothing$ *for every section* u;

(2) *the operator* $\widetilde{L} = \varkappa_2^* \circ L \circ (\varkappa_1^{-1})\colon \mathbb{R}_0^\infty(V, \mathbb{R}^l) \to \mathbb{R}_0^\infty(V, \mathbb{R}^r)$ *in the local coordinate system* $(V; x^1, \dots, x^k)$ *associated with the foliation has the form*

$$\widetilde{L} = \sum_{\langle \alpha, \rho \rangle \leqslant m} a_\alpha(x) \partial_x^\alpha,$$

where $\varkappa_1\colon V\times\mathbb{R}^l\to\mathcal{R}_1$ *and* $\varkappa_2\colon V\times\mathbb{R}^r\to\mathcal{R}_2$ *are canonical mappings,* $a_\alpha(x)$ *are* $r\times l$*-matrices.*

This theorem is proved in [1].

If the weighted order of L does not exceed m and there exist families of functions $\{a_1^j,\dots,a_{p(j)}^j\}\subset\mathcal{E}_j$ $(j=1,\dots,k)$ such that $\rho_1 p(1)+\dots+\rho_k p(k)=m$ and

$$(\partial_{a_1^1}\cdots\partial_{a_{p(k)}^k}L)\mathbb{R}_0^\infty(U,\mathcal{R}_1)|_U\neq 0,$$

then the number m is called the *weighted order* $\operatorname{ord}_{\mathrm{w}}L$ of the operator L in the domain U.

Let $\Phi=\{\varphi_1,\dots,\varphi_k\}$ be a family of smooth functions, where $\varphi_j\in\mathcal{E}_j(M)$. We define the mapping

$$\widehat{d}\colon\bigoplus_{j=1}^k\mathcal{E}_j(M)\to\widehat{\Lambda}^1(M),\qquad \widehat{d}_x\Phi=\bigoplus_{j=1}^k(d_x\varphi_j/d_x\mathcal{E}_{j+1}(M))\in\widehat{T}_x^*U.$$

Now assume that $\mathcal{R}_j$ is a locally trivial vector bundle with base M and fiber a complex vector space $(j=1,2)$. The construction of the weighted differential operator developed above can be applied to $\mathbb{C}$-homomorphisms

$$L\colon\mathbb{C}^\infty(M,\mathcal{R}_1)\to\mathbb{C}^\infty(M,\mathcal{R}_2),$$

because $\mathbb{R}\subset\mathbb{C}$.

The next fact allows us to introduce the notion of the leading part of a weighted principal symbol. It plays an important role in questions of solvability and smoothness of solutions.

PROPOSITION 4. *Let the manifold* M *be regular with respect to the foliation* (1). *Then there exists a* $\mathbb{C}$*-homomorphism*

$$\sigma(L)\colon\pi^*(\mathcal{R}_1)\to\pi^*(\mathcal{R}_2),$$

where $\pi\colon T^*M\to M$ *is the canonical projection, such that*

$$L(e^{i\Phi_\lambda}u)=e^{i\Phi_\lambda}(\lambda^m\sigma(L,\widehat{d}_x\Phi)u+o(\lambda^m)),\qquad \Phi_\lambda=\sum_{j=1}^k\lambda^{\rho_j}\varphi_j.$$

The construction of asymptotic solutions requires a more precise formula for the action of the anisotropic operator on oscillatory functions. Let $\mu=\inf(\rho_j)$ and $T^*_{\mathrm{ext}}(M)$ be the sum of cotangent bundles

$$T^*_{\mathrm{ext}}(M)=\bigoplus_{j=1}^k T^*M_j.$$

A *principal weighted symbol* $\mathcal{H}(L,d\Phi,\lambda^{-1})$ is a polynomial of degree $\mu-1$ with respect to λ^{-1} defined by the congruence

$$\mathcal{H}(L,d\Phi,\lambda^{-1})u=\lambda^{-m}e^{-i\Phi_\lambda}Le^{i\Phi_\lambda}u\mod(\lambda^{-m}),$$

where $d\Phi = (x, d\varphi_1, \dots, d\varphi_k) \in T^*_{\text{ext}}(M)$. Its coefficients are $\mathbb{C}$-homomorphisms $a\colon \pi^*_{\text{ext}}(\mathcal{R}_1) \to \pi^*_{\text{ext}}(\mathcal{R}_2)$, where $\pi_{\text{ext}}\colon T^*_{\text{ext}}(M) \to M$ is the canonical projection. The fact that the symbol $\mathcal{H}(L, d\Phi, \lambda^{-1})$ is well defined follows from Theorem 1. It is evident that

$$\sigma(L, \widehat{d\Phi}) = \mathcal{H}(L, d\Phi, 0).$$

In parallel with the principal weighted symbol below, we need an analog of the transport operator. It is a $\mathbb{C}$-linear differential operator of the first order whose coefficients are polynomials of degree $\mu - 1$ in λ^{-1} with values in $\mathbb{C}$-homomorphisms $a\colon \pi^*_{\text{ext}}(\mathcal{R}_1) \to \pi^*_{\text{ext}}(\mathcal{R}_2)$. More precisely, this operator is defined by the congruence

$$D(L, x, \xi, \lambda^{-1}) = \sum_{l=1}^{k} \sum_{j=1}^{n_l} \mathcal{H}_{\xi^l_j}(L, x, \xi, \lambda^{-1}) \lambda^{\mu-\rho_l} D_{x^l_j} \mod (\lambda^{-\mu})$$

in the associated local coordinates. We have the following

THEOREM 2. *Let the manifold M be regular with respect to the foliation* (1), $\operatorname{ord}_{\mathrm{w}} L = m$. *Then*

$$L(e^{i\Phi_\lambda}u) = e^{i\Phi_\lambda}\big(\lambda^m \mathcal{H}(L, d\Phi, \lambda^{-1})u + \lambda^{m-\mu}(D(L, d\Phi, \lambda^{-1}) + r(L, d\Phi, d^2\Phi, \lambda^{-1}))u + O(\lambda^{m-2\mu})u\big),$$

where r is a $\mathbb{C}$-linear mapping of fibers (when x and Φ are fixed) and the remainder $O(\lambda^{m-2\mu})$ represents a $\mathbb{C}$-linear differential operator whose coefficients are polynomials of degree $m - 2\mu$ in λ with values in $\mathbb{C}$-homomorphisms $\mathcal{R}_1 \to \mathcal{R}_2$.

4. Foliations of manifolds and representations of the multiplicative group $\mathbb{R}_+$ in the fibers of the cotangent bundle. Let x_0 be a regular point. The weight ρ_j corresponds to the manifold M_j from the foliation (1). The weight ρ_j is a positive number and $\rho_j < \rho_{j+1}$ for $j = 1, \dots, k-1$. In accordance with Proposition 3, there exists a local coordinate system $(U; x^1, \dots, x^k)$ associated with the foliation (1). This local coordinates induces the coordinate system (x, ∂_x) on the open set $T^*U \subset T^*M$. We define an action of the multiplicative group $\mathbb{R}_+$ in the coordinates (x, ∂_x) in following way

$$\lambda \circ (x(v), \partial_x(v)) = (x(v), \lambda^{\rho_1}\partial_{x^1}(v), \dots, \lambda^{\rho_k}\partial_{x^k}(v)), \tag{7}$$

where $v \in T^*U$. Because in the general case there are many associated local coordinates at every regular point, the representation of the group $\mathbb{R}_+$ is not uniquely defined by the weights and the foliation. The representation constructed above is local.

In fact a weighting for the investigation of anisotropic operators is used in the local coordinate approach. The analysis is developed in a fixed coordinate system. Now we study the connection between representations of the group $\mathbb{R}_+$ in the fibers of the cotangent bundle and the existence of associated coordinates. Let $\mathfrak{g}_\lambda\colon T^*U \to T^*U$ be a fiber-preserving representation of the group $\mathbb{R}_+$. Let $\mathfrak{A}$ be an infinitesimal operator and $\mathfrak{A}'$ the adjoint operator. We have

PROPOSITION 5. *If the representation of the group $\mathbb{R}_+$ in T^*V with respect to the coordinate system (X, ∂_x) is of the form*

$$\mathfrak{g}_\lambda(x, \xi) = (x, \lambda^{\mu_1}\xi_1, \dots, \lambda^{\mu_n}\xi_n)$$

with constant weights μ_j, where $\xi_j = \partial_{x_j}(v)$, $v \in T^*V$, then

1) *the characteristic polynomial* $\det(\mathfrak{A}(x) - vI)$ *does not depend on $x \in V$;*
2) *for every $x \in V$ and every $j = 1, \dots, n$ the covector $d_x x_j \in T_x^*V$ is an eigenvector of the linear operator $\mathfrak{A}(x)\colon T_x^*V \to T_x^*V$.*

So if in some local coordinate system of the given point the representation has the form (7), then the fibers of the cotangent bundle decompose into the direct sum of eigenvector spaces of an infinitesimal operator, and moreover the corresponding eigenvalues are locally constant and real.

Now let us find sufficient conditions for the representation to be of form (7). In the case in question, this is equivalent to the integrability of an operator field. We shall need the bracket of vector fields X and Y given by

$$[X, Y]_{\mathfrak{A}} = [\mathfrak{A}^t X, \mathfrak{A}^t Y] - \mathfrak{A}^t[\mathfrak{A}^t X, Y] - \mathfrak{A}^t[X, \mathfrak{A}^t Y] + (\mathfrak{A}^t)^2[X, Y],$$

where $[\ ,\ ]$ is the Lie bracket of vector fields.

The following result follows from the Frobenius theorem.

THEOREM 3. *Let the representation $\mathfrak{g}_\lambda$ satisfy the conditions:*

1) *the roots of the characteristic polynomial* $\det(\mathfrak{A}(x) - vI)$ *are real and constant;*
2) *the root spaces of the infinitesimal operator consist of eigenvectors.*

Then $[\ ,\ ]_{\mathfrak{A}} \equiv 0$ in a neighborhood U if and only if at every point of U there is a local coordinate system in which the representation has the form (7).

This theorem is proved in [1].

The existence of a coordinate system associated with the representation is reduced to finding a coordinate system for which the differentials of the coordinate functions are eigenvectors.

The condition in the theorem requiring the roots to be constant may be weakened. Let μ_j be a root of the characteristic polynomial. We suppose that all root spaces of the operator field $\mathfrak{A}\colon T^*U \to T^*U$ in some neighborhood U of a point x_0 consist of eigenvectors have constant dimension. We denote by V_j the subbundle of T^*U generated by the root space of the eigenvalue μ_j.

THEOREM 4. *Let $[\ ,\ ]_{\mathfrak{A}} \equiv 0$ in U and $d\mu_j \in V_j$ for all j. Then for every point $x_0 \in U$ one can find local coordinates in which the representation $\mathfrak{g}_\lambda$ has the form* (7).

PROOF. Let J_j be the ideal generated by sections of V_j in the exterior algebra $\Lambda(M)$. It is sufficient to show that for all j the Frobenius integrability conditions $dJ_j \subset J_j$ are fulfilled. For every 2-form η and for every j we define the 2-form

$$S_{j,\eta}(X, Y) = \eta((\mathfrak{A}^t - \mu_j)X, (\mathfrak{A}^t - \mu_j)Y),$$

where X, Y are vector fields. The following properties are valid:

(1) if $\eta \in J_j$, then $S_{j,\eta}(X, Y) = 0$;

(2) if $w \in J_l$ and $w' \in J_k$, then

$$S_{j,w\wedge w'}(\ ,\ ) = (\mu_j - \mu_l)(\mu_j - \mu_k) w \wedge w';$$

(3) if $w \in V_j$, then

$$S_{j,dw}(X, Y) = -w([X, Y]_{\mathfrak{A}}) - w(X)\, d\mu_j((\mathfrak{A}^t - \mu_j)Y) + w(Y)\, d\mu_j((\mathfrak{A}^t - \mu_j)X).$$

Consequently, if the assumptions of the theorem are fulfilled, then for every 1-form $w \in V_j$ the identity $S_{j,dw}(X, Y) \equiv 0$ is valid for all vector fields X and Y. The 2-form dw is a sum of two 2-forms. One of them belongs to the ideal J_j and other 2-form ζ is the sum of forms of type $w_l \wedge w_k$, where $w_l \in J_l$, $w_k \in J_k$ and $\mu_j \neq \mu_l$, $\mu_j \neq \mu_k$.

We have $S_{j,dw} = S_{j,\zeta}$ from property (1). From this relation and from property (2) we get $\zeta = 0$. So we have $dw \in J_j$.

5. Linear functionals. Let $\mathbb{C}^\infty(M)$ be the ring of complex infinitely differentiable functions on the manifold M and $\mathbb{C}_0^\infty(M)$ be its subring consisting of functions with compact support. Let $\mathcal{R}$ be a locally trivial vector bundle with base M whose fiber is a complex vector space of finite dimension r. We denote by $\mathbb{C}_0^\infty(M, \mathcal{R})$ the $\mathbb{C}^\infty(M)$-module of C^∞ sections of the bundle $\mathcal{R}$ with compact supports. The set of linear continuous (in the ordinary topology of $\mathbb{C}_0^\infty(M, \mathcal{R})$) is denoted by $D'(M, \mathcal{R})$.

Typical examples of bundles $\mathcal{R}$ are bundles of densities, bundles of differential forms, and their various tensor products. In the particular case when $\mathbb{C}_0^\infty(M, \mathcal{R}) = \Lambda_0^n(M)$, $n = \dim(M)$, we have the canonical imbedding of the space $\mathbb{C}^\infty(M)$ in the space of linear functionals

$$\langle f, v\rangle = \int fv,$$

where $f \in \mathbb{C}^\infty(M)$ and $v \in \Lambda_0^n(M)$. In this case the space of linear functionals is called the *space of distributions.*

Below we suppose that M is regular and denote by π the canonical projection $\pi\colon T^*M \to M$.

DEFINITION 2. We say that a point $(x^0, \xi^0) \in \widehat{T}^*M$ does not belong to the weighted wave front $\mathrm{WF_w}(u)$ of a linear functional $u \in D'(M, \mathcal{R})$ if there exists a neighborhood W of the point (x^0, ξ^0) such that for every variety $\mathcal{B}$ consisting of collections $\Phi = \{\varphi_1, \dots, \varphi_k\}$ satisfying the inclusion

$$\bigcup_{x\in\pi(W)} (x, \widehat{d_x\Phi}) \subset W$$

and bounded in

$$\bigoplus_{j=1}^k \mathcal{E}_j(\overline{\pi(W)}),$$

every bounded variety $\mathcal{B}_0$ in $\mathbb{C}_0^\infty(\pi(W), \mathcal{R})$, and every natural number N, there exists a constant C for which the inequality

$$|u(ve^{-i\Phi_\lambda})| \leqslant C(1+\lambda)^{-N}$$

is satisfied whenever $\Phi \in \mathcal{B}$, $v \in \mathcal{B}_0$, and $\lambda \in \mathbb{R}_+$.

Let $\{e_1, \dots, e_r\}$ be a basis of C^∞-sections of the bundle $\mathcal{R}$ in neighborhood $U \subset M$. For every $l = 1, \dots, r$ we can define a linear functional on $\mathbb{C}_0^\infty(U)$ by setting $u_l(v) = u(ve_l)$.

THEOREM 5. $\mathrm{WF_w}(u|_U) = \bigcup_{l=1}^r \mathrm{WF}(u_l)$.

One can find the proof in [1].

Usually, the definition of a weighted wave front is introduced in fixed coordinates and is connected with a fixed representation of the group $\mathbb{R}_+$. Now we discuss the relationship between these two approaches. More detailed information can be found in [1].

Let $(U, \varkappa) = (U; x^1, \dots, x^k)$ be an associated local coordinate system, where $\varkappa\colon U \to \mathbb{R}^n$ and $n = \dim(M)$. The mapping $\varkappa$ induces an isomorphism

$$\widehat{\varkappa}\colon D'(U, \mathcal{R}) \to D'(\varkappa(U), \mathcal{R}_1),$$

where the bundle $\mathcal{R}_1$ has base $\varkappa(U)$ induced by the mapping $\varkappa^{-1}$. The foliation of the neighborhood U generates a foliation of $\varkappa(U)$. We assume that the weights are preserved. The mapping $\varkappa$ is extended to an isomorphism

$$\widetilde{\varkappa}\colon \widehat{T}^*(U) \to \widehat{T}^*(\varkappa(U)).$$

The definitions imply

PROPOSITION 6. *Let $u \in D'(U, \mathcal{R})$. Then $\widetilde{\varkappa}\,\mathrm{WF_w}(u) = \mathrm{WF_w}(\widehat{\varkappa}(u))$.*

At every point $x \in U$ the following isomorphism is defined:

$$\varkappa_j^+ : d_x\mathcal{E}_j(U)/d_x\mathcal{E}_{j+1}(U) \to \mathbb{R}^{n_j},$$

where $j = 1, \dots, k$. It can be extended to the isomorphism

$$\varkappa^+ : \widehat{T}^*(U) \to \varkappa(U) \bigoplus_{j=1}^{k} \mathbb{R}^{n_j} (\cong T^*(\varkappa(U))).$$

DEFINITION 3. We say that a point $(x^0, \xi^0) \in T^*(\varkappa(U))$ does not belong to a weighted (with respect to the action of the group $\mathbb{R}_+$) wave front $\mathrm{WF}_{\mathrm{w},M}(u)$ of the linear functional $u \in D'(\varkappa(U))$ if there exists a neighborhood W_1 of the point ξ^0 and a section $\psi(x) \in C_0^\infty(\varkappa(U), \mathcal{R}_1)$, $\psi(x^0) \neq 0$, such that for every natural number N one can find a constant C for which the inequality

$$\left| u\left(\psi \exp\left(-i \sum_{j=1}^{n} \lambda^{\mu_j} x_j \xi_j \right)\right)\right| \leqslant C(1+\lambda)^{-N}$$

is satisfied for all $\xi \in W_1$, $\lambda \in \mathbb{R}_+$.

THEOREM 6. *Let* $u \in D'(U, \mathcal{R})$. *Then* $\varkappa^t \, \mathrm{WF}_{\mathrm{w}}(u) = \mathrm{WF}_{\mathrm{w}, M}(\widehat{\varkappa}(u))$.

Now we suppose that in a neighborhood U two coordinate systems associated with the filtration are defined. So in $T^* U$ two representations of the group $\mathbb{R}$ are defined.

COROLLARY. *If* $u \in D'(U, \mathcal{R})$, *then*

$$\mathrm{WF}_{\mathrm{w}, M_2}(\widehat{\varkappa}_2 u) = \varkappa_2^+ \circ (\varkappa_1^+)^{-1}(\mathrm{WF}_{\mathrm{w}, M_1}(\widehat{\varkappa}_1 u)).$$

This relation connects wave fronts in different local coordinate systems associated with a fixed foliation and a fixed sequence of weights. One can find the proofs in [1].

References

1. N. A. Shananin, *On the weighting of derivatives*, Mat. Zametki **51** (1992), no. 6, 115–125; English transl. in Math. Notes **51** (1992).

Translated by THE AUTHOR

9-3 RAMENKI AVE, APT. #558 MOSCOW 117607, RUSSIA

Amer. Math. Soc. Transl.
(2) Vol. **167**, 1995

Higher Symmetry Algebra Structures and Local Equivalences of Euler–Darboux Equations

VALERY E. SHEMARULIN

ABSTRACT. The structural properties of higher symmetry algebras for Euler–Darboux equations $\mathcal{Y}_n$ with the parameter $n \notin \mathbb{Z}$ are examined. The dimensions of the invariant solution spaces for these equations are found. It is proved that for $n, m \notin \mathbb{Z}$ the equality $n(n+1) = m(m+1)$ is a necessary and sufficient condition for the local equivalence of the equations $\mathcal{Y}_n$ and $\mathcal{Y}_m$ and for their symmetry algebras (both contact and higher ones) to be isomorphic. When this condition is met, all local equivalences of the equations $\mathcal{Y}_n$ and $\mathcal{Y}_m$ are explicitly described; this is also done for all algebra isomorphisms of subalgebras of their higher symmetries that are direct complements to the infinite-dimensional abelian ideals consisting of smooth solutions of these equations (under additional restrictions).

§1. Introduction

The paper is devoted to the study of the structural properties of the algebras $\operatorname{Sym}\mathcal{Y}_n$ for higher (local) symmetries of the Euler–Darboux equations

$$\mathcal{Y}_n = \{F_n(u) \equiv u_{\xi\eta} - n(\xi+\eta)^{-1}(u_\xi + u_\eta) = 0\}, \qquad n \in \mathbb{R}$$

with the parameter $n \notin \mathbb{Z}$ and to the local classification of the equations $\mathcal{Y}_n$ over the parameter n. The author described the algebras $\operatorname{Sym}\mathcal{Y}_n$ in [1, 2]. The studies of the $\operatorname{Sym}\mathcal{Y}_n$ algebra structure reduce essentially to the investigation of the structure of the subalgebra $\operatorname{NSym}\mathcal{Y}_n$ which is the direct complement to the infinite-dimensional abelian ideal consisting of smooth solutions of the equation $\mathcal{Y}_n$. Therefore the major part of this paper is devoted to the structural studies of the algebra $\operatorname{NSym}\mathcal{Y}_n$. The most significant results of this study include the decomposition of the algebra $\operatorname{NSym}\mathcal{Y}_n$ into an infinite direct sum of odd-dimensional irreducible $sl(2, \mathbb{R})$-modules and the representation of this algebra by linear ordinary differential operators. An important aspect of the study relates to the evaluation of dimensions of the space of $R(u)$-invariant solutions of the equation $\mathcal{Y}_n$ for $R(u) \in \operatorname{NSym}\mathcal{Y}_n$. Using the data on the dimension of invariant solution spaces, we prove that for $n, m \notin \mathbb{Z}$ the equality $n(n+1) = m(m+1)$ is a necessary and sufficient condition for the local equivalence of the equations

1991 *Mathematics Subject Classification*. Primary 35Q05.

Key words and phrases. Euler–Darboux equation, recursion operators, higher symmetry algebras, local equivalences, higher symmetry algebra isomorphisms, invariant solution spaces.

$\mathcal{Y}_n$ and $\mathcal{Y}_m$ and for the existence of an isomorphism of their symmetry algebras, both contact and higher ones. When this condition is met, all local equivalences between $\mathcal{Y}_n$ and $\mathcal{Y}_m$ are explicitly described; this is also done for all $\mathrm{NSym}\,\mathcal{Y}_n$ and $\mathrm{NSym}\,\mathcal{Y}_m$ subalgebras isomorphisms, when some additional restrictions are imposed. Note that in [3] the equivalence criterion is obtained for a special class of point transformations of one class of linear hyperbolic second-order equations with two independent variables; this was done using the Laplace invariants of these equations. Applying this criterion to the equations $\mathcal{Y}_n$ and $\mathcal{Y}_m$ yields the condition $n(n+1) = m(m+1)$, coinciding with the one found in the present paper. This agrees quite well with the fact that each local $\mathcal{Y}_n$ — $\mathcal{Y}_m$ equivalence described here belongs to the class of point equivalences considered in [3]. The main results obtained here were announced in [4]. The paper uses the methods of modern differential equation geometry developed in [5, 6].

All algebras and linear spaces are considered over the field $\mathbb{R}$ of real numbers. If not stated otherwise, for the parameter n of the equations $\mathcal{Y}_n$ we have $n \notin \mathbb{Z}$. The notation corresponds to that adopted in [2]. Let us recall a part of this notation and formulate some results from [1, 2] which will be used later.

Let $\pi\colon \mathbb{R}^2 \times \mathbb{R} \to \mathbb{R}^2$ be a trivial bundle whose coordinates in the base $\mathbb{R}^2$ are independent variables ξ and η and the coordinate in the fiber $\mathbb{R}$ is the dependent variable u. Let $J^l = J^l(\pi)$ and $J^\infty = J^\infty(\pi)$ be the manifolds of l-jets and infinite jets of smooth functions in the variables ξ and η (local sections of bundle π), $\mathcal{Y}_{n,\infty}$ the infinite prolongation of the equation $\mathcal{Y}_n$ considered as a submanifold in J^2. A special system of coordinates

$$\xi,\ \eta,\ u,\quad u_{m\xi,\,l\eta},\quad m \geqslant 0,\ l \geqslant 0\quad (u_{0\xi,\,0\eta} \equiv u)$$

exists on J^∞ such that for any $\varphi(\xi,\eta) \in C^\infty(J^2)$

$$j_\infty(\varphi)^*(u_{m\xi,\,l\eta}) = \frac{\partial^{m+l}\varphi(\xi,\eta)}{\partial\xi^m\partial\eta^l},\qquad m \geqslant 0,\ l \geqslant 0,$$

where $j_\infty(\varphi)\colon \mathbb{R}^2\{(\xi,\eta)\} \to J^\infty$ is the section corresponding to the infinite jet of the function φ. As usual, $C^\infty(J^l)$ denotes the algebra of smooth functions on J^l. The same coordinates restricted by $m + l \leqslant q$ will be used on J^q, $q \geqslant 0$. We take the variables

$$\xi,\ \eta,\ u,\ u_\xi,\ u_\eta,\ \ldots,\ u_{m\xi},\ u_{m\eta},\ \ldots,\qquad m = 1, 2, \ldots \tag{1}$$

as internal coordinates on the manifold $\mathcal{Y}_{n,\infty}$. The symmetries of the equation $\mathcal{Y}_n$ are identified with the functions φ (generating functions) in the variables (1) that are the solutions of the equation

$$\bar{l}_{F_n}(\varphi) = 0,\qquad l_{F_n} = D_\xi \circ D_\eta - n(\xi+\eta)^{-1}(D_\xi + D_\eta),$$

where D_ξ and D_η are the operators of total differentiation with respect to ξ and η respectively. The line over the operator or function symbol denotes the restrictions to $\mathcal{Y}_{n,\infty}$. The operators

$$\square = D_\xi - D_\eta,\quad \sigma_1 = \xi D_\xi + \eta D_\eta - nI,\quad \tau = \xi^2 D_\xi - \eta^2 D_\eta + n(\eta - \xi)I$$

are recursion operators for the equations $\mathcal{Y}_n$ and $\bar{l}_{F_n}(\varphi) = 0$ (I is the identical operator).

Direct calculations can easily establish the following commutation relations

$$[\square, \sigma_1] = \square, \quad [\square, \tau] = 2\sigma_1, \quad [\sigma_1, \tau] = \tau, \tag{2}$$

owing to which the algebra $\{\square, \sigma_1, \tau\}$ generated by the operators $\square$, σ_1, and τ is isomorphic to Lie algebra sl_2^0, inverse (anti-isomorphic) to the classical algebra $sl_2 = sl(2, \mathbb{R})$ of real traceless (2×2)-matrices: $\{\square, \sigma_1, \tau\} \cong sl_2^0$. This yields

$$\{\square(u), \sigma_1(u), \tau(u)\} \cong sl_2.$$

The symbol $\{S\}$ denotes the linear span of the family of functions S. Let A_n^∞ be the linear space of smooth solutions of the equation $\mathcal{Y}_n$ and let $\operatorname{Symc}\mathcal{Y}_n$ be the contact symmetry algebra of the equation $\mathcal{Y}_n$,

$$\operatorname{NSymc}\mathcal{Y}_n = \{u, \square(u), \sigma_1(u), \tau(u)\} \cong \{u\} \oplus sl_2.$$

It is easily seen that A_n^∞ is the infinite-dimensional abelian ideal of the algebra $\operatorname{Sym}\mathcal{Y}_n$ and $\operatorname{NSymc}\mathcal{Y}_n$ is a subalgebra in $\operatorname{Symc}\mathcal{Y}_n$. For $n \neq 0, -1$, the algebra of contact symmetries of the equation $\mathcal{Y}_n$ is a semidirect sum:

$$\operatorname{Symc}\mathcal{Y}_n = A_n^\infty \oplus \operatorname{NSymc}\mathcal{Y}_n.$$

For all real $n, m \neq 0, -1$, $\operatorname{NSymc}\mathcal{Y}_n$ and $\operatorname{NSymc}\mathcal{Y}_m$ are isomorphic as Lie algebras. For $n \notin \mathbb{Z}$ a similar semidirect expansion exists for the higher symmetry algebra:

$$\operatorname{Sym}\mathcal{Y}_n = A_n^\infty \oplus \operatorname{NSym}\mathcal{Y}_n.$$

Here $\operatorname{NSym}\mathcal{Y}_n = \{u, \square_j^i(u), 0 \leqslant j \leqslant 2i, i = 1, 2, \dots\}$ is a subalgebra in $\operatorname{Sym}\mathcal{Y}_n$,

$$\square_0^i = \square^i, \quad \square_j^i = [\dots[\square^i, \underbrace{\tau], \dots, \tau]}_{j}, \qquad i, j = 0, 1, \dots,$$

$\square^i$ is the i-degree of the operator $\square$, $[\square^i, \tau]$ is the ordinary commutator of operators.

§2. Linear bases of the algebra $\operatorname{NSym}\mathcal{Y}_n$. Some isomorphisms

Because of the above-mentioned direct sum decomposition, the investigation of the algebra structure of $\operatorname{Sym}\mathcal{Y}_n$ reduces essentially to the study of its subalgebra $\operatorname{NSym}\mathcal{Y}_n$. Therefore, our next objective is to study the structure of the algebra $\operatorname{NSym}\mathcal{Y}_n$. This study relies on two special bases in the space $\operatorname{NSym}\mathcal{Y}_n$ that are determined by the recursion operators. The first of these bases is described in the following theorem, which is easily derived from Theorems 2 and 3 in [1].

THEOREM 1. *The family of functions*

$$\{\overline{\tau}^i \circ \overline{\Box}^j(u)\,,\ \overline{\tau}^i \circ \overline{\sigma}_1 \circ \overline{\Box}^j(u)\,;\ i\,,\,j \geqslant 0\} \tag{3}$$

is a linear basis of the algebra $\operatorname{NSym}\mathcal{Y}_n$.

The basis (3) is used to prove some simple facts about $\operatorname{NSym}\mathcal{Y}_n$. The main role for further studies of the algebra $\operatorname{NSym}\mathcal{Y}_n$ is still played by the basis

$$\{\overline{\Box}^i_j(u)\,;\ i \geqslant 0\,,\ 0 \leqslant j \leqslant 2i\}. \tag{4}$$

The meaning of Proposition 11 from [2] is that the family (4) is a linear basis in $\operatorname{NSym}\mathcal{Y}_n$.

For brevity introduce the notation $L = sl_2^0$. Let $U(L)$ be the universal enveloping algebra of the Lie algebra L [7, 8]. Let us take standard monomials $\tau^i\sigma_1^k\Box^j$, $i\,,\,k\,,\,j \geqslant 0$ [7], as the basis of the linear space $U(L)$. Since each element $\rho \in U(L)$ can be uniquely represented as a sum of standard monomials, it follows that the following linear mapping is well defined

$$\lambda_n\colon U(L) \to \operatorname{NSym}\mathcal{Y}_n\,, \qquad \lambda_n(\rho) = \overline{\rho}(u)\,, \tag{5$'$}$$

$$\overline{\rho} = \sum_{i,k,j\geqslant 0} \alpha_{i,k,j}\overline{\tau}^i \circ \overline{\sigma}_1^k \circ \overline{\Box}^j\,, \quad \text{if } \rho = \sum_{i,k,j\geqslant 0} \alpha_{i,k,j}\tau^i\sigma_1^k\Box^j\,,$$

$\alpha_{i,k,j} = \operatorname{const}(i\,,\,k\,,\,j)$. The commutation relations (2) make almost obvious the following

LEMMA 1. *For arbitrary elements* ρ_1 *and* ρ_2 *of the algebra* $U(L)$, *their product* $\rho_1\rho_2$ *in* $U(L)$ *(i.e., the result of reducing the product* $\rho_1\rho_2$ *to a sum of standard monomials) considered as a differential operator in* $C^\infty(J^\infty)$ *coincides with the ordinary composition of the differential operators* ρ_1 *and* ρ_2: $\rho_1\rho_2 = \rho_1 \circ \rho_2$.

Using Lemma 1, we can easily prove

THEOREM 2. *The mapping* λ_n $(5')$ *is an epimorphism of linear spaces with kernel* $\operatorname{Ker}\lambda_n = (\Delta_n)$, *where* (Δ_n) *is the two-sided principal ideal generated by the element*

$$\Delta_n = \sigma_1^2 - \tau\Box - \sigma_1 - n(n+1)I\,,$$

I *is the unit element in* $U(L)$. *Further,* $\lambda_n(\rho_1\rho_2) = \overline{\rho}_1(\overline{\rho}_2(u))$, *and the induced mapping*

$$\lambda_n\colon [U(L)/(\Delta_n)]_L \cong \operatorname{NSym}\mathcal{Y}_n\,, \qquad \lambda_n([\rho]) = \overline{\rho}(u) \tag{5}$$

of the Lie algebra $[U(L)/(\Delta_n)]_L$ *of the associative algebra* $U(L)/(\Delta_n)$ *into the algebra* $\operatorname{NSym}\mathcal{Y}_n$ *is an anti-isomorphism of Lie algebras. (Here* $[\rho]$ *denotes the coset of the element* $\rho \in U(L)$ *with respect to the ideal* (Δ_n).)

COROLLARY 1. *The families of elements*

$$\{[\tau^i\Box^j]\,,\ [\tau^i\sigma_1\Box^j]\,;\ i\,,\,j \geqslant 0\}\,, \qquad \{[\Box^i_j]\,;\ i \geqslant 0\,,\ 0 \leqslant j \leqslant 2i\}$$

are linear bases of the algebra $[U(L)/(\Delta_n)]_L$.

COROLLARY 2. *The linear mapping*

$$(\widetilde{\mu}_n)_*\colon \operatorname{NSym}\mathcal{Y}_n \to \operatorname{NSym}\mathcal{Y}_{-(n+1)}$$

with its action on the elements of the linear basis of the algebra $\operatorname{NSym}\mathcal{Y}_n$ *defined by the formula*

$$(\widetilde{\mu}_n)_*(\overline{\tau}^i \circ \overline{\sigma}_1^k \circ \overline{\Box}^j(u)) = \overline{\tau}^i \circ \overline{\sigma}_1^k \circ \overline{\Box}^j(v), \qquad i, j \geqslant 0, \ k = 0, 1$$

is an isomorphism of Lie algebras.

Note that the isomorphism $(\widetilde{\mu}_n)_*$ coincides with the restriction to $\operatorname{NSym}\mathcal{Y}_n$ of the isomorphism of complete symmetry algebras, namely the isomorphism induced by the equivalence

$$\widetilde{\mu}_n\colon (\xi, \eta, u) \to (\xi, \eta, v), \qquad v = (\xi + \eta)^{-(2n+1)}u \tag{6}$$

of the equations $\mathcal{Y}_n$ and $\mathcal{Y}_{-(n+1)}$. For any real n, the mapping $\widetilde{\mu}_n$ is an equivalence of the equations $\mathcal{Y}_n$ and $\mathcal{Y}_{-(n+1)}$, since it can be easily seen that the function $u(\xi, \eta)$ is the solution of the equation $\mathcal{Y}_n$ if and only if $v \equiv (\xi+\eta)^{-(2n+1)}u(\xi, \eta)$ is the solution of the equation $\mathcal{Y}_{-(n+1)}$.

Let $A_{(3)}$ be the associative subalgebra generated by the operators I, $\Box$, σ_1, τ in the algebra of all linear $\mathcal{C}$-differential operators containing the differentiations D_ξ and D_η. Define the linear mapping

$$\lambda_n^1\colon U(L) \to A_{(3)} \tag{7}$$

by specifying its action on the standard monomials by the formula $\lambda_n^1(\tau^i\sigma_1^k\Box^j) = \tau^i \circ \sigma_1^k \circ \Box^j$. Let

$$\Delta_n^1 = \sigma_1^2 - \tau \circ \Box - \sigma_1 - n(n+1)I = (\xi+\eta)^2 l_{F_n} \in A_{(3)}. \tag{8}$$

Using Lemma 1 and simple considerations we prove

PROPOSITION 1. *The mapping*

$$\lambda_n^1\colon U(L)/(\Delta_n) \to A_{(3)}/(\Delta_n^1)$$

induced by the mapping (7) *is an isomorphism of associative algebras for any real* n.

According to Theorem 2 and Proposition 1,

$$\lambda_n^{(1)} = \lambda_n \circ (\lambda_n^1)^{-1}\colon [A_{(3)}/(\Delta_n^1)]_L \to \operatorname{NSym}\mathcal{Y}_n$$

is the anti-isomorphism of Lie algebras which coincides, due to the relation (5), with the well-known anti-isomorphism defined by matching $\varphi \leftrightarrow l_\varphi$ for the functions $\varphi \in C^\infty(J^l)$ that are linear in u_σ (see (13)). Here l_φ is the universal linearization operator.

§3. Commutation formulas for recursion operators. Expansion of NSym $\mathcal{Y}_n$ into a direct sum of odd-dimensional irreducible sl_2-modules

Using the commutation relations (2), we prove the following formulas by induction:

$$\begin{gathered}
\Box^i \circ \sigma_1 = \sigma_1 \circ \Box^i + i\,\Box^i , \\
\Box^i \circ \tau = \tau \circ \Box^i + 2i\sigma_1 \circ \Box^{i-1} + i\,(i-1)\,\Box^{i-1} , \\
\sigma_1^i \circ \tau = \tau \circ (\sigma_1 + I)^i = \sum_{j=0}^{i} \binom{i}{j} \tau \circ \sigma_1^j , \qquad i \geqslant 0, \\
\sigma_1 \circ \tau^i = \tau^i \circ (\sigma_1 + iI) , \\
\Box \circ \tau^i = \tau^i \circ \Box + 2i\tau^{i-1} \circ \sigma_1 + i\,(i-1)\,\tau^{i-1} ,
\end{gathered} \tag{9}$$

which then are used to prove the following statement by induction.

THEOREM 3. *For any integer $i \geqslant 1$, the relations*

$$\widetilde{\Box}^i_{2i+1} = \underbrace{[\dots[\widetilde{\Box}^i , \widetilde{\tau}]\dots\widetilde{\tau}]}_{2i+1} = \widetilde{0} , \qquad \underbrace{[\dots[\widetilde{\sigma}_1\widetilde{\Box}^{i-1} , \widetilde{\tau}]\dots\widetilde{\tau}]}_{2i} = \widetilde{0}$$

are valid in $U(L)/(\Delta_n)$ algebra.

Hereafter the symbol $\widetilde{\nabla}$ will denote the coset of the element $\nabla \in U(L)$ with respect to the ideal (Δ_n).

COROLLARY. *For any integer $i \geqslant 0$ and any $\lambda > 0$ we have*

$$\widetilde{\Box}^i_{2i+\lambda} = \widetilde{0} , \qquad \overline{\Box}^i_{2i+\lambda}(u) = 0 . \tag{10}$$

The equalities (10) explain why the basis (4) of the space NSym $\mathcal{Y}_n$ does not contain the elements $\overline{\Box}^i_j(u)$ with $j \geqslant 2i+1$. Simple considerations using the formulas (9) and (2) lead to important commutation relations

$$\begin{aligned}
&[\Box^i_j , \Box] = -j\,(2i - (j-1))\,\Box^i_{j-1} , && 1 \leqslant j \leqslant 2i , \\
&[\Box^i_j , \sigma_1] = (i-j)\,\Box^i_j , && 0 \leqslant j \leqslant 2i , \\
&[\Box^i_j , \tau] = \Box^i_{j+1} , && 0 \leqslant j \leqslant 2i ,\ i \geqslant 1 , \qquad (11)\\
&[\Box^i_1 , \Box^j] = -2ij\Box^{i+j-1} , && i , j \geqslant 0 . \qquad (12)
\end{aligned}$$

For the functions φ and ψ linear with respect to the variables u_σ we can easily verify that

$$l_{\{\varphi , \psi\}} = -[l_\varphi , l_\psi] . \tag{13}$$

Here $\{\varphi, \psi\}$ is the higher Jacobi bracket for the functions φ and ψ. Therefore, the relations (11) obviously yield the relations for the higher Jacobi bracket

$$
(14) \quad \begin{aligned}
&\{\square(u), \overline{\square}^i_j(u)\} = -j(2i-(j-1))\overline{\square}^i_{j-1}(u), && 1 \leqslant j \leqslant 2i, \\
&\{\square(u), \overline{\square}^i_0(u)\} = 0, && \\
&\{\sigma_1(u), \overline{\square}^i_j(u)\} = (i-j)\overline{\square}^i_j(u), && 0 \leqslant j \leqslant 2i, \\
&\{\tau(u), \overline{\square}^i_j(u)\} = \overline{\square}^i_{j+1}(u), && 0 \leqslant j \leqslant 2i.
\end{aligned} \qquad i \geqslant 1,
$$

Let M_i be a linear subspace in $\operatorname{NSym}\mathcal{Y}_n$ with the basis $\overline{\square}^i_j(u)$, $0 \leqslant j \leqslant 2i$. It is clear from (14) that for any i the subspace M_i is an irreducible two-sided sl_2-module. Since for each integer $i > 0$ only one (up to an isomorphism) irreducible sl_2-module of dimension i exists [7], the following result holds.

PROPOSITION 2. *The space* $\operatorname{NSym}\mathcal{Y}_n$ *is an infinite direct sum of all odd-dimensional irreducible* sl_2*-modules*:

$$
\operatorname{NSym}\mathcal{Y}_n = \{u\} \oplus \left(\bigoplus_{i=1}^{\infty} M_i\right),
$$

where $M_i = \{\overline{\square}^i_j(u),\ 0 \leqslant j \leqslant 2i\}$ *is an irreducible* sl_2*-module of dimension* $(2i+1)$.

§4. On ideals and subalgebras in $\operatorname{Sym}\mathcal{Y}_n$ and $\operatorname{NSym}\mathcal{Y}_n$. Canonical grading of the algebra $\operatorname{NSym}\mathcal{Y}_n$

THEOREM 4. *Let J be an ideal of* $\operatorname{NSym}\mathcal{Y}_n$ *which contains at least one nonzero element $g \neq cu$, $c \in \mathbb{R}$. Then there exists an integer $i_0 \geqslant 1$ such that each element $\overline{\square}^i_j(u)$. $i \geqslant i_0$, $0 \leqslant j \leqslant 2i$ belongs to the ideal J and each element $g \in J$ has the form*

$$
g = \alpha_0 u + \sum_{i=i_0}^{M}\sum_{j=0}^{2i} \alpha_i^j \overline{\square}^i_j(u), \qquad \alpha_0, \alpha_i^j = \text{const}.
$$

The proof is based on relations (14) and (12). Using (12), we can easily derive the following from Theorem 4.

THEOREM 5. *The algebra* $\operatorname{NSym}\mathcal{Y}_n$ *does not contain nonzero abelian ideals other than the trivial one-dimensional ideal $\{u\}$. In particular, $\{u\}$ is the center of the algebra* $\operatorname{NSym}\mathcal{Y}_n$.

Theorem 5 and Theorem 8 proved below yield

THEOREM 6. *For $n \notin \mathbb{Z}$ the ideal A_n^∞ is the only maximum abelian ideal of the algebra* $\operatorname{Sym}\mathcal{Y}_n$. *In particular, each abelian ideal of the algebra* $\operatorname{Sym}\mathcal{Y}_n$ *is contained in A_n^∞.*

The general result relative to the structure of ideals in $\operatorname{Sym}\mathcal{Y}_n$ is given by the following easy theorem.

THEOREM 7. *Let $n \notin \mathbb{Z}$. Then each ideal I in the algebra $\mathrm{Sym}\,\mathcal{Y}_n$ is a semidirect sum $I = A_I \oplus I_1$ of the abelian ideal $A_I = A_n^\infty \cap I$ of the algebra $\mathrm{Sym}\,\mathcal{Y}_n$ and the ideal $I_1 = \mathrm{NSym}\,\mathcal{Y}_n \cap I$ of the algebra $\mathrm{NSym}\,\mathcal{Y}_n$. If $I \neq 0$, then $A_I \neq 0$. For A_I and I_1 the following property holds*

$$(a \in A_n^\infty, g_1 \in I_1) \implies \{a, g_1\} \in A_I. \tag{15}$$

Vice versa, if $A_I \subset A_n^\infty$ is an ideal in $\mathrm{Sym}\,\mathcal{Y}_n$, I_1 is an ideal in $\mathrm{NSym}\,\mathcal{Y}_n$ and property (15) *holds for A_I and I_1, then $I = A_I \oplus I_1$ is an ideal in the algebra $\mathrm{Sym}\,\mathcal{Y}_n$.*

REMARK 1. The same arguments prove that the statements from Theorems 4–7 remain valid if $\mathrm{NSym}\,\mathcal{Y}_n$ and $\mathrm{Sym}\,\mathcal{Y}_n$ are replaced by $\mathrm{NSymc}\,\mathcal{Y}_n$ and $\mathrm{Symc}\,\mathcal{Y}_n$, respectively.

Consider the following three special abelian subalgebras in $\mathrm{NSym}\,\mathcal{Y}_n$

$$L_\square = \{\overline{\square}^i(u),\ i \geqslant 0\}, \quad L_{\sigma_1} = \{\overline{\sigma}_1^i(u),\ i \geqslant 0\}, \quad L_\tau = \{\overline{\tau}^i(u),\ i \geqslant 0\}.$$

By the first of the relations (14), it is obvious that each element $\varphi \in \mathrm{NSym}\,\mathcal{Y}_n$ commuting with $\square(u)$ belongs to $L_\square$. Therefore the following proposition holds.

PROPOSITION 3. *$L_\square$ is the maximal abelian subalgebra of $\mathrm{NSym}\,\mathcal{Y}_n$. Furthermore, $L_\square$ can be characterized as the only maximal abelian subalgebra in $\mathrm{NSym}\,\mathcal{Y}_n$ containing the element $\square(u)$.*

The following lemma can be easily proved by using relations (14) and (9).

LEMMA 2. *For any integer $i \geqslant 0$*

$$\overline{\tau}^i(u) = \beta_i \overline{\square}_{2i}^i(u), \qquad \beta_i = \mathrm{const}(i) \neq 0. \tag{16}$$

Lemma 2 yields

PROPOSITION 4. *L_τ is the maximal abelian subalgebra of $\mathrm{NSym}\,\mathcal{Y}_n$ and*

$$L_\tau = \{\overline{\square}_{2i}^i(u),\ i \geqslant 0\}.$$

Furthermore, L_τ can be characterized as the only maximal abelian subalgebra in $\mathrm{NSym}\,\mathcal{Y}_n$ containing the element $\tau(u)$.

The element $\sigma_1(u)$ determines the operator

$$\widetilde{\sigma}_1\colon \mathrm{NSym}\,\mathcal{Y}_n \to \mathrm{NSym}\,\mathcal{Y}_n$$

acting according to the formula $\widetilde{\sigma}_1(\overline{\nabla}(u)) = \{\sigma_1(u), \overline{\nabla}(u)\}$, $\overline{\nabla}(u) \in \mathrm{NSym}\,\mathcal{Y}_n$. Let R_m be the linear subspace in $\mathrm{NSym}\,\mathcal{Y}_n$ with the basis $\overline{\square}_i^{i+m}(u)$:

$$R_m = \{\overline{\square}_i^{i+m}(u);\ i \geqslant 0,\ i+m \geqslant 0\}, \qquad m \in \mathbb{Z}.$$

By the second of the relations (14), the next proposition is almost obvious.

PROPOSITION 5. *The set of eigenvalues of the operator* $\widetilde{\sigma}_1$ *coincides with* $\mathbb{Z}$. R_m *is the eigenspace of the operator* $\widetilde{\sigma}_1$ *corresponding to the eigenvalue* m, *and the dimension is* $\dim R_m = \infty$ *for each* $m \in \mathbb{Z}$.

It follows from the Jacobi identity that R_0 is a subalgebra in $\operatorname{NSym} \mathcal{Y}_n$. Obviously, $L_{\sigma_1} \subset R_0$, so that

$$\overline{\sigma}_1^i(u) = \sum_{k=0}^{i} \alpha_k^i \overline{\square}_k^k(u), \qquad \alpha_k^i \in \mathbb{R}, \ \alpha_i^i \neq 0. \tag{17}$$

By solving the previous equality with respect to $\overline{\square}_i^i(u)$, we can easily prove by induction that the following formula is also valid:

$$\overline{\square}_i^i(u) = \sum_{k=0}^{i} \gamma_k^i \overline{\sigma}_1^k(u), \qquad \gamma_k^i \in \mathbb{R}, \ \gamma_i^i \neq 0.$$

Therefore $R_0 \subset L_{\sigma_1}$. Since each element $\varphi \in \operatorname{NSym} \mathcal{Y}_n$ commuting with $\sigma_1(u)$ is contained in R_0, this proves

PROPOSITION 6. L_{σ_1} *is the maximal abelian subalgebra of* $\operatorname{NSym} \mathcal{Y}_n$ *and*

$$L_{\sigma_1} = R_0 = \{\overline{\square}_i^i(u), \ i \geqslant 0\}.$$

Furthermore, L_{σ_1} *can be characterized as the only maximal abelian subalgebra in* $\operatorname{NSym} \mathcal{Y}_n$ *containing the element* $\sigma_1(u)$.

The symbol $\{R_m, R_l\}$ will denote, as usual, the linear span of the commutators $\{a, b\}$ of elements $a \in R_m$, $b \in R_l$. Since the commutation with $\sigma_1(u)$ is a differentiation of Lie algebras, for any $m, l \in \mathbb{Z}$, we have the inclusion

$$\{R_m, R_l\} \subset R_{m+l}. \tag{18}$$

Besides, it is obvious that

$$\operatorname{NSym} \mathcal{Y}_n = \bigoplus_{m \in \mathbb{Z}} R_m.$$

Hence the following statement holds.

PROPOSITION 7. *The algebra* $\operatorname{NSym} \mathcal{Y}_n$ *is* $\mathbb{Z}$*-graded by the subspaces* R_m, $m \in \mathbb{Z}$.

The filtration is canonically determined using the above grading. For example, let $\mathcal{F}_m = \bigoplus_{\lambda \leqslant m} R_\lambda$, $\lambda, m \in \mathbb{Z}$, then (18) immediately yields $\{\mathcal{F}_m, \mathcal{F}_l\} \subset \mathcal{F}_{m+l}$. Hence the sequence

$$\cdots \subset \mathcal{F}_{-l} \subset \cdots \subset \mathcal{F}_{-2} \subset \mathcal{F}_{-1} \subset \mathcal{F}_0 \subset \mathcal{F}_1 \subset \mathcal{F}_2 \subset \cdots \subset \mathcal{F}_l \subset \ldots$$

of subspaces $\mathcal{F}_m$, $m \in \mathbb{Z}$ forms an increasing filtration of $\operatorname{NSym} \mathcal{Y}_n$. Similarly, subalgebras in $\operatorname{NSym} \mathcal{Y}_n$ can also be constructed. The validity of the relations

$$\{\tau(u), R_m\} = R_{m-1} \ \text{ with } m \leqslant 0, \qquad \{\square(u), R_m\} = R_{m+1} \ \text{ with } m \geqslant 0$$

can be easily verified. In particular,

$$\left.\begin{aligned} \underbrace{\{\tau(u), \ldots \{\tau(u)}_{l}, R_0\} \ldots\} &= R_{-l}, \\ \underbrace{\{\square(u), \ldots \{\square(u)}_{l}, R_0\} \ldots\} &= R_l, \end{aligned}\right\} \quad l \geqslant 1.$$

§5. The dimension of the space of $R(u)$-invariant solutions of the equation $\mathcal{Y}_n$; $R(u) \in \operatorname{NSym} \mathcal{Y}_n$

The dimensions of the spaces of $R(u)$-invariant solutions for the equation $\mathcal{Y}_n$ ($\mathcal{Y}_m$) must be known further to find conditions for the algebras $\operatorname{Sym} \mathcal{Y}_n$ and $\operatorname{Sym} \mathcal{Y}_m$ to be isomorphic in the case of $R(u) \in \operatorname{NSym} \mathcal{Y}_n$ ($R(u) \in \operatorname{NSym} \mathcal{Y}_m$) and $n, m \notin \mathbb{Z}$. As is shown in [2], for an arbitrary element $R(u) \in \operatorname{NSym} \mathcal{Y}_n$, the operator R has the form

$$R = \sum_{i=1}^{l} [B_i(\xi, \eta) D_\xi^i + C_i(\xi, \eta) D_\eta^i] + A_1(\xi, \eta) I, \qquad l \geqslant 0, \tag{19}$$

where $B_i(\xi, \eta)$, $C_i(\xi, \eta)$, $A_1(\xi, \eta)$ are functions analytic (almost everywhere) with respect to ξ and η; $B_l(\xi, \eta) \equiv B_l(\xi)$, $C_l(\xi, \eta) \equiv C_l(\eta)$, $B_l(\xi)$ and $C_l(\eta)$ are polynomials of degree $\leqslant 2l$, $C_l(\eta) \equiv (-1)^l B_l(-\eta) \not\equiv 0$. The $R(u)$-invariant solutions of $\mathcal{Y}_n$ are the solutions of the following system of equations [5, 6]:

$$\mathcal{R}_l : \begin{cases} \varphi_{\xi\eta} - n(\xi + \eta)^{-1}(\varphi_\xi + \varphi_\eta) = 0, \\ R(\varphi) = 0. \end{cases}$$

Thus, the space of such solutions coincides with the kernel of the operator $R|_{A_n^\infty}$, which is the restriction of the operator R to A_n^∞. The order of R will be denoted by $\deg R$. The main result of this section is the following theorem.

THEOREM 8. *Let $R(u) \in \operatorname{NSym} \mathcal{Y}_n$ and the order of the operator R be equal to l. Then each $R(u)$-invariant solution of the equation $\mathcal{Y}_n$ is analytic and for the dimension of the space of $R(u)$-invariant solutions we have*

$$\dim \operatorname{Ker}(R|_{A_n^\infty}) = 2 \deg R = 2l. \tag{20}$$

PROOF. The case $l = 0$ is quite trivial. Consider the case $l = 1$. The system $\mathcal{R}_1$ has "poor" formal properties: it is not formally integrable for an "arbitrary" operator R of the first order. Thus, instead of $\mathcal{R}_1$, we consider the equivalent system

$$\widetilde{\mathcal{R}}_1 : \begin{cases} \varphi_{\xi\eta} - n(\xi + \eta)^{-1}(\varphi_\xi + \varphi_\eta) = 0, \\ R(\varphi) = 0, \quad [R(\varphi)]_\xi = 0, \quad [R(\varphi)]_\eta = 0. \end{cases}$$

Note that in the two final equations of $\widetilde{\mathcal{R}}_1$ the mixed derivative $\varphi_{\xi\eta}$ is replaced by its expression obtained from the first equation. Whenever $B_l(\xi) C_l(\eta) \neq 0$, the symbol $\widetilde{G}_1$ of $\widetilde{\mathcal{R}}_1$ is zero: $\widetilde{G}_1 = 0$. The fact that $R(u)$ is a symmetry of $\mathcal{Y}_n$ easily yields that the natural projection $\pi_2^3 \colon \widetilde{\mathcal{R}}_{1+1} \to \widetilde{\mathcal{R}}_1$ is surjective; here $\widetilde{\mathcal{R}}_{1+1}$ is the first prolongation of $\widetilde{\mathcal{R}}_1$. By excluding singular points, if any, we conclude that this projection is an epimorphism almost everywhere. Hence, according to the involutivity criterion [9], $\widetilde{\mathcal{R}}_1$ is an involutive analytic system so that the Cartan–Kachler theorem can be applied to show that the assertion of the theorem is valid for $l = 1$. The case $l \geqslant 2$ is examined similarly, except that we must take as $\widetilde{\mathcal{R}}_l$ the system obtained by adding all derivatives up to the order $(l - 1)$ inclusive of the first equation of $\mathcal{R}_l$ to the system $\{R(\varphi) = 0, [R(\varphi)]_\xi = 0, [R(\varphi)]_\eta = 0\}$.

§6. The condition for the algebras Sym $\mathcal{Y}_n$ and Sym $\mathcal{Y}_m$ to be isomorphic and the equations $\mathcal{Y}_n$ and $\mathcal{Y}_m$ to be locally equivalent, $n, m \notin \mathbb{Z}$. Examples of nonlocal equivalences of $\mathcal{Y}_n$ and $\mathcal{Y}_m$

First we prove some auxiliary results. The classical real Lie algebra sl_2 is the algebra of traceless 2×2-matrices. Take the matrices

$$X_1 = \begin{pmatrix} 0 & 0 \\ -1 & 0 \end{pmatrix}, \qquad X_2 = \begin{pmatrix} -1/2 & 0 \\ 0 & 1/2 \end{pmatrix}, \qquad X_3 = \begin{pmatrix} 0 & 1 \\ 0 & 0 \end{pmatrix}$$

as a basis in sl_2. For brevity introduce the notation

$$\square^u \equiv \square(u), \qquad \sigma_1^u \equiv \sigma_1(u), \qquad \tau^u \equiv \tau(u).$$

We can easily verify that the assignment

$$X_1 \to \square^u, \qquad X_2 \to \sigma_1^u, \qquad X_3 \to \tau^u$$

linearly extended to the whole matrix algebra sl_2 is an isomorphism of the Lie algebras sl_2 and $\{\square^u, \sigma_1^u, \tau^u\} \subset \mathrm{NSym}\,\mathcal{Y}_n$. Below this isomorphism will be used to identify these algebras; in the same way, the elements $\square^u$, σ_1^u, τ^u will be identified, whenever necessary, with the corresponding matrices:

$$\square^u \equiv X_1, \qquad \sigma_1^u \equiv X_2, \qquad \tau^u \equiv X_3.$$

Obviously, the mapping $g: X \to A^{-1}XA$ is an automorphism of the algebra sl_2 for any nondegenerate real matrix A. By setting $A^{-1} = \begin{pmatrix} c & a \\ d & b \end{pmatrix}$ and evaluating the products $A^{-1}X_iA$, we easily find the formulas for the action of the automorphism g on basis elements of the algebra sl_2:

$$\begin{aligned} g(\square^u) &= A^{-1}\square^u A = \frac{b^2}{\Delta}\square^u + 2\frac{ba}{\Delta}\sigma_1^u + \frac{a^2}{\Delta}\tau^u, \\ g(\sigma_1^u) &= A^{-1}\sigma_1^u A = \frac{bd}{\Delta}\square^u + \frac{bc+ad}{\Delta}\sigma_1^u + \frac{ca}{\Delta}\tau^u, \qquad \Delta = cb - ad, \\ g(\tau^u) &= A^{-1}\tau^u A = \frac{d^2}{\Delta}\square^u + 2\frac{dc}{\Delta}\sigma_1^u + \frac{c^2}{\Delta}\tau^u. \end{aligned} \tag{21}$$

PROPOSITION 8. *The automorphisms group of the real algebra sl_2 consists of mappings of the form*

$$g: X \to A^{-1}XA, \tag{22}$$

where $A^{-1} = \begin{pmatrix} c & a \\ d & b \end{pmatrix}$ is a nondegenerate real matrix. The automorphism $g_1: X \to B^{-1}XB$ coincides with (22) if and only if $B = \rho A$, where ρ is a nonzero real number.

Indeed, let $g_C: sl_2^C \to sl_2^C$ be the complexification of the real automorphism g (22). Each automorphism of the algebra sl_2^C is of the form of (22), with nondegenerate complex matrix $A^{-1} = \begin{pmatrix} c & a \\ d & b \end{pmatrix}$ [7]. Thus formulas (21) are valid for g_C. Since g_C and g coincide on sl_2, all coefficients in (21) are real. This yields $A^{-1} = e^{i\varphi}\begin{pmatrix} \gamma & \alpha \\ \delta & \beta \end{pmatrix}$, $\varphi, \alpha, \beta, \gamma, \delta \in \mathbb{R}$. The second statement also follows trivially from properties of automorphisms of the algebra sl_2^C. This proves the proposition.

The contact vector fields X_{φ_i} on J^1, which are the lifts of vector fields defined on J^0, correspond to the point symmetries

$$\varphi_0 = u, \quad \varphi_1 = -\Box(u), \quad \varphi_2 = -\sigma(u) = -\xi u_\xi - \eta u_\eta, \quad \varphi_3 = \tau(u)$$

of the equation $\mathcal{Y}_n$. By integrating X_{φ_i}, we obtain the following one-parameter groups of finite symmetries (self-equivalences) for $\mathcal{Y}_n$, correspondingly:

$$G_0^t : \begin{cases} \xi \to \xi_1 = \xi, \\ \eta \to \eta_1 = \eta, \\ v \to u = e^t v, \end{cases} \qquad G_1^t : \begin{cases} \xi \to \xi_1 = \xi + t, \\ \eta \to \eta_1 = \eta - t, \\ v \to u = v, \end{cases} \qquad G_2^t : \begin{cases} \xi \to \xi_1 = e^t\xi, \\ \eta \to \eta_1 = e^t\eta, \\ v \to u = v, \end{cases}$$

$$G_3^t : \begin{cases} \xi \to \xi_1 = \xi(t\xi + 1)^{-1}, \\ \eta \to \eta_1 = -\eta(t\eta - 1)^{-1}, \\ v \to u = (t\xi + 1)^{-n}(t\eta - 1)^{-n} v. \end{cases}$$

For brevity, we omit the rule for the transformation of the variables u_ξ and u_η under G_i^t. Obviously, the transformations

$$g^- : \begin{cases} \xi \to \xi_1 = -\xi, \\ \eta \to \eta_1 = -\eta, \\ v \to u = v, \end{cases} \qquad G_0^- : \begin{cases} \xi \to \xi_1 = \xi, \\ \eta \to \eta_1 = \eta, \\ v \to u = -v \end{cases}$$

are also symmetries of $\mathcal{Y}_n$. The transformations G_i^t, $i = 1, 2, 3$ with any $t \in \mathbb{R}$, G_0^- and g^- generate the group of linear-fractional type self-equivalences of $\mathcal{Y}_n$:

$$\mu : J^0 \to J^0, \qquad \mu((\xi, \eta, v)) = (\xi_1, \eta_1, u),$$

$$\xi_1 = \frac{c\xi + d}{a\xi + b}, \qquad \eta_1 = \frac{-c\eta + d}{a\eta - b}, \qquad u = s_\mu \frac{v}{(a\xi + b)^n (a\eta - b)^n}, \tag{23}$$
$$a, b, c, d, s_\mu \in \mathbb{R}, \quad s_\mu \neq 0, \quad \Delta = cb - ad \neq 0.$$

Note that G_i^t, $i = 1, 2, 3$, $t \in \mathbb{R}$ generate transformations of the form (23) with determinant $\Delta > 0$. Since μ is a self-equivalence of $\mathcal{Y}_n$, formulas (23) yield immediately the following known result [10]:

if $u(\xi, \eta)$ is the solution of $\mathcal{Y}_n$, then $v(\xi, \eta) \equiv (a\xi + b)^n (a\eta - b)^n u(\xi_1, \eta_1)$ with ξ_1, η_1 defined by the formulas (23) is also the solution of $\mathcal{Y}_n$ for any $a, b, c, d \in \mathbb{R}$, $\Delta = cb - ad \neq 0$.

Set $s_\mu = 1$ in (23) (this being sufficient for our further considerations). Let $U_1^u = du - u_{\xi_1} d\xi_1 - u_{\eta_1} d\eta_1$, $U_1^v = dv - v_\xi \, d\xi - v_\eta \, d\eta$ be universal 1-forms. It can be easily verified that $\mu^*(U_1^u) = (a\xi + b)^{-n}(a\eta - b)^{-n} U_1^v$. Therefore, if the contact vector fields X_f and X_φ are μ-connected, $\mu_*(X_f) = X_\varphi$, then the relation for their generating functions

$$f = (a\xi + b)^n (a\eta - b)^n \mu^*(\varphi) \tag{24}$$

holds. In particular, for $\varphi = u$ it is obvious that $f = v$. Relation (24) can be rewritten in the form $f = (\mu^{-1})_*(\varphi)$. By extending the action of μ to the first derivatives, we can easily obtain formulas for the transformation of the basis elements of the algebra sl_2 (considered as a direct summand in $\operatorname{Symc} \mathcal{Y}_n$) under the action of the isomorphism

$$(\mu^{-1})_* \colon \operatorname{Symc} \mathcal{Y}_n^u \to \operatorname{Symc} \mathcal{Y}_n^v$$

induced by the self-equivalence μ^{-1}. Let $i_{v,u} \colon \operatorname{Symc} \mathcal{Y}_n^v \to \operatorname{Symc} \mathcal{Y}_n^u$ be the mapping acting according to the rule

$$i_{v,u}(\varphi(\xi, \eta, v, v_\xi, v_\eta)) = \varphi(\xi_1, \eta_1, u, u_{\xi_1}, u_{\eta_1}).$$

The above considerations show that the equation $\mathcal{Y}_n$ has the standard form $v_{\xi\eta} - n(\xi + \eta)^{-1}(v_\xi + v_\eta) = 0$ in the coordinates $(\xi, \eta, v; v_{i\xi, j\eta})$ as well as in the coordinates $(\xi_1, \eta_1, u; u_{i\xi_1, j\eta_1})$. Therefore, $i_{v,u}$ is an isomorphism of Lie algebras. For the composition of isomorphisms $\nu_* = i_{v,u} \circ (\mu^{-1})_*$ the following formulas hold

$$\begin{aligned}
\nu_*(\square^u) &= b^2\Delta^{-1}\square^u + 2ab\Delta^{-1}\sigma_1^u + a^2\Delta^{-1}\tau^u, \\
\nu_*(\sigma_1^u) &= bd\Delta^{-1}\square^u + (bc + ad)\Delta^{-1}\sigma_1^u + ac\Delta^{-1}\tau^u, \\
\nu_*(\tau^u) &= d^2\Delta^{-1}\square^u + 2cd\Delta^{-1}\sigma_1^u + c^2\Delta^{-1}\tau^u;
\end{aligned}$$

they coincide with formulas (21) for the automorphism g of the algebra $sl_2^u = \{\square^u, \sigma_1^u, \tau^u\}$. Since the matrix $\begin{pmatrix} c & d \\ a & b \end{pmatrix}$ of the linear-fractional transformation (23) for the variable ξ is transposed to the matrix $A^{-1} = \begin{pmatrix} c & a \\ d & b \end{pmatrix}$ of the automorphism (21), the above considerations and Proposition 8 yield

THEOREM 9. *Each automorphism of the algebra sl_2 (considered as a direct summand in $\operatorname{Symc} \mathcal{Y}_n$) is "induced" by some self-equivalence μ (23) of the equation $\mathcal{Y}_n$. The mapping $\mu \to \nu_*$ above constructed of the transformations group (23) with $s_\mu = 1$ into the automorphisms group of the algebra sl_2 is an anti-isomorphism.*

Now let

$$g \colon \operatorname{Symc} \mathcal{Y}_n \to \operatorname{Symc} \mathcal{Y}_m, \qquad n, m \neq 0, -1 \tag{25}$$

be an isomorphism of Lie algebras. Since A_n^∞ (A_m^∞) is the only maximal abelian ideal in $\operatorname{Symc} \mathcal{Y}_n$ ($\operatorname{Symc} \mathcal{Y}_m$) and $\{u\}$ ($\{v\}$) is the center of the algebra $\operatorname{NSymc} \mathcal{Y}_n$

(NSymc $\mathcal{Y}_m$), it is easily seen from the standard decomposition $\operatorname{Symc}\mathcal{Y}_i = A_i^\infty \oplus \{u_i\} \oplus sl_2^{(i)}$, $i = n, m$, that the induced isomorphism of Lie quotient algebras is defined:

$$g_N : sl_2^{(n)} \to sl_2^{(m)},$$

where $sl_2^{(n)}$ and $sl_2^{(m)}$ denote the algebra sl_2 with its standard generators depending on n and m parameters, respectively. The linear mapping of the quotient algebras

$$i_{m,n}: sl_2^{(m)} \to sl_2^{(n)}, \quad i_{m,n}(\widetilde{\square^v}) = \widetilde{\square^u}, \quad i_{m,n}(\widetilde{\sigma_1^v}) = \widetilde{\sigma_1^u}, \quad i_{m,n}(\widetilde{\tau^v}) = \widetilde{\tau^u}$$

is obviously an isomorphism of Lie algebras (here the symbol $\widetilde{\varphi}$ denotes the class of the element $\varphi \in \operatorname{Symc}\mathcal{Y}_i$ over the ideal $A_i^\infty \oplus \{u_i\}$, $i = n, m$, $u_n \equiv u$, $u_m \equiv v$). The composition $\tilde{g}_N = g_N \circ i_{m,n}$ is an automorphism of the algebra $sl_2^{(m)}$ and

$$\text{(26)} \qquad \tilde{g}_N(\widetilde{\square^v}) = g_N(\widetilde{\square^u}), \quad \tilde{g}_N(\widetilde{\sigma_1^v}) = g_N(\widetilde{\sigma_1^u}), \quad \tilde{g}_N(\widetilde{\tau^v}) = g_N(\widetilde{\tau^u}).$$

According to Theorem 9, the automorphism $\tilde{g}_N^{-1}$ is "induced" by some self-equivalence $\mu: \mathcal{Y}_m \to \mathcal{Y}_m$ of the form (23). Let g_m be the automorphism of the complete algebra $\operatorname{Symc}\mathcal{Y}_m$, the former being "induced" by this equivalence μ and let $g_m|_{sl_2^{(m)}}: sl_2^{(m)} \to sl_2^{(m)}$ be the corresponding quotient automorphism. Since $g_m|_{sl_2^{(m)}} = \tilde{g}_N^{-1}$, (26) yields

$$\text{(27)} \quad g_m|_{sl_2^{(m)}} \circ g_N(\widetilde{\square^u}) = \widetilde{\square^v}, \quad g_m|_{sl_2^{(m)}} \circ g_N(\widetilde{\sigma_1^u}) = \widetilde{\sigma_1^v}, \quad g_m|_{sl_2^{(m)}} \circ g_N(\widetilde{\tau^u}) = \tau^v.$$

Let $\tilde{g} = g_m \circ g$. Using the commutation relations for the elements $\square^v$, σ_1^v, τ^v of the basis of the algebra $sl_2^{(m)}$ similar to relations (2), we obtain from (27) that

$$\text{(28)} \qquad \begin{gathered} \tilde{g}(\square^u) = \square^v + a_1, \quad \tilde{g}(\sigma_1^u) = \sigma_1^v + a_2, \quad \tilde{g}(\tau^u) = \tau^v + a_3, \\ a_i \in A_m^\infty, \quad i = 1, 2, 3. \end{gathered}$$

Thus the following is proved.

PROPOSITION 9. *Let $n, m \neq 0, -1$. If the isomorphism g (25) of Lie algebras exists, the isomorphism*

$$\text{(29)} \qquad \tilde{g}: \operatorname{Symc}\mathcal{Y}_n \to \operatorname{Symc}\mathcal{Y}_m$$

of Lie algebras also exists for which the relations (28) are satisfied and which can be represented as the composition $\tilde{g} = g_m \circ g$, where g_m is the automorphism of $\operatorname{Symc}\mathcal{Y}_m$ algebra "induced" by some self-equivalence μ of the form (23) of the equation $\mathcal{Y}_m$.

Let $\tilde{g}$ be the isomorphism (29) satisfying the relations (28) and $\tilde{g}(1) = \tilde{1}$. Obviously, $\tilde{1} \in \operatorname{Ker}\square|_{A_m^\infty}$ follows from the equalities $\{1, \square(u)\} = \square(1) = 0$ and (28). By evaluating $\square(v)$-invariant solutions of $\mathcal{Y}_m$ in the standard way [3], we easily find

$$\text{(30)} \qquad \operatorname{Ker}\square|_{A_m^\infty} = \begin{cases} c_1 + c_2(\xi + \eta)^{2m+1}, & \text{if } m \neq -1/2, \\ c_1 + c_2 \ln|\xi + \eta|, & \text{if } m = -1/2, \end{cases} \qquad c_1, c_2 \in \mathbb{R}.$$

For clarity, assume that $m \neq -1/2$. Then $\widetilde{1} = \alpha \cdot 1 + \beta(\xi+\eta)^{2m+1}$ for some $\alpha, \beta \in \mathbb{R}$. Since $\{1, \sigma_1(u)\} = \sigma_1(1) = -n \cdot 1$, we have

$$\widetilde{\sigma}_1(\widetilde{1}) = -n \cdot \widetilde{1} = -n\alpha - n\beta(\xi+\eta)^{2m+1} \tag{31}$$

by virtue of (28) ($\widetilde{\sigma}_1 = \xi D_\xi + \eta D_\eta - mI$). On the other hand, we obtain

$$\widetilde{\sigma}_1(\widetilde{1}) = -m\alpha + (m+1)\beta(\xi+\eta)^{2m+1} \tag{32}$$

by acting by the operator $\widetilde{\sigma}_1$ on $\widetilde{1}$ directly. It follows from (31) and (32) that $n = m$ or $n = -(m+1)$. Equivalently, this can be written as $n(n+1) = m(m+1)$. In the case $m = -1/2$ the same result is obtained. This proves

THEOREM 10. *Let $n, m \neq 0, -1$. Then the isomorphism (25) of the contact symmetry algebras $\operatorname{Symc}\mathcal{Y}_n$ and $\operatorname{Symc}\mathcal{Y}_m$ of the equations $\mathcal{Y}_n$ and $\mathcal{Y}_m$ exists only if the condition $n(n+1) = m(m+1)$ is satisfied (that is, for $m = n$ or $m = -(n+1)$).*

COROLLARY. *If $n, m \neq 0, -1$, then a local contact equivalence of the equations $\mathcal{Y}_n$ and $\mathcal{Y}_m$ exists only when $n(n+1) = m(m+1)$.*

REMARK 2. In fact, the above considerations prove only that the condition $n(n+1) = m(m+1)$ is necessary. The sufficiency follows from the fact that the contact equivalence of $\mathcal{Y}_n$ and $\mathcal{Y}_m$ and corresponding isomorphism of the algebras $\operatorname{Symc}\mathcal{Y}_n$ and $\operatorname{Symc}\mathcal{Y}_m$ are provided by the global point equivalence $\widetilde{\mu}_n$ of the form (6) for $m = -(n+1)$. The same follows from the equivalence of the equations $\mathcal{Y}_n$ and $\widetilde{\mathcal{Y}}_{n(n+1)} = \{v_{\xi\eta} - n(n+1)(\xi+\eta)^{-2}v = 0\}$ realized by substituting $u = (\xi+\eta)^n v$.

Now let us see when the isomorphism

$$g\colon \operatorname{Sym}\mathcal{Y}_n \to \operatorname{Sym}\mathcal{Y}_m, \qquad n, m \notin \mathbb{Z} \tag{33}$$

of higher symmetry algebras and the local equivalence of $\mathcal{Y}_n$ and $\mathcal{Y}_m$ exist. According to Theorem 6,

$$g(A_n^\infty) = A_m^\infty \tag{34}$$

for any isomorphism (33). Let $R(u) \in \operatorname{NSym}\mathcal{Y}_n$ and

$$g(R(u)) = \widetilde{R}(v) + \widetilde{\alpha}, \qquad \widetilde{R}(v) \in \operatorname{NSym}\mathcal{Y}_m, \ \widetilde{\alpha} \in A_m^\infty. \tag{35}$$

It can be easily seen that $g(R(a)) = \widetilde{R}(g(a))$ for any $a \in A_n^\infty$ as given by (34) and that g is an isomorphism of Lie algebras. Hence the equivalence

$$a \in \operatorname{Ker} R|_{A_n^\infty} \iff g(a) \in \operatorname{Ker}\widetilde{R}|_{A_m^\infty}$$

and the relation $g(\operatorname{Ker} R|_{A_n^\infty}) = \operatorname{Ker}\widetilde{R}|_{A_m^\infty}$ hold. This, according to Theorem 8, yields

THEOREM 11. *Let $n, m \notin \mathbb{Z}$ and g be an isomorphism* (33) *of Lie algebras. Then the orders of the operators R and $\widetilde{R}$ related by* (35) *are equal*: $\deg R = \deg \widetilde{R}$.

COROLLARY. *If the conditions of Theorem* 11 *are satisfied, then the equivalence* $\deg R = 1 \iff \deg \widetilde{R} = 1$ *holds and hence*

$$g(\operatorname{Symc} \mathcal{Y}_n) = \operatorname{Symc} \mathcal{Y}_m .$$

Therefore, the theorem below follows immediately from Theorem 10.

THEOREM 12. *Let $n, m \notin \mathbb{Z}$. Then the isomorphism* (33) *of the higher symmetry algebras* $\operatorname{Sym} \mathcal{Y}_n$ *and* $\operatorname{Sym} \mathcal{Y}_m$ *of the equations $\mathcal{Y}_n$ and $\mathcal{Y}_m$ exists only if the condition $n(n+1) = m(m+1)$ is satisfied.*

COROLLARY. *When $n, m \notin \mathbb{Z}$, the local equivalence of the equations $\mathcal{Y}_n$ and $\mathcal{Y}_m$ occurs only in the case $n(n+1) = m(m+1)$.*

In this connection it is interesting that for some pairs (n, m) parameters satisfying the requirements $n, m \notin \mathbb{Z}$, $n(n+1) \neq m(m+1)$ for nonlocal equivalences between $\mathcal{Y}_n$ and $\mathcal{Y}_m$ equations can be indicated. As a classical example, let us point out the transformation associated with the change of Euler-to-mass Lagrange variables in the equations describing one-dimensional plane isentropic flows of polytropic gas. Note that this transformation belongs to the class of so-called independent variable transformations with respect to the solution [11, 12]. In the case of Euler variables, this system turns out to be equivalent to a single Euler–Darboux equation $\mathcal{Y}_n$ with parameter $n = (k-3)(k-1)^{-1}/2$, where k is the gas adiabatic exponent, and in the case of Lagrangian variables it is equivalent to the equation $\mathcal{Y}_{n'}$ with parameter $n' = n-1$. If $n \neq 0$, then obviously $n(n+1) \neq n'(n'+1)$. Hence, the corresponding equivalence of $\mathcal{Y}_n$ and $\mathcal{Y}_{n'}$ is nonlocal for $n \notin \mathbb{Z}$. Another example is given by the transformation of independent variables with respect to the solution reported in [13] (Ustinov transformation). This transformation provides a nonlocal equivalence of Euler variable equations describing one-dimensional flows of ideal gas of adiabatic exponent k with Euler equations describing the flows of ideal gas of adiabatic exponent $k_1 = k^{-1}$. Nonlocal equivalence of the equations $\mathcal{Y}_n$ and $\mathcal{Y}_{n_1}$ with parameters $n = (k-3)(k-1)^{-1}/2$ and $n_1 = 2-n$ corresponds to this transformation. Clearly, $n(n+1) \neq n_1(n_1+1)$ for $n \neq 1$.

§7. Description of the complete set of local equivalences of the equations $\mathcal{Y}_n$ and $\mathcal{Y}_m$, $n, m \notin \mathbb{Z}$

As usual, a transformation $\mu: \mathcal{Y}_n \to \mathcal{Y}_m$ (possibly local) preserving the contact structure on the equations $\mathcal{Y}_n$ and $\mathcal{Y}_m$ will be called an equivalence of these equations.

THEOREM 13. *Let $n \neq 0, -1$ and $\mu: \mathcal{Y}_n \to \mathcal{Y}_n$ be a self-equivalence of the equation $\mathcal{Y}_n$ inducing the automorphism $\mu_*: \operatorname{Sym} \mathcal{Y}_n \to \operatorname{Sym} \mathcal{Y}_n$ such that $\mu_*(\operatorname{Symc} \mathcal{Y}_n) = \operatorname{Symc} \mathcal{Y}_n$ and the corresponding automorphism*

$$\widetilde{\mu}_*: \{u\} \oplus sl_2 \to \{u\} \oplus sl_2 \tag{36}$$

of the quotient algebra $\{u\} \oplus sl_2$ *of the algebra* $\operatorname{Symc} \mathcal{Y}_n$ *over the ideal* A_n^∞ *is identical*: $\widetilde{\mu}_* = \mathrm{id}_{\{u\}\oplus sl_2}$. *Then* μ *is the restriction to* $\mathcal{Y}_n$ *of the lift of the composition of point transformations*:

$$\mu = \mu_{\lambda, a}^{i} \circ (\mu^-)^j, \qquad i, j = 0, 1,$$

(37)
$$\mu^- : J^0 \to J^0, \quad \mu^-((\xi, \eta, u)) = (\xi_1, \eta_1, u_1), \quad \xi_1 = -\eta, \quad \eta_1 = -\xi, \quad u_1 = u,$$

(38)
$$\mu_{\lambda, a} : J^0 \to J^0, \quad \mu_{\lambda, a}((\xi, \eta, u)) = (\xi, \eta, u_1), \quad u_1 = \lambda u + a, \quad \lambda \in \mathbb{R}, \ \lambda \neq 0, \ a \in A_n^\infty.$$

PROOF. Denote the induced automorphism μ_* of the algebra $\operatorname{Symc} \mathcal{Y}_n$ by g:

$$g = \mu_* : \operatorname{Symc} \mathcal{Y}_n \to \operatorname{Symc} \mathcal{Y}_n.$$

Obviously, $g^{-1}(\varphi) = \psi \mu^*(\varphi)$, $\psi \in C^\infty(J^\infty)$, $\psi \neq 0$ for any $\varphi \in \operatorname{Symc} \mathcal{Y}_n$. In particular, $g^{-1}(1) = \psi$. Let $\sigma = \xi D_\xi + \eta D_\eta$. By evaluating $\sigma(u)$- and $\tau(u)$-invariant solutions of $\mathcal{Y}_n$, we find

(39)
$$\operatorname{Ker} \sigma|_{A_n^\infty} = c_1 + c_2 \int (1 + x)^{2n} x^{-(n+1)} dx, \qquad x = \xi/\eta, \ c_1, c_2 \in \mathbb{R},$$

$$\operatorname{Ker} \tau|_{A_n^\infty} = \begin{cases} (\xi\eta)^n [c_1 + c_2(\xi^{-1} + \eta^{-1})^{2n+1}], & \text{if } n \neq -1/2, \\ (\xi\eta)^{-1/2} [c_1 + c_2 \ln|\xi^{-1} + \eta^{-1}|], & \text{if } n = -1/2, \end{cases} \qquad c_1, c_2 \in \mathbb{R}.$$

Since $\Box(1) = \sigma(1) = 0$ and g is an automorphism of Lie algebras, $\psi = g^{-1}(1) \in \operatorname{Ker} \Box|_{A_n^\infty} \cap \operatorname{Ker} \sigma|_{A_n^\infty}$. Therefore (30) and (39) yield $\psi = \alpha \in \mathbb{R}$, $\alpha \neq 0$ and

(40)
$$g^{-1}(\varphi) = \alpha \mu^*(\varphi), \qquad \varphi \in \operatorname{Symc} \mathcal{Y}_n.$$

As $g^{-1}(u_1) = u + a$ for some function $a \in A_n^\infty$, then $u_1 = \alpha^{-1}(u + a)$. $g^{-1}(A_n^\infty) = A_n^\infty$ so that $\mu^*(\varphi) \in A_n^\infty$ according to (40) for any $\varphi \in A_n^\infty$. In particular, using formulas (30) for $\operatorname{Ker} \Box|_{A_n^\infty}$ and (39) for $\operatorname{Ker} \tau|_{A_n^\infty}$, we obtain

$$\mu((\xi, \eta, u)) = (\xi_1, \eta_1, u_1), \qquad \xi_1 \equiv \xi_1(\xi, \eta), \ \eta_1 \equiv \eta_1(\xi, \eta),$$

or $\xi \equiv \xi(\xi_1, \eta_1)$, $\eta \equiv \eta(\xi_1, \eta_1)$. It is easily found that

(41)
$$\xi = \xi_1 + \varphi_1(\xi_1 + \eta_1), \qquad \eta = \eta_1 + \varphi_2(\xi_1 + \eta_1)$$

from $g^{-1}(\Box(u_1)) = \Box(u) + b_1$, $b_1 \in A_n^\infty$. Note that the Jacobian of the transformation (41) satisfies $\partial(\xi, \eta)/\partial(\xi_1, \eta_1) \equiv 1 + \varphi_1' + \varphi_2' \not\equiv 0$. Similarly, $\xi + \eta = (1 + c_0)(\xi_1 + \eta_1)$, $c_0 \in \mathbb{R}$, $c_0 \neq -1$ is easily obtained from $g(\sigma(u)) = \sigma(u_1) + a_2$, $a_2 \in A_n^\infty$. Hence, the relation $g(\tau(u)) = \tau(u_1) + a_3$, $a_3 \in A_n^\infty$ easily imply that $\varphi_1(x) \equiv \varphi_2(x) \equiv (c_0/2)x$ and thus

$$\xi = (1 + c_0/2)\xi_1 + (c_0/2)\eta_1, \qquad \eta = (c_0/2)\xi_1 + (1 + c_0/2)\eta_1.$$

Finally, $c_0 = 0$ or $c_0 = -2$ are obtained from the same condition $g(\tau(u)) = \tau(u_1) + a_3$. The transformations

$$\begin{cases} \xi = \xi_1, \\ \eta = \eta_1 \end{cases} \quad \text{and} \quad \begin{cases} \xi = -\eta_1, \\ \eta = -\xi_1 \end{cases}$$

correspond to these values of the constant c_0. Theorem 13 is proved.

Now let $\mu\colon \mathcal{Y}_n \to \mathcal{Y}_n$ be the self-equivalence inducing the automorphism $\widetilde{\mu}_*\colon \{u\} \oplus sl_2 \to \{u\} \oplus sl_2$ of the quotient algebra $\{u\} \oplus sl_2$ identical on sl_2 and $\widetilde{\mu}_*(u) = \lambda u_1$, $\lambda \in \mathbb{R}$, $\lambda \neq 0$, that is $\mu_*(u) = \lambda(u_1 + a)$, $a \in A_n^\infty$ (the notation is the same as in Theorem 13). As in the proof of Theorem 13, it is easily seen that $\xi_1 \equiv \xi_1(\xi, \eta)$, $\eta_1 \equiv \eta_1(\xi, \eta)$. Then it can be easily shown that $\lambda = 1$. Therefore, $\widetilde{\mu}_* = \mathrm{id}_{\{u\}\oplus sl_2}$ and Theorem 13 yields the following result.

THEOREM 14. *Let $n \neq 0, -1$ and $\mu\colon \mathcal{Y}_n \to \mathcal{Y}_n$ be self-equivalence of the equation $\mathcal{Y}_n$ inducing the automorphism $\mu_*\colon \mathrm{Sym}\,\mathcal{Y}_n \to \mathrm{Sym}\,\mathcal{Y}_n$ such that $\mu_*(\mathrm{Symc}\,\mathcal{Y}_n) = \mathrm{Symc}\,\mathcal{Y}_n$ and the corresponding automorphism*

$$\widetilde{\widetilde{\mu}}_*\colon sl_2 \to sl_2$$

of the quotient algebra sl_2 of $\mathrm{Symc}\,\mathcal{Y}_n$ with respect to the ideal $A_n^\infty \oplus \{u\}$ is identical: $\widetilde{\widetilde{\mu}}_* = \mathrm{id}_{sl_2}$. Then μ is the restriction to $\mathcal{Y}_n$ of the lift of the composition of point transformations:

$$\mu = \mu^i_{\lambda, a} \circ (\mu^-)^j, \qquad i, j = 0, 1, \ \lambda \in \mathbb{R}, \ \lambda \neq 0, \tag{42}$$

where μ^- and $\mu_{\lambda, a}$ are the transformations (37) *and* (38), *respectively. Each transformation* (42) *induces an automorphism $\widetilde{\mu}_* = \mathrm{id}_{\{u\}\oplus sl_2}$ for any $\lambda \neq 0$.*

By Corollary of Theorem 11, $g(\mathrm{Symc}\,\mathcal{Y}_n) = \mathrm{Symc}\,\mathcal{Y}_n$ for each automorphism g of the algebra $\mathrm{Sym}\,\mathcal{Y}_n$ with $n \notin \mathbb{Z}$. On the other hand, according to Theorem 9, each automorphism of the algebra sl_2 is induced by some equivalence of the form (23). Any equivalence (23) leaves the function u (as an element of $\mathrm{Symc}\,\mathcal{Y}_n$ algebra) "fixed". By Theorem 14, this results in the following

THEOREM 15. *For $n \notin \mathbb{Z}$, the complete group of self-equivalences $\mu_n\colon \mathcal{Y}_n \to \mathcal{Y}_n$ of the equation $\mathcal{Y}_n$ consists of the restrictions to $\mathcal{Y}_n$ of transformations of the form*

$$\begin{gathered} \mu_n = \mu^k_{(abcd)} \circ \mu^i_{\lambda, a} \circ (\mu^-)^j, \\ i, j, k = 0, 1, \ a, b, c, d, \lambda \in \mathbb{R}, \ \Delta = bc - ad \neq 0, \ \lambda \neq 0, \end{gathered} \tag{43}$$

where $\mu_{(abcd)}$, μ^-, and $\mu_{\lambda, a}$ are the transformations (23), (37), *and* (38), *respectively.*

Again, according to the Corollary of Theorem 11, $g(\mathrm{Symc}\,\mathcal{Y}_n) = \mathrm{Symc}\,\mathcal{Y}_m$ for any isomorphism $g\colon \mathrm{Sym}\,\mathcal{Y}_n \to \mathrm{Sym}\,\mathcal{Y}_m$ of Lie algebras with $n, m \notin \mathbb{Z}$. Therefore, Theorem 15 yields the following

THEOREM 16. *For $n \notin \mathbb{Z}$, the complete set of local equivalences $\mu\colon \mathcal{Y}_n \to \mathcal{Y}_{-(n+1)}$ of the equations $\mathcal{Y}_n$ and $\mathcal{Y}_{-(n+1)}$ consists of transformations of the form*

$$\mu = \widetilde{\mu}_n \circ \mu_n \,, \tag{44}$$

where μ_n is a self-equivalence of the form (43) *of the equation $\mathcal{Y}_n$ and $\widetilde{\mu}_n\colon \mathcal{Y}_n \to \mathcal{Y}_{-(n+1)}$ is the equivalence defined by the relation* (6).

If $n, m \notin \mathbb{Z}$ and $n \neq m$, then the local equivalence $\mu\colon \mathcal{Y}_n \to \mathcal{Y}_m$ exists only in the case $m = -(n+1)$. Thus Theorems 15 and 16 describe all possible local equivalences of the equations $\mathcal{Y}_n$ and $\mathcal{Y}_m$ with $n, m \notin \mathbb{Z}$.

§8. Representation of the algebra $[U(L)/(\Delta_n)]_L$ by linear ordinary differential operators

As it was shown in [10], for $-1 < n < 0$, $n \neq -1/2$, the general solution of the equation $\mathcal{Y}_n$ is expressed by the formula

$$\begin{aligned} &u(\xi, \eta) = J_1(-n, -n) + J_2(-n, -n)\,, \\ &J_1(-n, -n) = \int_0^1 \mu(\theta) t^{-n-1}(1-t)^{-n-1} dt\,, \\ &J_2(-n, -n) = (\xi+\eta)^{2n+1} \int_0^1 \nu(\theta) t^n (1-t)^n dt\,, \qquad \theta = \xi - (\xi+\eta)t\,, \end{aligned} \tag{45}$$

where $\mu(\theta)$ and $\nu(\theta)$ are arbitrary twice continuously differentiable functions. Since only smooth solutions of $\mathcal{Y}_n$ are considered here, let us consider $\mu(\theta), \nu(\theta) \in C^\infty(\mathbb{R}^1)$. Obviously, $u(\xi, \eta) = 0$ if and only if $\mu(\theta) = \nu(\theta) = 0$ and the correspondence $u \leftrightarrow (\mu, \nu)$ is a linear isomorphism. Therefore each solution $u(\xi, \eta)$ (45) will be denoted by a pair of its determining functions μ and ν:

$$u(\xi, \eta) = (\mu(\theta), \nu(\theta))\,. \tag{46}$$

Since $\square$, σ_1, τ are the recursion operators for the equation $\mathcal{Y}_n$, their action on the functions (46) gives functions of the same form. Simple but cumbersome transformations lead to the following formulas:

$$\begin{aligned} \square(u) &= \square((\mu, \nu)) = (\mu', \nu')\,, \\ \sigma_1(u) &= \sigma_1((\mu, \nu)) = (\theta\mu' - n\mu, \theta\nu' + (n+1)\nu)\,, \\ \tau(u) &= \tau((\mu, \nu)) = (\theta^2\mu' - 2n\theta\mu, \theta^2\nu' + 2(n+1)\theta\nu)\,, \end{aligned} \tag{47}$$

where $\mu' = d\mu/d\theta$, $\nu' = d\nu/d\theta$. This results in two representations of the algebra $A_{(3)}$ generated by the operators I, $\square$, σ_1, τ in the algebra of linear ordinary differential operators acting in the space of smooth functions defined on $\mathbb{R}^1$. According to formula (8), $\Delta_n^1(u) \equiv (\xi+\eta)^2 l_{F_n}(u) \equiv 0$ for each function $u(\xi, \eta)$ of the form (45). Hence for the classes $\widetilde{\square}$, $\widetilde{\sigma}_1$, $\widetilde{\tau}$ of the algebra $A_{(3)}/(\Delta_n^1)$ the same relations (47), which define the representations of this algebra, hold. Since the algebra $A_{(3)}/(\Delta_n^1)$ is identified with the algebra $U(L)/(\Delta_n)$ (see Proposition 1), formulas (47) define a pair of representations of $U(L)/(\Delta_n)$. For clarity, consider one of these representations which will be denoted by π (for brevity, the images of elements under the representation will be denoted by the same symbols as the elements themselves, but without tildes):

$$\pi\colon [U(L)/(\Delta_n)]_L \to L_{(n)} = \{I\,,\, \square\,,\, \sigma_1\,,\, \tau\}\,,$$

here $L_{(n)}$ is Lie algebra of the associative algebra generated by the compositions of the elements $\square$, σ_1, τ,

$$\begin{aligned}
&\pi(\widetilde{\square}) = \square = D_\theta\,, && \pi(\widetilde{I}) = I\,,\\
(48)\qquad &\pi(\widetilde{\sigma}_1) = \sigma_1 = \theta D_\theta - nI\,, && \\
&\pi(\widetilde{\tau}) = \tau = \theta^2 D_\theta - 2n\theta I\,, && D_\theta = d/d\theta\,.
\end{aligned}$$

If $\widetilde{R} \in U(L)/(\Delta_n)$ and $R(u) = \lambda_n(\widetilde{R}) \in \mathrm{NSym}\,\mathcal{Y}_n$, where λ_n is the anti-isomorphism (5), then R has the form (19) and according to Theorem 8 we have $\dim\mathrm{Ker}(R|_{A_n^\infty}) = 2\deg R$. Therefore, if $\pi(\widetilde{R}) = 0$ (that is $R|_{A_{n,1}^\infty} = 0$, where $A_{n,1}^\infty = \{(\mu\,,\,0) : \mu \in C^\infty(\mathbb{R}^1)\} \subset A_n^\infty$), then $R = 0$ and hence the representation π is faithful.

REMARK 3. The restrictions $-1 < n < 0$, $n \neq -1/2$ are redundant for the objectives of this paper. Essentially it is sufficient that for any smooth functions $\mu(\theta)$ and $\nu(\theta)$ integrals $J_1(-n\,,\,-n)$ and $J_2(-n\,,\,-n)$ (see (45)) converge and as functions of the variables ξ and η are solutions of the equation $\mathcal{Y}_n$ for $n < 0$ and $n > -1$, respectively. According to formulas (47), it is obvious that the subspaces of solutions $\{(\mu\,,\,0) : \mu \in C^\infty(\mathbb{R}^1)\}$ and $\{(0\,,\,\nu) : \nu \in C^\infty(\mathbb{R}^1)\}$ are invariant with respect to the operators $\square$, σ_1, and τ. Therefore, the representation π is defined for each $n < 0$, the second representation is defined for each $n > -1$. The representation π is obviously faithful for each $n < 0$, $n \notin \mathbb{Z}$.

8.1. Some identities in $[U(L)/(\Delta_n)]_L$. It follows from (48) that

$$(49)\qquad \pi(\widetilde{\square}^i_j) = \square^i_j = [\underbrace{\ldots[\square^i\,,\,\tau]\ldots\tau}_{j}]\,.$$

In this section the elements $\square^i_j$ will be regarded as elements of the algebra $L_{(n)}$. Since the representation π is faithful, the same commutation relations are satisfied in $[U(L)/(\square_n)]_L$ and in $L_{(n)}$. In particular, identities (11) are valid in $L_{(n)}$. Obviously,

$$(50)\qquad \square^i_j = \sum_{l=0}^{i} p^i_{j\,,\,i-l}(\theta)\, D_\theta^{i-l}\,,$$

where $p^i_{j\,,\,i-l}(\theta)$ is a polynomial of degree $\leqslant 2j$. Let us evaluate the commutator $[\square^i_j\,,\,\sigma_1]$ using the representation (50) and substitute the result into the second of the relations (11): $[\square^i_j\,,\,\sigma_1] = (i-j)\square^i_j$. We obtain the equations $\theta\cdot(p^i_{j\,,\,i-l})' = (j-l)\,p^i_{j\,,\,i-l}$. Hence, $p^i_{j\,,\,i-l}(\theta) = c^i_{j\,,\,j-l}\theta^{j-l}$, $c^i_{j\,,\,j-l} \in \mathbb{R}$, and

$$(51)\qquad \square^i_j = \sum_{l=0}^{i} c^i_{j\,,\,j-l}\theta^{j-l} D_\theta^{i-l}\,, \qquad c^i_{j\,,\,j-l} \in \mathbb{R}\,.$$

Similarly, by evaluating the commutator of the operator $\Box^i_j$ of the form (51) with τ and comparing the result with $\Box^i_{j+1}$, we find that

(52)

$$
\begin{aligned}
c^i_{j+1\,,\,j+1} &= (2i - j)\, c^i_{j\,,\,j}\,, \\
c^i_{j+1\,,\,0} &= (i - j)(i - (j+1) - 2n)\, c^i_{j\,,\,0}\,, \\
c^i_{j+1\,,\,j+1-l} &= (2i - j - l)\, c^i_{j\,,\,j-l} + (i - l + 1)(i - l - 2n)\, c^i_{j\,,\,j-(l-1)}\,, \qquad 1 \leqslant l \leqslant j.
\end{aligned}
$$

The relation

$$
(53) \qquad -(j+1)(2i - j)\, c^i_{j\,,\,j-l} = (l - j - 1)\, c^i_{j+1\,,\,j+1-l}\,, \qquad 0 \leqslant l \leqslant j
$$

is obtained in the same way from the relation $[\Box^i_{j+1}\,, \Box] = -(j+1)(2i - j)\Box^i_j$. Obviously, $\Box^i_0 = \Box^i = D^i_\theta$. Therefore, $c^i_{0\,,\,0} = 1$ and the first two relations (52) easily give

$$
(54) \qquad c^i_{j\,,\,j} = \prod_{\lambda=0}^{j-1} (2i - \lambda)\,, \qquad c^i_{j\,,\,0} = \prod_{\lambda=0}^{j-1} [(i - \lambda)(i - 1 - 2n - \lambda)]\,.
$$

Relation (53) and the last of the relations (52) obviously imply

$$
c^i_{j\,,\,j-l} = \frac{(i - 2n - l)(i - (l-1))(j - (l-1))}{l\,(2i - l + 1)}\, c^i_{j\,,\,j-(l-1)}\,, \qquad 1 \leqslant l \leqslant j\,,
$$

which finally leads to

$$
(55) \quad c^i_{j\,,\,j-l} = \binom{i - 2n - 1}{l} \prod_{\lambda=0}^{l-1} [(i - \lambda)(j - \lambda)] \prod_{\lambda=1}^{j-1} (2i - \lambda)\,, \qquad 1 \leqslant l \leqslant j - 1\,.
$$

The formulas (51), (54), and (55) give explicit expressions for the operators $\Box^i_j$. The following identities, which are important for further considerations

(56)

$$
\begin{aligned}
\frac{5}{24}[\Box^2_3\,, \Box^2] &= -\Box^3_2 + [4n(n+1) - 3]\Box\,, \\
\frac{7}{3\cdot 96}[\Box^3_4\,, \Box^2] &= -\frac{5}{24}\Box^4_3 + [3n(n+1) - 6]\Box^2_1\,, \\
\frac{2i+1}{12i(i-1)}[\Box^i_3\,, \Box^2] &= -\frac{2i-1}{i+1}\Box^{i+1}_2 - (i - 1 - 2n)(i + 1 + 2n)\Box^{i-1}\,, \qquad i > 1\,,
\end{aligned}
$$

can be obtained by simple but cumbersome transformations by evaluating the commutators $[\Box^2_3\,, \Box^2]$, $[\Box^3_4\,, \Box^2]$, and $[\Box^1_3\,, \Box^2]$ using the above explicit expressions.

The identities (56) are valid in $L_{(n)}$ for any real n. Since the representation π is defined and faithful for $n < 0$, $n \notin \mathbb{Z}$ (see Remark 3), the same identities (56) hold in the algebra $[U(L)/(\Delta_n)]_L$ for each $n < 0$, $n \notin \mathbb{Z}$, as well. By Theorem 2 and its Corollaries, the identities (56) are valid also in $[U(L)/(\Delta_{n'})]_L$, $n' = -(n+1)$. Since $n(n+1) = n'(n'+1)$ and $(i - 1 - 2n)(i + 1 + 2n) = i^2 - 1 - 4n(n+1)$, the replacement of n by n' does not change the right-hand sides in (56). Since $n' > -1$ for $n < 0$, we have

PROPOSITION 10. *For each* $n \notin \mathbb{Z}$, *the identities* (56) *are valid in the algebra* $[U(L)/(\Delta_n)]_L$.

The proof is quite similar for the identity

$$
\begin{aligned}
(57)\quad & (2/3)\, i\,(2i-1)(2i+1)(2i+2)(2i+3)(2i+4)[\square^i_5, \square^3] \\
& = -20i^2(2i-1)^2(2i-2)(2i-3)(2i-4)\,\square^{i+2}_4 - K_i\square^i_2 - \Omega_i\square^{i-2}, \\
& \qquad\qquad i \geqslant 3,\ n \notin \mathbb{Z},
\end{aligned}
$$

where K_i and Ω_i are polynomials with respect to i and n, and the degrees of K_i and Ω_i with respect to n are 2 and 4, respectively. The explicit form of K_i is not needed, and the representation

$$
(58)\qquad \Omega_i = 960i^2(i-1)(2i+2)(2i+3)(2i+4)(2i-4)\,n^4 + \mathcal{P}_3(i, n)
$$

holds for Ω_i. Here $\mathcal{P}_3(i, n)$ is a polynomial of third degree with respect to n.

8.2. "Adjoint" automorphism and variational subalgebra. Obviously, the equalities

$$
(59)\qquad \square^* = -\square', \qquad \sigma_1^* = -\sigma_1', \qquad \tau^* = -\tau'
$$

are valid for the adjoint operators, where $\square'$, σ_1', and τ' denote the operators obtained from $\square$, σ_1, and τ (see (48)) by replacing the parameter n with $n' = -(n+1)$. If we set

$$
(60)\qquad \lambda_n^T : L_{(n)} \to L_{(n')}, \qquad \lambda_n^T(\Delta) = -\Delta^*, \qquad n' = -(n+1),
$$

then the relation $\lambda_n^T([\Delta, \nabla]) = [\lambda_n^T(\Delta), \lambda_n^T(\nabla)]$ can be easily verified. Thus λ_n^T is an isomorphism of Lie algebras. Moreover, it can be easily seen that

$$
(61)\qquad \lambda_n^T(\square^i_j) = (-1)^{i-1}(\square')^i_j, \qquad i \geqslant 0,\ 0 \leqslant j \leqslant 2i.
$$

Consider the commutative diagram

$$
\begin{array}{ccc}
\mathrm{NSym}\,\mathcal{Y}_n & \overset{\lambda^T_{n,N}}{\dashrightarrow} & \mathrm{NSym}\,\mathcal{Y}_{n'} \\
{\scriptstyle i_n}\downarrow & \nwarrow{\scriptstyle j_n^{-1}} & \downarrow{\scriptstyle i_{n'}} \\
L_{(n)} & \underset{\lambda^T_n}{\longrightarrow} & L_{(n')},
\end{array}
$$

where $i_n(\overline{\square}^i_j(u)) = \square^i_j$, $i_{n'}(\overline{\square}^i_j(v)) = (\square')^i_j$, $j_n(\overline{\square}^i_j(u)) = (\square')^i_j$ and $\lambda^T_{n,N}$ is determined from condition that the diagram can be completed to a commutative one. Note that i_n and $i_{n'}$ are "induced" by the representation π (48) and j_n by the second representation

$$
\pi_2 : [U(L)/(\Delta_n)]_L \to L_{(n')}
$$

defined by formulas (47). Since i_n, $i_{n'}$, j_n are anti-isomorphisms,

$$\lambda^T_{n,a} = j_n^{-1} \circ \lambda^T_n \circ i_n \quad \text{and} \quad \lambda^T_{n,N} = i_{n'}^{-1} \circ \lambda^T_n \circ i_n$$

are respectively an automorphism and an isomorphism of Lie algebras. $\lambda^T_{n,a}$ ($\lambda^T_{n,N}$) can be naturally called the "adjoint" automorphism (isomorphism). It obviously follows from (61) that

$$\begin{aligned} &\lambda^T_{n,a}(\overline{\square}^i_j(u)) = (-1)^{i-1}\overline{\square}^i_j(u), \\ &\lambda^T_{n,N}(\overline{\square}^i_j(u)) = (-1)^{i-1}\overline{\square}^i_j(v), \end{aligned} \qquad i \geqslant 0, \ 0 \leqslant j \leqslant 2i. \tag{62}$$

In the case when $n \notin \mathbb{Z}$, we have

$$\{\overline{\square}^i_j(u), \overline{\square}^k_m(u)\} = \sum_{\lambda=0}^{l+k-1} \sum_{\mu=0}^{2\lambda} \alpha^\mu_\lambda \overline{\square}^\lambda_\mu(u), \qquad \alpha^\mu_\lambda \in \mathbb{R}. \tag{63}$$

Acting on the equality (63) by the automorphism $\lambda^T_{n,a}$ and using formula (62), we obtain

$$\alpha^\mu_\lambda \neq 0 \implies i + k + \lambda \equiv 1 \pmod 2.$$

Combined with the inclusion (18), this proves the following

PROPOSITION 11. *If $n \notin \mathbb{Z}$, the only nonzero coefficients α^μ_λ are those for which $i + k + \lambda \equiv 1 \pmod 2$ and $\lambda - \mu = (i + k) - (j + m)$ for any elements $\overline{\square}^i_j(u)$, $\overline{\square}^k_m(u) \in \mathrm{NSym}\,\mathcal{Y}_n$ in formula (63) for the higher Jacobi bracket of these elements.*

COROLLARY. *The linear subspace in $\mathrm{NSym}\,\mathcal{Y}_n$, $n \notin \mathbb{Z}$ generated by the elements*

$$\overline{\square}^i_j(u), \qquad i = 2m + 1, \ m = 0, 1, 2, \dots, \ 0 \leqslant j \leqslant 2i \tag{64}$$

is a subalgebra.

The subalgebra in $\mathrm{NSym}\,\mathcal{Y}_n$ generated by the elements (64) consists of variational symmetries (see [1, 2]) and will be further called *variational*. According to formula (62), the variational subalgebra coincides with the stationary subalgebra (that is the subalgebra consisting of fixed elements) of the "adjoint" automorphism $\lambda^T_{n,a}$.

THEOREM 17. *The direct sum $A_{2N+1} \oplus \{u\}$ of the variational subalgebra A_{2N+1} and the ideal $\{u\}$ is the maximal (non-abelian) subalgebra in $\mathrm{NSym}\,\mathcal{Y}_n$ for almost all $n \notin \mathbb{Z}$ (excluding, possibly, a countable set of values).*

PROOF. Let A be a subalgebra in $\mathrm{NSym}\,\mathcal{Y}_n$ such that $A_{2N+1} \oplus \{u\} \subset A$ and the element $\Delta(u) \in A \backslash A_{2N+1} \oplus \{u\}$ exists. Commuting $\Delta(u)$ with $\square(u)$ and $\tau(u)$ as many times as necessary, we find that $\overline{\square}^{2i_0}(u) \in A$ for some $i_0 \geqslant 1$. According to formula (12), $\{\overline{\square}^{2i_0}(u), \overline{\square}^3_1(u)\} = -12 i_0 \overline{\square}^{2i_0+2}(u) \in A$. Hence $\overline{\square}^{2i}(u) \in A$, and thus $\overline{\square}^{2i}_j(u) \in A$ for all $i \geqslant i_0$, $0 \leqslant j \leqslant 2i$. If $i_0 \geqslant 2$, then according to (57)–(58) $\overline{\square}^{2i_0-2}(u) \in A$ for almost all $n \notin \mathbb{Z}$, except, possibly, for the four roots of the equation $\Omega_{2i_0} = 0$. Thus, for almost all $n \notin \mathbb{Z}$, excluding, possibly, a finite set of values, $\overline{\square}^{2i}(u) \in A$ for all $i \geqslant 1$ and $A = \mathrm{NSym}\,\mathcal{Y}_n$. The theorem is proved.

8.3. On the uniqueness property for the nontrivial ideal in NSym $\mathcal{Y}_n$. Theorem 4 and the third relation in (56) yield the following

PROPOSITION 12. *Suppose $n \notin \mathbb{Z}$, $n \neq (\pm i - 1)/2$ for all integers $i > 1$, and J is an ideal of the algebra* NSym $\mathcal{Y}_n$ *containing at least one nonzero element $g \neq cu$, $c \in \mathbb{R}$. Then J contains all the elements*

$$\overline{\square}^i_j(u), \qquad i \geqslant 0,\ 0 \leqslant j \leqslant 2i \tag{65}$$

of the basis of the algebra NSym $\mathcal{Y}_n$ *except, possibly, the element u.*

LEMMA 3. *If $i+k = j+m$, then the element u does not appear in the expansion of the commutator $\{\overline{\square}^i_j(u), \overline{\square}^k_m(u)\}$ with respect to the basis* (65).

According to Proposition 6, $\{\overline{\square}^i_i(u), \overline{\square}^k_k(u)\} = 0$. So it is sufficient to examine the case of $j < i$, $m > k$. Since commutation with $\overline{\square}^i_j(u)$ is a differentiation, it can be easily seen that $\{\overline{\square}^i_j(u), \overline{\square}^k_m(u)\} = -\{\overline{\square}^i_{j+1}(u), \overline{\square}^k_{m-1}(u)\} + \dots$, where dots denote terms that do not contain the element u in their expansion with respect to the basis (65). Proceeding in this way, after $(i-j)$ steps we obtain the equality $\{\overline{\square}^i_j(u), \overline{\square}^k_m(u)\} = \pm\{\overline{\square}^i_i(u), \overline{\square}^k_k(u)\} + \dots = 0 + \dots$ which proves the lemma.

Let

$$J_1 = \{\overline{\square}^i_j(u),\ i \geqslant 1,\ 0 \leqslant j \leqslant 2i\} \tag{66}$$

be the linear span of the family of functions $\overline{\square}^i_j(u)$ with $i \geqslant 1$. If $i+k \neq j+m$, then according to Proposition 11, $\lambda - \mu \neq 0$ for each λ and μ in the representation (63) of the Jacobi bracket; hence $\{\overline{\square}^i_j(u), \overline{\square}^k_m(u)\} \in J_1$. When $i+k = j+m$, $\{\overline{\square}^i_j(u), \overline{\square}^k_m(u)\} \in J_1$ by Lemma 3. This proves

PROPOSITION 13. *For $n \notin \mathbb{Z}$, the linear space J_1 (66) is an ideal in* NSym $\mathcal{Y}_n$.

Thus, J_1 can be identified with the quotient algebra NSym $\mathcal{Y}_n/\{u\}$ and Theorem 17 yields

COROLLARY. *For almost all $n \notin \mathbb{Z}$ (excluding, possibly, a countable set of values) the variational subalgebra A_{2N+1} is a maximal subalgebra in J_1.*

The following theorem immediately results from Propositions 12 and 13.

THEOREM 18. *Suppose $n \notin \mathbb{Z}$, $n \neq (\pm i - 1)/2$ for all integer $i \geqslant 1$ and J is the nonzero eigenideal of the algebra* NSym $\mathcal{Y}_n$. *Then either $J = \{u\}$ or $J = J_1$, where J_1 is the ideal defined by* (66).

Hence, for almost all $n \notin \mathbb{Z}$, J_1 is essentially the only nontrivial ideal in NSym $\mathcal{Y}_n$.

§9. Description of the complete set of isomorphisms g of the algebras NSym $\mathcal{Y}_n$ and NSym $\mathcal{Y}_m$ satisfying the restriction $g(\text{NSymc}\,\mathcal{Y}_n) = \text{NSymc}\,\mathcal{Y}_m$ for $n, m \notin \mathbb{Z}$, $n, m \neq (\pm i - 1)/2$, $i = 2, 4, 6, \dots$

Here we consider the isomorphisms

$$g\colon \text{NSym}\,\mathcal{Y}_n \to \text{NSym}\,\mathcal{Y}_m, \qquad n, m \notin \mathbb{Z} \tag{67}$$

of Lie algebras for which $g(\mathrm{NSymc}\,\mathcal{Y}_n) = \mathrm{NSymc}\,\mathcal{Y}_m$. This class includes, for example, the isomorphisms g induced by isomorphisms $\tilde{g}\colon \mathrm{Sym}\,\mathcal{Y}_n \to \mathrm{Sym}\,\mathcal{Y}_m$ of the complete higher symmetry algebras (see the Corollary to Theorem 11).

We shall begin by examining the isomorphisms g (67) for which

$$g(\square^u) = \square^v, \quad g(\sigma_1^u) = \sigma_1^v, \quad g(\tau^u) = \tau^v, \qquad g(u) = \lambda v, \ \lambda = \mathrm{const} \neq 0. \tag{68}$$

The notation used here corresponds to that adopted in §6.

LEMMA 4. *Let $n, m \notin \mathbb{Z}$, $n, m \neq (\pm i - 1)/2$, $i = 2, 4, 6, \ldots$, and g be the isomorphism* (67) *of Lie algebras for which the relations* (68) *hold. Then*

$$g(\overline{\square}^2(u)) = c_2 \overline{\square}^2(v), \qquad c_2 = \mathrm{const} \neq 0. \tag{69}$$

PROOF. It follows from Proposition 3 and Theorem 18 that $g(L_\square u) = L_\square v$ and $g(J_1) = J_1$. We have $\overline{\square}^2(u) \in L_\square u \cap J_1$; hence $g(\overline{\square}^2(u)) \equiv \rho(v) \subset L_\square v \cap J_1$ and $\rho(v) = \sum_{i=1}^N c_i \overline{\square}^i(v)$, $c_i \in \mathbb{R}$. Since $\overline{\square}_5^2(u) = 0$, it follows that $\rho(v) = c_1 \square(v) + c_2 \overline{\square}^2(v)$. Finally, the second relation in (14) yields $c_1 = 0$. Relation (69) is proved.

PROPOSITION 14. *Let $n, m \notin \mathbb{Z}$, $n, m \neq (\pm i - 1)/2$, $i = 2, 4, 6, \ldots$, and g be an isomorphism* (67) *of Lie algebras for which relations* (68) *hold. Then*

$$g(\overline{\square}_j^i(u)) = c_2^{i-1} \overline{\square}_j^i(v), \qquad i \geqslant 1, \ 0 \leqslant j \leqslant 2i, \ c_2 = \mathrm{const} \neq 0. \tag{70}$$

Indeed, using Lemma 4 and relations (12), it is easily proved by induction that $g(\overline{\square}^i(u)) = c_2^{i-1} \overline{\square}^i(v)$, $i \geqslant 1$. Together with the definition

$$\overline{\square}_j^i(u) = \underbrace{\{\tau^u, \ldots \{\tau^u}_{j}, \overline{\square}^i(u)\} \ldots\},$$

this trivially yields the desired relation (70). If the conditions of Proposition 14 are satisfied, the constant c_2 is easily calculated in the following manner. In the algebras $\mathrm{NSym}\,\mathcal{Y}_n$ and $\mathrm{NSym}\,\mathcal{Y}_m$, the analogs of the relations (56) hold. In particular, in $\mathrm{NSym}\,\mathcal{Y}_n$

$$(5/24)\{\overline{\square}^2(u), \overline{\square}_3^2(u)\} = -\overline{\square}_2^3(u) + [4n(n+1) - 3]\square(u). \tag{71}$$

We find $c_2^2 = (4n(n+1) - 3)/(4m(m+1) - 3)$ by acting on (71) by the isomorphism g and comparing the result with the analog of relation (71) for $\mathrm{NSym}\,\mathcal{Y}_m$. Similarly, the second of the relations (56) leads to the equality $c_2^2 = (n(n+1) - 2)/(m(m+1) - 2)$. Now $n(n+1) = m(m+1)$, $c_2^2 = 1$, as can be obviously obtained by comparing the resulting expressions for c_2^2. The same result can be derived from the last of the relations (56) in a similar way. Thus the following proposition is proved.

PROPOSITION 15. *If the requirements of Proposition* 14 *are satisfied, then* $n(n+1) = m(m+1)$ *and the equality* $c_2^2 = 1$ *for the constant* c_2 *from* (70) *holds.*

Note that the already known relation $n(n+1) = m(m+1)$ is obtained as a necessary (and obviously, sufficient) condition for the existence of isomorphisms of the types under consideration. By formulas (62), the "adjoint" automorphism $\lambda^T_{n,a}$ (isomorphism $\lambda^T_{n,N}$) is an example of an automorphism (isomorphism in the case $m = -(n+1)$) satisfying relations (68) with constant $\lambda = 1$, for which $c_2 = -1$. Thus the previous considerations yield

THEOREM 19. *Let* $n \notin \mathbb{Z}$, $n \neq (\pm i - 1)/2$, $i = 2, 4, 6, \dots$, *and* g *be the automorphism of the algebra* $\mathrm{NSym}\,\mathcal{Y}_n$ *inducing the identical automorphism on* $\mathrm{NSymc}\,\mathcal{Y}_n$. *Then* g *is either an identical automorphism or the "adjoint" automorphism* $\lambda^T_{n,a}$ (62).

Denote by i_λ the automorphism

$$
(72) \qquad \begin{gathered} i_\lambda\colon \mathrm{NSym}\,\mathcal{Y}_n \to \mathrm{NSym}\,\mathcal{Y}_n\,, \qquad i_\lambda(\Box^i_j(u)) = \Box^i_j(u)\,, \\ i \geqslant 1\,,\ 0 \leqslant j \leqslant 2i\,,\ i_\lambda(u) = \lambda u\,,\ \lambda = \mathrm{const} \neq 0\,. \end{gathered}
$$

Obviously, the set $G = \{i_\lambda : \lambda \in \mathbb{R},\ \lambda \neq 0\}$ is an abelian group isomorphic to the multiplicative group of the field of real numbers. Now let g be an arbitrary automorphism of the algebra $\mathrm{NSym}\,\mathcal{Y}_n$ with $g(\mathrm{NSymc}\,\mathcal{Y}_n) = \mathrm{NSymc}\,\mathcal{Y}_n$. Then g induces some automorphism g_s of the algebra sl_2 identified with the quotient algebra $\mathrm{NSymc}\,\mathcal{Y}_n/\{u\}$. By Theorem 9, g_s is induced by some self-equivalence μ (23) of the equation $\mathcal{Y}_n$, i.e., $g_s = \mu_*|_{sl_2}$. Note that $\mu_*(\mathrm{NSymc}\,\mathcal{Y}_n) = \mathrm{NSymc}\,\mathcal{Y}_n$. Clearly, $g \circ \mu_*^{-1}|_{\mathrm{NSymc}\,\mathcal{Y}_n} = i_\lambda$ for some $\lambda \neq 0$. Therefore, Theorem 19 easily yields

THEOREM 20. *For* $n \notin \mathbb{Z}$, $n \neq (\pm i - 1)/2$, $i = 2, 4, 6, \dots$, *the complete group of automorphisms* g *of the algebra* $\mathrm{NSym}\,\mathcal{Y}_n$ *satisfying the restriction* $g(\mathrm{NSymc}\,\mathcal{Y}_n) = \mathrm{NSymc}\,\mathcal{Y}_n$ *consists of automorphisms of the form*

$$
(73) \qquad g_n = i_\lambda \circ (\lambda^T_{n,a})^i \circ \mu^j_*\,, \qquad i, j = 0, 1,\ \lambda \in \mathbb{R},\ \lambda \neq 0\,,
$$

where i_λ, $\lambda^T_{n,a}$ *are automorphisms defined by formulas* (72) *and* (62), μ_* *is the automorphism induced by the self-equivalence* μ (23) *of the equation* $\mathcal{Y}_n$.

Theorem 20 implies

THEOREM 21. *For* $n \notin \mathbb{Z}$, $n \neq (\pm i - 1)/2$, $i = 2, 4, 6, \dots$, *the complete set of isomorphisms* $g\colon \mathrm{NSym}\,\mathcal{Y}_n \to \mathrm{NSym}\,\mathcal{Y}_{-(n+1)}$ *satisfying the restriction* $g(\mathrm{NSymc}\,\mathcal{Y}_n) = \mathrm{NSymc}\,\mathcal{Y}_{-(n+1)}$ *consists of isomorphisms of the form*

$$
g = (\widetilde{\mu}_n)_* \circ g_n\,,
$$

where g_n *is an automorphism of the form* (73) *of the algebra* $\mathrm{NSym}\,\mathcal{Y}_n$ *and* $(\widetilde{\mu}_n)_*$ *is the isomorphism induced by the self-equivalence* $\widetilde{\mu}_n\colon \mathcal{Y}_n \to \mathcal{Y}_{-(n+1)}$ *defined by relation* (6).

Recall that $(\widetilde{\mu}_n)_*(\Box^i_j(u)) = \Box^i_j(v)$, $i \geqslant 0$, $0 \leqslant j \leqslant 2i$. According to Proposition 15, Theorems 20 and 21 describe all possible isomorphisms $g\colon \mathrm{NSym}\,\mathcal{Y}_n \to \mathrm{NSym}\,\mathcal{Y}_m$ satisfying the condition $g(\mathrm{NSymc}\,\mathcal{Y}_n) = \mathrm{NSymc}\,\mathcal{Y}_m$ for $n, m \notin \mathbb{Z}$, $n, m \neq (\pm i - 1)/2$, $i = 2, 4, 6, \dots$.

References

1. V. E. Shemarulin, *Higher symmetries and conservation laws for the equation of one-dimensional plane isentropic flows of a polytropic gas*, Mat. Zametki **47** (1990), no. 3, 138–140; English transl. in Math. Notes **47** (1990).
2. ———, *Higher symmetries and conservation laws of Euler–Darboux Equations. Geometry and Differential Equations*, World Scientific, Singapore, 1993 (to appear).
3. L. V. Ovsyannikov, *Group analysis of differential equations*, "Nauka", Moscow, 1978; English transl., Academic Press, New York, 1982.
4. V. E. Shemarulin, *Higher symmetry algebra structures and local equivalences of Euler–Darboux equations*, Dokl. Akad. Nauk SSSR **330** (1993), no. 1, 24–27; English transl. in Soviet Math. Dokl. **47** (1993).
5. A. M. Vinogradov, *Local symmetries and conservation laws*, Acta Appl. Math. **2** (1984), no. 1, 21–78.
6. A. M. Vinogradov, I. S. Krasilshchik, and V. V. Lychagin, *Introduction to geometry of nonlinear differential equations*, "Nauka", Moscow, 1986; English transl., *Geometry of jet spaces and nonlinear partial differential equations*, Gordon and Breach, New York, 1986.
7. N. Jacobson, *Lie algebras*, Interscience, New York and London, 1961.
8. J. Dixmier, *Algèbres enveloppantes*, Bordas, Paris, Bruxelles, and Montreal, 1974.
9. J. F. Pommaret, *Systems of partial differential equations and Lie pseudogroups*, Gordon and Breach, New York, London, and Paris, 1978.
10. S. G. Samko, A. A. Kilbas, and O. I. Marichev, *Fractional order integrals and derivatives and some of their applications*, "Nauka i Tekhnika", Minsk, 1987. (Russian)
11. B. L. Rozhdestvenskiĭ and N. N. Yanenko, *Systems of quasilinear equations and their applications to gas Dynamics*, "Nauka", Moscow, 1968. (Russian)
12. E. V. Ferapontov, *Transformations with respect to solution and their invariants*, Differentsial'nye Uravneniya (1989), no. 7, 1256–1265; English transl. in Differential Equations (1989), no. 7.
13. M. D. Ustinov, *Transformation and some solutions of perfect gas motion equations*, Izv. Akad. Nauk SSSR Ser. Mekh. Zhidk. Gaza (1966), no. 3, 68–74; English transl. in Fluid Mechanics (1966), no. 3.

Translated by THE AUTHOR

ALL-RUSSIA RESEARCH INSTITUTE OF EXPERIMENTAL PHYSICS (VNIIEF), 607200, ARZAMAS-16, RUSSIA,

E-mail address: otd8.1323@rfnc.nnov.su

Amer. Math. Soc. Transl.
(2) Vol. **167**, 1995

Hyperbolicity and Multivalued Solutions of Monge–Ampère Equations

D. V. TUNITSKY

The notion of multivalued solution is well known. It appears in the work of G. Monge, J. Pfaff, S. Lie, E. Cartan, and others. Nevertheless, there are no known results on the solvability of any wide set of partial differential equations of order higher than 1 in the class of multivalued solutions. The paper deals with the solvability of Monge–Ampère equations with two independent variables in the class of multivalued solutions. A definition of hyperbolicity that enlarges the class of hyperbolic equations and includes some equations with multiple characteristics into this set is formulated in this paper. Hyperbolic equations in this sense are reduced to corresponding diagonal systems of quasilinear equations. The unique local solvability of the initial value problem for hyperbolic Monge–Ampère equations in the class of multivalued solutions is proved on basis of this reduction. The local contact equivalence of any hyperbolic Monge–Ampère equation to the corresponding quasilinear equation is proved as a consequence of the unique solvability.

§1. Multivalued solutions

Consider the Monge–Ampère equation

$$A + Bz_{xx} + 2Cz_{xy} + Dz_{yy} + E(z_{xx}z_{yy} - z_{xy}^2) = 0. \tag{1.1}$$

The coefficients A, B, C, D, and E depend of x, y, z_x, z_y, and z, where $z = z(x, y)$ is the unknown function. Let A, B, C, D, and E be functions with continuous third derivatives such that

$$A^2 + B^2 + C^2 + D^2 + E^2 \neq 0. \tag{1.2}$$

Suppose z is a twice continuously differentiable function in a certain domain S of the xy plane satisfying equation (1.1). Put

$$p = z_x, \qquad q = z_y. \tag{1.3}$$

1991 *Mathematics Subject Classification*. Primary 35L70.

Expressions (1.3) and

$$x = x, \quad y = y, \quad z = z(x, y) \tag{1.4}$$

define in the space $\mathbb{R}^5$ with coordinates x, y, p, q, and z a two-dimensional surface $\sigma\colon S \to \mathbb{R}^5$. Consider another parameterization of this surface

$$x = x(u, v), \quad y = y(u, v), \quad p = p(u, v), \quad q = q(u, v), \quad z = z(u, v). \tag{1.5}$$

It follows from (1.5), (1.3)–(1.4) and the implicit function theorem that

$$\begin{gathered} z_{xx} = \frac{1}{\Delta}\begin{vmatrix} p_u & y_u \\ p_v & y_v \end{vmatrix}, \qquad z_{yy} = \frac{1}{\Delta}\begin{vmatrix} x_u & q_u \\ x_v & q_v \end{vmatrix}, \\ z_{xy} = \frac{1}{2\Delta}\left(\begin{vmatrix} x_u & p_u \\ x_v & p_v \end{vmatrix} + \begin{vmatrix} q_u & y_u \\ q_v & y_v \end{vmatrix}\right), \qquad z_{xx}z_{yy} - z_{xy}^2 = \frac{1}{\Delta}\begin{vmatrix} p_u & q_u \\ p_v & q_v \end{vmatrix}, \end{gathered} \tag{1.6}$$

where $\Delta = \begin{vmatrix} x_u & y_u \\ x_v & y_v \end{vmatrix}$. In (1.1) replace the second derivatives using (1.6). Since $\Delta \neq 0$, we have

$$\begin{aligned} A\begin{vmatrix} x_u & y_u \\ x_v & y_v \end{vmatrix} + B\begin{vmatrix} p_u & y_u \\ p_v & y_v \end{vmatrix} + C\left(\begin{vmatrix} x_u & p_u \\ x_v & p_v \end{vmatrix} + \begin{vmatrix} q_u & y_u \\ q_v & y_v \end{vmatrix}\right) & \\ + D\begin{vmatrix} x_u & q_u \\ x_v & q_v \end{vmatrix} + E\begin{vmatrix} p_u & q_u \\ p_v & q_v \end{vmatrix} &= 0. \end{aligned} \tag{1.7}$$

It follows from (1.3) that

$$z_u - px_u - qx_u, \qquad z_v - px_v - qx_v. \tag{1.8}$$

So we see that the vector-function $\sigma(x, y)$ satisfies the equalities

$$\sigma^*\omega_0 = 0, \qquad \sigma^*\omega_2 = 0, \tag{1.9}$$

where

$$\omega_0 = dz - p\,dx - q\,dy,$$
$$\omega_2 = A\,dx \wedge dy + B\,dp \wedge dy + C(dx \wedge dp + dq \wedge dy) + D\,dx \wedge dq + E\,dp \wedge dq.$$

It is well known that any differential equation can be written as a system of exterior differential equations on the corresponding manifold [1, item 79]. In this case the system (1.9) in space $\mathbb{R}^5$ corresponds to (1.1) equation (cf. [2]).

The differential form ω_0 defines a contact structure in the space $\mathbb{R}^5$ and ω_2 corresponds to the Monge–Ampère equation (1.1). The inverse operator σ^* commutes with the exterior differentiation operator [3, 2.33]. Therefore, $\sigma^*(\omega_1) = 0$ follows from the first relation (1.9), where

$$\omega_1 = d\omega_0 = dx \wedge dp + dy \wedge dq.$$

REMARK 1. The form ω_2 that corresponds to the left-hand side of equation (1.1) is chosen in such a way that there is no dz factor in it and the coefficients of the monomials $dx \wedge dy$ and $dq \wedge dy$ are equal to each other. Such forms are called *effective* [2, p. 140]. A simple algebraic criteria of effectiveness in this case are the conditions

$$\partial/\partial z \lrcorner\, \omega = 0, \qquad \omega \wedge \omega_1 = 0. \tag{1.10}$$

Here $\lrcorner$ is the interior product of a differential form and a vector field, $\wedge$ is the exterior product of differential forms.

Below all manifolds are assumed to be connected, Hausdorff, paracompact, and C^∞-smooth.

DEFINITION 1. Let S be a two-dimensional manifold. We shall say that an imbedded C^k-smooth submanifold (S, σ)

$$\sigma : S \to \mathbb{R}^5 \tag{1.11}$$

is a C^k-smooth *multivalued solution* of equation (1.1) if conditions (1.9) are fulfilled (cf. [4]).

§2. Hyperbolicity

The Pfaffian of the effective differential form ω_2,

$$\mathrm{Pf}(\omega_2) = -C^2 + BD - AE \tag{2.1}$$

coincides with the discriminant of the characteristic form of equation (1.1) calculated using the solution z, exact up to sign [5, Chapter 5, App. 1, §2]. Let the function (2.1) be nonnegative

$$\mathrm{Pf}(\omega_2) \leqslant 0. \tag{2.2}$$

Put $\Delta = \sqrt{-\mathrm{Pf}(\omega_2)}$, $\lambda_j = (-1)^{3-j}\Delta$, and suppose Δ has continuous derivatives of third order in the entire space $\mathbb{R}^5$.

A two-dimensional vector subbundle H_j of the tangent bundle $T\mathbb{R}^5$ is said to be a *characteristical distribution* if its elements ξ are solutions of the system of characteristic equations

$$\xi \lrcorner\, \omega_0 = 0, \qquad \xi \lrcorner\, (\omega_2 - \lambda_j \omega_1) = 0. \tag{2.3}$$

One can prove that the characteristic distribution H_j is generated by vector fields

$$\begin{aligned}
e_{j1} &= E\left(\frac{\partial}{\partial x} + p\frac{\partial}{\partial z}\right) - D\frac{\partial}{\partial p} + (C - \lambda_j)\frac{\partial}{\partial q}, \\
e_{j2} &= E\left(\frac{\partial}{\partial y} + q\frac{\partial}{\partial z}\right) + (C + \lambda_j)\frac{\partial}{\partial p} - B\frac{\partial}{\partial q}, \\
e_{j3} &= B\left(\frac{\partial}{\partial x} + p\frac{\partial}{\partial z}\right) + (C - \lambda_j)\left(\frac{\partial}{\partial y} + q\frac{\partial}{\partial z}\right) - A\frac{\partial}{\partial p}, \\
e_{j4} &= (C + \lambda_j)\left(\frac{\partial}{\partial x} + p\frac{\partial}{\partial z}\right) + D\left(\frac{\partial}{\partial y} + q\frac{\partial}{\partial z}\right) - A\frac{\partial}{\partial q}.
\end{aligned} \tag{2.4}$$

Among the vector fields (2.4) exactly two are linearly independent at any point of space $\mathbb{R}^5$. This follows from the definition of λ_j and inequality (1.2). For example, e_{j1} and e_{j2} are independent at the points where the coefficient E does not vanish. It is known that any vector bundle over $\mathbb{R}^5$ is trivial [6, Chapter 4, 2.5]. This means that in $\mathbb{R}^5$ for each $j = 1, 2$ there exist two linearly independent C^3-smooth vector fields

$$\xi_{j1}, \ \xi_{j2}, \tag{2.5}$$

that generate characteristical distribution H_j at each point of the space $\mathbb{R}^5$.

A *characteristic codistribution* H_j^* is defined in a dual way as the set of linear forms that are solutions of the equation

$$\omega \wedge (\omega_2 - \lambda_j \omega_1) = 0. \tag{2.6}$$

It can be proved that the characteristic codistribution H_j^* is two-dimensional and is generated by C^3-smooth linear differential forms ω_{3-j1} and ω_{3-j2} at each point of $\mathbb{R}^5$, where

$$\omega_{j1} = \xi_{j1} \lrcorner\, \omega_1, \qquad \omega_{j2} = \xi_{j2} \lrcorner\, \omega_1.$$

DEFINITION 2. Equation (1.1) is called *hyperbolic* if the inequality (2.2) is satisfied fulfilled in $\mathbb{R}^5$ and there exist C^2-smooth differential 5-forms Ω_{ji} such that

$$d\omega_{ji} \wedge \omega_{j1} \wedge \omega_{j2} \wedge \omega_0 = \Delta \Omega_{ji}, \qquad j, i = 1, 2. \tag{2.7}$$

It is obvious that Definition 2 is independent of the choice of the generators (2.5).

REMARK 2. We shall call equation (1.1) *strictly hyperbolic* if inequality (2.2) is fulfilled with the strict inequality sign. Obviously any strictly hyperbolic equation (1.1) is hyperbolic in the sense of Definition 2.

The following results hold.

LEMMA 2.1. *Suppose ω is a differential m-form and $a_1, \dots, a_n$ are independent C^k-smooth linear differential forms in space $\mathbb{R}^5$. Then the C^k-smooth differential $(m-1)$-forms $\beta_1, \dots, \beta_n$ such that*

$$\omega = \alpha_1 \wedge \beta_1 + \dots + \alpha_n \wedge \beta_n,$$

is true exist if and only if $\omega \wedge \alpha_1 \wedge \dots \wedge \alpha_n = 0$.

PROOF. See [1, item 17].

LEMMA 2.2. *Equation (1.1) is hyperbolic if and only if for certain C^2-smooth functions a_{ji} we have the representation*

$$d\omega_{ji} \wedge \omega_{j1} \wedge \omega_{j2} \wedge \omega_0 = a_{ji} \omega_1 \wedge \omega_{j1} \wedge \omega_{j2} \wedge \omega_0. \tag{2.8}$$

PROOF. By definition (2.6) and Lemma 2.1, we have

$$\omega_{3-j1} \wedge \omega_{3-j2} = r_j(\omega_2 - \lambda_j \omega_1), \qquad j = 1, 2, \tag{2.9}$$

for a certain nonzero C^3-smooth function r_j. It follows from (2.9), (2.8), and the second effectiveness condition (1.10) that (2.7) holds. Now suppose that condition (2.7) holds. Then $\Omega_{ji} = r_{ji}\omega_i \wedge \omega_1 \wedge \omega_0$ for a certain C^2-smooth function r_{ji}. Therefore (2.8) follows from (2.7), (2.9), and the second condition (1.10).

Lemma 2.2 is proved.

REMARK 3. Lemmas 2.2 and 2.1 imply the following criteria of hyperbolicity. Equation (1.1) is hyperbolic if and only if the representations

$$d\omega_{ji} = b^1_{ji} \wedge \omega_{j1} + b^2_{ji} \wedge \omega_{j2} + a_{ji} \wedge \omega_0 + c_{ji}\omega_1 \tag{2.10}$$

hold for $j, i = 1, 2$ and certain C^2-smooth differential forms b^1_{ji}, b^2_{ji}, a_{ji}, and c_{ji}.

It can be proved that if $E^2 + B^2 = 1$, then conditions (2.7) are equivalent to the congruences

$$\begin{gathered} e_1(\Delta) \equiv 0 \mod \Delta, \qquad e_2(\Delta) \equiv 0 \mod \Delta, \\ Be_1(E) - Ee_1(B) - e_2(C) \equiv 0 \mod \Delta, \\ e_1C + e_2(ED + BA) - C^2(Be_2(E) - Ee_2(B)) \equiv 0 \mod \Delta. \end{gathered} \tag{2.11}$$

Here

$$e_1 = \frac{\partial}{\partial x} + p\frac{\partial}{\partial z} + BC\left(\frac{\partial}{\partial y} + q\frac{\partial}{\partial z}\right) - (ED + BA)\frac{\partial}{\partial p} + EC\frac{\partial}{\partial q},$$

$$e_2 = E\left(\frac{\partial}{\partial y} + q\frac{\partial}{\partial z}\right) + C\frac{\partial}{\partial p} - B\frac{\partial}{\partial q}.$$

Congruency modulo Δ means that there exists a twice continuously differentiable function such that left part of the congruence can be represented as the product of this function and Δ. Thus $\mathrm{mod}\,\Delta = \{\Delta f \mid f \in C^2(\mathbb{R}^5)\}$ and congruences (2.11) imply divisibility of the left parts of this congruences by Δ in the class of C^2-smooth functions.

Examples of equations that are hyperbolic in the sense of Definition 2 but not strictly hyperbolic are presented below.

EXAMPLES. 1) $z_{xx}z_{yy} - z^2_{xy} = -a^2(z_x, z_y, z - xz_x - yz_y)$,

2) $z_{xx} = a^2(y, z_x, z - xz_x)z_{yy}$,

3) $z_{xx} = (a(y)z_y)_y$.

Here α is any nonnegative three times continuously differentiable function, $\Delta = a$, and the conditions of strict hyperbolicity break down at points where $a = 0$.

4) Equations (1.1) with constant coefficients A, B, C, D, E, and zero Pfaffian (2.1). Such equations appear in certain problems of geophysics [7].

5) Equations of regular isometric immersions of two-dimensional metrics into three-dimensional Euclidean spaces by means of the Darboux method.

6) The second order equation which is linear with respect to derivatives,

$$a_0 + a_1z_x + a_2z_y + z_{xx} + 2Cz_{xy} + Dz_{yy} = 0, \tag{2.12}$$

where a_1, a_2, C, D are functions of x and y, a_0 is a function of x, y, and z, is hyperbolic if

$$\Delta^2 = C^2 - D \geqslant 0, \qquad \Delta_x + C\Delta_y \equiv 0 \mod \Delta, \\ C_x + CC_y + a_1 C - a_2 \equiv 0 \mod \Delta. \tag{2.13}$$

Strict hyperbolicity is violated at the points where $\Delta = 0$. Known conditions for nonstrict hyperbolicity for linear equations are the Levi conditions [8] (see also [5, Chapter 5, §8]). These conditions ensure the possibility of representation of equation (2.12) in the form

$$\left(\frac{\partial}{\partial x} + \tau_i \frac{\partial}{\partial y}\right)\left(\frac{\partial}{\partial x} + \tau_{3-i} \frac{\partial}{\partial y}\right) z + b_1 \left(\frac{\partial}{\partial x} + \tau_1 \frac{\partial}{\partial y}\right) z + b_2 \left(\frac{\partial}{\partial x} + \tau_2 \frac{\partial}{\partial y}\right) z + a_0 = 0$$

either for $i = 1$ or for $i = 2$. Here $\tau_1 = C + \Delta$, $\tau_2 = C - \Delta$, and b_1, b_2 are certain smooth functions of x and y. It is easy to prove that the condition of representability is reduced to the congruency

$$\tau_{3-ix} + C\tau_{3-iy} + a_1 C - a_2 \equiv 0 \mod \Delta \tag{2.14}$$

either for $i = 1$ or for $i = 2$. It is obvious that the congruencies (2.13) are equivalent to the congruencies (2.14) for $i = 1$ and $i = 2$ together. Thus we can say that relations (2.11) give a nonlinear analog of the Levi conditions.

§3. Initial value problem and contact equivalence

Consider an initial value for equation (1.1) in the form of an imbedded *initial curve*

$$l\colon (a, b) \to \mathbb{R}^5 \tag{3.1}$$

that is of class $C^3(a, b)$, $-\infty \leqslant a < b \leqslant +\infty$, and for which

$$\omega_0(\dot{l}(t)) = 0. \tag{3.2}$$

Suppose this curve to be *free*, i.e.,

$$\dot{l}(t) \notin H_j(l(t)), \qquad j = 1, 2, \quad a < t < b. \tag{3.3}$$

We say that a solution (1.11) of equation (1.1) is a *solution of initial value problem* defined by free initial curve (3.1) if

$$l(t) \subset \sigma(S), \qquad a < t < b. \tag{3.4}$$

A solution (S, σ) is called *nonsingular* if it is not tangent to characteristic distributions at any point, i.e.,

$$\sigma_*(T_s S) \neq H_j(\sigma(s)), \qquad s \in S, \ j = 1, 2.$$

LEMMA 3.1. *Let* (S, σ) *be a nonsingular solution of the hyperbolic equation* (1.1). *Then the linear space*

$$\sigma_*(T_sS) \cap H_j(\sigma(s)) \tag{3.5}$$

is one-dimensional for any $s \in S$ *and* $j = 1, 2$.

PROOF. By the assumption, (S, σ) is a nonsingular solution of equation (1.1). Therefore the dimension of the space (3.5) is at most 1. It follows from (1.9) and (2.9) that the forms $\sigma^*(\omega_{j1})$ and $\sigma^*(\omega_{j2})$ are linearly dependent for $i = 1, 2$. In other words, if e_1, e_2 is a basis of the space T_sS, then

$$\omega_1(a_j\xi_{j1} + b_j\xi_{j2}, e_k) = 0$$

for the vector fields (2.5), $k = 1, 2$, and corresponding coefficients a_j and b_j. It follows from this and the first equality (2.3) that the vector $(a_j\xi_{j1} + b_j\xi_{j2})(\sigma(s))$ is tangent to the solution (S, σ), i.e., this vector belongs to the space $\sigma_*(T_sS)$.

Lemma 3.1 is proved.

LEMMA 3.2. *Consider manifolds* M, N, P, *a map* $\psi: N \to M$, *and an imbedding* $\varphi: P \to M$ *of class* C^k. *Suppose that* ψ *passes through the submanifold* (P, φ), *i.e.,* $\psi(N) \subset \varphi(P)$. *Then there exists a unique map* $\psi_0: N \to P$ *of class* C^k *such that* $\varphi \circ \psi_0 = \psi$.

PROOF. See [3, 1.32].

LEMMA 3.3. *Let* (S, σ) *be a nonsingular solution of equation* (1.1). *Then the smooth one-dimensional distributions* h_1 *and* h_2 *are uniquely defined on the manifold* S *and we have*

$$\sigma_*(h_j)(s) \subset H_j(\sigma(s)), \qquad s \in S.$$

PROOF. The existence and uniqueness of the distributions follow from Lemma 3.1. While proving this lemma we found that for any point $s \in S$ there exist a neighborhood U that trivializes the tangent bundle TS and smooth functions $a_j, b_j: U \to \mathbb{R}^5$, defined in this neighborhood such that the map

$$\psi_j: U \ni s \mapsto (a_j\xi_{j1} + b_j\xi_{j2})(\sigma(s)) \in T\mathbb{R}^5 \tag{3.6}$$

is smooth. This map passes through the imbedded submanifold $\sigma_*: TU \to T\mathbb{R}^5$. Applying Lemma 3.2 to the map (3.6) and submanifold (TU, σ_*), we obtain the smoothness of the distributions h_j.

Lemma 3.3 is proved.

By the Frobenius theorem [3, 1.63] there is a unique connected one-dimensional integral manifold $\gamma_j(s)$, maximal with respect to inclusion, of the distribution h_j through any point $s \in S$. This manifold is called a *characteristic* of the nonsingular solution (S, σ) of equation (1.1). A nonsingular solution (S, σ) of the initial value problem (1.1), (3.4) is said to be *defined* if the intersection of $\sigma(\gamma_j(s))$ and the initial curve (3.1) consists of exactly one point for $j = 1, 2$ and any point $s \in S$.

The main results of this paper are the following two theorems.

THEOREM 1. *Let equation* (1.1) *be hyperbolic. Then there exists an open neighborhood* V *of the initial curve* (3.1) *in the space* $\mathbb{R}^5$ *such that the following results are true.*

(Existence.) *There exists a simply connected manifold* S *and an imbedding* (1.11) *such that the submanifold* (S, σ) *is contained in the neighborhood* V *(i.e.,* $\sigma(S) \subset V$*) and this submanifold is a defined solution of the initial value problem* (1.1), (3.4) *of class* C^2.

(Uniqueness.) *Let* $\widetilde{S}$ *be a two-dimensional simply connected manifold and* $\widetilde{\sigma}\colon \widetilde{S} \to \mathbb{R}^5$ *be a defined solution of the problem* (1.1), (3.4) *of class* C^2 *that is contained in* V. *Then* $\widetilde{\sigma}(\widetilde{S}) \subseteq \sigma(S)$. *Thus the solution* (S, σ) *is the largest with respect to inclusion simply connected defined solution of the initial value problem* (1.1), (3.4) *of class* C^2 *in the neighborhood* V.

The proof of this theorem is given below.

REMARK 4. The input data of Theorem 1 is of class C^3 while the solution (S, σ) is of class C^2. This effect of degradation of smoothness of the solution by one is due to the multiplicity of characteristics of equation (1.1). This effect is impossible if all the characteristics of equation (1.1) are distinct, i.e., this equation is strictly hyperbolic [9, 10].

THEOREM 2. *Let equation* (1.1) *be hyperbolic and* m *be an arbitrary point of the space* $\mathbb{R}^5$. *Then there exists a contact diffeomorphism* f *(i.e.,* $f^*(\omega_0) \wedge \omega_0 = 0$*) of a certain neighborhood of the point* m *such that leaves* $f(m) = m$ *and*

$$\begin{aligned} &f^*(\omega_2) - \omega_0 \wedge (\partial/\partial z \lrcorner f^*(\omega_2)) \\ &\qquad = a\, dx \wedge dy + \beta\, dp \wedge dy + \gamma\,(dx \wedge dp + dq \wedge dy) + \delta\, dx \wedge dq, \end{aligned}$$

where α, β, γ, *and* δ *are certain smooth coefficients.*

The proof of this theorem is analogous to that of the well-known result of Sophus Lie on the contact classification of analytical strictly hyperbolic Monge–Ampère equations [11]. Theorem 1 must be applied instead of the Cauchy–Kovalevskaya theorem.

In fact Theorem 2 means that by using a suitable substitution of independent and dependent variables any hyperbolic Monge–Ampère equation can be reduced to the corresponding second order quasilinear equation.

§4. Diagonal system

The main step in the proof of the Theorem 1 is reducing the hyperbolic equation (1.1) to the corresponding diagonal system (4.11) of two first order quasilinear equations in $\mathbb{R}^5$. Certain auxiliary propositions must be proved to perform this reduction.

LEMMA 4.1. *Let* (S, σ) *be an imbedded submanifold of the space* $\mathbb{R}^5$ *and*

$$\xi_j = \xi_{j1} \cos u_j + \xi_{j2} \sin u_j, \qquad j = 1, 2, \tag{4.1}$$

where the smooth functions u_j *are defined in a certain neighborhood* W *of the set* $\sigma(S)$ *and* ξ_{ji} *are the vector fields* (2.5). *Suppose that the following conditions are satisfied.*

(a) *The set* $\sigma(S)$ *contains the initial curve* (3.1)–(3.3) *and*

$$\omega_1(\dot{l}(t), \xi_j(l(t))) = 0, \qquad j = 1, 2.$$

(b) *Any point* m *of the set* $\sigma(S)$ *can be joined to the initial curve by a segment of an integral curve of the vector field* ξ_1 *that entirely belongs to* $\sigma(S)$.

(c) *In a certain neighborhood of the set* $\sigma(S)$ *for* $j = 1, 2$ *there exist continuous functions* a^j *and linear differential forms* c_0^j, c_1^j, c_2^j *such that*

$$L_{\xi_j}\omega_1 = a^j\omega_1 + c_0^j \wedge \omega_0 + c_1^j \wedge (\xi_1 \lrcorner\, \omega_1) + c_2^j \wedge (\xi_2 \lrcorner\, \omega_1)$$

at any point of the set $\sigma(S)$.

Then (S, σ) *is a solution of equation* (1.1) *and the vector fields* ξ_1 *and* ξ_2 *are tangent to the submanifold* (S, σ).

PROOF. Consider an arbitrary point m of the set $\sigma(S)$. By condition (b), this point and the initial curve l can be joined by an integral curve $\gamma(\tau)$, $\tau_0 \leqslant \tau \leqslant \tau_1$, of the vector field ξ_1. To be definite, let $\gamma(\tau_0) \in l$ and $\gamma(\tau_1) = m$. It can be proved that in a certain neighborhood of the curve $\gamma(\tau)$ there exists a smooth vector field g that is tangent to the submanifold (S, σ), coincides with $\dot{l}(t)$ at points of the curve l,

$$g(l(t)) = \dot{l}(t), \tag{4.2}$$

and is noncollinear to the vector fields ξ_1 and ξ_2. Put

$$\begin{aligned} w_0(\tau) &= \omega_0(g)(\gamma(\tau)), \\ w_1(\tau) &= \omega_1(\xi_1, g)(\gamma(\tau)), \\ w_2(\tau) &= \omega_1(\xi_2, g)(\gamma(\tau)). \end{aligned} \tag{4.3}$$

It follows from (4.2) and condition (a) that

$$\omega_0(\tau_0) = 0, \quad \omega_1(\tau_0) = 0, \quad \omega_2(\tau_0) = 0. \tag{4.4}$$

According to well-known properties of the Lie derivative [3, 2.25] and the Lie bracket of vector fields [3, 1.55], we have

$$L_{\xi_1}(\omega_1(\xi_j, g)) = \omega_1([\xi_1, \xi_j], g) + \omega_1(\xi_j, [\xi_1, g]) + (L_{\xi_1}\omega_1)(\xi_j, g), \tag{4.5}$$

$$[\xi_1, g] = \alpha_1\xi_1 + \alpha_2 g, \tag{4.6}$$

where α_1 and α_2 are continuous coefficients. It can be proved that

$$\omega_0(\xi_j) = 0, \qquad \omega_1(\xi_1, \xi_2) = 0. \tag{4.7}$$

It follows from the representation (4.5), condition (c) and relations (4.6), (4.7) that

$$dw_j/ds = a_{j0}w_0 + a_{j1}w_1 + a_{j2}w_2, \qquad dw_0/ds = w_1. \tag{4.8}$$

Here $j = 1, 2$ and a_{ji} are continuous coefficients. The theorem on the uniqueness of solutions and relations (4.4), (4.8) guarantee that the functions (4.3) vanish, i.e.,

$$(4.9) \qquad \omega_0(g) = 0, \quad \omega_1(\xi_1, g) = 0, \quad \omega_1(\xi_2, g) = 0, \qquad \tau_0 \leqslant \tau \leqslant \tau_1.$$

By the construction (4.1), the vector ξ_j belongs to the characteristic distribution H_j. Hence, by definition (2.3),

$$(4.10) \qquad \omega_2(\xi_j, g) = \lambda_j \omega_1(\xi_j, g) = 0, \qquad j = 1, 2.$$

Since the vector field g is noncollinear to ξ_1 and ξ_2 by construction, it follows from (4.7), (4.9), and (4.10) that

$$\sigma^*(\omega_0) = 0, \quad \sigma^*(\omega_1) = 0, \quad \sigma^*(\omega_2) = 0,$$

i.e., (S, σ) is a solution of equation (1.1).

If the vector fields ξ_1 and ξ_2 are collinear at a point m of the set $\sigma(S)$, then according to condition (b) they are tangent to the submanifold (S, σ). If these vector fields are noncollinear at a point of the set $\sigma(S)$, then the linear forms ω_0, $\xi_1 \lrcorner \omega_1$, $\xi_2 \lrcorner \omega_1$ are linearly independent at this point. The vector subspace of $T_m(\mathbb{R}^5)$ that annihilates forms is two-dimensional. It follows from (4.7) and (4.9) that vector fields ξ_1, g and the vector ξ_2 are contained in this space, i.e., ξ_2 is complanar to the vectors ξ_1 and g that are tangent to the submanifold (S, σ). Hence, ξ_2 is also tangent to (S, σ).

Lemma 4.1 is proved.

Direct calculations prove that the following proposition is true.

LEMMA 4.2. *Condition* (c) *from Lemma* 4.1 *is true if and only if*

$$(4.11) \quad \xi_{3-j} \lrcorner (du_j + b_{j1}^2 \cos^2 u_j + (b_{j2}^2 - b_{j1}^1) \cos u_j \sin u_j - b_{j2}^1 \sin^2 u_j) = 0,$$

at points of the set $\sigma(S)$ *for* $j = 1, 2$, *where* b_{ji}^k, $i, k = 1, 2$, *are the linear forms from the hyperbolicity conditions* (2.10).

LEMMA 4.3. *Let* S *be a simply connected manifold and* (S, σ) *be a nonsingular solution of class* C^2 *of equation* (1.1). *Then in* $\mathbb{R}^5$ *there exist an open neighborhood* W *of the set* $\sigma(S)$ *and smooth functions* u_1 *and* u_2 *defined in* W *such that the vector fields* (4.1) *are contained in the image* $\sigma_*(TS)$ *of the tangent bundle* TS *at all points of the set* $\sigma(S)$, *i.e.,*

$$(4.12) \qquad (\xi_{j1} \cos u_j + \xi_{j2} \sin u_j)(\sigma(s)) \in \sigma_*(T_s S), \qquad s \in S, \ j = 1, 2.$$

PROOF. This lemma is true because S is simply connected. Elementary properties of trigonometric functions and standard facts of topology [6, Chapter 4, 2.5] and cohomology theory [12, §§17, 15] must be used to obtain the detailed proof of the lemma.

LEMMA 4.4. *Suppose that all the conditions of Lemma* 4.3 *are fulfilled and let each pair of functions* (u_{11}, u_{21}) *and* (u_{12}, u_{22}) *satisfy the assertion of this lemma. Then there exists an integer* n_j *such that*

$$(u_{j1} - u_{j2})|_{\sigma(S)} = \pi n_j, \qquad j = 1, 2.$$

The proof of this lemma is based on elementary properties of trigonometric functions.

LEMMA 4.5. *There exist an open neighborhood* W *of the initial curve* (3.1)–(3.3) *in* $\mathbb{R}^5$ *and functions* u_j, $j = 1, 2$, *of class* $C^2(W)$ *such that for the vector fields* ξ_1 *and* ξ_2 *defined by the expression* (4.1) *the following relations*

$$\omega_1(\dot{l}(t), \xi_j(l(t))) = 0, \qquad \omega_2(\dot{l}(t), \xi_j(l(t))) = 0 \tag{4.13}$$

hold. If $\widetilde{u}_j$ *are other functions that satisfy these conditions, then they differ from* u_j *by a constant summand of the form* πn_j, *where* n_j *is an integer,* $j = 1, 2$.

The proof is similar to those of Lemmas 4.3 and 4.4.

According to Lemma 4.3, there exists an open neighborhood W of the set $\sigma(S)$ in $\mathbb{R}^5$ and functions u_1 and u_2 of class $C^1(W)$ such that the vector fields ξ_1 and ξ_2 defined by (4.1) satisfy inclusion (4.12) at all points of the set $\sigma(S)$ for $j = 1, 2$. The following proposition is true for these vectors.

LEMMA 4.6. *Let* S *be a simply connected manifold and* (S, σ) *be a nonsingular solution of equation* (1.1) *of class* C^2, *and let the smooth functions* u_1 *and* u_2 *be determined according to Lemma* 4.3. *Then for the vector fields* ξ_1 *and* ξ_2 *defined by expression* (4.1) *there exist*:

1) *an open neighborhood* W_+ *of the set* $\sigma(S_+)$ *in* $\mathbb{R}^5$, *where*

$$S_+ = \{s \in S \mid \Delta(\sigma(s)) = 0\},$$

and continuous linear differential forms a_0^j, a_1^j, *and* a_2^j *on* W_+ *such that*

$$L_{\xi_j}\omega_1 = a_0^j \wedge \omega_0 + a_1^j \wedge (\xi_1 \lrcorner\, \omega_1) + a_2^j \wedge (\xi_2 \lrcorner\, \omega_1) \tag{4.14}$$

at points of the set $\sigma(S_+)$;

2) *an open neighborhood* W_0 *of the set* $\sigma(S_0)$ *in* $\mathbb{R}^5$, *where*

$$S_0 = \{s \in S \mid \Delta(\sigma(s)) = 0\},$$

a continuous function c^j, *and linear differential forms* c_0^j *and* c_1^j *on* W_0 *such that*

$$L_{\xi_j}\omega_1 = c^j\omega_1 + c_0^j \wedge \omega_0 + c_1^j \wedge (\xi_{3-j} \lrcorner\, \omega_1) \tag{4.15}$$

at points of the set $\sigma(S_0)$.

PROOF. 1) It can be proved that if the inequality $\operatorname{Pf}(\omega_2)|_m < 0$ is true at a point $m \in \mathbb{R}^5$, then

$$(H_1 \cap H_2)|_m = 0. \tag{4.16}$$

Thus if $\Delta > 0$, then the vector fields ξ_1 and ξ_2 are noncollinear. Let $m \in \sigma(S_+)$. Then in a certain neighborhood U_m of this point there exist smooth vector fields ξ_3^m and ξ_4^m such that ξ_1, ξ_2, ξ_3^m, and ξ_4^m form a basis of $\ker \omega_0$ at every point of the neighborhood U_m. Hence, the smooth linear forms $\beta_0 = \omega_0$, $\beta_i^m = \xi_i^m \lrcorner \omega_1$, $i = 1, 2, 3, 4$, form a basis of T_n^* at every point $n \in U_m$ and well-defined representation

$$L_{\xi_j}\omega_1 = a_{m0}^j \wedge \omega_0 + a_{m1}^j \wedge (\xi_1 \lrcorner \omega_1) + a_{m2}^j \wedge (\xi_2 \lrcorner \omega_1) + \omega_{m3}^{j4}\beta_3^m \wedge \beta_4^m, \tag{4.17}$$

holds in the neighborhood U_m, where

$$a_{m0}^j = \sum_{b=1}^4 \omega_{m0}^{jb}\beta_b^m, \quad a_{m1}^j = \sum_{b=2}^4 \omega_{m1}^{jb}\beta_b^m, \quad a_{m2}^j = \sum_{b=3}^4 \omega_{m2}^{jb}\beta_b^m,$$

and ω_{mi}^{jb} are well-defined continuous functions on U_m for $j = 1, 2$, $i = 1, 2, 3$, $i < b \leqslant 4$. Since (S, σ) is a solution of equation (1.1), it follows from inclusion (4.12) that

$$\sigma^*(L_{\xi_j}\omega_1) = 0, \qquad j = 1, 2, \; s \in S.$$

This equality and inclusion (4.12) imply that

$$\omega_{m3}^{j4}|_{\sigma(S_+)\cap U_m} = 0 \tag{4.18}$$

[1, item 17]. For every point $m \in \sigma(S_+)$ let us construct the corresponding neighborhood U_m as it was done above and put

$$W_+ = \bigcup_{m\in\sigma(S_+)} U_m.$$

The system of sets $\{U_m\}_{m\in\sigma(S_+)}$ forms an open covering of the set W_+. It is known that for W_+ there exists a smooth partition of unity $\{\varphi_m\}_{m\in\sigma(S_+)}$ subordinate to this covering [3, 1.11]. Let us define the linear differential forms

$$a_i^j = \sum_{m\in\sigma(S_+)} \varphi_m a_{mi}^j, \qquad j = 1, 2, \; i = 0, 1, 2. \tag{4.19}$$

By definition of a partition of unity and according to relations (4.17) and (4.18), the forms (4.19) are continuous in W_+ and the representation (4.14) holds at every point of the set $\sigma(S_+)$.

The case 2) is analogous to the case 1).

Lemma 4.6 is proved.

LEMMA 4.7. *Let all the conditions of Lemma* 4.6 *be satisfied. Then equations* (4.11) *hold at every point of the set* $\sigma(S)$ *for* $j = 1, 2$.

PROOF. By Lemma 4.6 the representations (4.14) and (4.15) are valid. The deduction of equations (4.11) from these representations is a direct calculation.

§5. Existence and uniqueness

Before starting the proof of Theorem 1, we need two auxiliary results. The first result is obvious.

LEMMA 5.1. *Let W be an open neighborhood of the initial curve* (3.1), *where the functions u_1 and u_2 are determined by Lemma* 4.5. *Then there exists a four-dimensional manifold L imbedded into W*

$$v\colon L \to W, \tag{5.1}$$

of class C^2 with the following properties.

(a) *The initial curve* (3.1) *belongs to $v(L)$: $l((a, b)) \subset v(L)$.*

(b) *The vector fields are transversal to the submanifold* (5.1), *i.e.,*

$$\xi_j(v(r)) \notin v_*(T_r L), \qquad r \in L, \ j = 1, 2.$$

Let us go on to the second result. In the neighborhood W consider functions u_1 and u_2 that satisfy Lemma 4.5. Denote the restrictions of these functions to the submanifold (5.1) by $\overline{u}_1$ and $\overline{u}_2$ respectively. Thus,

$$u_j|_{v(L)} = \overline{u}_j, \qquad j = 1, 2. \tag{5.2}$$

LEMMA 5.2. *There exists a neighborhood V of the submanifold* (5.1) *in $\mathbb{R}^5$ constructed according to Lemma* 5.1 *that has the following properties.*

(a) *There exist functions u_1 and u_2 of class $C^2(V)$ that are solutions of the system* (4.11) *in V with initial values* (5.2).

(b) *The neighborhood V is a domain of definition of the solution (u_1, u_2) mentioned in* (a) *of the initial value problem* (5.2) *and the initial curve* (3.1) *is closed in the relative topology of V.*

PROOF. To construct of the solution (u_1, u_2) of problem (5.2), (4.11) we use the method of successive approximations and the method of characteristics [13, 14].

PROOF OF THEOREM 1. (*Existence.*) Let us take the open subset of $\mathbb{R}^5$ from Lemma 5.2 as the set V. Properties (a) and (b) of the lemma are fulfilled for this subset. A characteristic curve $\gamma_j(t; m)$ is a solution of the problem

$$\dot{\gamma}_j(t; m) = \xi_j(\gamma_j(t; m)), \qquad \gamma_j(0; m) = m. \tag{5.3}$$

Here m is in V, $j = 1, 2$, and ξ_j are the vector fields determined by relation (4.1), where u_j are functions of class $C^2(V)$ defined in V according to Lemma 5.2. Define the set S_0 by the expression

$$S_0 = \{m \in V \mid \text{there exists } t, \tau \in \mathbb{R}^1 \text{ such that } m = \gamma_1(t; l(\tau))\}, \tag{5.4}$$

where l is the initial curve (3.1). Define the map σ_0 as

$$\sigma_0 = \iota, \tag{5.5}$$

where $\imath$ is the inclusion map of the subset S_0 into the set V. By definition (5.4), the set S_0 is a topological space with the topology induced by the inclusion into V. The map σ_0 defined by expression (5.5) is a homeomorphism for this reason.

We can define the structure of a smooth submanifold on S_0 that agrees with the topology induced from V and such a structure is unique [3, 1.33]. Therefore (S_0, σ_0) is a submanifold of class C^2 imbedded into V. It is known [6, Chapter 2, 2.10] that S_0 is C^2-diffeomorphic to a C^∞-manifold S, i.e., there exists a C^2-diffeomorphism $\rho\colon S \to S_0$. Put

$$\sigma = \sigma_0 \circ \rho\colon S \to \mathbb{R}^5. \tag{5.6}$$

The submanifold (S, σ) is closed in the topology of the space V because the set $l((a, b))$ is closed in this topology by Lemma 5.2(b).

The initial value (5.2) is defined according to Lemma 4.5. Therefore the submanifold (S, σ) is a solution of equation (1.1) according to Lemmas 5.2, 4.1, and 4.2. It follows from the construction (5.4) that it satisfies the initial value (3.4). Thus the submanifold (5.6) is a solution of the initial value problem (1.1), (3.4) of class C^2.

By Lemma 5.2(b), the set V is the domain of definition of the solution (u_1, u_2) of the initial value problem (5.2). Therefore, since $\sigma(S)$ is closed in the topology of V, it follows from Lemma 4.1 that (S, σ) is a well-defined solution of the Monge–Ampère equation (1.1). This fact and the simple connectedness of the initial curve (3.1) imply that the manifold S is simply connected.

To complete the proof of the existence part of Theorem 1, it remains to prove that the solution (S, σ) is nonsingular. Let m be an arbitrary point contained in the set $\sigma(S)$. Consider two cases.

Case 1. Let $\operatorname{Pf}(\omega_2) < 0$ at the point $m \in \sigma(S)$. Then, on the one hand, $\xi_j(m) \in H_j(m)$, for $j = 1, 2$ and, according to the above, $\xi_j(m) \in T_m S$, on the other. Therefore the solution is nonsingular at the point m according to equality (4.16) because in the converse case the dimension of the tangent space $T_m S$ is less than two while dimension of the manifold S is exactly equal to two.

Case 2. Let $\operatorname{Pf}(\omega_2) = 0$ at the point under consideration. The proof is similar.

The existence of the solution is proved.

(*Uniqueness.*) Suppose that the assertion of Theorem 1 on the uniqueness of the solution is false. Then there exists another solution $(\widetilde{S}, \widetilde{\sigma})$ of equation (1.1) that possesses all the properties from Theorem 1 on the existence of a solution, and $\widetilde{\sigma}(\widetilde{S})$ is not contained in $\sigma(S)$. By Lemma 4.7 (see also Lemmas 4.3 and 4.6), this follows from the fact that there is a neighborhood $\widetilde{V} \subset V$ of the set $\widetilde{\sigma}(\widetilde{S})$ and smooth functions $\widetilde{u}_1$ and $\widetilde{u}_2$ defined in it such that

$$\widetilde{\xi}_{3-j} \lrcorner (d\widetilde{u}_j + b^2_{j1}\cos^2\widetilde{u}_2 + (b^2_{j2} - b^1_{j1})\cos\widetilde{u}_j \sin\widetilde{u}_j - b^1_{j2}\sin^2\widetilde{u}_j) = 0, \tag{5.7}$$

at points of the set $\widetilde{\sigma}(\widetilde{S})$, where $\widetilde{\xi}_j = \xi_{j1}\cos\widetilde{u}_j + \xi_{j2}\sin\widetilde{u}_j$ (cf. (4.1)).

By the assumptions, $l((a, b)) \subset \widetilde{\sigma}(\widetilde{S})$. Therefore conditions (4.13) are necessarily fulfilled for the vectors $\widetilde{\xi}_j$. By Lemma 4.5, $(\widetilde{u}_j - u_j)|_{l((a, b))} = \pi n_j$, where n_j is an integer constant. If $n_j \neq 0$, then substitute $\widetilde{u}_j$ by $\widetilde{u}_j - \pi n_j$. It is obvious that $\widetilde{u}_1$ and $\widetilde{u}_2$ still satisfy (5.7). Thus we have

$$u_j|_{l((a, b))} = \widetilde{u}_j|_{l((a, b))}, \qquad j = 1, 2. \tag{5.8}$$

Let us subtract equations (4.11) from (5.7). We obtain

(5.9)

$$\begin{aligned}
\tilde{\xi}_{3-j} \lrcorner\, d(\tilde{u}_j - u_j) &= (\tilde{\xi}_{3-j} - \xi_{3-j}) \lrcorner\, (du_j + b_{j1}^2 \cos^2 u_j + (b_{j2}^2 - b_{j1}^1)\cos u_j \sin u_j - b_{j2}^1 \sin^2 u_j) \\
&\quad - \tilde{\xi}_{3-j} \lrcorner\, b_{j1}^2(\cos \tilde{u}_j + \cos u_j)(\cos \tilde{u}_j - \cos u_j) \\
&\quad - \tilde{\xi}_{3-j} \lrcorner\, (b_{j2}^2 - b_{j1}^1)(\sin(2\tilde{u}_j) - \sin(2u_j))/2 \\
&\quad + \tilde{\xi}_{3-j} \lrcorner\, b_{j2}^1(\sin \tilde{u}_j + \sin u_j)(\sin \tilde{u}_j - \sin u_j).
\end{aligned}$$

Return to definition (5.3) and define a curve $\tilde{\gamma}_j(t\,;m)$ in a similar way:

$$\dot{\tilde{\gamma}}_j(t\,;m) = \tilde{\xi}_j(\tilde{\gamma}_j(t\,;m)), \qquad \tilde{\gamma}_j(0;m) = 0. \tag{5.10}$$

Since by assumption $(\widetilde{S}, \tilde{\sigma})$ is a well-defined solution of the Monge–Ampère equation (1.1), it is possible to define smooth functions

$$\begin{gathered}
\tilde{t}_j : \tilde{\sigma}(\widetilde{S}) \to R^1, \qquad \tilde{\pi}_j : \tilde{\sigma}(\widetilde{S}) \to l((a,b)), \\
\tilde{\gamma}_j(\tilde{t}_j(m);m) \in l((a,b)), \quad \tilde{\pi}_j(m) = \tilde{\gamma}_j(\tilde{t}_j(m);m), \qquad j = 1, 2, \\
d\tilde{\pi}_j|_{\tilde{\sigma}(r)}(\xi) \neq 0
\end{gathered} \tag{5.11}$$

for all vectors $\xi \in \tilde{\sigma}_*(T_r\widetilde{S})$ noncollinear to the vectors $\tilde{\xi}_j|_{\tilde{\sigma}(r)}$.

It follows from the finite increment formula that equation (5.9) can be reduced to an ordinary linear equation with unknown function $\tilde{u}_j - u_j$. Integrate the resulting equation along the curve $\tilde{\gamma}_{3-j}(t\,;m)$ from 0 to t according to the Cauchy formula. From the fact that the solution $(\widetilde{S}, \tilde{\sigma})$ is well defined and from equalities (5.8) it follows that

$$\begin{aligned}
(\tilde{u}_j - u_j)(\tilde{\gamma}_{3-j}(t\,;m)) = \int_0^t &(\tilde{u}_{3-j} - u_{3-j})\, G_j(\tilde{\gamma}_{3-j}(\tau\,;m)) \\
&\times \exp \int_\tau^0 R_j(\tilde{\gamma}_{3-j}(\tau_1\,;m))\, d\tau_1 \\
&\times \exp \int_0^t R_j(\tilde{\gamma}_{3-j}(\tau_1\,;m))\, d\tau_1\, d\tau
\end{aligned} \tag{5.12}$$

for $m \in l((a,b))$ and $j = 1, 2$. Here G_j and R_j are certain continuous functions in $\widetilde{V}$. From (5.12) we deduce

(5.13)

$$\begin{aligned}
&(\tilde{u}_j - u_j)(\tilde{\gamma}_{3-j}(t\,;m)) \\
&\quad = \int_0^t G_j(\tilde{\gamma}_{3-j}(\tau\,;m))\, d\tau \exp \int_\tau^t R_j(\tilde{\gamma}_{3-j}(\tau_1\,;m))\, d\tau_1 \\
&\qquad \times \left(\int_0^{-\tilde{t}_j(\tilde{\gamma}_{3-j}(\tau\,;m))} (\tilde{u}_j - u_j)\, G_{3-j}(\tilde{\gamma}_j(\tau_2\,;\tilde{\pi}_j(\tilde{\gamma}_{3-j}(\tau\,;m)))) \right) \\
&\qquad \times \left(\exp \int_{\tau_2}^{-\tilde{t}_j(\tilde{\gamma}_{3-j}(\tau\,;m))} R_{3-j}(\tilde{\gamma}_j(\tau_3\,;\tilde{\pi}_j(\tilde{\gamma}_{3-j}(t\,;m))))\, d\tau_3)\, d\tau_2 \right) d\tau,
\end{aligned}$$

where the notation from (5.11) and the fact that $\widetilde{\gamma}_j(t\,;\,m_1) = m_2$ implies $m_1 = \widetilde{\gamma}_j(-t\,;\,m_2)$ are used [3, 1.48]. Using the known uniqueness theorem and (5.13), we obtain

$$\widetilde{u}_j(m) = u_j(m)\,, \qquad j = 1\,,\,2\,,\;\; m \in \widetilde{\sigma}(\widetilde{S})\,. \tag{5.14}$$

By equalities (5.14) and definitions (5.10) and (5.3), we get

$$\widetilde{\gamma}_j(t\,;\,m) = \gamma_j(t\,;\,m)\,, \qquad j = 1\,,\,2\,,\;\; m \in \widetilde{\sigma}(\widetilde{S})\,,\;\; 0 \leqslant t \leqslant \tilde{t}_j(m)\,.$$

Since $(\widetilde{S}\,,\,\widetilde{\sigma})$ is a well-defined solution by assumption, it follows from (5.15) and the construction (5.4)–(5.6) that $\widetilde{\sigma}(\widetilde{S}) \subseteq \sigma(S)$. Since this contradicts the initial assumption, the last inclusion proves the uniqueness of the solution.

Theorem 1 is proved.

References

1. E. Cartan, *Les systèmes différentiels extérieurs et leurs applications scientifiques*, Hermann, Paris, 1946.
2. V. Lychagin, *Contact geometry and nonlinear second order differential equations*, Uspekhi Mat. Nauk **34** (1979), no. 1, 101–171; English transl., Russian Math. Surveys **34** (1979), 149–180.
3. W. F. Warner, *Foundations of differentiable manifolds and Lie groups*, Springer-Verlag, Berlin and Heidelberg, 1983.
4. A. M. Vinogradov, *Multivalued solutions and classification principle of nonlinear differential equations*, Dokl. Akad. Nauk SSSR **210** (1973), no. 1, 11–14; English transl. in Soviet Math. Dokl. **14** (1973).
5. R. Courant, *Methods ofmathematical physics. Vol* II: *Partial differential equations*, Interscience, New York, 1962.
6. M. W. Hirsch, *Differential topology*, Springer-Verlag, Berlin and Heidelberg, 1976.
7. E. R. Rozendorn, *An approximated solution of wind and pressure balance equation for anti-cyclones on tropical and subtropical latitude*, Dokl. Akad. Nauk SSSR **253** (1980), no. 3, 584–587; English transl. in Soviet Math. Dokl. **22** (1980).
8. E. E. Levi, *Sul problema di Cauchy per le equazion lineari in due variabli a caratteristice reali,* I–II, Rend. Ist. Lombardo Ser. 2 **41** (1908), 409–421, 691–712.
9. D. V. Tunitsky, *Cauchy problem for hyperbolic Monge–Ampère equations*, Izv. Akad. Nauk SSSR Ser. Mat. **57** (1993), no. 4, 174–191; English transl. in Math. USSR-Izv. **43** (1994).
10. ———, *On the Cauchy problem for Monge–Ampère equations of hyperbolic type*, Mat. Zametki **51** (1992), no. 6, 80–90; English transl. in Math. Notes **51** (1992).
11. V. V. Lychagin and V. N. Rubtsov, *On Sophus Lie's theorems for Monge–Ampère equations*, Dokl. Akad. Nauk BSSR **27** (1983), no. 5, 396-398. (Russian)
12. R. Bott and L. W. Tu, *Differential forms in algebraic topology*, Springer-Verlag, Berlin and Heidelberg, 1982.
13. A. Douglis, *Some existence theorems for hyperbolic systems of partial differential equations*, Comm. Pure Appl. Math. **5** (1952), no. 2, 119–154.
14. P. Hartman and A. Wintner, *On hyperbolic partial differential equations*, Amer. J. Math. **74** (1952), no. 4, 834–864.

Translated by THE AUTHOR

9, ul. Plushcheva, korp. 3, apt. 19, Moscow, 111524 Russia

Amer. Math. Soc. Transl.
(2) Vol. **167**, 1995

Singularities of Solutions of the Maxwell–Dirac Equation

L. ZILBERGLEIT

ABSTRACT. The Maxwell–Dirac equations are studied from a general point of view as the natural equations on the Grassmann algebra of exterior differential forms on pseudo-Riemann space. The complex structure of electromagnetism is induced by the Hodge operator on differential forms in this approach. Singularities of the Maxwell–Dirac equation's solutions (particles) are defined by the complex vector fiber bundle over the characteristic manifold. Fibers of this fiber bundle are the kernels of the symbol of the corresponding operator. The transport operator coordinating with complex structure acts in this fiber bundle. The example is singularities and the transport operator in the neighborhood of the point gravitational particle (Schwarzschild metric).

Introduction

The purpose of this paper to study the Maxwell and Dirac equations from the general invariant point of view. As we shall see in the subsequent discussion we simply deal with electromagnetic theory. Therefore the notion of electromagnetic field acquires paramount significance. The treatment of this notion allows us to choose two nonequivalent approaches to the study of electrodynamics. In the first classical approach, the electromagnetic field is regarded as the aggregate of two vector fields (electric and magnetic) on the physical space. The second invariant approach studies 6-component antisymmetric 2-tensor on Minkowski space, i.e., an element of the space of differential 2-forms (see [1–3]).

The second approach, besides being mathematically attractive, has a sound physical foundation, because its main proposition is a paraphrase of the fact that all fundamental experiments with electric and magnetic fields naturally lead to the notion of orientation of 2-dimensional surface elements in space–time (i.e., to differential 2-forms): the motion of a conductor in a magnetic field is characterized by the value of the current that it induces, i.e., to each 2-dimensional surface element spanning the spatial and temporal coordinates a number is assigned. If two motionless conductors with current interact, then a number is assigned to a 2-dimensional spatial oriented surface element. Let us add that the consideration of an electromagnetic field as a 2-form leads (in contrast with the classical situation)

1991 *Mathematics Subject Classification*. Primary 35Q60.

to the natural appearance of a complex structure: the "source" of complexity in electrodynamic equations is the Hodge operator $\#$ on differential 2-forms. Moreover, if we consider the space of all differential forms on Minkowski space, then the objects obtained are called spinors in classical physics. This suggests that the Maxwell and Dirac equations put together are in fact one equation. This thought is confirmed by direct examination in local cartesian coordinates. This Maxwell–Dirac operator coincides with the operator $d+\delta$, where d is the external de Rham differential, δ is the codifferential, $\delta = \#^{-1}d\#$. Finally, forms are natural objects from the geometrical point of view: by transformations of coordinates allowed in Minkowski space, they also transform as electric and magnetic fields (respectively, as Ψ-functions).

The appearance of this article, to a great extent, is the result of many discussions with V. V. Lychagin. The author takes this opportunity to thank him for his permanent support.

§1. Linear algebra in Minkowski space

The object of study in this section will be external forms. The analysis of the algebra of external forms on a metric vector space leads to the specification of natural structures in this algebra: the complex and spinor structures.

1.1. Let E be a vector space over the field Q, $\dim E = n$, and let E^* be the dual space to the space E.

We denote by e_θ and i_X (or $X \lrcorner$) linear operators of external and internal multiplications:

$$\begin{gathered} i_X\colon \Lambda^i(E^*) \to \Lambda^{i-1}(E^*), \\ [i_X(\omega)](X_1, \dots, X_{n-1}) = \omega(X, X_1, \dots, X_{n-1}), \\ X, X_1, \dots, X_{n-1} \in E, \qquad \omega \in \Lambda^i(E^*), \\ e_\theta\colon \Lambda^i(E^*) \to \Lambda^{i+1}(E^*), \quad e_\theta(\omega) = \theta \wedge \omega, \qquad \theta \in E^*. \end{gathered}$$

They naturally arise as the symbols of the external differential and codifferential operators. The main properties of the operators i_X and e_θ are

$$e_{\theta_1}e_{\theta_2} + e_{\theta_2}e_{\theta_1} = 0, \qquad i_{X_1}i_{X_2} + i_{X_2}i_{X_1} = 0.$$

The operators e_θ, i_X generate operators e_ω, i_U, $\omega \in \Lambda^\bullet(E^*)$, $U \in \Lambda^\bullet(E)$ satisfying the following conditions:

$$\begin{aligned} e_{\omega_1\wedge\omega_2} &= e_{\omega_1} \circ e_{\omega_2}, & \omega_i &\in \Lambda^\bullet(E^*), \quad i = 1, 2, \\ i_{U_1\wedge U_2} &= i_{U_1} \circ i_{U_2}, & U_i &\in \Lambda^\bullet(E^*), \quad i = 1, 2. \end{aligned}$$

On decomposable forms these operators have the form

$$\begin{aligned} e_{\theta_1\wedge\dots\wedge\theta_k} &= e_{\theta_1} \circ \dots \circ e_{\theta_k}, & \theta_i &\in E^*, \quad i = 1, \dots, k, \\ i_{X_1\wedge\dots\wedge X_k} &= i_{X_1} \circ \dots \circ i_{X_k}, & X_i &\in E, \quad i = 1, \dots, k. \end{aligned}$$

1.2. Let $\langle\ ,\ \rangle$ be the pairing between E and E^*,

$$\langle\ ,\ \rangle \colon E \times E^* \to Q.$$

Recall that the operator $A_g \colon E \to E^*$ defines a metric with signature l on the space E:

$$g(e_1, e_2) = \langle A_g e_1, e_2 \rangle, \qquad e_1, e_2 \in E$$

if the form g is symmetric and the number of negative eigenvalues in the canonic form of g is equal to l. The choice of the form g is equivalent to the choice of the operator A_g, but the operator A_g is more convenient because the metric on the space E^* may be defined by means of the operator A_g^{-1} and the metric on the vector spaces $\Lambda^k E^*$ may be defined by means of external degrees of the operator A_g^{-1}.

On elements of the space $\Lambda^k E$ this metric is defined by the formula

$$g(\zeta, \xi) = \langle A_g \zeta, \xi \rangle, \tag{1.2.1}$$

where $\zeta, \xi \in \Lambda^k E$. The sign of this metric on the space $\Lambda^n(E)$ coincides with the sign of the expression $(-1)^\nu$, $\nu = [n/2] + l$. We introduce the following notation:

$$\Lambda^\cdot(E) = \bigoplus_k \Lambda^k(E), \qquad \hat{p} = A_g^{-1}(p), \quad p \in \Lambda^\cdot(E^*),$$

$$\Lambda^{\mathrm{ev}}(E) = \bigoplus_{k \equiv 0 \bmod 2} \Lambda^k(E), \qquad \Lambda^{\mathrm{od}}(E) = \bigoplus_{k \equiv 1 \bmod 2} \Lambda^k(E).$$

In this notation we obtain

$$g(V, \theta) = \widehat{V} \lrcorner\, \theta = \hat{\theta} \lrcorner\, V, \qquad \theta, V \in \Lambda^k E^*.$$

As a consequence, we obtain a relationship between the operators i and e and the metric g:

$$i_{\hat{\theta}_1} e_{\theta_2} + e_{\theta_2} i_{\hat{\theta}_1} = g(\theta_1, \theta_2)\, \mathbb{1}, \qquad \theta_1, \theta_2 \in E^*.$$

1.3. Let us fix the volume form $\Omega \in \Lambda^n(E^*)$, $g(\Omega, \Omega) = (-1)^\nu$. The form Ω defines the operator $\# \in \operatorname{Aut}(\Lambda^\cdot(E^*))$ so that following diagram

$$\begin{array}{ccc} \Lambda^k(E^*) & \xrightarrow{A_g^{-1}} & \Lambda^k(E) \\ & \overset{\#}{\searrow} & \big\downarrow \lrcorner\, \Omega \\ & & \Lambda^{n-k}(E^*) \end{array}$$

is commutative.

The operator $\#$ on decomposable forms can be expressed as

$$\#(\theta_1 \wedge \cdots \wedge \theta_k) = i_{\hat{\theta}_1} \circ \cdots \circ i_{\hat{\theta}_k}(\Omega), \qquad \theta_i \in E^*, \ i = 1, \ldots, k.$$

PROPOSITION 1.3.1. *The homomorphism* $\#$ *satisfies the following conditions*:
1°. $\# e_\theta = i_{\hat\theta}\#$, $\theta \in E^*$.
2°. $\#^2|_{\Lambda^i(E^*)} = (-1)^\nu \mathbb{1}_{\Lambda^i(E^*)}$.

PROOF. Statement 1° of Proposition 1.3.1 is a trivial consequence of definition of $\#$. It suffices to verify statement 2° of Proposition 1.3.1 on decomposable forms. Consider the form $\theta_1 \wedge \cdots \wedge \theta_k$, $\theta_i \in E^*$, $g(\theta_i, \theta_j) = 0$, $i \neq j$, $i, j = 1, \ldots, k$ and present the volume form Ω as $\Omega = \theta_1 \wedge \cdots \wedge \theta_k \wedge \theta_{k+1} \wedge \cdots \wedge \theta_n$, $g(\theta_i, \theta_j) = 0$, $i \neq j$, $i, j = 1, \ldots, n$. Then

$$\begin{aligned}
\#^2(\theta_1 \wedge \cdots \wedge \theta_k) &= \#(i_{\hat\theta_1} \circ \cdots \circ i_{\hat\theta_k}\Omega) \\
&= \#[(-1)^{k(k-1)/2} g(\theta_1, \theta_1) \cdots g(\theta_k, \theta_k)\theta_{k+1} \wedge \cdots \wedge \theta_n] \\
&= (-1)^{k(k-1)/2} g(\theta_1, \theta_1) \cdots g(\theta_k, \theta_k) i_{\hat\theta_{k+1}} \cdots i_{\hat\theta_n}\Omega \\
&= (-1)^{k(k-1)/2} g(\theta_1, \theta_1) \cdots g(\theta_k, \theta_k) \\
&\qquad \times (-1)^{(n-k)(n-k-1)/2 + k(n-k)} g(\theta_{k+1}, \theta_{k+1}) \cdots g(\theta_n, \theta_n)\theta_1 \wedge \cdots \wedge \theta_k \\
&= (-1)^\nu \theta_1 \wedge \cdots \wedge \theta_k. \quad \square
\end{aligned}$$

The second statement of Proposition 1.3.1 leads to the following very important theorem.

THEOREM 1.3.1. *For odd* ν *the operator* $\#$ *defines a complex structure on the space* $\Lambda^\cdot(E^*)$.

COROLLARY. $\# i_{\hat\omega} = e_\omega \#$ *for* $\omega \in \Lambda^\cdot(E^*)$.

PROOF. From the definition of the operator $\#$, we get $\# w = i_{\hat\omega}(\Omega)$. Therefore

$$\# e_w v = \#(\omega \wedge v) = i_{\widehat{\omega\wedge v}}(\Omega) = i_{\hat\omega} \circ i_{\hat v}(\Omega) = i_{\hat\omega}\#(v). \quad \square$$

From the definition of metric structure, it follows that

$$g(U, V) = i_{\hat U}(V), \qquad U, V \in \Lambda^k(E^*). \tag{1.3.2}$$

In the sequel we shall suppose that the number ν is odd.

1.4. A Clifford structure exists on the space $\Lambda^\cdot(E^*)$. It is determined by the symbol of the Dirac operator $d + \delta$ (see 2.1), which is equal to the sum of the external and internal multiplication operators:

$$B_\theta \colon \Lambda^\cdot(E^*) \to \Lambda^\cdot(E^*), \quad B_\theta = e_\theta + i_\theta, \qquad \theta \in E^*.$$

The operator B_θ satisfies the following conditions:

$$\begin{aligned}
B_{\theta_1} \circ B_{\theta_2} + B_{\theta_2} \circ B_{\theta_1} &= 2g(\theta_1, \theta_2)\mathbb{1}, &\quad \theta_1, \theta_2 \in E^*, \\
[B_\theta]^2 &= g(\theta, \theta)\mathbb{1}, &\quad \theta \in E^*, \\
B_\theta\# &= \# B_\theta, &\quad \theta \in E^*.
\end{aligned}$$

The operator B_θ, $\theta \in E^*$ defines the Clifford algebra $C(E^*, g)$ corresponding to the form g. In other words, we have the linear mapping

$$\rho \colon E^* \to \operatorname{End}_{\mathbb{C}} \Lambda^\cdot(E^*), \quad \theta \mapsto B_\theta$$

for which $[\rho(\theta)]^2 = g(\theta, \theta)\mathbb{1}$.

Note that the mapping ρ can be naturally extended to the linear mapping

$$\rho \colon \Lambda^\cdot(E^*) \to \operatorname{End}_{\mathbb{C}} \Lambda^\cdot(E^*).$$

§2. Structure of the kernel bundle for the symbols of the Dirac–Maxwell operators

2.1. In this section we introduce the Dirac and Maxwell operators and study their symbols and the structures associated with them. The principal attention will be focused on the fiber bundle of kernels of the symbols of Dirac–Maxwell operators. This point of view is founded on the analysis of the corpuscular–wave dualism, which leads to the following mathematical situation.

Let the solution ω of the equation

$$\nabla\omega = g \tag{2.1.1}$$

have a singularity on the wave front $s = 0$. Here ∇ is a first order differential operator on the module $\Lambda^{\cdot}(M)$ (for example, the Dirac–Maxwell operator). We are interested in the behavior of the solution ω in the neighborhood of the surface $s = 0$ (we suppose that $d_q s \neq 0$ if $s(q) = 0$). We look for the solution in the form of the composition of progressing waves

$$\omega_s = \omega_s^0 + f_1(s)\omega_s^1 + f_2(s)\omega_s^2 + \dots , \tag{2.1.2}$$

where $df_i/ds = f_{i-1}$, $\omega_s^i \in \Lambda^{\cdot}(M)$. By substituting this series in equation (2.1.1), we get the following equation as the first condition on the "function" ω_s^1 (the coefficient of f_1):

$$\sigma_{ds}(\nabla)\omega_s^1 = 0, \tag{2.1.3}$$

where $\sigma_{ds}(\nabla)\colon \Lambda^{\cdot}(M) \to \Lambda^{\cdot}(M)$ is the symbol of the operator ∇, $\sigma_{ds}(\nabla)\omega = \nabla(s\omega) - s\nabla(\omega)$, $s \in C^\infty(M)$.

Equation (2.1.3) has a nontrivial solution if $\operatorname{Ker}\sigma_{ds}(\nabla) \neq 0$. This means that the surface $s = 0$ is a characteristic surface. In other words, the section

$$l_{ds}\colon M \to T^*M, \quad q \mapsto (q, d_q s)$$

to points of the surface $s = 0$ assigns points of the set

$$\operatorname{Char}\nabla = \{(q, p) \in T^*M \mid \operatorname{Ker}\sigma_p(\nabla) \neq 0,\ p \in T_q^*M \setminus 0\}.$$

We assume that $\operatorname{Char}\nabla$ is a smooth manifold.

This allows us to introduce the universal object on the space T^*M. This is the fiber bundle of kernels over the space $\operatorname{Char}\nabla$.

The fiber of this bundle $K(\nabla)$ over a point $(q, p) \in \operatorname{Char}\nabla$ is the vector space $\operatorname{Ker}\sigma_p(\nabla)$, $p \in T_q^*M$. The fiber bundle $K(\nabla)$ is naturally included in the following commutative diagram:

$$\begin{array}{ccccc}
K(\nabla) & \hookrightarrow & \pi^*(\Lambda^{\cdot}T^*M) & \longrightarrow & \Lambda^{\cdot}T^*M \\
\downarrow & & \downarrow & & \downarrow \\
\operatorname{Char}(\nabla) & \hookrightarrow & T^*M & \xrightarrow{\pi} & M
\end{array}$$

The universality of $K(\nabla)$ means that we can "pull down" kernels from this fiber bundle to every characteristic surface $s = 0$ by using the appropriate sections.

The fiber bundle $K(\nabla)$ inherits the grading of the bundle $\Lambda^{\cdot}T^*M$. Therefore, together with the bundle $K(\nabla)$ we may introduce the fiber bundles $K^i(\nabla)$ with fiber $\operatorname{Ker}\sigma_p(\nabla) \cap \Lambda^i T^*M$ over the point $(q, p) \in \operatorname{Char}(\nabla)$. We also introduce the fiber bundle $C(\nabla)$ over the base $\operatorname{Char}(\nabla)$ with fiber $\operatorname{Coker}\sigma_p(\nabla)$ over the point $(q, p) \in \operatorname{Char}(\nabla)$.

Note that from the physical point of view the fact that the principal part of the singularity of a solution (particle) belongs to the kernel of the symbol means that it is created from waves satisfying certain degeneracy conditions.

2.2. Let M^n be an oriented manifold, g be the metric on the manifold M with signature l, and Ω be the volume form on the manifold M, $g(\Omega, \Omega) = -1$.

From the mathematical point of view, the operators of external differentiation and codifferentiation are natural first order differential operators on the manifold M. A pleasant (but by no means unexpected) fact consists in that these operators are natural from the physical point of view. Therefore, from both points of view (physical and mathematical) it is natural to consider the following first order differential operators on the manifold M:

1. The de Rham external differential

$$d\colon \Lambda^i(M) \to \Lambda^{i+1}(M).$$

2. The codifferential

$$\delta\colon \Lambda^i(M) \to \Lambda^{i-1}(M), \qquad \delta = \#^{-1}d\#.$$

3. The Dirac operator

$$D = d + \delta\colon \Lambda^{\mathrm{ev}}(M) \to \Lambda^{\mathrm{od}}(M).$$

4. The Maxwell operator, i.e., the Dirac operator restricted to 2-forms ($n = 4$, $l = 3$):

$$\Delta = D|_{\Lambda^2(M)}\colon \Lambda^2(M) \to \Lambda^1(M) \oplus \Lambda^3(M),$$
$$\Delta(\omega) = \delta\omega \oplus d\omega, \qquad \omega \in \Lambda^2(M).$$

The Maxwell operator is not convenient because it is overdetermined. Since these operators mainly interest us from the point of view of the behavior of their solution's singularities, we shall considerably simplify the situation if we define the operator $\overline{\Delta}\colon \Lambda^2(M) \to \Lambda^2(M)$, which is equivalent to the operator Δ as a solution of the Cauchy problem.

The Cauchy problem for the Maxwell equation on the manifold M can be stated in the following way.

Consider the function $t \in C^\infty(M)$, $g(dt, dt) > 0$. It is necessary to find a form $\omega \in \Lambda^2(M)$ that coincides with the well-known 2-form ω_0 at points of the manifold $M_0 = (t = 0) \cap M$: $\omega_q = \omega_{0,q}$ for $q \in M_0$. In addition, the form ω is the solution of the Maxwell equation: $\Delta\omega = 0$. It should be noted that the comparison of the forms ω and ω_0 is "pointwise". Further on, we shall also use the restriction of the form ω to the manifold M_0, which we denote by $\omega|_t = 0$. Solvability conditions of the Cauchy problem for the Maxwell equation have the form:

$$d(\omega|_{t=0}) = 0, \quad d(\#\omega|_{t=0}) = 0. \tag{2.2.1}$$

PROPOSITION 2.2.1. *The form* $\omega \in \Lambda^2(M)$ *satisfies the Cauchy problem*:

$$\text{(I)} \qquad \Delta\omega = 0\,, \qquad \omega_q = \omega_{0,q}\,, \quad q \in M_0$$

if and only if the form ω *satisfies the Cauchy problem*

$$\text{(II)} \qquad \overline{\Delta}\omega = 0\,, \qquad \omega_q = \omega_{0,q}\,, \quad q \in M_0\,,$$

where $\overline{\Delta} = i_{\widehat{dt}}d - e_{dt}\delta$, *and conditions* (2.2.1) *hold.*

PROOF. Let ω be the solution of problem (II) satisfying conditions (2.2.1). Then $(i_{\widehat{dt}}d - e_{dt}\delta)\omega = 0$. By using the operator $e_{dt}i_{\widehat{dt}}$, we obtain $e_{dt}\,\delta\omega = 0$ and therefore $i_{\widehat{dt}}\,d\omega = 0$.

For the proof of the relations $d\omega = 0$ and $\delta\omega = 0$, it suffices to check that $L_{\widehat{dt}}(d\omega) = 0$ and $L_{\widehat{dt}}(\#\delta\omega) = 0$. But $L_{\widehat{dt}}(d\omega) = d\,(i_{\widehat{dt}}\,d\omega) = 0$, $L_{\widehat{dt}}(\#\delta\omega) = d\,(i_{\widehat{dt}}\#\delta\omega) = d\,(\#e_{dt}\delta\omega) = 0$. The second part of the proof is obvious. □

Let us calculate the symbols of the operators considered above. We have

1. $\sigma_{d\lambda}(d)(\omega) = d\lambda \wedge \omega\,, \qquad \lambda \in C^\infty(M)\,,$
   $\sigma_p(d) = e_p\,, \qquad p \in T_q^*M\,,$
2. $\sigma_{d\lambda}(\delta)(\omega) = \delta\,(\lambda\omega) - \lambda\delta\omega = \#^{-1}[d\,(\lambda\#\omega) - \lambda d\,(\#\omega)]$
   $= \#^{-1}e_{d\lambda}\#\omega = i_{\widehat{d\lambda}}\omega\,, \qquad \lambda \in C^\infty(M)\,,$
   $\sigma_p(\delta) = i_{\hat{p}}\,, \qquad p \in T_q^*M\,,$
3. $\sigma_p(D) = e_p + i_{\hat{p}} = B_p\,, \qquad p \in T_q^*M\,,$
4. $\sigma_p(\Delta) = i_{\hat{p}} \oplus e_p\,, \qquad p \in T_q^*M\,,$
5. $\sigma_p(\overline{\Delta}) = i_{\widehat{dt}}e_p - e_{dt}i_{\hat{p}} = i_{\hat{p}}e_{dt} - e_p i_{\widehat{dt}}\,, \qquad p \in T_q^*M\,.$

2.3. Now we turn to the study of the symbols of differential Dirac and Maxwell operators.

I. The Dirac operator: $\sigma_p(D)\omega = B_p\omega\,,\ p \in T_q^*M\,,\ \omega \in \Lambda^{\mathrm{ev}}(M)$.

We use the definitions

$$\mathrm{Char}_q\,D = \{p \in T_q^*M \setminus 0 \mid g\,(p\,,\,p) = 0\}\,,$$
$$V_p = \{X \in T_qM \mid p(X) = 0\,,\ p \in \mathrm{Char}_q\,D\}\,.$$

PROPOSITION 2.3.1. *The operator* B_p, $p \in T_q^*M$ *has a kernel if and only if* $p \in \mathrm{Char}_q\,D$. *In this case* $\mathrm{Ker}\,B_p = \mathrm{Im}\,B_p$.

PROOF. If the operator B_p , $p \in T_q^*M$ has a kernel in the space $\Lambda^\cdot(T_q^*M)$, then it is obvious that $p \in \mathrm{Char}_q\,D$. Also, the inclusion $\mathrm{Im}\,B_p \subset \mathrm{Ker}\,B_p$, $p \in \mathrm{Char}_q\,D$ is obvious. We study the space $\mathrm{Ker}\,B_p$ for the set $\Lambda^\cdot(V_p^*)$. From the condition $B_p\varpi = 0$, $\varpi \in \Lambda^\cdot(V_p^*)$ it follows that $G_p\varpi = 0$, $G_p = i_{\hat{p}}e_p$ and hence $\varpi \in \Lambda^\cdot(E_p^*)$, $E_p = V_p/\hat{p}$. Therefore $B_p\varpi = e_p\varpi$ and $\mathrm{Ker}\,B_p|_{\Lambda^\cdot(V_p^*)} = 0$. We have the following relations:

$$\dim \mathrm{Ker}\,B_p + \dim \mathrm{Im}\,B_p = 2^n \quad \text{and} \quad \mathrm{Ker}\,B_p \supset \mathrm{Im}\,B_p \supset \mathrm{Im}\,B_p|_{\Lambda^\cdot(V_p^*)}\,.$$

As proved above, $\dim \operatorname{Im} B_p|_{\Lambda^{\bullet}(V_p^*)} = \dim \Lambda^{\bullet}(V_p^*) = 2^{n-1}$. Therefore $\operatorname{Ker} B_p = \operatorname{Im} B_p = \operatorname{Im} B_p|_{\Lambda^{\bullet}(V_p^*)}$. $\square$

From Proposition 2.3.1 and from the properties of the operators B_p it follows that

PROPOSITION 2.3.2. *The fiber bundle* $K(D)$ *is a* 2^{n-3}*-dimensional complex fiber bundle with a fiber* $\Lambda^{\mathrm{od}}(V_p)$ *at the point* $p \in T_q^*M$, *the complex structure being defined by the operator* $\#$.

Now we study the fiber bundle $K^i(D)$. Conditions which define an element of the fiber ω_i of this bundle are $e_p\omega_i = 0$, $i_{\hat{p}}\omega_i = 0$. Therefore $\omega_i = e_p\theta_{i-1}$, $\theta_k \in \Lambda^k(E_p^*)$, $E_p = \operatorname{Ker} p/\hat{p}$. It is obvious that if $\omega_i \in K_p^i$, then $\#\omega_i \in K_p^{n-i}$. Thus we have

PROPOSITION 2.3.3. 1°. $\dim K^i(D) = \begin{pmatrix} i-1 \\ n-2 \end{pmatrix}$.

2°. $\#\colon C^\infty(K^i(D)) \to C^\infty(K^{n-i}(D))$.

For any fiber bundle ζ, we use the notation $C^\infty(\zeta)$ for the module of smooth sections of the bundle ζ.

The description of the elements $K_p^i(D)$ permits us to define the isomorphism

$$k_p\colon K_p^i(D) \xrightarrow{\sim} \Lambda^{i-1}(E_p^*), \qquad e_p\theta_{i-1} \mapsto \theta_{i-1}.$$

Thus we may define the operator

$$I\colon \Lambda^k(E_p^*) \to \Lambda^{n-k-2}(E_p^*), \qquad I^2 = -\mathbb{1},$$

so that the following diagram

$$\begin{array}{ccc} K_p^i & \xrightarrow{\#} & K_p^{n-i} \\ {\scriptstyle k_p}\downarrow & & \downarrow{\scriptstyle k_p} \\ \Lambda^{i-1}(E_p^*) & \xrightarrow{I} & \Lambda^{n-i-1}(E_p^*) \end{array}$$

is commutative. Hence we have

PROPOSITION 2.3.4. *The space* $\Lambda^{\bullet}(E_p^*)$, $p \in \operatorname{Char} D$ *is the* 2^{n-3}*-dimensional complex plane where the complex structure is defined by the operator* I.

PROPOSITION 2.3.5. *The operator* I *satisfies the following conditions*:

1°. $I^2 = -\mathbb{1}_{\Lambda^{\bullet}(E_p^*)}$.

2°. $e_{\theta}I = Ii_{\widehat{v}}$, $v \in \Lambda^{\bullet}(E_p^*)$.

PROOF. Condition 1° is a simple consequence of the definition of the operator I. Let us prove 2°.

Let $\omega_{j+1} \in K_p^{j+1}$, i.e. $\omega_{j+1} = e_p\theta_j$, $\theta_j \in \Lambda^j(E_p^*)$. Then

$$i_{\widehat{v}}\#e_p\theta_j = \#e_v e_p\theta_j = (-1)^k\#e_p e_v\theta_j = (-1)^k e_p I e_v \theta_j.$$

On the other hand, $i_{\widehat{v}}\#e_p\theta_j = i_{\widehat{v}}e_p I\theta_j = (-1)^k e_p i_v I\theta_j$. $\square$

COROLLARY. *For* $\theta_1, \theta_2 \in \Lambda^k(E_p^*)$ *we have*

$$i_{\hat{\theta}_1}\theta_2 = I^{-1} e_{\theta_1} I \theta_2 .$$

Similarly to the fiber bundle $K^i(D)$ let us introduce the fiber bundle $\mathfrak{L}^i(D)$ over the set $\operatorname{Char} D$ with fiber $\Lambda^i(E_p^*)$ over the point $(q, p) \in \operatorname{Char} D$.

PROPOSITION 2.3.6. 1°. $\dim \mathfrak{L}^i(D) = \binom{i}{n-2}$.
2°. $I\colon C^\infty(\mathfrak{L}^i(D)) \to C^\infty(\mathfrak{L}^{n-i-2}(D))$, $I^2 = -\mathbb{1}$.

II. The modified Maxwell operator:

$$\sigma_p(\overline{\Delta})\omega = (i_{\widehat{dt}} e_p - e_{dt} i_{\hat{p}})\omega, \qquad \omega \in \Lambda^2(M).$$

The structure of the kernel fiber bundle is determined by the following statement.

LEMMA 2.3.1. *We have*

$$\operatorname{Ker} \sigma_p(\overline{\Delta}) = \{\omega \in \Lambda^2(M) \mid e_p\omega = 0,\ i_{\hat{p}}\omega = 0,\ p^2 = 0\}.$$

PROOF. The form ω belongs to the space $\operatorname{Ker} \sigma_p(\overline{\Delta})$ if

$$(i_{\widehat{dt}} e_p - e_{dt} i_{\hat{p}})\omega = 0. \tag{2.3.1}$$

We apply the operator $i_{\widehat{dt}}$ to (2.3.1). We obtain $-i_{\hat{p}}\omega \cdot g(dt, dt) + e_{dt} i_{\widehat{dt}} i_{\hat{p}}\omega = 0$. By applying the operator e_{dt}, we get $e_{dt} i_{\hat{p}}\omega = 0$, $i_{\widehat{dt}} e_p\omega = 0$. Hence $i_{\hat{p}}\omega = \lambda_1\, dt$, $e_p\omega = \lambda_2 \# dt$. We apply the operator $i_{\hat{p}}$ to the first relation and the operator e_p to the second one. Since $g(p, dt) \neq 0$, we obtain $\lambda_1 = \lambda_2 = 0$. □

Let us describe the structure of the space $\operatorname{Ker} \sigma_p(\overline{\Delta})$, $p^2 = 0$.

From $e_p\omega = 0$ it follows that $\omega = e_p\theta$, $\theta \in T^*M/\mathbb{R}p$. From $i_{\hat{p}}\omega = 0$ it follows that $i_{\hat{p}} e_p\theta = p^2\theta - g(\theta, p)p = 0$ that is $g(p, p) = 0$, $g(\theta, p) = 0$. Therefore,

$$\operatorname{Ker} \sigma_p(\overline{\Delta}) = \{\omega \in \Lambda^2(T_q^*M) \mid \omega = e_p\theta,\ i_{\hat{p}}\theta = 0,\ g(p, p) = 0\}.$$

Geometrically this is meant that

$$K_p = \operatorname{Ker} \sigma_p(\overline{\Delta}) \overset{k_p}{\rightarrow} E_p^*[\operatorname{Ker} p/\hat{p}]^*, \quad (q, p) \in \operatorname{Char}(\overline{\Delta}), \qquad e_p\theta \mapsto \theta \bmod p .$$

It is obvious that k_p is well defined and is an isomorphism of 2-dimensional spaces. As a consequence we obtain the following statement.

PROPOSITION 2.3.7. *The space* k_p *is a complex line in the space* $\Lambda^2 T_q^*M$, $p \in T_q^*M$, $(q, p) \in \operatorname{Char}(\overline{\Delta})$.

PROOF. The statement follows from Lemma 2.3.1. If $\omega \in K_p$ then

$$e_p \#\omega = \# i_{\hat{p}}\omega = 0, \qquad i_{\hat{p}}\#\omega = \# e_p\omega = 0. \quad \square$$

COROLLARY. *The space E_p^* is the complex line.*

PROOF. By definition of the isomorphism k_p, there is an operator I such that the following diagram

$$\begin{array}{ccc} K_p & \xrightarrow{\#} & K_p \\ {\scriptstyle k_p}\downarrow & & \downarrow{\scriptstyle k_p} \\ E_p^* & \xrightarrow{I} & E_p^* \end{array}$$

is commutative. It is obvious that $I^2 = -\mathbb{1}_{E_p^*}$. The operator I determines the complex structure on the space E_p^*. □

By analogy with the foregoing, one may define the fiber bundle $\mathfrak{L}(\overline{\Delta})$ over the base $\mathrm{Char}(\overline{\Delta})$ with fiber E_p^* over the point $(q, p) \in \mathrm{Char}(\overline{\Delta})$.

PROPOSITION 2.3.8. *The fiber bundle $K(\overline{\Delta})$ (respectively $\mathfrak{L}(\overline{\Delta})$) is a 1-dimensional complex fiber bundle, the complex structure being defined by the operator # (respectively the operator I).*

Note that k_p defines the isomorphism $k\colon C^\infty(K(\nabla)) \rightleftarrows C^\infty(\mathfrak{L}(\nabla))$, $\nabla = D, \overline{\Delta}$.

2.4. Next, let us consider a useful identification of the spaces $\mathrm{Ker}\,\sigma_p(\nabla)$ and $\mathrm{Coker}\,\sigma_p(\nabla)$ for the operators $\nabla = D, \overline{\Delta}$. For this purpose we construct operator

$$G_p\colon \Lambda^{\cdot}(M) \to K_p(\nabla), \qquad p \in \mathrm{Char}_q(\nabla)$$

satisfying the condition

$$\mathrm{Ker}\, G_p = \mathrm{Im}\,\sigma_p(\nabla), \qquad p \in \mathrm{Char}_q(\nabla).$$

As a consequence we defined of the operator $G\colon C^\infty(K(\nabla)) \rightleftarrows C^\infty(K(\nabla))$. We describe the operator G_p for Dirac and Maxwell operators.

1. *The Dirac operator.* Proposition 2.3.1 implies that $G_p = B_p$.

2. *The modified Maxwell operator.* Let us prove that the operator $G_p = i_{\hat{p}} e_p$ satisfies all the necessary conditions.

LEMMA 2.4.1. $G_p \circ \sigma_p(\overline{\Delta}) = 0$, $p \in \mathrm{Char}(\overline{\Delta})$.

PROOF. Using the results of 2.2, we obtain:

$$G_p \circ \sigma_p(\overline{\Delta}) = i_{\hat{p}} e_p (e_{dt} i_{\hat{p}} - i_{\widehat{dt}} e_p) = -(p, dt) e_p i_{\hat{p}} - (p, dt) i_{\hat{p}} e_p = 0. \quad \square$$

LEMMA 2.4.2. $\mathrm{Ker}\, G_p = \mathrm{Im}\,\sigma_p(\overline{\Delta})$, $p \in \mathrm{Char}(\overline{\Delta})$.

PROOF. Lemma 2.4.1 implies that $\mathrm{Im}\,\sigma_p(\overline{\Delta}) \subset \mathrm{Ker}\, G_p$, $p \in \mathrm{Char}(\overline{\Delta})$. Therefore it is suffices to prove that $\dim \mathrm{Ker}\, G_p = \dim \mathrm{Im}\,\sigma_p(\overline{\Delta})$. We have $\dim \mathrm{Ker}\,\sigma_p(\overline{\Delta}) = 2$, so that $\dim \mathrm{Im}\,\sigma_p(\overline{\Delta}) = 4$. Consider $\mathrm{Ker}\, G_p = \{\omega \in \Lambda^2(M) \mid e_p i_{\hat{p}} \omega = 0\}$. Conditions on the space $\mathrm{Ker}\, G_p$ may be rewritten as $i_{\hat{p}}\omega = \lambda p$ or $\omega = e_p\theta + \omega^0$, $i_{\hat{p}}\omega^0 = 0$, $\theta \in \Lambda^1(M)/\mathbb{R}p$. The forms $e_p\theta$ generate a 3-dimensional subspace of the space $\Lambda^2(M)$. A similar statement is true for the forms $\omega^0 \in \Lambda^2(M)$, $i_{\hat{p}}\omega^0 = 0$. These subspaces intersect in a 2-dimensional subspace $\mathrm{Ker}\,\sigma_p(\overline{\Delta})$. Therefore $\dim \mathrm{Ker}\, G_p = 3 + 3 - 2 = 4$. □

§3. The transport operator in Maxwell–Dirac theory

3.1. In this section we consider the transport of the principal part of the singularity of a solution. The operator that defines this evolution is called the transport operator (see [7]). For this purpose we return to subsection 2.1. As it is shown there, the principal part of the singularity of the solution ω_s^1 belongs to $\operatorname{Ker}\sigma_{ds}(\nabla)$. In order to give more exact information about ω_s^1, we consider the next term in equation (2.1.1) after the substitution of the series (2.1.2), i.e., the coefficient of the function f_1:

$$\nabla(\omega_s^1) + \sigma_{ds}(\nabla)\omega_s^2 = 0 \quad \text{or} \quad \nabla(\omega_s^1) \in \operatorname{Im}\sigma_{ds}(\nabla).$$

Note that this condition depends on the value of ω_s^1 on the surface $s = 0$ only and does not depend upon the prolongation of ω_s^1 to the neighborhood of the surface $s = 0$. Indeed, in order to apply the operator ∇, it is necessary to prolong ω_s^1 to the neighborhood of the surface $s = 0$ in some way, i.e., consider the element $\omega_s^1 + sv$ instead of ω_s^1. Then $\nabla(\omega_s^1 + sv) = \nabla(\omega_s^1) + \sigma_{ds}(\nabla)v + s\nabla(v)$ or

$$[\nabla(\omega_s^1 + sv) - \nabla(\omega_s^1)] \subset \operatorname{Im}\sigma_{ds}(\nabla)$$

if $S(q) = 0$. Thus we obtain a well-defined transport operator

$$\nabla_1 \colon C^\infty(K(\nabla)) \to C^\infty(C(\nabla)),$$
$$\nabla_1(\omega)(q\,,\, d_q s) = [\nabla(\omega) \mid \operatorname{Im}\sigma_{ds}(\nabla)], \qquad s(q) = 0.$$

The operator $G(\nabla)$ constructed in 2.4 allows us to define the transport operator $[\nabla_1]$ as an operator taking $C^\infty(K(\nabla))$ to itself. We define it from the following commutative diagram:

$$\text{(3.1.1)} \qquad \begin{array}{ccc} C^\infty(K(\nabla)) & \xrightarrow{\ \nabla_1\ } & C^\infty(C(\nabla)) \\ & \searrow^{[\nabla_1]} & \downarrow{\scriptstyle G(\nabla)} \\ & & C^\infty(K(\nabla)) \end{array}$$

The operator $[\nabla_1]$ preserves the grading of the fiber bundle $K(\nabla)$:

$$[\nabla_1] \colon C^\infty(K^i(\nabla)) \to C^\infty(K^i(\nabla)).$$

Therefore using the isomorphism k (see 2.3), we can define the transport operator $\langle\nabla_1\rangle$ on sections $C^\infty(\mathfrak{L}^i(\nabla))$ by putting $k \circ [\nabla_1] = \langle\nabla_1\rangle \circ k$.

Now we describe the structure of the transport operators. Let $M_0 = \{h = 0\}$ be the characteristic surface.

I°. The modified Maxwell operator

$$\text{(3.1.2)} \quad [\Delta_1](\omega)(q\,,\, d_q h) = i_{\widehat{dh}} e_{dh}(i_{\widehat{dt}} d - e_{dt}\delta)\omega = g\,(dh\,,\, dt)[L_{\widehat{dh}} + \#^{-1} L_{\widehat{dh}} \#]\omega\,,$$

where $\omega \in C^\infty(K(\overline{\Delta}))$, $\omega = dh \wedge \theta_{dh}$, $\theta_{dh} \in C^\infty(\mathfrak{L}(\overline{\Delta}))$,

$$\text{(3.1.2}'\text{)} \qquad \langle\Delta_1\rangle(\theta_{dh})(q\,,\, d_q h) = g\,(dh\,,\, dt)[L_{\widehat{dh}} + I^{-1} L_{\widehat{dh}} I].$$

PROPOSITION 3.1.1. 1°. *The operator*

$$[\overline{\Delta}_1](\omega)(q, p) = g(p, dt)[L_{\hat{p}} + \#^{-1}L_{\hat{p}}\#]\omega,$$
$$\omega \in C^\infty(K(\overline{\Delta})), \quad (q, p) \in \mathrm{Char}(\overline{\Delta})$$

(*respectively*

$$\langle\overline{\Delta}_1\rangle(\theta)(q, p) = g(p, dt)[L_{\hat{p}} + I^{-1}L_{\hat{p}}I],$$
$$\theta \in C^\infty(\mathfrak{L}(\overline{\Delta})), \quad (q, p) \in \mathrm{Char}(\overline{\Delta}))$$

takes the set of sections of the fiber bundle $K(\overline{\Delta})$ (*respectively* $\mathfrak{L}(\overline{\Delta})$) *to itself and preserves the complex structure of this fiber bundle.*

2°. *The operators* $[\overline{\Delta}_1]$ *and* $\langle\overline{\Delta}_1\rangle$ *cover the vector field* $\hat{p}$ *at the point* $(q, p) \in \mathrm{Char}(\overline{\Delta})$, *i.e., we have*

$$[\Delta_1](f\omega)(q, p) = 2g(p, dt)\,\hat{p}(f)\omega + f[\Delta_1](\omega),$$
$$\langle\Delta_1\rangle(f\theta)(q, p) = 2g(p, dt)\,\hat{p}(f)\theta + f\langle\Delta_1\rangle(\theta),$$

where $\omega \in C^\infty(K(\overline{\Delta}))$, $\theta \in C^\infty(\mathfrak{L}(\overline{\Delta}))$, $f \in C^\infty(M)$.

II°. The Dirac operator.

$$(3.1.3)\qquad \begin{aligned} [D_1](\omega)(q, d_qh) &= \sum_i i_{\widehat{dh}}(d\omega_i + \delta\omega_{i+2}) + e_{dh}(d\omega_{i-2} + \delta\omega_i) \\ &= \sum_i (L_{\widehat{dh}} + \#^{-1}L_{\widehat{dh}}\#)\omega_i, \end{aligned}$$

where $\omega_i \in C^\infty(K^i(D))$, $(q, d_qh) \in \mathrm{Char}(D)$. Accordingly,

$$(3.1.3')\qquad \langle D_1\rangle(\theta)(q, d_qh) = \sum_i (L_{\widehat{dh}} + I^{-1}L_{\widehat{dh}}I)\theta_i,$$

where $\theta_i \in C^\infty(\mathfrak{L}^i(D))$.

As a consequence, we obtain the following statement.

PROPOSITION 3.1.2. 1°. *The operator* $[D_1] = L_{\hat{p}} + \#^{-1}L_{\hat{p}}\#$ (*respectively* $\langle D_1\rangle = L_{\hat{p}} + I^{-1}L_{\hat{p}}I$) *at a point* $(q, p) \in \mathrm{Char}\,D$ *acts in sections of the fiber bundle* $K(D)$ (*respectively* $\mathfrak{L}(D)$) *and preserves the grading of this fiber bundle as well as its complex structure*:

$$\#[D_1] = [D_1]\#, \qquad I\langle D_1\rangle = \langle D_1\rangle I.$$

2°. *The operators* $[D_1]$ *and* $\langle D_1\rangle$ *cover the vector field* $\hat{p}$ *at the point* $(q, p) \in \mathrm{Char}(D)$, *i.e., we have*

$$[D_1](f\omega) = 2\hat{p}(f)\omega + f[D_1](\omega), \qquad \langle D_1\rangle(f\theta) = 2\hat{p}(f)\theta + f\langle D_1\rangle(\theta),$$

where $\omega \in C^\infty(K(D))$, $\theta \in C^\infty(\mathfrak{L}(D))$, $f \in C^\infty(M)$.

3.2. Any characteristic manifold $M_0 = \{q \in M \mid h(q) = 0, d_q h \neq 0\}$ defines the homogeneous Lagrangian manifold $N_{dh} \subset \operatorname{Char} D$, $N_{dh} = \{(q, d_q h) \mid h(q) = 0\}$, $\dim \operatorname{Ker} \pi_* T_q N_{dh} = 1$, where $\pi\colon T^*M \to M$.

Having in mind the construction of the transport operator in the neighborhood of points of the caustic, we carry out all the constructions on a arbitrary homogeneous Lagrangian manifold. For this purpose we need to extend the transport operator acting in sections of the fiber bundle $\mathfrak{L}$ to a neighborhood of the singularity of the Lagrangian manifold's projection on the manifold M. The prolongation will depend on the construction of the singularity, and concretely it is only necessary to prolong the operator I. For this purpose we recall the structure of the cotangent space.

Let ρ be the universal 1-form on the cotangent space: $\rho_{(q,p)}(X) = p(\pi_* X)$, where $(q, p) \in T^*M$, $X \in T_{(q,p)}(T^*M)$.

The form ρ defines a symplectic space structure on the space T^*M, i.e., the nonsingular 2-form $d\rho$. The form $d\rho$ defines a natural isomorphism between the spaces $T^*(T^*M)$ and $T(T^*M)$:

$$X_\varphi \lrcorner\, d\rho = \varphi, \quad \overline{X}_\varphi = \varphi \in T^*(T^*M), \quad X_\varphi \in T(T^*M).$$

The natural generalization of the concept of characteristic manifold is the conic Lagrangian manifold.

DEFINITION 3.2.1. The manifold $N \subset T^*N$ is called the *conic Lagrangian manifold* associated with the differential operator ∇ if

1°. $\rho|_N = 0$.

2°. $N \subset \operatorname{Char}(\nabla)$.

The term "conic" is related to the fact that this manifold is always tangent to the field X_ρ: $(X_\rho \lrcorner\, d\rho)|_N = \rho|_N = 0$ and the one-parameter group associated with the field X_ρ acts as the group of dilatations on the fiber bundle T^*M: $(q, p) \to (q, e^t p)$.

We shall use the following fact from the theory of the Lagrangian manifolds. At any point $z = (q, p) \in N$ we have the equalities

$$\operatorname{Ker} \pi_*|_{T_z N} = \operatorname{Ann} \operatorname{Im} \pi_*|_{T_z N} = (\operatorname{Im} \pi_*|_{T_z N})^\perp,$$

where $\perp$ denotes the orthogonal complement with respect to the metric g. Let us prove this statement.

Take a vector $X \in \operatorname{Ker} \pi_*|_{T_z N}$, i.e., suppose that X is a vertical vector from the space $T_z N$. Then $\overline{X}$ is a horizontal 1-form and $\overline{X} \in \operatorname{Ann}(\operatorname{Im} \pi_*|_{T_z N})$, since for any vector $Y \in \operatorname{Im} \pi_*|_{T_z N}$ we have $\overline{X}(Y) = d\rho(X, Y') = 0$, where $Y' \in T_z N$, $\pi_*(Y') = Y$.

Using the metric g, identify the spaces $\operatorname{Ann}(\operatorname{Im} \pi_*|_{T_z N})$ and $(\operatorname{Im} \pi_*|_{T_z N})^\perp$. We compare the 1-form $\varphi \in \operatorname{Ann}(\operatorname{Im} \pi_*|_{T_z N})$ with the vector $\widehat{\varphi}$. Since $\varphi(X) = g(\widehat{\varphi}, X) = 0$ for any $X \in (\operatorname{Im} \pi_*|_{T_z N})$, we obtain $\widehat{\varphi} \in (\operatorname{Im} \pi_*|_{T_z N})^\perp$.

Further we suppose that $l = n - 1$. Introduce the notation $E^M_{q,p} = \operatorname{Ker} p/\hat{p}$, $E^N_{q,p} = T_{(q,p)}N/(X_p, X_{dH})$, $H = g(p, p)$. The vector X_p is vertical and the vector X_{dH} is horizontal and covers the vector $\hat{p}$ under the projection π_*. Therefore the following relations is true:

$$\begin{aligned}(\operatorname{Ker}\pi_*|_{E^N_{q,p}}) &= (\operatorname{Ker}\pi_*|_{T_{q,p}N})/X_p \simeq \operatorname{Ann}(\operatorname{Im}\pi_*|_{T_{q,p}N})/p \\ &\simeq (\operatorname{Im}\pi_*|_{T_{q,p}N})^{\perp}/\hat{p} \simeq (\operatorname{Im}\pi_*|_{T_{q,p}N})^{\perp}|_{E^M_{p,q}}.\end{aligned}$$

The last relation requires an explanation. By Definition 3.2.1, $\operatorname{Im}\pi_*|_{T_{p,q}N} \in \operatorname{Ker} p$ and since $\pi_*(X_{dH}) = \hat{p}$, we obtain $\hat{p} \in \operatorname{Im}\pi_*|_{T_{q,p}N}$. Therefore $g(\widehat{\varphi}, \hat{p}) = p(\widehat{\varphi}) = 0$ for any $\widehat{\varphi} \in (\operatorname{Im}\pi_*|_{T_{q,p}N})^{\perp}$.

Let us define the operator

$$\underline{I}\colon \Lambda^k[(E^N_{q,p})^*] \to \Lambda^{n-k-2}[(E^N_{q,p})^*]$$

so that $\underline{I}^2 = -\mathbb{1}$ and $\underline{I}$ coincides with the operator I constructed above outside the singularities of the projection of the conic Lagrangian manifold.

It suffices to construct the operator $\underline{I}$ in the neighborhood of Lagrangian manifold's singularities. For this purpose consider the following commutative diagram:

$$\begin{array}{ccccccccc} 0 & \longrightarrow & \operatorname{Ker}\pi_*|_{E^N_{q,p}} & \longrightarrow & E^N_{q,p} & \longrightarrow & \operatorname{Im}\pi_*|_{E^N_{q,p}} & \longrightarrow & 0 \\ & & {\scriptstyle\alpha}\downarrow\wr & & & & {\scriptstyle\alpha}\downarrow\wr & & \\ 0 & \longleftarrow & (\operatorname{Im}\pi_*|_{T_{q,p}N})^{\perp}|_{E^M_{q,p}} & \longrightarrow & E^M_{q,p} & \longrightarrow & (\operatorname{Im}\pi_*|_{T_{q,p}N})|_{E^N_{q,p}} & \longleftarrow & 0 \end{array} \tag{3.2.1}$$

Let us construct the isomorphism from the space $E^M_{q,p}$ to the space $E^N_{q,p}$. For this purpose note that the second row in the diagram is split by the metric g:

$$E^M_{q,p} \simeq (\operatorname{Im}\pi_*|_{T_{q,p}N})|_{E^M_{q,p}} \oplus (\operatorname{Im}\pi_*|_{T_{q,p}N})^{\perp}|_{E^M_{q,p}},$$

since $(\operatorname{Im}\pi_*|_{T_{q,p}N}) \cap (\operatorname{Im}\pi_*|_{T_{q,p}N})^{\perp} = (\hat{p})$.

Let us split the first row. Using the metric on the manifold M, one may define a unique pseudo-Riemannian connection on the fiber bundle T^*M (see [6]). This implies the existence of a natural decomposition of the space $T_z(T^*M)$ in the direct sum of the horizontal space and the space tangent to the fiber. The required splitting is defined by the projection on the space tangent to the fiber:

$$E^N_{q,p} = (\operatorname{Ker}\pi_*|_{E^N_{q,p}}) \oplus (\operatorname{Im}\pi_*|_{E^N_{q,p}}).$$

The splitting thus constructed allows us to define the isomorphism $\alpha\colon E^N_{q,p} \to E^M_{q,p}$ and the operator $\underline{I} = \alpha^{-1} \circ I \circ \alpha$. They satisfy all the needed conditions. We can then define the fiber bundle $\mathfrak{L}^N$ over the homogeneous Lagrangian manifold N with fiber $\Lambda^{\boldsymbol{\cdot}}[(E^N_{q,p})^*]$ over any point $(q, p) \in N$.

PROPOSITION 3.2.1. *The fiber bundle $\mathfrak{L}^N$ is a 2^{n-3}-dimensional complex fiber bundle, the complex structure being defined by the operator $\underline{I}$.*

3.3. Let us construct the transport operator

$$(\nabla_1)\colon C^\infty(\mathfrak{L}^N) \to C^\infty(\mathfrak{L}^N)$$

that preserves the grading and the complex structure of the fiber bundle $\mathfrak{L}^N$ and coincides with the operator $\langle\nabla_1\rangle$ outside the caustic points.

1. The modified Maxwell operator

$$(\overline{\Delta}_1) = g(p\,,\,dt)[L_{X_{dH}} + \underline{I}^{-1}L_{X_{dH}}\underline{I}]\,. \tag{3.3.1}$$

2. The Dirac operator

$$(D_1) = L_{X_{dH}} + \underline{I}^{-1}L_{X_{dH}}\underline{I}\,. \tag{3.3.2}$$

Therefore we have

THEOREM 3.3.1. *The transport operators preserve the complex structure defined by the operator $\underline{I}$:*

$$(\overline{\Delta}_1)\circ\underline{I} = \underline{I}\circ(\overline{\Delta}_1)\,, \qquad (D_1)\circ\underline{I} = \underline{I}\circ(D_1)\,.$$

As a consequence, we obtain the following statement.

THEOREM 3.3.2. *The transport operator covers the Hamiltonian vector field, i.e.,*

$$(\overline{\Delta}_1)(f\omega) = 2g(p\,,\,dt)\,L_{X_{dH}}(f)\,\omega + f(\overline{\Delta}_1)(\omega)\,,$$
$$(D_1)(f\theta) = 2L_{X_{dH}}(f)\,\theta + f(D_1)(\theta)\,,$$

where $\omega \in C^\infty(\mathfrak{L}(\overline{\Delta}))$, $\theta \in C^\infty(\mathfrak{L}(D))$, $f \in C^\infty(N)$.

The statements of Theorems 3.3.1 and 3.3.2 are direct consequences of formulas (3.3.1) and (3.3.2).

Appendix

1. The transport operator for the Maxwell equation in local coordinates. Let us choose some (local) basis α_i in the 2-forms $\Lambda^2(M)$. Any form $\omega \in \Lambda^2(M)$ is then expressed as $\omega = \sum_i \omega^i\alpha_i$. Conditions for having $\omega \in \operatorname{Ker}\sigma_p(\overline{\Delta})$ then have the form:

$$\omega^i(\hat{p} \lrcorner \alpha_i) = 0\,, \quad \omega^i(\hat{p} \lrcorner \#\alpha_i) = 0\,, \quad p \in \operatorname{Char}_q(\overline{\Delta})\,, \quad \hat{p} \in T_qM\,,$$
$$\hat{p} = A_g^{-1}(p)\,, \quad \hat{p} = \sum_k p_k X_k\,,$$

where X_k is the local basis in the space T_qM.

Therefore the transport operator may be described in the following way:

$$\begin{aligned}(1.1)\qquad \Delta_1\omega &= g(p, dt)[L_{\hat{p}} + \#^{-1}L_{\hat{p}}\#]\omega \\ &= g(p, dt)\sum_i [2\hat{p}(\omega^i) + \omega^i(L_{\hat{p}} + \#^{-1}L_{\hat{p}}\#)]\alpha_i .\end{aligned}$$

Similarly the transport operator may be calculated in local coordinates on sections $C^\infty(\mathfrak{L}(\overline{\Delta}))$. Here the fiber of the fiber bundle $\mathfrak{L}(\overline{\Delta})$ is the space $\Lambda^1[(E^M_{q,p})^*]$ at the point $(q, p) \in \mathrm{Char}(\overline{\Delta})$.

Let e_i be the local basis in the space $C^\infty[\mathfrak{L}(\overline{\Delta})]$. Any section $\theta \in C^\infty(\mathfrak{L}(\overline{\Delta}))$ has the form: $\theta = \sum_i \theta^i e_i$. Conditions on the section θ are the following ones:

$$\sum_i \theta^i p_i = 0, \qquad \sum_i (I\theta)^i p_i = 0.$$

As before, simple calculations lead us to the following expression for the transport operator

$$(1.2)\qquad \langle\Delta_1\rangle(\theta) = g(p, dt)\sum_i [2L_{\hat{p}}(\theta^i) + \theta^i(L_{\hat{p}} + I^{-1}L_{\hat{p}}I)]e_i .$$

Both equations (1.1) and (1.2) are equivalent to systems of ordinary differential equations on functions ω^i and θ^i with Lie differentiation along the vector $\hat{p}$.

2. The transport operator for the Maxwell equation in the Schwarzschild metric. The Schwarzschild metric defines the geometry of empty space surrounding a gravitational pointwise particle (see [11]). It has the form

$$g = \left(1 - \frac{2m}{r} - \frac{\Lambda}{3}r^2\right)dt^2 - \left(1 - \frac{2m}{r} - \frac{\Lambda}{3}r^2\right)^{-1} dr^2 - r^2 d\theta^2 - r^2\sin^2\theta\, d\varphi^2 ,$$

where m is the mass of the particle and Λ is the cosmological constant. Calculating the operators $\#$ and $\hat{}$, we obtain the following expression for the transport operators in coordinates

$$\begin{aligned}(\alpha_1, \alpha_2, \alpha_3, \alpha_4, \alpha_5, \alpha_6) &= (dr\wedge d\theta, dr\wedge d\varphi, dr\wedge dt, d\varphi\wedge dt, d\theta\wedge dt, d\theta\wedge d\varphi): \\ [\Delta_1]\omega = p_t\Bigg\{ &\sum_{k=1}^{6}([2\hat{p}(\omega^k) + \omega^k\hat{p}(\ln\gamma_k)]\alpha_k \\ &+ \sum_{j=1}^{6}[-L^j_k(\omega) + L^{c(j)}_{c(k)}(\omega)\gamma_{c(k)}\gamma_{c(j)}]\alpha_j)\Bigg\},\end{aligned}$$

where $L \in \mathrm{Diff}_1(\Lambda^2 T^* M)$, $\omega^k \alpha_k \mapsto \sum_{j=1}^6 L_k^j(\omega)\alpha_j$, $(L_k^j(\omega))$ is the matrix

$$\begin{pmatrix}
p_r\frac{\partial\omega^1}{\partial r}+p_\theta\frac{\partial\omega^1}{\partial\theta} & p_\theta\frac{\partial\omega_1}{\partial\varphi} & p_\theta\frac{\partial\omega^1}{\partial t} & 0 & -p_r\frac{\partial\omega^1}{\partial t} & -p_r\frac{\partial\omega^1}{\partial\varphi} \\
p_\theta\frac{\partial\omega^2}{\partial\theta} & p_r\frac{\partial\omega^2}{\partial r}+p_\varphi\frac{\partial\omega^2}{\partial\varphi} & p_\varphi\frac{\partial\omega^2}{\partial t} & -p_r\frac{\partial\omega^2}{\partial t} & 0 & p_r\frac{\partial\omega^2}{\partial\theta} \\
p_t\frac{\partial\omega^3}{\partial\theta} & p_t\frac{\partial\omega^3}{\partial\varphi} & p_r\frac{\partial\omega^3}{\partial r}+p_t\frac{\partial\omega^3}{\partial t} & p_r\frac{\partial\omega^3}{\partial\varphi} & p_r\frac{\partial\omega^3}{\partial\theta} & 0 \\
0 & -p_t\frac{\partial\omega^4}{\partial r} & p_\varphi\frac{\partial\omega^4}{\partial r} & p_\varphi\frac{\partial\omega^4}{\partial\varphi}+p_t\frac{\partial\omega^4}{\partial t} & p_\varphi\frac{\partial\omega^4}{\partial\theta} & -p_t\frac{\partial\omega^4}{\partial\theta} \\
-p_t\frac{\partial\omega^5}{\partial r} & 0 & p_\theta\frac{\partial\omega^5}{\partial r} & p_\theta\frac{\partial\omega^5}{\partial\varphi} & p_\theta\frac{\partial\omega^5}{\partial\theta}+p_t\frac{\partial\omega^5}{\partial t} & p_t\frac{\partial\omega^5}{\partial\varphi} \\
-p_\varphi\frac{\partial\omega^6}{\partial r} & p_\theta\frac{\partial\omega^6}{\partial r} & 0 & -p_\theta\frac{\partial\omega^6}{\partial t} & p_\varphi\frac{\partial\omega^6}{\partial t} & p_\theta\frac{\partial\omega^6}{\partial\theta}+p_\varphi\frac{\partial\omega^6}{\partial\varphi}
\end{pmatrix},$$

and

$$c(k) = (k+3) \bmod 6, \qquad \hat{p} = p_r\frac{\partial}{\partial r} + p_\theta\frac{\partial}{\partial\theta} + p_\varphi\frac{\partial}{\partial\varphi} + p_t\frac{\partial}{\partial t},$$

$$\gamma_1 = -(1 - 2m/r - (\Lambda/3)r^2)\sin\theta, \qquad \gamma_2 = \frac{1 - 2m/r - (\Lambda/3)r^2}{\sin\theta},$$

$$\gamma_3 = -r^2\sin\theta, \qquad \gamma_{3+i} = -1/\gamma_i, \quad i = 1, 2, 3.$$

References

1. A. Sommerfeld, *Vorlesungen über Theoretische Physik. Bd.* III. *Electrodynamik*, Leipzig, 1949.
2. A. Einstein, *Eine neue formale Deutung der Maxwellschen Feldgleichungen der Elektrodynamik*, Sitzungsber. Preus. Akad. Wiss. **1** (1916), 184–188.
3. E. Cartan, *Leçons sur les invariants intégraux*, Hermann, Paris, 1922.
4. P. K. Rashevskiĭ, *Theory of spinors*, Uspekhi Mat. Nauk **10** (1955), no. 2, 3–110. (Russian)
5. N. Bourbaki, *Eléments de mathématique. Première partie: les structures foundamentales de l'analyse. Livre* II, *Algèbre*, Hermann, Paris, 1953.
6. S. Helgason, *Differential geometry and symmetric spaces*, Academic Press, New York and London, 1962.
7. V. V. Lychagin, *About the singularities of a differential equation's solutions*, Dokl. Akad. Nauk SSSR; English transl., Soviet Math. Dokl. **21** (1980), no. 2, 523–528.
8. R. K. Luneburg, *Mathematical theory of optics*, Univ. California Press, Berkeley, CA, 1964.
9. M. Kline and I. Kay, *Electromagnetic theory and geometrical optics*, Interscience, New York, 1965.
10. M. A. Lichnerowicz, *Laplacien sur une variété riemanniene et spineurs*, Atti. Accad. Naz. Lincei **33** (1963), no. 5.
11. R. C. Tolman, *Relativity thermodynamics and cosmology*, Clarendon Press, Oxford, 1969.
12. L. V. Zilbergleit, *About the singularities of the solutions of the Maxwell-Dirac equation*, Preprint N858-B86 (1986), VINITI, 153–182. (Russian)

Translated by A. B. SOSSINSKY

GOLUBINSKAYA ST., 7–2, APT. 554, 117574, MOSCOW, RUSSIA

Amer. Math. Soc. Transl.
(2) Vol. **167**, 1995

Characteristic Classes of Monge–Ampère Equations

L. ZILBERGLEIT

ABSTRACT. Using a spectral sequence, we compute characteristic classes for multivalued solutions of the Monge–Ampère equation. These classes are generated by Grassmannians consisting of planes tangent to solutions of Monge–Ampère equations. In the two-dimensional nondegenerate case, these Grassmannians coincide with the torus T^2 or the projective line $\mathbb{C}P^1$, depending on the type of Sp-orbits. In the three-dimensional case the result is more refined and there are characteristic classes that cannot be computed in terms of the Stiefel–Whitney classes of the tautological fiber bundle over the Grassmannian. In dimensions greater than 3, we describe the Sp-invariant partition of Monge–Ampère equations by nonlinearity degree. Using this partition, we obtain conditions of coincidence of characteristic classes of multivalued solutions of Monge–Ampère equations with absolute characteristic classes generated by absolute Grassmannians of Lagrangian planes.

Introduction

Among the second order differential equations on a smooth manifold M, $\dim M = n$ we consider the important class of Monge–Ampère equations. This class is essentially of geometric nature and is defined by n-forms $\omega \in \Lambda^n_{\mathrm{ef}}(J^1M)$ [1]. It is closed with respect to the action of the contact transformations group. Tangent planes to the solutions of this equations define Grassmannians, which are the fibers of a smooth vector bundle in the general case.

The cohomology of the Grassmannians associated with Monge–Ampère equations defines certain topological invariants on singularities of the solutions of the Monge–Ampère equation ([2–4]). The meaning of these classes is the following: the first class is responsible for the construction of the discontinuous solutions from the multivalued ones (as the asymptotic solutions do in the theory of Maslov's canonic operator [5, 6]); the classes of codimension one are responsible for the solvability of the boundary value problems; classes of codimension greater than one are responsible for the solvability of Sobolev problems with data on submanifolds of codimension greater than one. The purpose of this article is the computation

1991 *Mathematics Subject Classification*. Primary 58G30.
Key words and phrases. Monge–Ampère equation, characteristic class, spectral sequence, degree of nonlinearity.

of characteristic classes for Monge–Ampère equations. Our method is based on the construction of a spectral sequence determined by the natural filtration in the Grassmannians associated with the given Monge–Ampère equations.

§1. Jet fiber bundles, Grassmannians, and the spectral sequence

Let μ_x be the ideal of the ring $C^\infty(M)$ associated with the point $x \in M$:

$$\mu_x = \{f \in C^\infty(M) \mid f(x) = 0\}.$$

The fiber bundle $\pi_k\colon J^k M \to M$ with fiber $J^k_x M = C^\infty(M)/\mu_x^{k+1} C^\infty(M)$ over the point $x \in M$ is called the *k-jet fiber bundle*. The image of a function $f \in C^\infty(M)$ in the fiber $J^k_x M$ will be denoted by $[f]^k_x$. The fiber bundle $J^1 M$ has a natural contact structure defined by the universal Cartan 1-form U_1 [1]. If $q_1, \dots, q_n$ are the local coordinates on the manifold M in a neighborhood of the point $x \in M$ and $u, p_1, \dots, p_n$ are the local coordinates in the fibers $J^1_x M$ over this neighborhood, then the form U_1 may be described as

$$U_1 = du - \sum_{i=1}^{n} p_i\, dq_i .$$

DEFINITION 1.1. A plane $L \subset T_{x_1} J^1 M$ is called *Lagrangian* if and only if
1°. $\dim L = n$,
2°. $U_1|_L = 0$,
3°. $dU_1|_L = 0$.

In other words, the plane L is the integral n-plane of the differential ideal determined by the 1-form U_1 (Cartan ideal).

An important example of Lagrangian planes are the planes $L(x_2)$, $x_2 \in J^2 M$. They are defined in the following way. For any function $f \in C^\infty(M)$ consider the natural section $S_{j_1(f)}$ of the fiber bundle $J^1 M$:

$$S_{j_1(f)}\colon M \to J^1 M, \qquad x \mapsto [f]^1_x .$$

Then the *plane* $L(x_2)$ is defined as $T_{x_1}[S_{j_1(f)}(M)]$, $x_1 = [f]^1_x$.

Any Lagrangian plane L decomposes in the direct sum $L = L_V \oplus L_h$, where $L_V = L \cap \pi_1^{-1}(x) = \operatorname{Ann}\zeta_x$, $\zeta_x = \pi_{1,*}L \subset T_x M$, and $L_h \subset L(x_2)$ for some $x_2 \in J^2 M$, $L_h = L \cap L(x_2)$. The relation $L \cap L(x_2) = L \cap L(x_2')$ is true if and only if $x_2 - x_2' \in T^*_x M \cdot \operatorname{Ann}\zeta_x$. Hence the subplane $L_h \subset L$ is uniquely defined by an element of the space $S^2 T^*_x M / T^*_x M \cdot \operatorname{Ann}\zeta_x \simeq S^2 \zeta^*_x$. Finally, the Lagrangian plane L is identified with $(\operatorname{Ann}\zeta_x, \theta)$, where $\zeta_x = \pi_{1,*}L \subset T_x M$, $\theta \in S^2 \zeta^*_x$ [2].

The *Grassmannian* $I(x_1)$ is by definition the set of Lagrangian planes at the point $x_1 \in J^1 M$, with the natural homogeneous space structure:

$$I(x_1) \simeq \operatorname{Sp}(2n)/\operatorname{GL}(n)$$

(see [7]). The Grassmannian $I(x_1)$ possesses the following natural filtration:

$$I(x_1) = X_n \supset X_{n-1} \supset \cdots \supset X_0 \supset X_{-1} = \varnothing, \tag{1}$$

where $X_i = \{L \in I(x_1) \mid \dim L \cap \pi_1^{-1}(x) \geqslant n - i\}$.

The graded objects associated to the filtration (1) are the following fiber bundles:

$$\varkappa\colon (X_p, X_{p-1}) \xrightarrow{(\mathbb{R}^{\binom{p+1}{2}}, \mathbb{R}_0^{\binom{p+1}{2}})} G_{n-p,p}, \quad \varkappa(L) = L_V.$$

THEOREM 1.1 [7]. *The filtration* (1) *defines a spectral sequence* $(E_r^{pq}(I), d_r^{pq})$ *that stabilizes at the first step and converges to the cohomology* $H^*(I(x_1), \mathbb{Z}_2)$. *A direct consequence of this spectral sequence is the relation*

$$H^*(I(x_1), \mathbb{Z}_2) \overset{(n)}{\simeq} \mathbb{Z}_2[W_1, \dots, W_n]/(W_1^2, \dots, W_n^2).$$

Here W_i, $i = 1, \dots, n$, are the Stiefel–Whitney classes of the tautological fiber bundle over the Grassmannian $I(x_1)$. We denote by the symbol $\overset{(n)}{\simeq}$ an isomorphism of graded rings up to degree n.

EXAMPLE. Let $n = 3$. Then the term $E_1^{pq}(I)$ of the spectral sequence is

q				
	0	0	0	0
	0	$\mathbb{Z}_2$	$\mathbb{Z}_2$	0
	0	$\mathbb{Z}_2$	$\mathbb{Z}_2$	0
	$\mathbb{Z}_2$	$\mathbb{Z}_2$	0	0
				p

We use the computation of the ring $H^*(I(x_1), \mathbb{Z}_2)$ performed by Borel [7] and Fuchs [8].

§2. Monge–Ampère equations, associated Grassmannians, and the spectral sequence

Monge–Ampère equations are second order differential equations naturally associated with the geometry of the fiber bundle of 1-jets [1].

The universal form U_1 and its differential dU_1 define the structure of an $SL(2)$-representation on the module $\Lambda^\cdot(J^1M)$. Primitive n-dimensional elements of this representation are called *effective forms* and are denoted by $\Lambda^n_{\mathrm{ef}}(J^1M)$. There is a one-to-one mapping between effective forms $\Lambda^n_{\mathrm{ef}}(J^1M)$ and second order differential Monge–Ampère operators:

$$\Delta_\omega\colon C^\infty(M) \to \Lambda^n(M), \quad f \mapsto S^*_{j_1(f)}\omega, \qquad \omega \in \Lambda^n_{\mathrm{ef}}(J^1M).$$

We denote by $\mathrm{Ct}(x_1)$ the group of germs of contact diffeomorphisms with fixed point $x_1 \in J^1M$. The action of group $\mathrm{Ct}(x_1)$ on Monge–Ampère operators is defined by

$$\alpha(\Delta_\omega) = \Delta_{\alpha^*(\omega)}, \qquad \alpha \in \mathrm{Ct}(x_1).$$

Suppose that a contact vector field X_f, $f(x_1) \neq 0$, $L_{X_f}(\Delta_\omega) = 0$ is given. Then the $\mathrm{Ct}(x_1)$-action on Monge–Ampère operators reduces to Sp-action.

We use the Sp-classification of orbits in general position on the space of primitive n-dimensional forms over $\mathbb{R}$ [9] (for the Sp-classification over $\mathbb{C}$, see [10, 11]).

We denote the symbol of the nonlinear differential operator Δ_ω at the point $x_2 \in J^2M$ by $\sigma_{x_2}(\omega)$:

$$\sigma_{x_2}(\omega): S^2T_x^*M \to \Lambda^n T_x^*M, \qquad \sigma_{x_2}(\omega)(\zeta^2) = dh \wedge S^*_{j_1(f)}[X_{dh} \lrcorner\, \omega].$$

Here $h \in C^\infty(M)$, $h(x) = 0$, $\zeta = d_xh$, $f \in C^\infty(M)$, $[f]_x^2 = x_2$, and X_{dh} is the dU_1-symplectic vector field determined by the differential dh.

DEFINITION 2.1. The covector $\zeta \in T_x^*M$, $\zeta \neq 0$, is said to be *characteristic* for the operator Δ_ω at the point $x_2 \in J^2M$ if the value of $\sigma_{x_2}(\omega)(\zeta^2)$ is zero.

We denote by $\operatorname{Char}(\omega)(x)$ the set of characteristic covectors at the point $x \in M$. If $\operatorname{Char}(\omega)(x) = 0$, then the operator Δ_ω is called *elliptic* at the point $x \in M$.

The set of Lagrangian planes at the point $x_1 \in J^1M$ belonging to the Monge–Ampère equation $E_\omega = \Delta_\omega^{-1}(0) \subset J^2M$ forms the Grassmannian $IE_\omega(x_1) \subset I(x_1)$ associated with the equation E_ω:

$$IE_\omega(x_1) = \{L \in I(x_1) : \omega_{x_1}|_L = 0\}.$$

We assume that the distribution $\chi: J^1M \to IE_\omega$, $x_1 \mapsto IE_\omega(x_1)$ is a smooth vector bundle. We have a natural filtration on the Grassmannian $IE_\omega(x_1)$:

$$IE_\omega(x_1) = X_n^\omega \supset X_{n-1}^\omega \supset \cdots \supset X_0^\omega \supset X_{-1}^\omega = \varnothing. \tag{2}$$

We assume that the graded objects associated to the filtration (2) determine smooth vector bundles:

$$\varkappa_\omega: (X_p^\omega, X_{p-1}^\omega) \xrightarrow{(F_p^\omega, F_p^{\omega,0})} B_p^\omega,$$

where

$$\varkappa_\omega(L) = L_V, \quad F_p^\omega|_{L_V} = \{L \in (X_p^\omega, X_{p-1}^\omega) \mid \varkappa_\omega(L) = L_V\},$$
$$F_p^{\omega,0} = \overline{F_p^\omega} \cap X_{p-1}^\omega.$$

THEOREM 2.1. *The filtration* (2) *defines a spectral sequence* $(E_r^{pq}(IE_\omega), d_r^{pq})$ *converging to the cohomology* $H^*(IE_\omega(x_1), \mathbb{Z}_2)$.

From the Leray–Hirsch theorem we obtain

THEOREM 2.2. *The cohomology* $H^*(IE_\omega, \mathbb{Z}_2)$ *is the module generated by the cohomology* $H^*(IE_\omega(x_1), \mathbb{Z}_2)$ *over the algebra* $H^*(M, \mathbb{Z}_2)$.

Finally, let us note that any class $W \in H^*(IE_\omega, \mathbb{Z}_2)$ defines the characteristic class $\overline{W}: \mathcal{L} \to H^*(\mathcal{L}, \mathbb{Z}_2)$ on the solution $\mathcal{L}$ of the equation E_ω as $\overline{W}(\mathcal{L}) = \tau_{\mathcal{L}}^*(W)$, where $\tau_{\mathcal{L}}: \mathcal{L} \to IE_\omega$, $x_1 \mapsto T_{x_1}\mathcal{L}$ is the tangent map.

§3. Grassmannians associated with Monge–Ampère equations in dimension 2

There is only one invariant of the Sp-action on Monge–Ampère operators in dimension 2 that defines Sp-orbits in general position (see [12]). This invariant is the Pfaffian $\mathrm{Pf}(\omega)$ for any Monge–Ampère's operator Δ_ω:

$$\omega \wedge \omega = \mathrm{Pf}(\omega)\, dU_1 \wedge dU_1 .$$

There are two types of orbits in general position of the Sp-action on Monge–Ampère operators:

1°. $\mathrm{Pf}(\omega)(x_1) > 0$: elliptic orbits;

2°. $\mathrm{Pf}(\omega)(x_1) < 0$: hyperbolic orbits.

THEOREM 3.1. *If the form ω_{x_1} belongs to an elliptic* Sp-*orbit, then there is a homeomorphism between the Grassmannian $IE_\omega(x_1)$ and the projective line $\mathbb{C}P^1$. If the form ω_{x_1} belongs to a hyperbolic* Sp-*orbit, then there is a homeomorphism between the Grassmannian $IE_\omega(x_1)$ and the torus T^2.*

PROOF. 1°. Let the form $\omega \in \Lambda^2_{\mathrm{ef}}(J^1M^2)$ belong to a hyperbolic orbit at the point $x_1 \in J^1M$; then this form can be reduced by Sp-transformations to the expression $dp_1 \wedge dq_1 - dp_2 \wedge dq_2$. Lagrangian planes from the Grassmannian $IE_\omega(x_1)$ are defined as solutions of the following system:

$$dp_1 \wedge dq_1 - dp_2 \wedge dq_2|_L = 0, \qquad dp_1 \wedge dq_1 + dp_2 \wedge dq_2|_L = 0.$$

This system can be transformed to the following form:

$$dp_1 \wedge dq_1|_L = 0, \quad dp_2 \wedge dq_2|_L = 0.$$

Therefore there are two nontrivial 2-vectors $(a_{11}, a_{12}) \in \mathbb{R}^2_0$ and $(a_{21}, a_{22}) \in \mathbb{R}^2_0$, which are solutions of the following system

$$a_{11}\, dp_1 + a_{12}\, dq_1|_L = 0, \qquad a_{21}\, dp_2 + a_{22}\, dq_2|_L = 0.$$

Therefore any plane $L \in IE_\omega(x_1)$ is defined by an element of $\mathbb{R}^2_0 \times \mathbb{R}^2_0 \simeq T^2$.

2°. If the form $\omega \in \Lambda^2_{\mathrm{ef}}(J^1M^2)$ belongs to an elliptic Sp-orbit at the point $x_1 \in J^1M$, then it can be expressed as $dp_1 \wedge dq_2 + dq_1 \wedge dp_2$. Therefore any plane $L \in IE_\omega(x_1)$ is defined as a solution of the system

$$dp_1 \wedge dq_2 + dq_1 \wedge dp_2|_L = 0, \qquad dp_1 \wedge dq_1 + dp_2 \wedge dq_2|_L = 0.$$

This system is equivalent to the following equation in the space $\mathbb{C}^2 = (\mathbb{Z}_1, \mathbb{Z}_2)$:

$$dz_1 \wedge dz_2 = 0, \quad \text{where } z_1 = q_1 + iq_2,\ z_2 = p_1 - ip_2 .$$

Hence any plane $L \in IE_\omega(x_1)$ is determined by the line

$$a_1\, dz_1 + a_2\, dz_2|_L = 0, \qquad (a_1, a_2) \in \mathbb{C}^2_0$$

and the Grassmannian $IE_\omega(x_1)$ is homeomorphic to the projective line $\mathbb{C}P^1$.

§4. Characteristic classes of Monge–Ampère equations in dimension 3

There exists an Sp-invariant on forms $\omega \in \Lambda^3_{\mathrm{ef}}(J^1M^3)$ in dimension 3, namely the quadratic form:

$$q\omega(X, Y) = -(1/4)\perp^2(X \lrcorner \omega) \wedge (Y \lrcorner \omega),$$

where $X, Y \in \operatorname{Ker} U_1, x_1$, $\perp$ is the substitution operator of the symplectic bivector dual to the 2-form dU_{1,x_1} [9].

The form $q\omega$ possesses the following descriptions:

1°. The form $q\omega$ is nondegenerate and the signature $\operatorname{sign} q\omega$ is equal to 0 or 2.

2°. The form $q\omega$ is degenerate and $q\omega \not\equiv 0$.

3°. The form $q\omega$ is 0.

In other words, this is the description of the various types of orbits of Sp-transformations on Monge–Ampère equations. For example, the form ω belongs to the first type of orbit if and only if this orbit does not contain linear representatives.

THEOREM 4.1. I. *Assume that the form $q\omega$ is nondegenerate.*

1°. *If* $\operatorname{sign} q\omega = 0$, *then*

$$H^*(IE_\omega(x_1), \mathbb{Z}_2) \overset{(3)}{\simeq} \mathbb{Z}_2[W_1, W_2, UU_2]/(W_1^2, W_1 \cdot UU_2).$$

2°. *If* $\operatorname{sign} q\omega = 2$, *then*

$$H^*(IE_\omega(x_1), \mathbb{Z}_2) \overset{(3)}{\simeq} \mathbb{Z}_2[W_1, W_2, UU_2]/(W_1^2).$$

II. *Assume that the form $q\omega$ is degenerate, but $q\omega \not\equiv 0$.*

1°. *If the form ω belongs to an elliptic orbit, then*

$$H^*(IE_\omega(x_1), \mathbb{Z}_2) \overset{(3)}{\simeq} H^*(I(x_1), \mathbb{Z}_2)/(W_1 \cdot W_2).$$

2°. *If the form ω belongs to a hyperbolic orbit, then*

$$H^*(IE_\omega(x_1), \mathbb{Z}_2) \overset{(3)}{\simeq} H^*(I(x_1), \mathbb{Z}_2).$$

PROOF. 1. If the form ω is nondegenerate and $\operatorname{sign} q\omega = 0$, then the orbit of Sp-transformations containing the form ω also contains the representative

$$\omega_1 = dp_1 \wedge dp_2 \wedge dp_3 - \lambda\, dq_1 \wedge dq_2 \wedge dq_3$$

(see [9]). Using the homotopy of λ to 1 and the Legendre transformation $p_3 \mapsto q_3$, $q_3 \mapsto -p_3$, we can reduce this representative to the form

$$\omega_1 = dp_1 \wedge dp_2 \wedge dq_3 + dq_1 \wedge dq_2 \wedge dp_3.$$

Now let us describe the term $E_1^{pq}(IE_{\omega_1})$ of our spectral sequence:

a. $p = 0$. The set $X_0^{\omega_1}$ consists of one three-dimensional vertical plane isomorphic to $T_x^* M$. Therefore the following relations hold

$$(X_0^{\omega_1}, X_{-1}^{\omega_1}) = *, \qquad E_1^{0q}(IE_{\omega_1}) = H^q(*, \mathbb{Z}_2).$$

b. $p = 1$. Assume that the plane L belongs to $(X_1^{\omega_1}, X_0^{\omega_1})$ and the 2-plane L_V is generated by the vectors X_{ζ_i}, $\zeta_i \in T_x^* M$, $i = 1, 2$. Then the form $(X_{\zeta_1} \wedge X_{\zeta_2}) \lrcorner\, \omega_{1, x_1}$ belongs to the space $T_x^* M$ and the following relation holds

$$(X_{\zeta_1} \wedge X_{\zeta_2}) \lrcorner\, \omega_{1, x_1}\big|_{\operatorname{Ker} \zeta_1 \cap \operatorname{Ker} \zeta_2} = 0.$$

Hence the bicovector $\zeta = \zeta_1 \wedge \zeta_2$ is the solution of the equation

$$\zeta \wedge (X_\zeta \lrcorner\, \omega_{1, x_1}) = 0,$$

namely $\zeta = dq_3 \wedge \theta$, $\theta \in (\operatorname{Ker} dq_3)^* \cap T_x^* M$.

Any such bicovector ζ is defined by the line passing through the vector θ in the space $(\operatorname{Ker} dq_3)^* \cap T_x^* M \simeq \mathbb{R}^2$. Therefore $B_1^{\omega_1}$ coincides with $\mathbb{R}P^1$. Since the form $X_\zeta \lrcorner\, \omega$ is horizontal, there are no additional conditions for the prolongation of the plane L_V to the plane L. This means that the fiber $F_1^{\omega_1}$ is equal to $\mathbb{R}^1$, as in the absolute case. Therefore the space $(X_1^{\omega_1}, X_0^{\omega_1})$ is a fiber bundle over $\mathbb{R}P^1$ with fiber $(\mathbb{R}^1, \mathbb{R}_0^1)$. Using the Thom isomorphism, we get:

$$E_1^{1q}(IE_{\omega_1}) = H^{1+q}([X_1^{\omega_1}, X_0^{\omega_1}], \mathbb{Z}_2) = H^q(\mathbb{R}P^1, \mathbb{Z}_2).$$

c. $p = 2$. Let the plane L belong to $(X_2^{\omega_1}, X_1^{\omega_1})$ and the line L_V be generated by the vector X_ζ, $\zeta \in T_x^* M$. In this case the 2-form $X_\zeta \lrcorner\, \omega_1$ defines a hyperplane in quadrics $S^2(\operatorname{Ker} \zeta)^*$. Each element of this hyperplane is the complement to the vertical line L_V to the plane in $(X_2^{\omega_1}, X_1^{\omega_1})$. Therefore the fiber $F_2^{\omega_1}$ is a hyperplane in the fiber $F_2 \simeq \mathbb{R}^3$ and $F_2^{\omega_1} \simeq \mathbb{R}^2$. The set $B_2^{\omega_1}$ is equivalent to the set $B_2 \simeq \mathbb{R}P^2$. Thus $(X_2^{\omega_1}, X_1^{\omega_1})$ is a fiber bundle with base $\mathbb{R}P^2$ and fiber $(\mathbb{R}^2, \mathbb{R}_0^2)$. Using the Thom isomorphism, we obtain:

$$E_1^{2q}(IE_{\omega_1}) = H^{2+q}([X_2^{\omega_1}, X_1^{\omega_1}], \mathbb{Z}_2) = H^q(\mathbb{R}P^2, \mathbb{Z}_2).$$

d. $p = 3$. In this case all planes L from $(X_3^{\omega_1}, X_2^{\omega_1})$ are horizontal and $B_3^{\omega_1} = *$. In the quadrics $S^2 T_x^* M \simeq \mathbb{R}^6$, the equation E_{ω_1} defines the surface

$$p_{33} = p_{12}^2 - p_{11} p_{22}.$$

Therefore the fiber $F_3^{\omega_1}$ is the graph of some function and $F_3^{\omega_1} \simeq \mathbb{R}^5$. Using the Thom isomorphism, we get

$$E_1^{3q}(IE_{\omega_1}) = H^{3+q}([X_3^{\omega_1}, X_2^{\omega_1}], \mathbb{Z}_2) = H^{q-2}(*, \mathbb{Z}_2).$$

Thus the term $E_1^{pq}(IE_{\omega_1})$ of our spectral sequence is of the form

q

0	0	0	0
0	0	$\mathbb{Z}_2$	$\mathbb{Z}_2$
0	$\mathbb{Z}_2$	$\mathbb{Z}_2$	0
$\mathbb{Z}_2$	$\mathbb{Z}_2$	$\mathbb{Z}_2$	0

p

The inclusion $IE_{\omega_1}(x_1) \subset I(x_1)$ induces the map $X_p^{\omega_1} \subset X_p$, $p = 0, 1, 2, 3$. We remark that the set $X_0^{\omega_1}$ coincides with the set X_0. We regard the map $(X_1^{\omega_1}, X_0^{\omega_1}) \subset (X_1, X_0)$ as fiber bundles morphism

$$\begin{array}{ccc} (X_1^{\omega_1}, X_0^{\omega_1}) & \hookrightarrow & (X_1, X_0) \\ {\scriptstyle (F_1^{\omega_1}, F_1^{\omega_1, 0})}\big\downarrow & & \big\downarrow{\scriptstyle (F_1, F_1^0)} \\ B_1^{\omega_1} & \hookrightarrow & B_1 \end{array}$$

Here $B_1^{\omega_1} = \mathbb{R}P^1$, $B_1 = \mathbb{R}P^2$ and the imbedding $B_1^{\omega_1} \subset B_1$ induces the isomorphism $H^1(B_1, \mathbb{Z}_2) \xrightarrow{\sim} H^1(B_1^{\omega_1}, \mathbb{Z}_2)$. Using the Thom isomorphism and the relation $F_1^{\omega_1} = F_1 \simeq \mathbb{R}^1$, we obtain

$$H^{k+1}([X_1, X_0], \mathbb{Z}_2) \xrightarrow{\sim} H^{k+1}([X_1^{\omega_1}, X_0^{\omega_1}], \mathbb{Z}_2), \qquad k \leqslant 1,$$

or $E_1^{1q}(I) \xrightarrow{\sim} E_1^{1q}(IE_{\omega_1})$, for $q \leqslant 1$. Since $E_1^{20}(I) = 0$, the mapping

$$(X_2^{\omega_1}, X_1^{\omega_1}) \hookrightarrow (X_2, X_1)$$

induces the monomorphism $E_1^{20}(I) \to E_1^{20}(IE_{\omega_1})$. Thus we have a morphism of spectral sequences $E_1(I) \to E_1(IE_{\omega_1})$ satisfying condition $\langle 1 \,|\, 1 \rangle$ (see the Appendix). It follows from Theorem A that condition $\langle 1 \,|\, \infty \rangle$ is true. Therefore

1°. The monomorphism $E_\infty^{11}(I) \to E_\infty^{11}(IE_{\omega_1})$ and the relation $E_\infty^{11}(I) = \mathbb{Z}_2$ imply

$$E_\infty^{11}(IE_{\omega_1}) = \mathbb{Z}_2, \quad d_1|_{E_1^{11}(IE_{\omega_1})} = 0, \quad E_\infty^{21}(IE_{\omega_1}) = E_1^{21}(IE_{\omega_1}).$$

2°. From condition $\langle 1 \,|\, 1 \rangle$, we obtain $d_1|_{E_1^{10}(IE_{\omega_1})} = 0$.

Thus $E_\infty^{20}(IE_{\omega_1}) = E_1^{20}(IE_{\omega_1})$. As the result, we see that the spectral sequence $(E_r^{pq}(IE_{\omega_1}), d_r^{pq})$, $p + q \leqslant 3$ stabilizes at the first step.

2. If the form $q\omega$ is nondegenerate and $\operatorname{sign} q\omega = 2$, then we can find a representative expressed as

$$\omega_2 = v^2 dp_1 \wedge dp_2 \wedge dp_3 + dq_1 \wedge dp_2 \wedge dq_3 + dp_1 \wedge dq_2 \wedge dq_3 - dq_1 \wedge dq_2 \wedge dp_3$$

in the orbit of Sp-transformations containing the form ω [9]. Homotoping v to 1 and using the Legendre transformation $p_3 \mapsto q_3$, $q_3 \mapsto -p_3$, we reduce this representative to the following form

$$\omega_2 = dp_1 \wedge dp_2 \wedge dq_3 - dq_1 \wedge dp_2 \wedge dp_3 - dp_1 \wedge dq_2 \wedge dp_3 - dq_1 \wedge dq_2 \wedge dq_3.$$

Let us describe the term $E_1^{pq}(IE_{\omega_2})$ of our spectral sequence.

a. $p = 0$. As in the previous case, the set $X_0^{\omega_2}$ consists of one three-dimensional vertical plane isomorphic to T_x^*M. Therefore the following relations hold:

$$(X_0^{\omega_2}, X_{-1}^{\omega_2}) = *, \qquad E_1^{0q}(IE_{\omega_2}) = H^q(*, \mathbb{Z}_2).$$

b. $p = 1$. Let the plane L belong to $(X_1^{\omega_2}, X_0^{\omega_2})$ and the 2-plane L_V be generated by the vectors X_{ζ_i}, $\zeta_i \in T_x^*M$, $i = 1, 2$. Denote $\zeta = \zeta_1 \wedge \zeta_2$, $X_\zeta = X_{\zeta_1} \wedge X_{\zeta_2}$. As in part 1 of the proof, the bicovector ζ is the solution of the equation $\zeta \wedge (X_\zeta \lrcorner\, \omega_{2, x_1}) = 0$. This equation determines a conic surface in $\Lambda^2 T_x^*M$. In the basis $(dq_1 \wedge dq_2, dq_1 \wedge dq_3, dq_2 \wedge dq_3)$ this equation is $\zeta_{(1)}^2 - \zeta_{(2)}^2 - \zeta_{(3)}^2 = 0$, $\zeta = (\zeta_{(1)}, \zeta_{(2)}, \zeta_{(3)})$. Therefore $B_1^{\omega_1}$ is topologically equivalent to S^1. As in part 1, using the Thom isomorphism we get

$$E_1^{1q}(IE_{\omega_2}) = H^{q+1}([X_1^{\omega_2}, X_0^{\omega_2}], \mathbb{Z}_2) = H^q(S^1, \mathbb{Z}_2).$$

c. $p = 2$. Repeating the argument from part 1, we see that $(X_2^{\omega_2}, X_1^{\omega_2})$ is a fiber bundle over $\mathbb{R}P^2$ with fiber $(\mathbb{R}^2, \mathbb{R}_0^2)$. Therefore we have $E_1^{2q}(IE_{\omega_2}) = H^q(\mathbb{R}P^2, \mathbb{Z}_2)$.

d. $p = 3$. In the quadrics $S^2T_x^*M \simeq \mathbb{R}^6$, the equation E_{ω_2} defines the following surface N:

$$(p_{11}p_{22} - p_{12}^2) - (p_{22}p_{33} - p_{32}^2) - (p_{11}p_{33} - p_{13}^2) = 1.$$

Using the change of variables

$$\begin{aligned} p_{11} &= (x_1 - x_4)/2 - x_5, & p_{12} &= x_6, \\ p_{22} &= (x_1 - x_4)/2 + x_5, & p_{32} &= x_2, \\ p_{33} &= -(3x_1 + 5x_4)/4, & p_{13} &= x_3, \end{aligned}$$

we transform this equation to the form $x_1^2 + x_2^2 + x_3^2 - x_4^2 - x_5^2 - x_6^2 = 1$. This equation defines a nontrivial vector bundle η over the sphere S^2 with fiber $\mathbb{R}^3$

$$\begin{gathered} \eta\colon N \xrightarrow{\mathbb{R}^3} S^2, \\ \eta(x_1, x_2, x_3, x_4, x_5, x_6) = (x_1/y, x_2/y, x_3/y), \\ y = (1 + x_4^2 + x_5^2 + x_6^2)^{1/2}, \\ \eta^{-1}(\alpha_1, \alpha_2, \alpha_3) = (\alpha_1 y, \alpha_2 y, \alpha_3 y, x_4, x_5, x_6) \simeq \mathbb{R}^3, \\ \alpha_1^2 + \alpha_2^2 + \alpha_3^2 = 1. \end{gathered}$$

Hence $(X_3^{\omega_2}, X_2^{\omega_2})$ is the fiber bundle over the sphere S^2 with fiber $(\mathbb{R}^3, \mathbb{R}_0^3)$. Using the Thom isomorphism, we obtain

$$E_1^{3q}(IE_{\omega_2}) = H^{q+3}([X_3^{\omega_2}, X_2^{\omega_2}], \mathbb{Z}_2) = H^q(S^2, \mathbb{Z}_2).$$

Therefore, the term $E_1^{pq}(IE_{\omega_2})$ has the form:

q

0	0	0	0
0	0	$\mathbb{Z}_2$	$\mathbb{Z}_2$
0	$\mathbb{Z}_2$	$\mathbb{Z}_2$	0
$\mathbb{Z}_2$	$\mathbb{Z}_2$	$\mathbb{Z}_2$	$\mathbb{Z}_2$

p

As in part 1, we see that the map $IE_{\omega_2}(x_1) \hookrightarrow I(x_1)$ induces the isomorphism $E_1^{1q}(I) \xrightarrow{\sim} E_1^{1q}(IE_{\omega_2})$ for $q \leqslant 1$ and the monomorphism $E_1^{20}(I) \to E_1^{20}(IE_{\omega_2})$. Thus we have a morphism of spectral sequences satisfying condition $\langle 1 \,|\, 1 \rangle$ (see the Appendix). It follows from Theorem A that condition $\langle 1 \,|\, \infty \rangle$ holds. From the monomorphism $E_\infty^{11}(I) \to E_\infty^{11}(IE_{\omega_2})$ and the relation $E_\infty^{11}(I) = \mathbb{Z}_2$, we see that

$$E_\infty^{11}(IE_{\omega_2}) = \mathbb{Z}_2, \qquad d_1|_{E_1^{11}(IE_{\omega_2})} = 0.$$

Thus the relation $E_\infty^{21}(IE_{\omega_2}) = E_\infty^{21}(I)$ holds. From condition $\langle 1 \,|\, 1 \rangle$ we obtain $d_1|_{E_1^{10}(IE_{\omega_2})} = 0$. Let us prove that $E_\infty^{30}(IE_{\omega_2}) \neq 0$. Consider the map

$$f : D_0^3 \to IE_{\omega_2}(x_1), \qquad D_0^3 = (\mathbb{R}^3, \mathbb{R}_0^3),$$
$$f(x_1, x_2, x_3) = (\alpha_1 y, \alpha_2 y, \alpha_3 y, x_1, x_2, x_3) \in N,$$
$$y = (1 + x_1^2 + x_2^2 + x_3^2)^{1/2}, \quad \alpha_1^2 + \alpha_2^2 + \alpha_3^2 = 1, \quad (x_1, x_2, x_3) \in D_0^3.$$

The map f takes the Thom class UU_3 of the fiber bundle $(X_3^{\omega_2}, X_2^{\omega_2})$ to the class f^*UU_3 on D_0^3. Since $f^*UU_3 \neq 0$, the class UU_3 is nonzero and therefore $E_\infty^{30}(IE_{\omega_2}) \neq 0$.

In this case the condition $d_1|_{E_1^{20}(IE_{\omega_2})} \neq 0$ is not valid and the spectral sequence $E(IE_{\omega_2})$ becomes stable at the first step.

3. If the form $q\omega$ is degenerate, $q\omega \not\equiv 0$, and the form ω belongs to the elliptic orbit, then we can find the representative

$$\omega_3 = dp_1 \wedge dq_2 \wedge dq_3 + dq_1 \wedge dp_2 \wedge dq_3 + dq_1 \wedge dq_2 \wedge dp_3$$

(see [9]). Let us describe the term $E_1^{pq}(IE_{\omega_3})$

a. $p = 0$. The set $X_0^{\omega_3}$ consists of one three-dimensional vertical plane isomorphic to T_x^*M, $X_0^{\omega_3} = X_0$, $(X_0^{\omega_3}, X_{-1}^{\omega_3}) = *$. Therefore we have $E_1^{0q}(IE_{\omega_3}) = H^q(*, \mathbb{Z}_2)$.

b. $p = 1$. Let the plane L belong to (X_1, X_0). Then the 2-plane L_V annihilates the form ω_3. Therefore the plane L belongs to $(X_1^{\omega_3}, X_0^{\omega_3})$. Thus $X_1^{\omega_3} = X_1$ and $(X_1^{\omega_3}, X_0^{\omega_3})$ is the fiber bundle over $B_1 = \mathbb{R}P^2$ with fiber $(\mathbb{R}^1, \mathbb{R}_0^1)$. Using the Thom isomorphism, we obtain:

$$E_1^{1q}(IE_{\omega_3}) = H^{q+1}([X_1, X_0], \mathbb{Z}_2) = H^q(\mathbb{R}P^2, \mathbb{Z}_2).$$

c. $p = 2$. Let the plane L belong to $(X_2^{\omega_3}, X_1^{\omega_3})$. Then the line L_V is generated by the vector X_ζ, $\zeta \in T_x^*M$. The form $X_\zeta \lrcorner\, \omega_{3,x_1}$ belongs to

$\Lambda^2 T_x^* M$. The covector ζ is the solution of the equation $\zeta \wedge (X_\zeta \lrcorner \omega_{3,x_1}) = 0$. Thus the operator Δ_{ω_3} is elliptic and this equation only has the trivial solution $\zeta \equiv 0$. Therefore $X_2^{\omega_3} = X_1^{\omega_3}$, and

$$E_1^{2q}(IE_{\omega_3}) = H^{q+2}([X_2^{\omega_3}, X_1^{\omega_3}], \mathbb{Z}_2) = 0.$$

d. $p = 3$. In this case $B_3^{\omega_3} = *$. In the quadrics $S^2 T_x^* M \simeq \mathbb{R}^6$, the equation E_{ω_3} defines a five-dimensional hyperplane. Therefore $(X_3^{\omega_3}, X_2^{\omega_3})$ is a fiber bundle over the point with fiber $(\mathbb{R}^5, \mathbb{R}^5_0)$. Using the Thom isomorphism, we get

$$E_1^{3q}(IE_{\omega_3}) = H^{q+3}([X_3^{\omega_3}, X_2^{\omega_3}], \mathbb{Z}_2) = H^{q-2}(*, \mathbb{Z}_2).$$

Thus the term $E_1^{pq}(IE_{\omega_3})$ has the form

q				
	0	0	0	0
	0	$\mathbb{Z}_2$	0	$\mathbb{Z}_2$
	0	$\mathbb{Z}_2$	0	0
	$\mathbb{Z}_2$	$\mathbb{Z}_2$	0	0
				p

From the relations $X_p^{\omega_3} = X_p$, $p = 0, 1$, and Theorem A, we deduce that the spectral sequence $E_r^{pq}(IE_{\omega_3})$, $p + q \leqslant 3$, stabilizes at the first step.

4. If the form $q\omega$ is degenerate, $q\omega \not\equiv 0$, and the form ω belongs to a hyperbolic orbit, then we can find the representative

$$\omega_4 = dp_1 \wedge dq_2 \wedge dq_3 + dq_1 \wedge dp_2 \wedge dq_3 - dq_1 \wedge dq_2 \wedge dp_3$$

(see [9]). Let us describe the term $E_1^{pq}(IE_{\omega_4})$.

a. $p = 0$. As in part 3.a, the set $X_0^{\omega_4}$ is the set $X_0 = *$ and therefore

$$E_1^{0q}(IE_{\omega_4}) = H^q(*, \mathbb{Z}_2).$$

b. $p = 1$. Repeating the argument of part 3.b, we obtain $X_1^{\omega_4} = X_1$ and therefore $(X_1^{\omega_4}, X_0^{\omega_4})$ is a fiber bundle over $B_1 = \mathbb{R}P^2$ with fiber $(\mathbb{R}^1, \mathbb{R}^1_0)$. Using the Thom isomorphism, we obtain

$$E_1^{1q}(IE_{\omega_4}) = H^{q+1}([X_1, X_0], \mathbb{Z}_2) = H^q(\mathbb{R}P^2, \mathbb{Z}_2).$$

c. $p = 2$. Let the plane L belong to $(X_2^{\omega_4}, X_1^{\omega_4})$. Then the line L_V is generated by the vector X_ζ, $\zeta \in T_x^* M$. Repeating the argument of part 3.c, we see that the covector ζ is the solution of the equation $\zeta \wedge (X_\zeta \lrcorner \omega_{4,x_1}) = 0$. Since the operator Δ_{ω_4} is hyperbolic, this equation defines the sphere S^1. Since the form $X_\zeta \lrcorner \omega_{4,x_1}$ belongs to the space $\Lambda^2 T_x^* M$, any solution X_ζ of the previous equation and any quadric $\theta \in S^2[\operatorname{Ker}\zeta]^*$ define the plane $L \in (X_2^{\omega_4}, X_1^{\omega_4})$. Therefore $(X_2^{\omega_4}, X_1^{\omega_4})$ is a fiber bundle over $B_2^{\omega_4} = S^1$ with fiber $(\mathbb{R}^3, \mathbb{R}^3_0)$. Using the Thom isomorphism, we obtain

$$E_1^{2q}(IE_{\omega_4}) = H^{q+2}([X_2^{\omega_4}, X_1^{\omega_4}], \mathbb{Z}_2) = H^{q-1}(S^1, \mathbb{Z}_2).$$

d. $p=3$. Repeating the argument of part 3.d, we see that $(X_3^{\omega_4}, X_2^{\omega_4})$ is a fiber bundle over the point with fiber $(\mathbb{R}^5, \mathbb{R}_0^5)$. Using the Thom isomorphism, we obtain

$$E_1^{3q}(IE_{\omega_4}) = H^{q+3}([X_3^{\omega_4}, X_2^{\omega_4}], \mathbb{Z}_2) = H^{q-2}(*, \mathbb{Z}_2).$$

Therefore the term $E_1^{pq}(IE_{\omega_4})$ is of the form

q

0	0	0	0
0	$\mathbb{Z}_2$	$\mathbb{Z}_2$	$\mathbb{Z}_2$
0	$\mathbb{Z}_2$	$\mathbb{Z}_2$	0
$\mathbb{Z}_2$	$\mathbb{Z}_2$	0	0

p

The mapping $IE_{\omega_4}(x_1) \hookrightarrow I(x_1)$ induces the mapping $X_p^{\omega_4} \hookrightarrow X_p$, $p=0,1,2,3$. The relations $X_p^{\omega_4} = X_p$, $p=0,1$ hold. We regard the mapping $(X_2^{\omega_4}, X_1^{\omega_4}) \hookrightarrow (X_2, X_1)$ as a morphism of fiber bundles

$$\begin{array}{ccc} (X_2^{\omega_4}, X_1^{\omega_4}) & \hookrightarrow & (X_2, X_1) \\ {\scriptstyle (F_2^{\omega_4}, F_2^{\omega_4, 0})}\downarrow & & \downarrow{\scriptstyle (F_2, F_2^0)} \\ B_2^{\omega_4} & \hookrightarrow & B_2 \end{array}$$

Since $B_2^{\omega_4} = S^1$ and $B_2 = \mathbb{R}P^2$, it follows that the imbedding $B_2^{\omega_4} \hookrightarrow B_2$ induces an isomorphism $H^k(B_2, \mathbb{Z}_2) \xrightarrow{\sim} H^k(B_2^{\omega_4}, \mathbb{Z}_2)$, $k \leqslant 1$. Since $F_2^{\omega_4} \simeq F_2 \simeq \mathbb{R}^3$, the Thom isomorphism implies

$$H^{k+3}([X_2, X_1], \mathbb{Z}_2) \xrightarrow{\sim} H^{k+3}([X_2^{\omega_4}, X_1^{\omega_4}], \mathbb{Z}_2), \qquad k \leqslant 1,$$

or $E_1^{2q}(I) \xrightarrow{\sim} E_1^{2q}(IE_{\omega_4})$, for $q \leqslant 1$. The map $(X_3^{\omega_4}, X_2^{\omega_4}) \hookrightarrow (X_3, X_2)$ induces an isomorphism $E_1^{3q}(I) \xrightarrow{\sim} E_1^{3q}(IE_{\omega_4})$, $q = 0, 1$ (all these terms are equal to zero).

Thus we have a mapping $E_1^{pq}(I) \to E_1^{pq}(IE_{\omega_4})$ satisfying condition $\langle 3 \,|\, 1 \rangle$. By Theorem A, condition $\langle 3 \,|\, \infty \rangle$ holds. Therefore we conclude that the spectral sequence $E_r^{pq}(IE_{\omega_4})$, $p+q \leqslant 3$ stabilizes at the first step and

$$H^*(IE_{\omega_4}(x_1), \mathbb{Z}_2) \overset{(3)}{\simeq} H^*(I(x_1), \mathbb{Z}_2).$$

§5. Characteristic classes of Monge–Ampère equations in arbitrary dimension

Keeping in mind the classification of Monge–Ampère equations, we define the degree of nonlinearity for these equations.

DEFINITION 5.1. The *degree of nonlinearity of the equation* E_ω, where $\omega \in \Lambda_{\mathrm{ef}}^n(J^1M)$, *with respect to the plane* $L \in IE_\omega(x_1)$ is the minimal integer k such that any element $a \in \Lambda^{k+1}L \subset \Lambda^{k+1}T_{x_1}(J^1M)$ annihilates the form ω: $a \lrcorner \omega_{x_1} = 0$.

The *degree of nonlinearity* $k_{x_1}(\omega)$ *of the equation* E_ω *at the point* $x_1 \in J^1M$ is the minimum among degrees of nonlinearity $k(L)$ with respect to the planes $L \in IE_\omega(x_1)$.

The degree of nonlinearity of the equation E_ω is an invariant of the Sp-transformations group. Using Sp-transformations, we can identify the plane $L \in IE_\omega(x_1)$, $k(L) = k_{x_1}(\omega)$, with the vertical n-plane tangent to the fiber $J^1_x M$.

If $k_{x_1}(\omega) = 1$, then the equation E_ω is Sp-equivalent to a quasilinear equation at the point $x_1 \in J^1 M$. The degree of nonlinearity is closely associated with the Sp-classification of Monge–Ampère equations. When $n = 2$, we always have $k_{x_1}(\omega) = 1$. When $n = 3$, for orbits in general position we have $k_{x_1}(\omega) = 3$ and for all other orbits, $k_{x_1}(\omega) = 1$ (see [9]).

THEOREM 5.1. *If* $k_{x_1}(\omega) = 1$ *and* $n > 3$, *then*

$$H^*(IE_\omega(x_1), \mathbb{Z}_2) \overset{(n)}{\simeq} H^*(I(x_1), \mathbb{Z}_2).$$

PROOF. We describe the term $E_1^{pq}(IE_\omega)$, $p + q \leqslant n + 1$. Since the equation E_ω is quasilinear it follows that the set of the filtration X_p^ω coincides with the set X_p: $X_p^\omega = X_p$, $p = 0, 1, \dots, n-2$. Let the plane L belong to $(X_{n-1}^\omega, X_{n-2}^\omega)$. Then the line L_V is generated by the vector X_ζ, $\zeta \in T_x^* M$. The form $X_\zeta \lrcorner \omega_{x_1}$ belongs to the space $\Lambda^{n-1} T_x^* M$ and $(X_\zeta \lrcorner \omega_{x_1})|_{\mathrm{Ker}\,\zeta} = 0$. Therefore the covector ζ is the solution of the equation $\zeta \wedge (X_\zeta \lrcorner \omega_{x_1}) = 0$. In other words, the set B_{n-1}^ω coincides with $\mathrm{Char}(\omega)(x)$. From the condition $X_\zeta \lrcorner \omega_{x_1} \in \Lambda^{n-1} T_x^* M$, we obtain $F_{n-1}^\omega = F_{n-1} \simeq \mathbb{R}^{l(n)}$, where $l(n) = n(n-1)/2$.

Thus $(X_{n-1}^\omega, X_{n-2}^\omega)$ is a fiber bundle with base $\mathrm{Char}(\omega)(x)$ whose fiber is $(\mathbb{R}^{l(n)}, \mathbb{R}_0^{l(n)})$. Using the Thom isomorphism, we obtain

$$E_1^{n-1,q}(IE_\omega) = H^{n+q-1}([X_{n-1}^\omega, X_{n-2}^\omega], \mathbb{Z}_2) = H^{n+q-l(n)-1}(\mathrm{Char}(\omega)(x), \mathbb{Z}_2).$$

Therefore $E_1^{n-1q}(IE_\omega)$ is equal to zero when $q = 0, 1, 2$ if $n > 3$.

Equation E_ω defines a hyperplane in the quadric $S^2\mathbb{R}^n$. Hence $(X_n^\omega, X_{n-1}^\omega)$ is a fiber bundle with fiber $(\mathbb{R}^{l(n)-1}, \mathbb{R}_0^{l(n)-1})$ over the point $B_n^\omega = *$.

Using the Thom isomorphism, we get

$$E_1^{nq}(IE_\omega) = H^{n+q}([X_n^\omega, X_{n-1}^\omega], \mathbb{Z}_2) = H^{n+q+1-l(n)}(*, \mathbb{Z}_2).$$

Thus $E_1^{nq}(IE_\omega) = 0$, $q = 0, 1$, $n > 3$, and the mapping $IE_\omega(x_1) \hookrightarrow I(x_1)$ determines isomorphisms

$$\begin{aligned} E_1^{pq}(I) &\xrightarrow{\sim} E_1^{pq}(IE_\omega), && p + q \leqslant n + 1,\ p = 0, 1, \dots, n-2, \\ E_1^{n-1q}(I) &\simeq E_1^{n-1q}(IE_\omega), && q = 0, 1, 2, \\ E_1^{nq}(I) &\simeq E_1^{nq}(IE_\omega), && q = 0, 1. \end{aligned}$$

Using Theorem A, we get the isomorphism $E_\infty^{pq}(I) \xrightarrow{\sim} E_\infty^{pq}(IE_\omega)$, for $p + q \leqslant n$.

THEOREM 5.2. *Suppose we have the following conditions*:

1° *the inequality* $n > k + [1 + \sqrt{8k+1}]/2$ *holds*;

2° *the algebraic manifolds* $(F_p^\omega, F_p^{\omega,0})$ *are n-acyclic when* $p = n - k + 2, \dots, n-1$ *and* $H^n([F_n^\omega, F_n^{\omega,0}], \mathbb{Z}_2) = 0$;

3° *the degree of nonlinearity of the equation* E_ω *is equal to* k.

Then

$$H^*(IE_\omega(x_1), \mathbb{Z}_2) \overset{(n)}{\simeq} H^*(I(x_1), \mathbb{Z}_2).$$

PROOF. Let us describe the term $E_1^{pq}(IE_\omega)$, $p+q \leqslant n+1$. Since $k_{x_1}(\omega) = k$, for the sets of filtration we have

$$X_p^\omega = X_p, \qquad p = 0, 1, \dots, n-k+1.$$

Let the plane L belong to $(X_{n-k}^\omega, X_{n-k-1}^\omega)$. Then the plane L_V is generated by $X_{\zeta_1}, \dots, X_{\zeta_k}$, $\zeta_i \in T_x^*M$, $i = 1, 2, \dots, k$. We introduce the following notation: $\zeta = \zeta_1 \wedge \dots \wedge \zeta_k$, $X_\zeta = X_{\zeta_1} \wedge \dots \wedge X_{\zeta_k}$. The form $X_\zeta \lrcorner\, \omega_{x_1}$ belongs to the space $\Lambda^{n-k}T_x^*M$ and $X_\zeta \lrcorner\, \omega_{x_1}|_{\bigcap_i \operatorname{Ker}\zeta_i} = 0$. Therefore the form ζ is the solution of the equation $\zeta \wedge (X_\zeta \lrcorner\, \omega_{x_1}) = 0$. This equation defines the set B_{n-k}^ω. Since the form $X_\zeta \lrcorner\, \omega_{x_1}$ belongs to the space $\Lambda^{n-k}T_x^*M$, we have $F_{n-k}^\omega = F_{n-k} \simeq \mathbb{R}^{l(n-k+1)}$.

Using the Thom isomorphism, we get

$$E_1^{n-kq}(IE_\omega) = H^{n+q-k}([X_{n-k}^\omega, X_{n-k-1}^\omega], \mathbb{Z}_2) = H^{n+q-k-l(n-k+1)}(B_{n-k}^\omega, \mathbb{Z}_2).$$

Therefore if $q = 0, 1, \dots, k$ and $n > k + [1+\sqrt{8k+1}]/2$, then $E_1^{n-kq}(IE_\omega) = 0$. Note that $B_p^\omega = B_p$, $p = n-k+1, \dots, n$.

Thus $(X_p^\omega, X_{p-1}^\omega)$ is a fiber bundle over the base B_p with fiber $(F_p^\omega, F_p^{\omega,0})$, $p = n-k+1, \dots, n$. Since $E_1^{pq}(IE_\omega) = H^{p+q}([X_p^\omega, X_{p-1}^\omega], \mathbb{Z}_2)$, the Leray–Hirsch theorem and condition 2° imply that $E_1^{pq}(IE_\omega) = 0$, $p+q \leqslant n$, $p = n-k+1, \dots, n$.

Thus the map $IE_\omega(x_1) \hookrightarrow I(x_1)$ determines the map $E_1^{pq}(I) \to E_1^{pq}(IE_\omega)$ satisfying condition $\langle n-1 \,|\, 1 \rangle$.

Using Theorem A and the isomorphisms $E_1^{pq}(I) \rightrightarrows E_1^{pq}(IE_\omega)$, $E_1^{pq}(I) \simeq E_\infty^{pq}(I)$, $p+q = n$, we get the isomorphism $E_\infty^{pq}(I) \rightrightarrows E_\infty^{pq}(IE_\omega)$, $p+q \leqslant n$.

REMARK. This theorem is similar to the well-known Lefschetz theorem about hyperplane sections [13].

Appendix. Comparison theorem for spectral sequences

We have used the following version of the comparison theorem for spectral sequences [14].

THEOREM A. *Let* A *and* A' *be differential modules and* f *be a mapping compatible with filtrations in* A *and* A', $f: A \to A'$. *We suppose that for some integer* k *the homomorphism* $f_k^*: E_k^{pq}(A) \to E_k^{pq}(A')$ *satisfies the following*

Condition $\langle n \,|\, k \rangle$.

1. *f_k^* is an isomorphism when $p+q \leqslant n$,*
2. *f_k^* is a monomorphism when $p+q = n+1$.*

Then for any integer $r \geqslant k$ the homomorphism $f_r^*\colon E_r^{pq}(A) \to E_r^{pq}(A')$ satisfies condition $\langle n \mid r \rangle$.

PROOF. We consider the complexes

$$M = \bigoplus_{p\,,\,p+q\leqslant n+1} E_k^{pq}(A) \quad \text{and} \quad L = \bigoplus_{p\,,\,p+q\leqslant n+1} E_k^{pq}(A')$$

and the cochain map $f_k^*\colon M \to L$. We also consider the exact sequence constructed by using the mapping cone of f_k:

$$0 \to L \to Cf_k^* \to M^+ \to 0. \tag{3}$$

Here $Cf_k^* = (L_i \oplus M_{i+1}\,,\,(d_L + f_k^*) \oplus (-d_M))\,,\ M^+ = (M_{i+1}\,,\,-d_M)$.

The proof of Theorem A follows from the next two lemmas.

LEMMA A.1 [14]. *The coboundary operator of the exact sequence* (3) *coincides with the mapping $Hf_k^*\colon HM \to HL$.*

LEMMA A.2. *Under the assumptions of Theorem* A, *the cone Cf_k^* is n-acyclic.*

PROOF OF LEMMA A.2. Let $(y\,,\,x)$ belong to $\operatorname{Ker} d_{Cf_k^*}$ $\dim y \leqslant n$. This means that $d_L y + f_k^*(x) = 0$ for $d_M x = 0$. Since the mapping f_k^* is an isomorphism up to dimension n, it follows that $y = f_k^*(x')$. Since the mapping f_k^* is a monomorphism up to dimension $n+1$, it follows that $d_M x' + x = 0$. Therefore $(y\,,\,x) = d_{Cf_k^*}(0\,,\,x')$ and $H^t(Cf_k^*) = 0\,,\ t \leqslant n$. □

Applying Lemmas A.1 and A.2 to the exact cohomology sequence (3), we obtain the proof of Theorem A.

COROLLARY. *The map $f\colon A \to A'$ induces the homomorphism $f^*\colon H(A) \to H(A')$, which is an isomorphism up to dimension* n *and is a monomorphism in dimension $n+1$.*

References

1. V. Lychagin, *Contact geometry and nonlinear second order differential equations*, Uspekhi Mat. Nauk **34** (1979), no. 1, 101–171; English transl., Russian Math. Surveys **34** (1979), 149–180.
2. ———, *Topological aspects of the geometric theory of differential equations*, Topology Application in Modern Analysis, Voronezh, 1985, pp. 106–123. (Russian)
3. L. V. Zilbergleit, *Characteristic classes of solutions of Monge–Ampère equations*, Uspekhi Mat. Nauk **39** (1984), no. 4, 159–160; English transl. in Russian Math. Surveys **39** (1984).
4. ———, *Topological invariants of geometric singularities of solutions of Monge–Ampère equations*, Izv. Vyssh. Uchebn. Zaved. Mat. **9** (1988), 30–37; English transl. in Soviet Math. (Iz. VUZ) **32** (1988).
5. V. P. Maslov, *Theory of perturbations and asymptotic methods*, Moscow State University, 1965. (Russian)
6. V. I. Arnold, *About the characteristic class appearing in the quantization condition*, Funktsional. Anal. i Prilozhen. **1** (1967), no. 1, 1–14; English transl. in Functional Anal. Appl. **1** (1967), no. 1.
7. A. Borel, *La cohomologie* mod 2 *de certains éspaces homogènes*, Comment. Math. Helv. **27** (1953), no. 2–3, 165–197.

8. D. B. Fuks, *About Maslov–Arnold characteristic classes*, Dokl. Akad. Nauk SSSR **178** (1968), no. 2, 303–306; English transl. in Soviet Math. Dokl. **9** (1968).
9. V. V. Lychagin and V. N. Rubtsov, *Local classification of Monge–Ampère differential equations*, Dokl. Akad. Nauk SSSR **272** (1983), no. 1, 34–38; English transl. in Soviet Math. Dokl. **28** (1983).
10. J. Igusa, *A classification of spinors up to dimension twelve*, Amer. J. Math. **92** (1970), no. 4, 997–1028.
11. V. L. Popov, *A classification of spinors of dimension fourteen*, Trudy Moskov. Mat. Obshch. **37** (1978), 173–217; English transl. in Trans. Moscow Math. Soc. **1978**, no. 1.
12. W. Slebodzinski, *Exterior forms and their applications*, PNW-Polish Scientific Publishers, Warsaw, 1970.
13. S. Lefschetz, *L'analysis situs et la géométrie algébrique*, Gauthier-Villars, Paris, 1924.
14. S. Maclane, *Homology*, Springer-Verlag, Berlin, Göttingen, and Heidelberg, 1963.

Translated by THE AUTHOR

7–2, Golubinskaya St., Apt. 554, Moscow, 117574 Russia